AF846322

Orexin and Sleep

Takeshi Sakurai · S.R. Pandi-Perumal
Jaime M. Monti
Editors

Orexin and Sleep

Molecular, Functional and Clinical Aspects

Springer

Editors
Takeshi Sakurai
Department of Molecular Neuroscience
and Integrative Physiology
Faculty of Medicine, Kanazawa University
Kanazawa, Ishikawa
Japan

Jaime M. Monti
Department of Pharmacology
and Therapeutics, School of Medicine
Clinics Hospital Montevideo
Montevideo
Uruguay

S.R. Pandi-Perumal
Somnogen Canada Inc.
Toronto, ON
Canada

ISBN 978-3-319-23077-1 ISBN 978-3-319-23078-8 (eBook)
DOI 10.1007/978-3-319-23078-8

Library of Congress Control Number: 2015948160

Springer Cham Heidelberg New York Dordrecht London
© Springer International Publishing Switzerland 2015
This work is subject to copyright. All rights are reserved by the Publisher, whether the whole or part of the material is concerned, specifically the rights of translation, reprinting, reuse of illustrations, recitation, broadcasting, reproduction on microfilms or in any other physical way, and transmission or information storage and retrieval, electronic adaptation, computer software, or by similar or dissimilar methodology now known or hereafter developed.
The use of general descriptive names, registered names, trademarks, service marks, etc. in this publication does not imply, even in the absence of a specific statement, that such names are exempt from the relevant protective laws and regulations and therefore free for general use.
The publisher, the authors and the editors are safe to assume that the advice and information in this book are believed to be true and accurate at the date of publication. Neither the publisher nor the authors or the editors give a warranty, express or implied, with respect to the material contained herein or for any errors or omissions that may have been made.

Printed on acid-free paper

Springer International Publishing AG Switzerland is part of Springer Science+Business Media
(www.springer.com)

Preface

Almost 17 years have passed since the discovery of orexin/hypocretin. Initially, the peptide was thought to be exclusively a regulator of feeding behavior. Soon after its discovery, however, it was proposed that orexin deficiency was a cause of narcolepsy in humans and other mammalian species, thus implicating orexin's potential role in the regulation of sleep and wakefulness. More recently, it has been suggested that orexin is an important modifier and regulator of emotion, energy homeostasis, reward, drug addiction, and arousal. Accumulating evidence has shown that more generally orexin neurons facilitate the body's sensing of external and internal environments and that accordingly they regulate vigilance states.

In recent years, orexin has received a great deal of attention as a potent endogenous, arousal-promoting peptide. The clinical implications of orexin's properties subsequently led to the development of several orexin receptor antagonists for the treatment of sleep disorders. In this respect, the orexin 1 and 2 receptor antagonist suvorexant has been recently made available for the treatment of insomnia disorders. Given the orexin system's broad range of functions, it has been suggested that orexin receptor antagonists might be also beneficial for treating a variety of conditions other than sleep disorders, including addictive, mood, and eating disorders.

In the present book, the authors discuss the physiological functions of orexins from various perspectives. The therapeutic potential of drugs that target orexin receptors is also discussed in detail. It is anticipated that these studies, which have used a number of different approaches, are expected to provide valuable insights into the physiological functions of the orexin system.

Japan — Takeshi Sakurai
Canada — S.R. Pandi-Perumal
Uruguay — Jaime M. Monti

Contents

Contributors

Antoine R. Adamantidis University of Bern, Department of Neurology, Inselspital University Hospital, Bern, Switzerland; McGill University, Department of Psychiatry, Douglas Institute, Montreal, Canada

Oscar Arias-Carrión Unidad de Trastornos del Movimiento y Sueño (TMS), Hospital General "Dr. Manuel Gea González", México, D.F., México

Paul Christian Baier Department of Psychiatry and Psychotherapy, Centre for Sleep Medicine, University Kiel, Kiel, Germany

Mino D.C. Belle Faculty of Life Sciences, University of Manchester, Manchester, UK

Paul J. Coleman Merck Co. Inc, West Point, PA, USA

Luis de Lecea Department of Psychiatry and Behavioral Sciences, Stanford University School of Medicine, Stanford, CA, USA

Hiromasa Funato Department of Anatomy, Toho University School of Medicine, Tokyo, Japan; International Institute for Integrative Sleep Medicine, University of Tsukuba, Ibaraki, Japan

Thomas C. Gent University of Bern, Department of Neurology, Inselspital University Hospital, Bern, Switzerland

Robert Göder Department of Psychiatry and Psychotherapy, Centre for Sleep Medicine, University Kiel, Kiel, Germany

Y. Audrey Hay Department of Physiology, Development and Neuroscience, University of Cambridge, Physiological Laboratory, Cambridge, UK

Andrea Herrera-Solís Unidad de Trastornos del Movimiento y Sueño (TMS), Hospital General "Dr. Manuel Gea González", México, D.F., México

Y. Hishikawa Department of Neuropsychiatry, Akita University School of Medicine, Akita, Japan

A. Imanishi Department of Neuropsychiatry, Akita University School of Medicine, Akita, Japan

Masaru Ishibashi Department of Physiology, New York Medical College, New York, USA

Ramsés Jiménez-Moreno Laboratorio de Neurociencias Moleculares e Integrativas, Escuela de Medicina, División Ciencias de la Salud, Universidad Anáhuac Mayab, Mérida, YUC, México

Alok Joshi Faculty of Life Sciences, University of Manchester, Manchester, UK

T. Kanbayashi Department of Neuropsychiatry, Akita University School of Medicine, Akita, Japan; International Institute for Integrative Sleep Medicine (WPI-IIIS), University of Tsukuba, Tsukuba, Japan

Mudasir Ahmad Khanday School of Life Sciences, Jawaharlal Nehru University, New Delhi, India

Leszek Kubin Department of Animal Biology 209E/VET, School of Veterinary Medicine, Philadelphia, PA, USA

Christopher S. Leonard Department of Physiology, New York Medical College, New York, USA

Birendra Nath Mallick School of Life Sciences, Jawaharlal Nehru University, New Delhi, India

Michihiro Mieda Department of Molecular Neuroscience and Integrative Physiology, Faculty of Medicine, Kanazawa University, Ishikawa, Kanazawa, Japan

Takatoshi Mochizuki Department of Neurology, Beth Israel Deaconess Medical Center, Boston, MA, USA

Jaime Monti Department of Pharmacology and Therapeutics, School of Medicine Clinics Hospital, University of the Republic, Montevideo, Uruguay

Eric Murillo-Rodríguez Laboratorio de Neurociencias Moleculares e Integrativas, Escuela de Medicina, División Ciencias de la Salud, Universidad Anáhuac Mayab, Mérida, YUC, México; Grupo de Investigación en Envejecimiento, División Ciencias de la Salud, Universidad Anáhuac Mayab, Mérida, YUC, México; Grupo de Investigación Desarrollos Tecnológicos para la Salud, División de Ingeniería y Ciencias Exactas, Universidad Anáhuac Mayab, Mérida, YUC, México

Seiji Nishino Sleep and Circadian Neurobiology Laboratory, Department of Psychiatry and Behavioral Sciences, Center for Narcolepsy, Stanford University School of Medicine, Palo Alto, CA, USA

Y. Ohmori Department of Neuropsychiatry, Akita University School of Medicine, Akita, Japan

M. Omokawa Department of Neuropsychiatry, Akita University School of Medicine, Akita, Japan

S.R. Pandi-Perumal Division of Health and Behavior, Department of Population Health, NYU Langone Medical Center, New York, NY, USA

Hugh D. Piggins Faculty of Life Sciences, University of Manchester, Manchester, UK

Alwin Poot-Aké Laboratorio de Neurociencias Moleculares e Integrativas, Escuela de Medicina, Divisi6n Ciencias de la Salud, Universidad Anáhuac Mayab, Mérida, YUC, México

John J. Renger Merck Co. Inc, West Point, PA, USA

Y. Sagawa Department of Neuropsychiatry, Akita University School of Medicine, Akita, Japan

Pradeep Sahota Harry S. Truman Memorial Veterans Hospital and Department of Neurology, University of Missouri, Columbia, MO, USA

Takeshi Sakurai Department of Molecular Neuroscience and Integrative Physiology, Faculty of Medicine, Kanazawa University, Kanazawa, Ishikawa, Japan

Toshiyasu Sasaoka Department of Clinical Pharmacology, University of Toyama, Toyama, Japan

M. Sato Department of Neuropsychiatry, Akita University School of Medicine, Akita, Japan

Rishi Sharma Harry S. Truman Memorial Veterans Hospital and Department of Neurology, University of Missouri, Columbia, MO, USA

T. Shimizu Department of Neuropsychiatry, Akita University School of Medicine, Akita, Japan; International Institute for Integrative Sleep Medicine (WPI-IIIS), University of Tsukuba, Tsukuba, Japan

Jerome M. Siegel Department of Psychiatry and Biobehavioral Sciences, University of California Los Angeles, Los Angeles, CA, USA; Veterans Administration Greater Los Angeles Healthcare System, Neurobiology Research (151A3), North Hills, CA, USA

Abhishek Singh School of Life Sciences, Jawaharlal Nehru University, New Delhi, India

Y. Takahashi Department of Neuropsychiatry, Akita University School of Medicine, Akita, Japan

Mahesh M. Thakkar Harry S. Truman Memorial Veterans Hospital and Department of Neurology, University of Missouri, Columbia, MO, USA

Thomas C. Thannickal Department of Psychiatry and Biobehavioral Sciences, University of California Los Angeles, Los Angeles, CA, USA; Veterans Administration Greater Los Angeles Healthcare System, Neurobiology Research (151A3), North Hills, CA, USA

Pablo Torterolo Laboratory of Sleep Neurobiology, Department of Physiology, School of Medicine, University of the Republic, Montevideo, Uruguay

Hiroshi Tsuneki Department of Clinical Pharmacology, University of Toyama, Toyama, Japan

Tomomi Tsunematsu Research Institute of Environmental Medicine, Nagoya University, Nagoya, Japan; Strathclyde Institute of Pharmacy and Biomedical Sciences, University of Strathclyde, Glasgow, UK

Jason M. Uslaner Merck Co. Inc, West Point, PA, USA

Tsutomu Wada Department of Clinical Pharmacology, University of Toyama, Toyama, Japan

Sara Lena Weinhold Department of Psychiatry and Psychotherapy, Centre for Sleep Medicine, University Kiel, Kiel, Germany

Chris J. Winrow Merck Co. Inc, West Point, PA, USA

KongFatt Wong-Lin Intelligent Systems Research Centre, University of Ulster, Magee Campus, Derry, Northern Ireland, UK

Akihiro Yamanaka Research Institute of Environmental Medicine, Nagoya University, Nagoya, Japan

Kyoko Yoshida-Court Department of Neurology, Beth Israel Deaconess Medical Center, Boston, MA, USA

History of Orexin Research

Takeshi Sakurai

Abstract Orexin A and orexin B (also known as hypocretin 1 and hypocretin 2) are hypothalamic neuropeptides that were discovered as endogenous cognate ligands for two orphan G-protein coupled receptors in 1998. Initially, these peptides were reported as regulators of feeding behavior (Sakurai et al. in Cell 92:573–585, 1998). Thereafter, several studies suggested that orexin deficiency causes narcolepsy in several mammalian species including humans, highlighting roles of this hypothalamic neuropeptide in the regulation of sleep and wakefulness (Sakurai in Nat Rev Neurosci 8:171–181, 2007). Studies of efferent and afferent systems of orexin-producing neurons have revealed that orexin neurons has close interactions with systems that regulate emotion, energy homeostasis, the reward system, and arousal (Boutrel et al. in Proc Natl Acad Sci USA 102:19168–19173, 2005; Yamanaka et al. in Neuron 38:701–713, 2003a; Akiyama et al. in Eur J Neurosci 20:3054–3062, 2004; Mieda et al. in J Neurosci 24:10493–10501, 2004; Sakurai et al. in Neuron 46:297–308, 2005; Yoshida et al. in J Comp Neurol 494:845–861, 2006; Harris et al. in Nature 437:556–559, 2005; Narita et al. in J Neurosci 26:398–405, 2006). Subsequent studies suggested that emotionally salient cues and contexts excite orexin neurons to promote arousal, and to support behavior. This system seems to be important to maintain the vigilance and arousal during doing various motivated and adaptive behaviors. Recently, suvorexant, a dual orexin receptor antagonist, has become clinically available for treatment of insomnia. In this chapter, I will overview the history of orexin research, highlighting some of the physiological roles of orexins.

T. Sakurai (✉)
Department of Molecular Neuroscience and Integrative Physiology, Faculty of Medicine, Kanazawa University, 13-1 Takaramachi, Kanazawa, Ishikawa 920-8640, Japan
e-mail: tsakurai@med.kanazawa-u.ac.jp

© Springer International Publishing Switzerland 2015

T. Sakurai et al. (eds.), *Orexin and Sleep*,
DOI 10.1007/978-3-319-23078-8_1

1 Orexin and Orexin Receptors

1.1 Identification of Orexin (Hypocretin)

In 1998, we identified and purified novel neuropeptides, orexin A and orexin B, from rat brain extracts as two endogenous peptide ligands for an orphan G-protein-coupled receptor (GPCR), called HFGAN-72 by means of a method called "reverse pharmacology" (Sakurai et al. 1998). We expressed numbers of GPCRs in HEK293 or CHO cells, and used them as then assay system to identify the cognitive ligands for each receptor. We successfully identified and purified two peptide ligands for HFGAN-72 from rat brains. Subsequent structural analysis and molecular cloning studies showed that both rat orexin A and orexin B are derived from a common precursor peptide, *prepro-orexin* (Fig. 1a).

Our structural analysis of purified peptides showed that orexin A is a 33-amino-acid peptide with an N-terminal pyroglutamyl residue, two intra-chain disulfide bonds, and C-terminal amidation. This structure is completely conserved among several mammalian species identified so far (human, rat, mouse, cow, sheep, dog and pig). Orexin B is a 28-amino-acid, C-terminally amidated linear peptide. There are several species differences in the structure of orexin B, although also highly conserved. The C-terminal half of orexin B is very similar to that of orexin A, whereas the N-terminal half is more variable.

An mRNA encoding the same precursor peptide was independently identified by de Lecea et al. as a hypothalamus-specific transcript (de Lecea et al. 1998), as described in detail in the separate chapter (de Lecea). They predicted that the transcript encoded a polypeptide precursor that is cleaved to form two neuropeptides, termed hypocretin-1 and hypocretin-2 (corresponding to orexin A and orexin B, respectively).

1.2 Orexin Receptors

Because an in vitro study suggested that orexin B has a much lower affinity to HFGAN72 than orexin A did, we had searched several EST data bases, and identified another subtype of orexin receptor, which we term orexin-2 receptor (OX2R), which appeared to have similar affinities to both orexin A and orexin B, and we re-named HFGAN72 as orexn-1 receptor (OX1R) (Fig. 1a). The actions of orexins are mediated via these two G-protein coupled receptors (GPCRs). OX1R has one-order-of-magnitude greater affinity for orexin A over orexin B. In contrast, OX2R binds both ligands with similar affinities (Sakurai et al. 1998) (Fig. 1b). *OX1R* and *OX2R* mRNAs exhibit a markedly different and basically complementary distributions, suggesting that these receptors have distinct physiological roles through different neuronal pathways (Marcus et al. 2001; Mieda et al. 2011). Recently, the structure of the human OX_2R bound to suvorexant was solved using

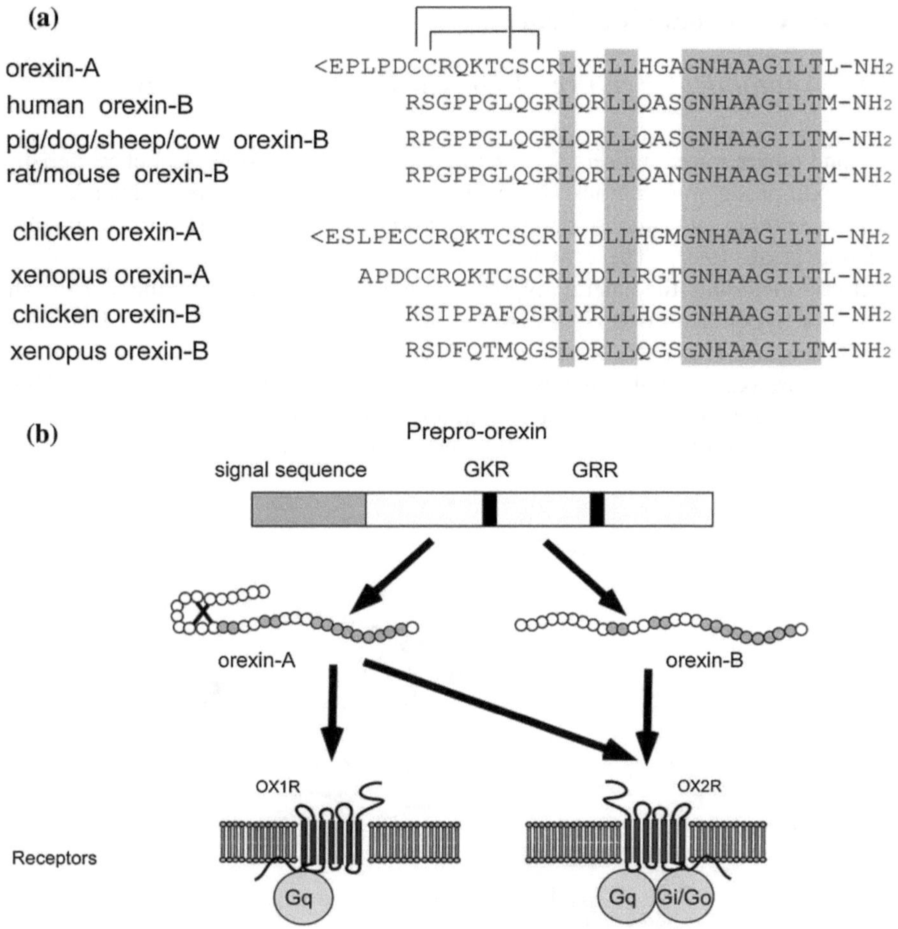

Fig. 1 Orexin and orexin receptors. **a** Structures of various species of orexin A and orexin B. The topology of the two intrachain disulfide bonds of orexin A is indicated above the sequence. *Shadows* indicate amino acid identity. Mammalian orexin A sequences thus far identified (human, rat, mouse, pig, dog, sheep, cow) are all identical. **b** Orexin A and orexin B are derived from a common precursor peptide, prepro-orexin. The actions of orexins are mediated via two G protein-coupled receptors named orexin-1 (OX1R) and orexin-2 (OX2R) receptors. OX1R is relatively selective for orexin A, whereas OX2R shows similar affinities for both orexin A and orexin B. OX1R is coupled to the Gq subclass of heterotrimeric G proteins, whereas OX2R couples to Gi/o and/or Gq in neuronal cell lines

lipid-mediated crystallization (Yin et al. 2014). The structure revealed that suvorexant adopts a π-stacked horseshoe-like conformation and binds to the receptor deep in the orthosteric pocket, stabilizing extracellular salt bridges and blocking transmembrane helix motions necessary for activation. Signal transduction system and physiological roles of these receptors are discussed in a separate chapter by Mieda et al.

2 Orexin-Producing Neurons

2.1 Anatomical Feature of Orexin-Producing Neurons

We and others raised antisera for orexin, and examined histological characteristics of orexin-producing neurons. Numbers of orexin neurons are estimated to be around 3,000 in rat or mouse brains, or 70,000 in human brains (Peyron et al. 1998; Nambu et al. 1999). These neurons are found exclusively in the hypothalamic regions, including the lateral hypothalamic area (LHA), perifornical area, and posterior hypothalamus (PH) (Peyron et al. 1998; Nambu et al. 1999; Date et al. 1999) (Fig. 2). However, orexin-imunoreactive fibers were observed in the almost entire neuroaxis excluding the cerebellum (Peyron et al. 1998; Nambu et al. 1999; Date et al. 1999). Especially dens staining of fibers were found in the paraventricular nucleus of the thalamus, several regions of the hypothalamus, including

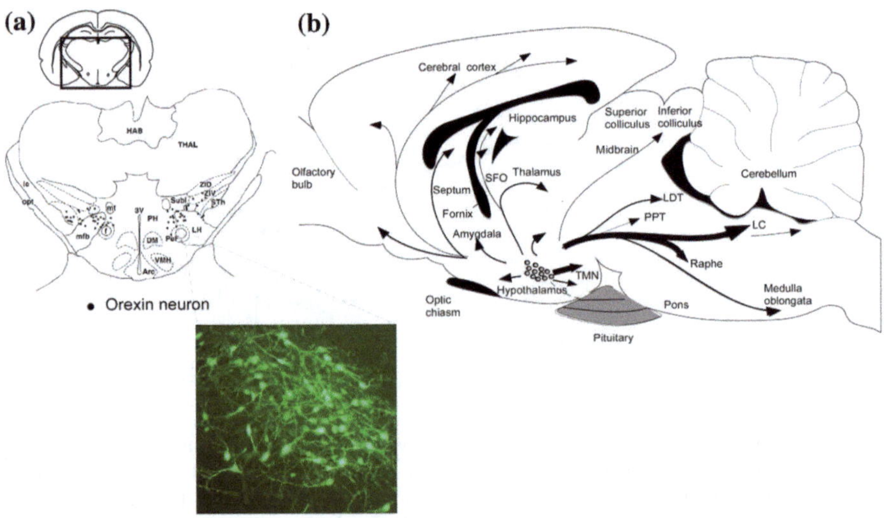

Fig. 2 Schematic drawing of coronal section and sagittal section of rat brain, summarizing the orexin neuronal system. **a** *Prepro-orexin* mRNA-containing neurons are shown in *black* superimposed upon anatomical structures of the hypo- and subthalamic areas. The *rectangle* designates the area schematized in the figure. Abbreviations: lateral hypothalamic area (*LH*), perifornical nucleus (*PeF*), posterior hypothalamic area (*PH*), subthalamic nucleus (*Sth*), subincertal nucleus (*SubI*), ventral zona incerta (*ZIV*). Additional landmarks include: thalamus (*THAL*), habenular complex (*HAB*), internal capsule (*ic*), optic tract (*opt*), mammillothalamic tract (mt), fornix (*f*), medial forebrain bundle (*mfb*), third ventricle (*3V*), arcuate hypothalamic nucleus (*Arc*), dorsomedial hypothalamic nucleus (*DM*), and ventromedial hypothalamic nucleus (*VMH*). *Inset* shows immunostaining image of orexin neurons. **b** Orexin neurons are found only in the lateral hypothalamic area and project to the entire central nervous system. The *thickness of arrows* represents relative abundance of projections. Abbreviations: third ventricle (*3V*), fourth ventricular (*4V*), tuberomammillary nucleus (*TMN*), locus coeruleus (*LC*), laterodorsal tegmental nucleus (*LDT*), pedunculopontine nucleus (*PPT*)

arcuate nucleus, ventromedial hypothalamus, and posterior hypothalamus, and most notably, monoaminergic nuclei in the hypothalamic/brain stem regions, such as the locus coeruleus (LC) (containing noradrenergic neurons), raphe nuclei (containing serotonergic neurons), tuberomammillary nucleus (TMN) (containing histaminergic neurons), and laterodorsal and pedunculopontine tegmental nuclei (LDT/PPT) (containing cholinergic neurons) (Nambu et al. 1999; Date et al. 1999; Peyron et al. 2000). The distribution of the orexin receptor mRNA was consistent with these projection sites, with differential expression of each subtype; within the brain, *OX1R* is most abundantly expressed in the LC, while *OX2R* is highly expressed in the TMN (Marcus et al. 2001). Both regions are important for the maintenance of arousal (Marcus et al. 2001). The raphe nuclei, LDT/PPT, and ventral tegmental area (VTA) contain both *OX1R* and *OX2R* (Marcus et al. 2001), although they are expressed in distinct neuronal populations in each region (Mieda et al. 2011). These observations suggest that these monoaminergic regions, implicated in the regulation of wakefulness, are major effector sites of orexins.

In vivo recording studies revealed changes of orexin neuronal activity across the sleep-wake cycle in rats or mice (Mileykovskiy et al. 2005; Lee et al. 2005; Takahashi et al. 2008). Basically, orexin neurons fire during active waking, decrease discharge during quiet waking, and virtually cease firing during both rapid eye movement (REM) and non-rapid eye movement (NREM) sleep. This firing pattern was consistent with earlier studies showing that Fos expression (a marker of neuronal activity) in orexin neurons in rats or mice is increased during the dark, active period in which the awake state is dominant (Estabrooke et al. 2001), and orexin levels in CSF peak during the dark period and decrease during the light period in which the sleep state is dominant (Yoshida et al. 2001).

2.2 *Regulation of Orexin Neurons*

2.2.1 Input and Output of Orexin Neurons

Orexin system is involved in the diverse functions. Knowledge about the regulatory mechanisms of orexin neurons is important for understanding the roles of orexin neurons in these functions (Fig. 3). Studies using anterograde and retrograde tracers suggest that orexin neurons receive abundant projections from the lateral septum, preoptic area, amygdala, bed nucleus of the stria terminalis (BNST), posterior/dorsomedial hypothalamus, and the raphe nuclei (Sakurai et al. 2005; Yoshida et al. 2006). Input and output of orexin neurons are described in detail in a chapter by Mochizuki.

Abundant input from the limbic system suggests it plays a part in the regulation of the firing rate of orexin neurons by conveying emotive factors to maintain arousal. The limbic input to orexin neurons might also be involved in the regulation of feeding behavior, because some of the affective content of the perception of food

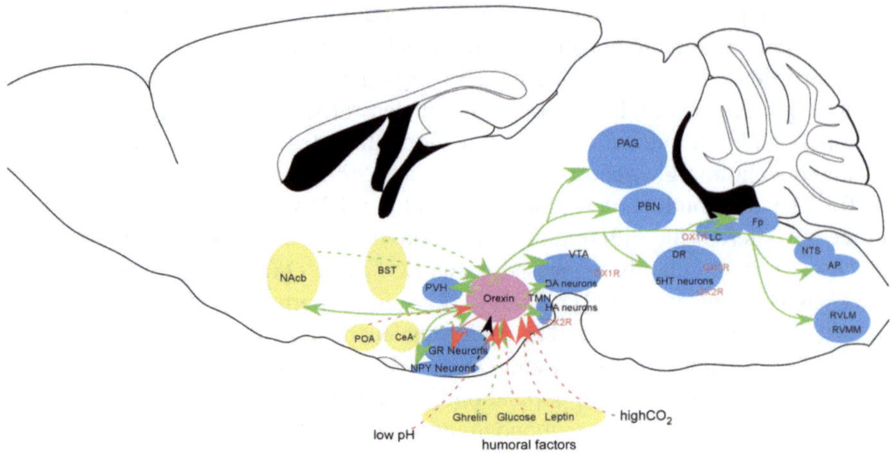

Fig. 3 Connections of orexin neurons with other regions. Orexin neurons in the lateral hypothalamic area (*LHA*) provide a link between the limbic system, energy homeostasis and the brain stem nuclei. Modified from Sakurai (2007). *Circles* show major target sites for orexins. Included in these are the locus coeruleus (LC, containing noradrenaline, NA), tuberomammillary nucleus (TMN, containing histamine, HA), raphe nuclei (Raphe, containing 5-HT), ventral tegmental area (VTA, containing dopamine, DA), and laterodorsal/pedunclopontine tegmental nuclei (PPT/LDT, containing acetylcholine, Ach). Orexin neurons promote wakefulness through the monoaminergic/cholinergic nuclei that are wake-active. Connection between dopaminergic centers and orexin neurons plays to modulate the reward systems. Input from the limbic system might be important to regulate the activity of orexin neurons upon emotional stimuli to evoke emotional arousal or fear-related responses. Sleep-active neurons in the POA send inhibitory influences to monoaminergic/cholinergic neurons and orexin neurons. Orexin neurons send both direct excitatory input to cholinergic neurons in the LDT/PPT and indirect inhibitory input to these cells through GABAergic local interneurons and GABAergic neurons in the substantia nigra pars reticulata (Takakusaki et al. 2005). Noradrenergic neurons in the LC and serotonergic neurons in the RN also send inhibitory influences to these cholinergic neurons. Blood glucose levels also affect the activity of orexin neurons through fluctuations of glucose levels in the CSF and vagal afferent. *NAc* nucleus accumbens; *PVH* paraventricular hypothalamic nucleus; *TMN* tuberomammilary nucleus; *LHA* lateral hypothalamic area; *DMH* dorsomedial hypothalamus; *ARC* arcuate nucleus; *VTA* ventral tegmental area; *SN* substantia nigra; *SCN* suprachiasmatic nucleus; *RN* raphe nucleus; *LC* locus coeruleus; *PPT* pedunculopontine tegmental nucleus; *LDT* laterodorsal tegmental nucleus

is thought to be processed in the amygdala and limbic system (Berthoud 2004). Interestingly, it is well known that food perception often evokes cataplexy in narcoleptic dogs (Reid et al. 1998), suggesting that orexin signaling is activated upon perception of food, and that this system is necessary to evoke normal feeding behavior.

Orexin neurons also receive abundant input from the preoptic area (POA), which is thought to play an important role in initiation and maintenance of sleep. This connection seems to be important for silencing orexin neurons during sleep (Saito et al. 2013).

2.2.2 Factors that Influence Firing of Orexin Neurons

In vitro electrophysilogical studies identified a number of factors that affect activity of orexin neurons (Table 1). In addition to the classical aminoacid neurotransmitters, glutamate and GABA, several factors were shown to influence the activity of orexin neurons. Both noradrenaline and serotonin inhibited orexin neurons through the activation of G protein-regulated inwardly rectifying K^+ (GIRK or Kir3) channels via α_2-adrenoceptors and $5HT_{1A}$-receptors, respectively (Li et al. 2002; Yamanaka et al. 2003b, 2006). The cholinergic agonist carbachol activated 27 %

Table 1 Factors that influence activity of orexin neurons

Excitation	Receptor involved
Glutamate	AMPAR, NMDAR mGluRs
Acetylcholine (muscarinic) (27 %)	M3
Orexin	OX_2R
Ghrelin	GHSR
Cholecystokinin	CCKA
Neurotensin	NTSR2 (unpublished data)
Vasopressin	V1a
Oxytocin	V1a
Glucagon-like peptide 1	ND
Corticotropin-releasing factor	CRFR1
Thyrotropin-releasing hormone	TRH1
BRS3 agonist	BRS3
ATP	P2X
H^+	ASIC1a
CO_2	ND
Mixture of amino acids	System-A amino acid transporters
Inhibition	
Glucose	Unknown
GABA	$GABA_A$, $GABA_B$
Glycine	Glycine receptor
Serotonin	$5HT_{1A}$
Noradrenaline	α_2
Dopamine	α_2
Acetylcholine (muscarinic) (6 %)	ND
Neuropeptide Y	Y_1
Enkephalin	μ opioid-R
Nociceptin	NOPR
Leptin	Ob-R
Adenosine	A_1
BRS3 agonist	BRS3

and inhibits 6 % of orexin neurons (Sakurai et al. 2005; Yamanaka et al. 2003b), whereas histamine had little effect on orexin neurons. Although orexin neurons do not express functional dopamine receptors, dopamine inhibited orexin neurons by acting through α_2-adrenoceptors (Yamanaka et al. 2003b, 2006). Several neuropeptides, including cholecystokinin (CCK-8S), neurotensin, oxytocin, and vasopressin, induced depolarization and excitation of orexin neurons (Tsujino et al. 2005). A synthetic surrogate ligand for an orphan receptor, BRS-3, also directly activated these neurons, although it also inhibits orexin neurons through activation of GABAergic interneurons (Furutani et al. 2010). Orexin itself was also shown to activate orexin neurons through OX2R, suggesting a positive feedback mechanism that maintain orexin neuronal activity (Yamanaka et al. 2010). Similarly, neurotensin was shown to co-localize with orexin neurons and excites orexin neurons (Furutani et al. 2013).

Interestingly, metabolic signals also seem to contribute to the regulation of orexin neurons. Decreasing the extracellular glucose concentration produced depolarization and increased the frequency of firing of orexin neurons, whereas increasing glucose concentration induced hyperpolarization and cessation of firing (Yamanaka et al. 2003a; Burdakov et al. 2005). Importantly, this mechanism is sensitive enough to monitor variations in glucose levels in cerebrospinal fluid reflecting those occurring physiologically between normal meals (Burdakov et al. 2005). These responses were shown to be mediated by tandem-pore K^+ (K_{2P}) channels (Burdakov et al. 2006).

In addition, a gut/stomach-derived hormone, ghrelin activated 60 % of dispersed orexin neurons with depolarization and an increase in firing frequency (Yamanaka et al. 2003a). By contrast, bath-application of leptin, an anorexigenic protein hormone secreted by adipocytes, was found to robustly inhibit most of the orexin neurons examined, causing hyperpolarization and a decrease in firing rate (Yamanaka et al. 2003a). These findings show that peripheral humoral factors that are related to energy metabolism influence the activity of orexin neurons (Sakurai et al. 1998; Yamanaka et al. 2003a; Willie et al. 2001). The ability of orexin neurons of sensing metabolic signals might play an important physiological role in the regulation of feeding behavior. This suggests that negative energy balance activates orexin neurons to increase arousal, thereby reinforcing food-seeking/feeding pathways. Consistently, orexin neurons-ablated mice failed to exhibit this fasting-induced arousal (Yamanaka et al. 2003a), suggesting that orexin neurons are necessary for evoking adaptive behavioral arousal during fasting. The mechanism that helps to ensure survival in nature clearly involves orexins. As discussed later, motivation toward food might also activate orexin neurons when animals are fasted.

3 Physiological Functions of Orexins

3.1 Orexin and Feeding

We initially reported orexins as factors that is involved in the regulation of feeding behavior (Sakurai et al. 1998). Orexin neurons are bilaterally and synmetorically distributed within the LHA and adjacent regions (Sakurai et al. 1998). The LHA has been thought to be the "feeding center", because lesions in this region caused anorexia, whereas electrical stimulation resulted in overeating and obesity in rats (Anand and Brobeck 1951). Indeed, we found an orexigenic effect of intracerebroventricular (icv) administration of orexin A and orexin B in rats (Sakurai et al. 1998), and this effect was subsequently confirmed in several species (Sakurai 2007b). Furthermore, central administration of an orexin antibody or an OX1R antagonist have been shown to decrease food intake (Yamada et al. 2000; Haynes et al. 2000a).

Intraperitoneal administration of the selective OX1R antagonist (1-SORA) SB-334867 or RNAi-mediated knockdown of the orexin gene reduced food intake in mice exposed to mild food restriction (Sharf et al. 2010a), and orexin-deficient mice showed decreased food intake (Willie et al. 2001; Hara et al. 2005). However, importantly, orexin signaling increases not only food intake but also energy expenditure, and a decrease in the overall orexin tone generally results in obesity (Funato et al. 2009; Hara et al. 2001). This is consistent with the findings that human narcolepsy patients show increased incidence of obesity (Hara et al. 2001). The role of orexins in body weight regulation is discussed in a separate chapter of this volume (Funato).

The orexin system may contribute to the regulation of energy homeostasis by integrating information regarding metabolic state and regulating wakefulness to support feeding behavior (Yamanaka et al. 2003a; Mieda and Sakurai 2012; Sakurai and Mieda 2011). Indeed, mice lacking orexin neurons do not show an increase in wakefulness or locomotor activity in response to starvation, unlike wild-type mice (Yamanaka et al. 2003a). Moreover, *prepro-orexin* mRNA is upregulated in fasted animals (Sakurai et al. 1998) and several studies reported that the firing rates of orexin neurons are influenced by glucose, triglycerides and amino acids (Yamanaka et al. 2003a; Burdakov et al. 2005b; Venner et al. 2011; Chang et al. 2004; Karnani et al. 2011). Orexin neurons were also shown to be innervated by neurons in the arcuate nucleus (which are primary sensors for plasma leptin levels) (Elias et al. 1998), and they are directly inhibited by leptin and excited by ghrelin (Yamanaka et al. 2003a). Together, these observations suggest that orexin neurons sense the animal's metabolic and nutritional status through both direct and indirect pathways, and integrate it in order to evoke a level of arousal necessary to promote food-seeking behavior in response to negative energy balance. In addition, motivation toward food might also contribute to activate orexin neurons.

As already mentioned, one of the possible mechanisms by which orexins promote feeding is that these factors increase arousal to secure feeding behavior. However, although the OX2R is thought to be a major player in the regulation of wakefulness, the studies using 1-SORA SB-334867 pointed to the importance of OX1R in the regulation of food seeking (Sharf et al. 2010a; Haynes et al. 2000b). This suggests that the orexin system influences food intake and wakefulness through at least partially different receptors and pathways.

Orexins might also directly affect the neuronal circuits in the hypothalamus that are implicated in the regulation of feeding behavior. They inhibit glucoreceptor neurons in the ventromedial hypothalamus (VMH), and excite neuropeptide Y (NPY) neurons in the arcuate nucleus and melanin-concentrating hormone (MCH) neurons in the LHA (van den Pol et al. 2004; Shiraishi et al. 2000; Yamanaka et al. 2000). Local injection of orexin in the paraventricular nucleus of the hypothalamus (PVN), DMH or LHA increased food intake in rats (Dube et al. 1999; Thorpe et al. 2003; Sweet et al. 1999). The area postrema and nucleus of the solitary tract (NTS) were also shown to be involved in orexin-mediated feeding (Baird et al. 2009; Thorpe and Kotz 2005). Together, orexin is likely to promote feeding by influencing multiple elements of the feeding circuitry.

Orexin-mediated feeding seems to be closely related with the reward system, in which OX1R has been also shown to be involved (Sharf et al. 2010a; Harris et al. 2005b; Choi et al. 2010). In rats, orexin increased the motivation to food-seeking when administered icv, especially for palatable food (Borgland et al. 2009, 2010; Thorpe et al. 2005). Furthermore, feeding behavior induced by administration of the mu-opioid receptor agonist DAMGO (D-Ala(2)-N-MePhe(4)-Gly-ol(5)-enkephalin) into the shell of the nucleus accumbens (NAc) was dependent on OX1R activation (Zheng et al. 2007), and intraperitoneal injection of the 1-SORA SB-334867 reduced high-fat food intake in food-restricted rats (Choi et al. 2010; Borgland et al. 2009; Nair et al. 2008).

A recent study showed that the number of Fos-immunoreactive orexin neurons in the hypothalamus increased in response to a chow-predictive (that is, conditioned) cue in rats (Petrovich et al. 2012). Similarly, the expectation of receiving a palatable food like chocolate increased numbers of Fos-positive hypothalamic orexin neurons (Choi et al. 2010). Numbers of Fos-positive orexin neurons in the LHA were increased after conditioned place preference training for a sweet cereal reward in rats (Harris et al. 2005a). Together, these findings indicate that orexin plays a role in food pursuit, especially when motivation towards food is high (e.g., when an animal is food-deprived or when foods are palatable), or when reward-conditioned cues are present. Input from the limbic system and NAc, which are thought to process the affective content of the perception of food (Berthoud 2004), might be involved in this function. Notably, food-related cues often evoke cataplexy in narcoleptic dogs (which are defective in orexin signaling) (Reid et al. 1998), suggesting that the perception of food normally induces orexin signaling, and that this signaling is necessary to elicit feeding behavior, including the maintenance of motor activity and wakefulness. Collectively, orexin neurons are likely to be excited by food-related cues and/or a low energy balance through neuronal connections with the limbic system and by factors that indicate a low energy balance to ensure feeding behavior.

3.2 *Orexin and Wakefulness*

The importance of orexins in the maintenance of wakefulness was highlighted by the findings that showed the involvement of the dysfunction of orexin signaling in a sleep disorder, narcolepsy. Clues that revealed the dysfunction of orexins are involved in narcolepsy initially came from animal models (Chemelli et al. 1999; Lin et al. 1999) (see other chapters). The link between orexin signaling and narcolepsy was subsequently supported by studies with human patients (Peyron et al. 2000; Thannickal et al. 2000).

Narcolepsy is a debilitating neurological disorder that affects approximately 1 in 2,000 individuals in the United States (Mignot 1998). A cardinal symptom of the disorder is excessive daytime sleepiness (an insurmountable urge to sleep), which often results in falling asleep at inappropriate times and situations ('sleep attack'). The latency for rapid eye movement (REM) sleep is notably reduced in narcolepsy patients, and the existence of 'sleep-onset REM periods' (i.e. REM-sleep is directly preceded by an awake period) is one of the diagnostic criteria for narcolepsy. Nocturnal sleep is also often disturbed by nocturnal wakening combined with the occurrence of hypnagogic hallucinations, vivid dreaming, and sleep paralysis. Narcolepsy patients often suffer from a condition called "cataplexy", which is a sudden weakening of muscle tone, ranging from jaw dropping and speech slurring to complete bilateral collapse of the postural muscles. Cataplexy is usually triggered by emotional stimuli. Unlike the sleep attack, consciousness is preserved during cataplexy.

A postmortem study of human narcoleptic brains showed no detectable levels of orexin peptides in the cortex and pons, in which normally orexinergic projections are found. In the hypothalamus, an 80–100 % reduction in the number of neurons containing detectable *prepro-orexin* mRNA or orexin-like immunoreactivity were found (Peyron et al. 2000; Thannickal et al. 2000). More than 90 % of patients with narcolepsy are shown to have decreased orexin A levels in the cerebrospinal fluid (Mignot et al. 2002). Relationship between orexin deficiency and narcolepsy is described in detail in other chapters (Nishino, Kanbayashi).

As already discussed, the projection pattern of orexin neurons and distributions of orexin receptor mRNAs suggested that main effector sites for orexin are monoaminergic/cholinergic neurons in the brainstem, and electrophysiological experiments also showed that firing rates of monoaminergic cells in these nuclei are increased by orexins. For instance, noradrenergic cells of the LC (Hagan et al. 1999; Horvath et al. 1999), dopaminergic cells of the VTA (Nakamura et al. 2000), serotonergic cells of the dirsal raphe (DR) (Liu et al. 2002; Brown et al. 2002), and histaminergic cells in the TMN (Yamanaka et al. 2002) were all shown to increase their firing rates by orexins. The firing rates of these monoaminergic neurons are well known to be associated with sleep/wakefulness states. They fire tonically during awake period, less during NREM sleep, and cease firing during REM sleep (Vanni-Mercier et al. 1984), displaying similar firing patterns with orexin neurons. These observations suggest that firing of these wake-active monoaminergic neurons mediates arousal are supported by orexins. Orexin neurons also project directly to the LDT/PPT, which contain cholinergic neurons. Some populations of these cholinergic neurons are implicated in the maintenance of wakefulness and REM sleep (W/REM-on neurons) (Shouse and Siegel 1992), whereas other populations are implicated in desynchronization of cerebral cortex and muscle atonia during REM sleep (REM-on neurons) (Shouse and Siegel 1992). Pharmacologically, a direct injection of orexin A into the LDT of cats results in an increased awake time and a decreased REM sleep time (Xi et al. 2001). In addition, several reports have shown that orexin induces long-lasting excitation of cholinergic neurons in the LDT (Takahashi et al. 2002). However, recent work also showed that orexin A inhibits cholinergic neurons in the PPT via activation of GABAergic local interneurons and GABAergic neurons in the substantia nigra pars reticulate (SNr), which send inhibitory projections to the PPT (Takakusaki et al. 2005). Since orexins show strong inhibitory effects on REM sleep, the indirect inhibition of cholinergic neurons of orexin might play an important role during REM sleep. In fact, we found robust expression of *OX1R* in the cholinergic neurons, and both receptor expressions in GABAergic neurons in these regions (Mieda et al. 2011). Collectively, these results indicate that in the LDT/PPT, orexin may directly activate W/REM-on cholinergic neurons through OX1R to facilitate wakefulness. Simultaneously, orexin is likely to activate GABAergic interneurons through both receptors to inhibit REM-on cholinergic neurons. Additionally, orexinergic activations of wake-active noradrenergic and serotonergic neurons in the LC and raphe nuclei, respectively, are likely to counteract activation of REM-on cholinergic neurons in the LDT/PPT during wakefulness (Sakurai 2007a; Pace-Schott and Hobson 2002). This is consistent with the fact that tricyclic antidepressants and serotonin-specific reuptake inhibitors are effective for treating cataplexy in narcoleptic patients.

Recent optogenetic studies revealed that orexin neurons are also glutamatergic (Schone et al. 2014). Glutamate-mediated fast and orexin-mediated slow neurotorasmission make it possible for orexin neurons to convert signals into transient and sustained signals in the same postsynaptic target neurons. The downstream pathways of orexin receptor-expressing neurons are described in more detail in a separate chapter (Mieda).

3.3 Orexin and Reward/Addiction

Besides feeding and arousal, orexin system also plays an important role in the reward system. The reward system is closely related to both feeding and wakefulness. Cues and contexts associated with rewards, including food, sex and drugs, increase the number of Fos-positive orexin neurons and *prepro-orexin* mRNA levels (Sakurai et al. 1998; Harris et al. 2005a; Di Sebastiano et al. 2011; Cason et al. 2010). Orexin neurons send dense projections to the ventral tegmental area (VTA), in which dopaminergic neurons that send innervations to the NAc are localized (Yoshida et al. 2006) (Fig. 2). The NAc in turn sends projections to orexin neurons, constituting a reciprocal link. Intracerebroventricular injection of orexins or local administration into the VTA can reinstate previously-extinguished drug-seeking or food-seeking behaviour in rodents (Boutrel et al. 2005; Harris et al. 2005a), and orexin neurons are activated during the behavioural expression of preferences for cues associated with reward (Harris et al. 2005a). The VTA expresses both OX1R and OX2R (Marcus et al. 2001), with dopaminergic neurons predominantly expressing OX1R, and orexin signaling in the VTA has been implicated in reinforcement and reward-related processes via actions on VTA dopamine neurons (Balcita-Pedicino and Sesack 2007) (Fig. 3).

An increasing body of work shows that orexin neurons also play a part in the behavioural presentation of addiction to drugs including cocaine, amphetamine, morphine, heroin, nicotine, ethanol and cannabinoids (Espana et al. 2011; Mahler et al. 2012; Martin-Fardon and Boutrel 2012). Generally, orexin seems to be involved in the modulation of highly-motivated reward seeking, especially when this seeking is triggered by external cues that are conditioned with the rewards. Although the 1-SORA SB-334867 did not affect the expression of cocaine and amphetamine-sensitization in animals tested immediately after training (Borgland et al. 2006), it blocked the expression of sensitization after a period of abstinence following amphetamine sensitization training, as did the DORA almorexant (Quarta et al. 2010; Winrow et al. 2010). This suggests that the OX1R is involved in the acquisition of sensitization. In another example, the 1-SORAs SB-334867 and GSK-1059865 attenuated the expression of cocaine and amphetamine-induced conditioned place preference in rats (Gozzi et al. 2011; Hutcheson et al. 2011; Sartor and Aston-Jones 2012).

Importantly, neither orexins nor their receptor antagonists affect self-adminstration of addictive drugs such as cocaine in rodents (Espana et al. 2010, 2011; Hutcheson et al. 2011), suggesting that orexins have no major role in the reinforcing or the priming effects of cocaine. However, orexin A has been shown to promote motivation in a study using rats in which high levels of effort were required for seeking addictive drugs in self-administration paradigms (Borgland et al. 2009). These results suggested that orexins plays an essential role in reward-seeking not by influencing the primary reinforcing or priming effects of rewards, but by supporting motivated behavior. Intraperitoneal injections or intra-VTA administration of the 1-SORA SB-334867 blocked reinstatement of cocaine seeking elicited by either discrete cues or contextual stimuli in rats (Smith et al. 2009, 2010; James et al. 2011), whereas intraperitoneal injection of the 2-SORA 4-PT did not affect cue-induced reinstatement (Smith et al. 2009). These observations suggest that orexin neurons might be activated by external reward-related stimuli and then send information to the VTA to induce reinstatement through the activation of OX1R.

Orexins also seem to play a role in addiction to drugs other than cocaine and amphetamine. For example, Fos expression was increased in orexin neurons in rats following acute nicotine administration or nicotine withdrawal (Pasumarthi et al. 2006; Plaza-Zabala et al. 2013), and nicotine withdrawal was decreased in orexin knockout mice (Plaza-Zabala et al. 2013). Prior intraperitoneal administrations of the 1-SORA SB-334867, but not the 2-SORA TCSOX229, attenuated nicotine withdrawal (Plaza-Zabala et al. 2013), as did intra-PVN infusion of the 1-SORA SB-334867 (Plaza-Zabala et al. 2013). Moreover, systemic administration of the DORA almorexant or the 1-SORA SB-334867 reduced nicotine self-administration (Hollander et al. 2008; LeSage et al. 2010).

Orexin knockout mice and wild-type mice that received the 1-SORA SB-334867 show reduced morphine withdrawal responses (Sharf et al. 2010a, b; Georgescu et al. 2003), suggesting that orexin may also be involved in opiate addiction. Furthermore, intraperitoneal injection of the 1-SORA SB-334867 reduced the expression of morphine-induced conditioned place preference in rats and mice (Harris et al. 2005a; Sharf et al. 2010b). Similarly, a report showed that orexin-deficient mice did not show morphine-induced conditioned-place preference and hyperlocomotion (Narita et al. 2006), although this finding is argued in another study (Sharf et al. 2010b). Action of orexin in the VTA is important for the expression of morphine-induced conditioned place preference (Richardson and Aston-Jones 2012). Moreover, in contrast with the lack of effect of the 1-SORA SB-334867 on cocaine self-administration, intraperitoneal delivery of the 1-SORA SB-334867 reduced heroin self-administration (Smith and Aston-Jones 2012).

3.4 Stress, Emotion and Emotional Memory

Emotional stimuli are known to evoke autonomic responses and arousal through neural connections between the amygdala and LHA, the region orexin neurons locate. Historically, the LHA has been recognized as the ‘defence area’, as electrical stimulation here can evoke aggressive behavior and the accompanying sympathetic activation (Hilton 1982). The orexin system has close functional relationships with systems that regulate emotion and stress response, including so-called “fight or flight” response.

Orexin system has been shown to be involved in emotion-induced changes in autonomic and neuroendocrine functions. Orexin neurons in the LHA receive innervations from limbic regions including the lateral septum, bed nucleus of the stria terminalis (BNST) and the amygdala (Sakurai et al. 2005; Yoshida et al. 2006) and send projections to monoaminergic and cholinergic regions in the brain stem, periaqueductal grey (PAG), parabrachial nucleus, NST, PVN, and the rostral ventrolateral and ventromedial medulla (RVLM and RVMM) (Peyron et al. 1998; Nambu et al. 1999). Thus, orexin neurons may link limbic structures with arousal and premotor autonomic centres (Fig. 2). Indeed, several studies showed that excitation of orexin neurons increases arousal along with increased sympathetic outflow in response to various physiological and emotional stimuli (Kayaba et al. 2003a; Zhang et al. 2006, 2009, 2010).

Notably, oral administration of a DORA, almorexant, inhibited cardiovascular responses to novelty and contextual fear (Furlong et al. 2009) without affecting responses to cold or restraint stress. Similarly, orexin knock-out mice have a decreased cardiovascular response to social stress but a normal response to a tail pinch (Kayaba et al. 2003a). Together with the finding that oral administration of the 1-SORA ACT335827 reduced the tachycardic response to social stress (Steiner et al. 2013), these data suggest that orexin more profoundly contributes to autonomic responses to psychological stressors than to physical stressors. Similarly, narcoleptic patients show reduced autonomic responses to emotional stimuli, especially aversive ones (Tucci et al. 2003), whereas they have a normal cardiovascular response to basic homeostatic challenges (such as head-up tilt, Valsalva manoeuver and cold pressor test). This suggests that orexin regulates the sympathetic nervous system primarily in response to salient emotional cues or contexts, rather than physical stress.

Several papers have suggested that dysregulation of the orexin system has been implicated in anxiety and panic-like behaviour in humans and rats (Johnson et al. 2010). For example, humans with panic anxiety have higher orexin levels in the cerebrospinal fluid compared to subjects without panic anxiety (Johnson et al. 2010), as do patients with post-traumatic stress disorder (PTSD) (Strawn and Geracioti 2008).

The limbic input to orexin neurons (Sakurai et al. 2005; Yoshida et al. 2006; Winsky-Sommerer et al. 2004) might regulate or modulate physiological responses to emotional and stressful stimuli. Indeed, the cardiovascular and locomotor responses that wild-type mice show after exposure to an intruder mouse are diminished in orexin-deficient mice (Kayaba et al. 2003b). Similarly, cardiovascular responses to air-jet stress were reduced in mice in which orexin neurons were genetically ablated (Zhang et al. 2006). Disinhibition of the amygdala or BNST through microinjections of a $GABA_A$ receptor antagonist increased Fos immunoreactivity in nuclei of orexin neurons and induced cardiorespiratory excitation in wild-type mice but not in mice lacking orexin neurons (Zhang et al. 2009). Together, these observations indicate that orexin neurons are excited by input from the amygdala and BNST (Zhang et al. 2009). Thus, it is possible that activation of orexin neurons by the limbic system maintains wakefulness during emotional arousal by conveying various emotional stimuli to orexin neurons.

The regulation of orexin neurons by the limbic system is also implicated in pathophysiology of cataplexy, as strong, generally positive emotional stimuli are well known to trigger this phenomenon in patients. Cholinergic neurons in the PPT have a role in REM sleep-related atonia (Shiromani et al. 1988) and are therefore likely to be implicated in cataplexy as well. Indeed, local injections of orexin into the PPT strongly inhibited REM-related atonia in cats (Takakusaki et al. 2005). Thus, excitatory input from the limbic system might increase orexin release in the PPT in order to sustain muscle tone that may be required to respond to salient situations.

Orexin neurons send abundant projections to the LC, and noradrenergic neurons strongly express OX1R (Marcus et al. 2001; Mieda et al. 2011). The orexin–LC pathway was shown to be involved in the formation of emotional memory in mice (Soya et al. 2013). Mice lacking OX1R displayed reduced freezing (a behavioral expression of fear) and reduced lateral amygdala activation (as measured by expression of the immediate-early gene *Zif268*) in response to cued- and contextual fear stimuli (Soya et al. 2013). Interestingly, this study showed that re-expression of OX1R in noradrenergic neurons in the LC by an AAV-mediated gene transfer in these mice restored both freezing time and lateral amygdala activation in the test phase of the cued fear conditioning procedure, but not in the contextual fear procedure. Mice lacking the OX2R ($Ox2r^{-/-}$ mice) also showed reduced freezing in the contextual fear test, but normal freezing in the cued fear test (Soya et al. 2013). This study thus suggested that OX1R, but not OX2R, plays a major role in the establishment of explicit cue-dependent emotional memory.

Icv infusion of the 1-SORA SB 334867 blocked the establishment of long-term fear memory, whereas infusion of the 2-SORA TCS-OX2-29 did not (Sears et al. 2013). Furthermore, blockade of OX1R signalling in the LC before conditioning, but not immediately after conditioning inhibited threat-memory formation. These findings suggest that OX1R signalling is important during the learning phase of fear memory formation. In human, individuals with narcolepsy showed reduced amygdala activity during aversive conditioning (Ponz et al. 2010), suggesting that orexin deficiency may result in impaired emotional learning.

A recent study using pharmacological MRI in rats found that amphetamine-induced activation in the extended amygdala, BNST and NAc, regions involved in emotion processing and emotional memory formation, was attenuated by administration of the 1-SORA GSK1059865, whereas activation of the frontal cortex and thalamus, regions that are involved in regulating arousal, was attenuated by the 2-SORA JNJ-1037049 (Gozzi et al. 2011). These findings suggest that OX2R plays a major role in maintenance of arousal, while OX1R is predominantly involved in processing emotive or reward information.

Together, these observations suggest that the orexins are important in the formation of cued fear memory. This is consistent with the observation that human narcolepsy patients show impairments in fear response acquisition as well as reduced amygdala activity relative to controls when exposed to aversively conditioned stimuli (Ponz et al. 2010; Khatami et al. 2007).

Stress response is also closely related with emotion, and orexin is also involved in this response. Early studies showed that icv administration of orexin results in increased CRH levels in the hypothalamus and in activation of the hypothalamus–pituitary–adrenal (HPA) axis (Al-Barazanji et al. 2001; Sakamoto et al. 2004) and, conversely, that orexin neurons are activated by CRH (Winsky-Sommerer et al. 2004, 2005). Indeed, an in vitro study, application of CRH depolarized the membrane potential and increased the firing rate in a subpopulation of orexin neurons by activating CRHR1 on these neurons (Winsky-Sommerer et al. 2004). These findings are in accordance with the reciprocal connections between CRH neurons in the PVN and orexin neurons in the LHA (Winsky-Sommerer et al. 2004) and suggest that orexin might play a role in activating the HPA axis in response to stress. Indeed, forced swim stress caused orexin neuron activation and increased plasma levels of adrenocorticotropic hormone (ACTH) (Chang et al. 2007), and icv administration of a 2-SORA reduced this ACTH response. Considering that the PVN expresses abundant OX2R mRNA (Marcus et al. 2001), these results suggest that stress increases orexin neuron firing to stimulate CRH neurons in the PVN through OX2R.

Numbers of Fos-positive orexin neurons were reported to be increased by a fear-conditioned cue, but not by restraint stress (Furlong et al. 2009). These observations again suggest that orexin neurons are activated by psychological stressors. This response seems to be an appropriate response to which requires proper vigilance level and attention to environmental cues.

4 Therapeutic Potential of Orexin Agonists/Antagonists

Several orexin receptor antagonists (DORAs and SORAs) are expected to become next-generation drugs for insomnia. To date, several orexin receptor antagonists with different pharmacological characteristics have been developed (Table 2), and a DORA suvorexant has been already clinically available for insomnia disorder in Japan and the U.S. OX2R selective antagonists were also shown to be effective in animal studies for inducing sleep (Dugovic et al. 2009; Etori et al. 2014). The clinical features of orexin receptor antagonists for insomnia treatment are discussed in detail in a separate chapter (Uslaner).

On the other hand, orexin agonists a promising candidate for narcolepsy treatment in the future. Given the broad range of function of the orexin system, these drugs might be also beneficial for treating a variety of conditions other than sleep disorders.

As discussed above, it has been shown that orexin mediates many behaviors associated with drug addiction in rodents due to its effects on the VTA (Harris et al. 2007). It is reasonable to speculate that orexin receptor antagonists might be effective for treating drug addiction. Recent report showed that orexin-1 receptor antagonist SB-334867 reduces the acquisition and expression of cocaine-conditioned reinforcement and the expression of amphetamine-conditioned reward, suggesting that OXR1-selective antagonists (1-SORA) have the potential to be a treatment for individuals struggling with drug relapse and dependency (Hutcheson et al. 2011). 1-SORA might also be effective for panic disorders, because they were shown to inhibit elevations of mean arterial pressure, heart rates and freezing responses in rat models of panic disorder (Johnson et al. 2010).

Table 2 Orexin receptor antagonists

	Compound	Affinity		Units	Reference
		OX1R	OX2R		
DORA	ACT-078573 (almorexant)	7.9 (human), 7.8 (rat)	8.1 (human), 7.8 (rat)	pIC_{50}	Brisbare-Roch et al. (2007)
DORA	MK-4305	9.26	9.46	pK_i	Cox et al. (2010)
OX1R SORA	SB-410220	7.7	nd	pK_i	Langmead et al. (2004)
OX1R SORA	SB-334867	7.2	nd	pK_i	Langmead et al. (2004)
OX1R SORA	SB-408124	7	nd	pK_i	Langmead et al. (2004)
OX1R SORA	[3H]SB-674042	8.3	nd	pK_d	Langmead et al. (2004)
OX1R SORA	SB-410220	8.1	6.3	pK_b	Langmead et al. (2004)
OX1R SORA	SB-334867	7.4	5.7	pK_b	Porter et al. (2001)
OX1R SORA	SB-408124	7.7	5.9	pK_b	Langmead et al. (2004)
OX1R SORA	SB-674042	9	6.9	pK_b	Langmead et al. (2004)
OX2R SORA	1-(2-bromo-phenyl)-3-((4S,5S)-2,2-dimethyl-4-phenyl-[1,3] dioxan-5-yl)-urea	5.3–6.1	6.8–7.1	pK_i	McAtee et al. (2004)
OX2R SORA	1-(2,4-dibromo-phenyl)-3-((4S,5S)-2,2-dimethyl-4-phenyl-[1,3]dioxan-5-yl)-urea (JNJ-10397049)	5.3–5.8	8.0–8.6	pK_i	McAtee et al. (2004)

5 Conclusion and Future Perspective

Proper arousal levels are necessary for executing any purposeful behavior that requires high motivation. These behaviors are usually triggered by external cues. These functions are closely related and are interconnected through the orexin system.

References

Akiyama M et al (2004) Reduced food anticipatory activity in genetically orexin (hypocretin) neuron-ablated mice. Eur J Neurosci 20:3054–3062

Al-Barazanji KA, Wilson S, Baker J, Jessop DS, Harbuz MS (2001) Central orexin-A activates hypothalamic-pituitary-adrenal axis and stimulates hypothalamic corticotropin releasing factor and arginine vasopressin neurones in conscious rats. J Neuroendocrinol 13:421–424

Anand BK, Brobeck JR (1951) Localization of a "feeding center" in the hypothalamus of the rat. Proc Soc Exp Biol Med 77:323–324

Baird JP et al (2009) Orexin-A hyperphagia: hindbrain participation in consummatory feeding responses. Endocrinology 150:1202–1216. doi:10.1210/en.2008-0293

Balcita-Pedicino JJ, Sesack SR (2007) Orexin axons in the rat ventral tegmental area synapse infrequently onto dopamine and gamma-aminobutyric acid neurons. J Comp Neurol 503:668–684

Berthoud HR (2004) Mind versus metabolism in the control of food intake and energy balance. Physiol Behav 81:781–793

Borgland SL, Taha SA, Sarti F, Fields HL, Bonci A (2006) Orexin A in the VTA is critical for the induction of synaptic plasticity and behavioral sensitization to cocaine. Neuron 49:589–601

Borgland SL et al (2009) Orexin A/hypocretin-1 selectively promotes motivation for positive reinforcers. J Neurosci 29:11215–11225. doi:10.1523/JNEUROSCI.6096-08.2009

Borgland SL, Ungless MA, Bonci A (2010) Convergent actions of orexin/hypocretin and CRF on dopamine neurons: emerging players in addiction. Brain Res 1314:139–144. doi:10.1016/j.brainres.2009.10.068

Boutrel B et al (2005) Role for hypocretin in mediating stress-induced reinstatement of cocaine-seeking behavior. Proc Natl Acad Sci USA 102:19168–19173

Brisbare-Roch C et al (2007) Promotion of sleep by targeting the orexin system in rats, dogs and humans. Nat Med 13:150–155. doi:10.1038/nm1544 nm1544 [pii]

Brown RE, Sergeeva OA, Eriksson KS, Haas HL (2002) Convergent excitation of dorsal raphe serotonin neurons by multiple arousal systems (orexin/hypocretin, histamine and noradrenaline). J Neurosci 22:8850–8859

Burdakov D, Gerasimenko O, Verkhratsky A (2005) Physiological changes in glucose differentially modulate the excitability of hypothalamic melanin-concentrating hormone and orexin neurons in situ. J Neurosci 25:2429–2433. doi:10.1523/JNEUROSCI.4925-04.2005 25/9/2429 [pii]

Burdakov D et al (2006) Tandem-pore K+ channels mediate inhibition of orexin neurons by glucose. Neuron 50:711–722

Cason AM et al (2010) Role of orexin/hypocretin in reward-seeking and addiction: implications for obesity. Physiol Behav 100:419–428. doi:10.1016/j.physbeh.2010.03.009

Chang GQ, Karatayev O, Davydova Z, Leibowitz SF (2004) Circulating triglycerides impact on orexigenic peptides and neuronal activity in hypothalamus. Endocrinology 145:3904–3912. doi:10.1210/en.2003-1582

Chang H et al (2007) Inhibitory effects of an orexin-2 receptor antagonist on orexin A- and stress-induced ACTH responses in conscious rats. Neurosci Res 57:462–466. doi:10.1016/j.neures.2006.11.009

Chemelli RM et al (1999) Narcolepsy in orexin knockout mice: molecular genetics of sleep regulation. Cell 98:437–451

Choi DL, Davis JF, Fitzgerald ME, Benoit SC (2010) The role of orexin-A in food motivation, reward-based feeding behavior and food-induced neuronal activation in rats. Neuroscience 167:11–20. doi:10.1016/j.neuroscience.2010.02.002

Cox CD et al (2010) Discovery of the dual orexin receptor antagonist [(7R)-4-(5-chloro-1,3-benzoxazol-2-yl)-7-methyl-1,4-diazepan-1-yl][5-methy l-2-(2H-1,2,3-triazol-2-yl) phenyl]methanone (MK-4305) for the treatment of insomnia. J Med Chem 53:5320–5332. doi:10.1021/jm100541c

Date Y et al (1999) Orexins, orexigenic hypothalamic peptides, interact with autonomic, neuroendocrine and neuroregulatory systems. Proc Natl Acad Sci USA 96:748–753

de Lecea L et al (1998) The hypocretins: hypothalamus-specific peptides with neuroexcitatory activity. Proc Natl Acad Sci USA 95:322–327

Di Sebastiano AR, Wilson-Perez HE, Lehman MN, Coolen LM (2011) Lesions of orexin neurons block conditioned place preference for sexual behavior in male rats. Horm Behav 59:1–8. doi:10.1016/j.yhbeh.2010.09.006

Dube MG, Kalra SP, Kalra PS (1999) Food intake elicited by central administration of orexins/hypocretins: identification of hypothalamic sites of action. Brain Res 842:473–477

Dugovic C et al (2009) Blockade of orexin-1 receptors attenuates orexin-2 receptor antagonism-induced sleep promotion in the rat. J Pharmacol Exp Ther 330:142–151. doi:10.1124/jpet.109.152009 jpet.109.152009 [pii]

Elias CF et al (1998) Chemically defined projections linking the mediobasal hypothalamus and the lateral hypothalamic area. J Comp Neurol 402:442–459

Espana RA et al (2010) The hypocretin-orexin system regulates cocaine self-administration via actions on the mesolimbic dopamine system. Eur J Neurosci 31:336–348. doi:10.1111/j.1460-9568.2009.07065.x

Espana RA, Melchior JR, Roberts DC, Jones SR (2011) Hypocretin 1/orexin A in the ventral tegmental area enhances dopamine responses to cocaine and promotes cocaine self-administration. Psychopharmacology 214:415–426. doi:10.1007/s00213-010-2048-8

Estabrooke IV et al (2001) Fos expression in orexin neurons varies with behavioral state. J Neurosci 21:1656–1662

Etori K, Saito YC, Tsujino N, Sakurai T (2014) Effects of a newly developed potent orexin-2 receptor-selective antagonist, compound 1 m, on sleep/wakefulness states in mice. Front Neurosci 8:8. doi:10.3389/fnins.2014.00008

Funato H et al (2009) Enhanced orexin receptor-2 signaling prevents diet-induced obesity and improves leptin sensitivity. Cell Metab 9:64–76. doi:10.1016/j.cmet.2008.10.010 S1550-4131(08)00351-3 [pii]

Furlong TM, Vianna DM, Liu L, Carrive P (2009) Hypocretin/orexin contributes to the expression of some but not all forms of stress and arousal. Eur J Neurosci 30:1603–1614. doi:10.1111/j.1460-9568.2009.06952.x

Furutani N, Hondo M, Tsujino N, Sakurai T (2010) Activation of bombesin receptor subtype-3 influences activity of orexin neurons by both direct and indirect pathways. J Mol Neurosci 42:106–111. doi:10.1007/s12031-010-9382-5

Furutani N et al (2013) Neurotensin co-expressed in orexin-producing neurons in the lateral hypothalamus plays an important role in regulation of sleep/wakefulness states. PLoS ONE 8:e62391. doi:10.1371/journal.pone.0062391

Georgescu D et al (2003) Involvement of the lateral hypothalamic peptide orexin in morphine dependence and withdrawal. J Neurosci 23:3106–3111. 23/8/3106 [pii]

Gozzi A et al (2011) Functional magnetic resonance imaging reveals different neural substrates for the effects of orexin-1 and orexin-2 receptor antagonists. PLoS ONE 6:e16406. doi:10.1371/journal.pone.0016406

Hagan JJ et al (1999) Orexin A activates locus coeruleus cell firing and increases arousal in the rat. Proc Natl Acad Sci USA 96:10911–10916

Hara J et al (2001) Genetic ablation of orexin neurons in mice results in narcolepsy, hypophagia, and obesity. Neuron 30:345–354. S0896-6273(01)00293-8 [pii]

Hara J, Yanagisawa M, Sakurai T (2005) Difference in obesity phenotype between orexin-knockout mice and orexin neuron-deficient mice with same genetic background and environmental conditions. Neurosci Lett 380:239–242. doi:10.1016/j.neulet.2005.01.046

Harris GC, Wimmer M, Aston-Jones G (2005) A role for lateral hypothalamic orexin neurons in reward seeking. Nature 437:556–559

Harris GC, Wimmer M, Randall-Thompson JF, Aston-Jones G (2007) Lateral hypothalamic orexin neurons are critically involved in learning to associate an environment with morphine reward. Behav Brain Res 183:43–51

Haynes AC et al (2000) A selective orexin-1 receptor antagonist reduces food consumption in male and female rats. Regul Pept 96:45–51

Hilton SM (1982) The defence-arousal system and its relevance for circulatory and respiratory control. J Exp Biol 100:159–174

Hollander JA, Lu Q, Cameron MD, Kamenecka TM, Kenny PJ (2008) Insular hypocretin transmission regulates nicotine reward. Proc Natl Acad Sci USA 105:19480–19485. doi:10.1073/pnas.0808023105

Horvath TL et al (1999) Hypocretin (orexin) activation and synaptic innervation of the locus coeruleus noradrenergic system. J Comp Neurol 415:145–159
Hutcheson DM et al (2011) Orexin-1 receptor antagonist SB-334867 reduces the acquisition and expression of cocaine-conditioned reinforcement and the expression of amphetamine-conditioned reward. Behav Pharmacol 22:173–181. doi:10.1097/FBP.0b013e328343d761
James MH et al (2011) Orexin-1 receptor signalling within the ventral tegmental area, but not the paraventricular thalamus, is critical to regulating cue-induced reinstatement of cocaine-seeking. Int J Neuropsychopharmacol/official scientific journal of the Collegium Internationale Neuropsychopharmacologicum 14:684–690. doi:10.1017/S1461145711000423
Johnson PL et al (2010) A key role for orexin in panic anxiety. Nat Med 16:111–115. doi:10.1038/nm.2075 nm.2075 [pii]
Karnani MM et al (2011) Activation of central orexin/hypocretin neurons by dietary amino acids. Neuron 72:616–629. doi:10.1016/j.neuron.2011.08.027
Kayaba Y et al (2003) Attenuated defense response and low basal blood pressure in orexin knockout mice. Am J Physiol Regul Integr Comp Physiol 285:R581–R593
Khatami R, Birkmann S, Bassetti CL (2007) Amygdala dysfunction in narcolepsy-cataplexy. J Sleep Res 16:226–229. doi:10.1111/j.1365-2869.2007.00587.x
Langmead CJ et al (2004) Characterisation of the binding of [3H]-SB-674042, a novel nonpeptide antagonist, to the human orexin-1 receptor. Br J Pharmacol 141:340–346. doi:10.1038/sj.bjp.0705610 sj.bjp.0705610 [pii]
Lee MG, Hassani OK, Jones BE (2005) Discharge of identified orexin/hypocretin neurons across the sleep-waking cycle. J Neurosci 25:6716–6720
LeSage MG, Perry JL, Kotz CM, Shelley D, Corrigall WA (2010) Nicotine self-administration in the rat: effects of hypocretin antagonists and changes in hypocretin mRNA. Psychopharmacology 209:203–212. doi:10.1007/s00213-010-1792-0
Li Y, Gao XB, Sakurai T, van den Pol AN (2002) Hypocretin/orexin excites hypocretin neurons via a local glutamate neuron-A potential mechanism for orchestrating the hypothalamic arousal system. Neuron 36:1169–1181
Lin L et al (1999) The sleep disorder canine narcolepsy is caused by a mutation in the hypocretin (orexin) receptor 2 gene. Cell 98:365–376
Liu RJ, van den Pol AN, Aghajanian GK (2002) Hypocretins (orexins) regulate serotonin neurons in the dorsal raphe nucleus by excitatory direct and inhibitory indirect actions. J Neurosci 22:9453–9464
Mahler SV, Smith RJ, Moorman DE, Sartor GC, Aston-Jones G (2012) Multiple roles for orexin/hypocretin in addiction. Prog Brain Res 198:79–121. doi:10.1016/B978-0-444-59489-1.00007-0
Marcus JN et al (2001) Differential expression of orexin receptors 1 and 2 in the rat brain. J Comp Neurol 435:6–25
Martin-Fardon R, Boutrel B (2012) Orexin/hypocretin (Orx/Hcrt) transmission and drug-seeking behavior: is the paraventricular nucleus of the thalamus (PVT) part of the drug seeking circuitry? Front Behav Neurosci 6:75. doi:10.3389/fnbeh.2012.00075
McAtee LC et al (2004) Novel substituted 4-phenyl-[1,3]dioxanes: potent and selective orexin receptor 2 (OX(2)R) antagonists. Bioorg Med Chem Lett 14:4225–4229. doi:10.1016/j.bmcl.2004.06.032 S0960894X0400767X [pii]
Mieda M, Sakurai T (2012) Overview of orexin/hypocretin system. Prog Brain Res 198:5–14. doi:10.1016/B978-0-444-59489-1.00002-1
Mieda M et al (2004) Orexin neurons function in an efferent pathway of a food-entrainable circadian oscillator in eliciting food-anticipatory activity and wakefulness. J Neurosci 24:10493–10501
Mieda M et al (2011) Differential roles of orexin receptor-1 and -2 in the regulation of non-REM and REM sleep. J Neurosci 31:6518–6526. doi:10.1523/JNEUROSCI.6506-10.2011 31/17/6518 [pii]
Mignot E (1998) Genetic and familial aspects of narcolepsy. Neurology 50:S16–S22

Mignot E et al (2002) The role of cerebrospinal fluid hypocretin measurement in the diagnosis of narcolepsy and other hypersomnias. Arch Neurol 59:1553–1562

Mileykovskiy BY, Kiyashchenko LI, Siegel JM (2005) Behavioral correlates of activity in identified hypocretin/orexin neurons. Neuron 46:787–798

Nair SG, Golden SA, Shaham Y (2008) Differential effects of the hypocretin 1 receptor antagonist SB 334867 on high-fat food self-administration and reinstatement of food seeking in rats. Br J Pharmacol 154:406–416. doi:10.1038/bjp.2008.3

Nakamura T et al (2000) Orexin-induced hyperlocomotion and stereotypy are mediated by the dopaminergic system. Brain Res 873:181–187

Nambu T et al (1999) Distribution of orexin neurons in the adult rat brain. Brain Res 827:243–260

Narita M et al (2006) Direct involvement of orexinergic systems in the activation of the mesolimbic dopamine pathway and related behaviors induced by morphine. J Neurosci 26:398–405

Pace-Schott EF, Hobson JA (2002) The neurobiology of sleep: genetics, cellular physiology and subcortical networks. Nat Rev Neurosci 3:591–605. doi:10.1038/nrn895 nrn895 [pii]

Pasumarthi RK, Reznikov LR, Fadel J (2006) Activation of orexin neurons by acute nicotine. Eur J Pharmacol 535:172–176. doi:10.1016/j.ejphar.2006.02.021

Petrovich GD, Hobin MP, Reppucci CJ (2012) Selective Fos induction in hypothalamic orexin/hypocretin, but not melanin-concentrating hormone neurons, by a learned food-cue that stimulates feeding in sated rats. Neuroscience 224:70–80. doi:10.1016/j.neuroscience.2012.08.036

Peyron C et al (1998) Neurons containing hypocretin (orexin) project to multiple neuronal systems. J Neurosci 18:9996–10015

Peyron C et al (2000) A mutation in a case of early onset narcolepsy and a generalized absence of hypocretin peptides in human narcoleptic brains. Nat Med 9:991–997

Plaza-Zabala A et al (2013) A role for hypocretin/orexin receptor-1 in cue-induced reinstatement of nicotine-seeking behavior. Neuropsychopharmacology 38:1724–1736. doi:10.1038/npp.2013.72

Ponz A et al (2010) Reduced amygdala activity during aversive conditioning in human narcolepsy. Ann Neurol 67:394–398. doi:10.1002/ana.21881

Porter RA et al (2001) 1,3-Biarylureas as selective non-peptide antagonists of the orexin-1 receptor. Bioorg Med Chem Lett 11:1907–1910. S0960894X01003432 [pii]

Quarta D, Valerio E, Hutcheson DM, Hedou G, Heidbreder C (2010) The orexin-1 receptor antagonist SB-334867 reduces amphetamine-evoked dopamine outflow in the shell of the nucleus accumbens and decreases the expression of amphetamine sensitization. Neurochem Int 56:11–15. doi:10.1016/j.neuint.2009.08.012

Reid MS et al (1998) Neuropharmacological characterization of basal forebrain cholinergic stimulated cataplexy in narcoleptic canines. Exp Neurol 151:89–104

Richardson KA, Aston-Jones G (2012) Lateral hypothalamic orexin/hypocretin neurons that project to ventral tegmental area are differentially activated with morphine preference. J Neurosci 32:3809–3817. doi:10.1523/JNEUROSCI.3917-11.2012

Saito YC et al (2013) GABAergic neurons in the preoptic area send direct inhibitory projections to orexin neurons. Front Neural Circuits 7:192. doi:10.3389/fncir.2013.00192

Sakamoto F, Yamada S, Ueta Y (2004) Centrally administered orexin-A activates corticotropin-releasing factor-containing neurons in the hypothalamic paraventricular nucleus and central amygdaloid nucleus of rats: possible involvement of central orexins on stress-activated central CRF neurons. Regul Pept 118:183–191

Sakurai T (2007) The neural circuit of orexin (hypocretin): maintaining sleep and wakefulness. Nat Rev Neurosci 8:171–181

Sakurai T, Mieda M (2011) Connectomics of orexin-producing neurons: interface of systems of emotion, energy homeostasis and arousal. Trends Pharmacol Sci 32:451–462. doi:10.1016/j.tips.2011.03.007 S0165-6147(11)00050-2 [pii]

Sakurai T et al (1998) Orexins and orexin receptors: a family of hypothalamic neuropeptides and G protein-coupled receptors that regulate feeding behavior. Cell 92:573–585

Sakurai T et al (2005) Input of orexin/hypocretin neurons revealed by a genetically encoded tracer in mice. Neuron 46:297–308. doi:10.1016/j.neuron.2005.03.010 S0896-6273(05)00205-9 [pii]

Sartor GC, Aston-Jones GS (2012) A septal-hypothalamic pathway drives orexin neurons, which is necessary for conditioned cocaine preference. J Neurosci 32:4623–4631. doi:10.1523/JNEUROSCI.4561-11.2012

Schone C, Apergis-Schoute J, Sakurai T, Adamantidis A, Burdakov D (2014) Coreleased orexin and glutamate evoke nonredundant spike outputs and computations in histamine neurons. Cell Rep. doi:10.1016/j.celrep.2014.03.055

Sears RM et al (2013) Orexin/hypocretin system modulates amygdala-dependent threat learning through the locus coeruleus. Proc Natl Acad Sci USA 110:20260–20265. doi:10.1073/pnas.1320325110

Sharf R et al (2010a) Orexin signaling via the orexin 1 receptor mediates operant responding for food reinforcement. Biol Psychiatry 67:753–760. doi:10.1016/j.biopsych.2009.12.035

Sharf R, Guarnieri DJ, Taylor JR, DiLeone RJ (2010b) Orexin mediates morphine place preference, but not morphine-induced hyperactivity or sensitization. Brain Res 1317:24–32. doi:10.1016/j.brainres.2009.12.035

Shiraishi T, Oomura Y, Sasaki K, Wayner MJ (2000) Effects of leptin and orexin-A on food intake and feeding related hypothalamic neurons. Physiol Behav 71:251–261

Shiromani PJ, Armstrong DM, Berkowitz A, Jeste DV, Gillin JC (1988) Distribution of choline acetyltransferase immunoreactive somata in the feline brainstem: implications for REM sleep generation. Sleep 11:1–16

Shouse MN, Siegel JM (1992) Pontine regulation of REM sleep components in cats: integrity of the pedunculopontine tegmentum (PPT) is important for phasic events but unnecessary for atonia during REM sleep. Brain Res 571:50–63

Smith RJ, Aston-Jones G (2012) Orexin/hypocretin 1 receptor antagonist reduces heroin self-administration and cue-induced heroin seeking. Eur J Neurosci 35:798–804. doi:10.1111/j.1460-9568.2012.08013.x

Smith RJ, See RE, Aston-Jones G (2009) Orexin/hypocretin signaling at the orexin 1 receptor regulates cue-elicited cocaine-seeking. Eur J Neurosci 30:493–503. doi:10.1111/j.1460-9568.2009.06844.x

Smith RJ, Tahsili-Fahadan P, Aston-Jones G (2010) Orexin/hypocretin is necessary for context-driven cocaine-seeking. Neuropharmacology 58:179–184. doi:10.1016/j.neuropharm.2009.06.042

Soya S et al (2013) Orexin receptor-1 in the locus coeruleus plays an important role in cue-dependent fear memory consolidation. J Neurosci 33:14549–14557. doi:10.1523/JNEUROSCI.1130-13.2013

Steiner MA et al (2013) Discovery and characterization of ACT-335827, an orally available, brain penetrant orexin receptor type 1 selective antagonist. ChemMedChem 8:898–903. doi:10.1002/cmdc.201300003

Strawn JR, Geracioti TD Jr (2008) Noradrenergic dysfunction and the psychopharmacology of posttraumatic stress disorder. Depress Anxiety 25:260–271. doi:10.1002/da.20292

Sweet DC, Levine AS, Billington CJ, Kotz CM (1999) Feeding response to central orexins. Brain Res 821:535–538

Takahashi K, Koyama Y, Kayama Y, Yamamoto M (2002) Effects of orexin on the laterodorsal tegmental neurones. Psychiatry Clin Neurosci 56:335–336

Takahashi K, Lin JS, Sakai K (2008) Neuronal activity of orexin and non-orexin waking-active neurons during wake-sleep states in the mouse. Neuroscience 153:860–870

Takakusaki K et al (2005) Orexinergic projections to the midbrain mediate alternation of emotional behavioral states from locomotion to cataplexy. J Physiol 568(Pt 3):1003–1020

Thannickal TC et al (2000) Reduced number of hypocretin neurons in human narcolepsy. Neuron 27:469–474

Thorpe AJ, Kotz CM (2005) Orexin A in the nucleus accumbens stimulates feeding and locomotor activity. Brain Res 1050:156–162. doi:10.1016/j.brainres.2005.05.045

Thorpe AJ, Mullett MA, Wang C, Kotz CM (2003) Peptides that regulate food intake: regional, metabolic, and circadian specificity of lateral hypothalamic orexin A feeding stimulation. Am J Physiol Regul Integr Comp Physiol 284:R1409–R1417. doi:10.1152/ajpregu.00344.2002

Thorpe AJ, Teske JA, Kotz CM (2005) Orexin A-induced feeding is augmented by caloric challenge. Am J Physiol Regul Integr Comp Physiol 289:R367–R372. doi:10.1152/ajpregu.00737.2004

Tsujino N et al (2005) Cholecystokinin activates orexin/hypocretin neurons through the cholecystokinin A receptor. J Neurosci 25:7459–7469

Tucci V et al (2003) Emotional information processing in patients with narcolepsy: a psychophysiologic investigation. Sleep 26:558–564

van den Pol AN, Acuna-Goycolea C, Clark KR, Ghosh PK (2004) Physiological properties of hypothalamic MCH neurons identified with selective expression of reporter gene after recombinant virus infection. Neuron 42:635–652

Vanni-Mercier G, Sakai K, Jouvet M (1984) Neurons specifiques de l'eveil dans l'hypothalamus posterieur du chat. CR Acad Sci III 298:195–200

Venner A et al (2011) Orexin neurons as conditional glucosensors: paradoxical regulation of sugar sensing by intracellular fuels. J Physiol 589:5701–5708. doi:10.1113/jphysiol.2011.217000

Willie JT, Chemelli RM, Sinton CM, Yanagisawa M (2001) To eat or to sleep? Orexin in the regulation of feeding and wakefulness. Annu Rev Neurosci 24:429–458

Winrow CJ et al (2010) Orexin receptor antagonism prevents transcriptional and behavioral plasticity resulting from stimulant exposure. Neuropharmacology 58:185–194. doi:10.1016/j.neuropharm.2009.07.008

Winsky-Sommerer R et al (2004) Interaction between the corticotropin-releasing factor system and hypocretins (orexins): a novel circuit mediating stress response. J Neurosci 24:11439–11448

Winsky-Sommerer R, Boutrel B, de Lecea L (2005) Stress and arousal: the corticotrophin-releasing factor/hypocretin circuitry. Mol Neurobiol 32:285–294

Xi M, Morales FR, Chase MH (2001) Effects on sleep and wakefulness of the injection of hypocretin-1 (orexin-A) into the laterodorsal tegmental nucleus of the cat. Brain Res 901:259–264

Yamada H, Okumura T, Motomura W, Kobayashi Y, Kohgo Y (2000) Inhibition of food intake by central injection of anti-orexin antibody in fasted rats. Biochem Biophys Res Commun 267:527–531. doi:10.1006/bbrc.1999.1998

Yamanaka A et al (2000) Orexin-induced food intake involves neuropeptide Y pathway. Brain Res 859:404–409. S0006899300020436 [pii]

Yamanaka A et al (2002) Orexins activate histaminergic neurons via the orexin 2 receptor. Biochem Biophys Res Commun 290:1237–1245

Yamanaka A et al (2003a) Hypothalamic orexin neurons regulate arousal according to energy balance in mice. Neuron 38:701–713

Yamanaka A, Muraki Y, Tsujino N, Goto K, Sakurai T (2003b) Regulation of orexin neurons by the monoaminergic and cholinergic systems. Biochem Biophys Res Commun 303:120–129

Yamanaka A et al (2006) Orexin neurons are directly and indirectly regulated by catecholamines in a complex manner. J Neurophysiol 96:284–298

Yamanaka A, Tabuchi S, Tsunematsu T, Fukazawa Y, Tominaga M (2010) Orexin directly excites orexin neurons through orexin 2 receptor. J Neurosci 30:12642–12652. doi:10.1523/JNEUROSCI.2120-10.2010 30/38/12642 [pii]

Yin J, Mobarec JC, Kolb P, Rosenbaum DM (2014) Crystal structure of the human OX orexin receptor bound to the insomnia drug suvorexant. Nature. doi:10.1038/nature14035

Yoshida Y et al (2001) Fluctuation of extracellular hypocretin-1 (orexin A) levels in the rat in relation to the light-dark cycle and sleep-wake activities. Eur J Neurosci 14:1075–1081

Yoshida K, McCormack S, Espana RA, Crocker A, Scammell TE (2006) Afferents to the orexin neurons of the rat brain. J Comp Neurol. 494:845–861

Zhang W, Sakurai T, Fukuda Y, Kuwaki T (2006) Orexin neuron-mediated skeletal muscle vasodilation and shift of baroreflex during defense response in mice. Am J Physiol Regul Integr Comp Physiol (in press)

Zhang W, Zhang N, Sakurai T, Kuwaki T (2009) Orexin neurons in the hypothalamus mediate cardiorespiratory responses induced by disinhibition of the amygdala and bed nucleus of the stria terminalis. Brain Res 1262:25–37. doi:10.1016/j.brainres.2009.01.022 S0006-8993(09)00125-5 [pii]

Zhang W et al (2010) Orexin neurons are indispensable for stress-induced thermogenesis in mice. J Physiol 588:4117–4129. doi:10.1113/jphysiol.2010.195099 jphysiol.2010.195099 [pii]

Zheng H, Patterson LM, Berthoud HR (2007) Orexin signaling in the ventral tegmental area is required for high-fat appetite induced by opioid stimulation of the nucleus accumbens. J Neurosci (the official journal of the Society for Neuroscience) 27:11075–11082. doi:10.1523/JNEUROSCI.3542-07.2007

The Hypocretin Story

Luis de Lecea

Abstract The main topic of this book stems from the discovery of two neuropeptides derived from the same precursor, conducted in parallel by two different groups (de Lecea et al. in Proc Natl Acad Sci USA 95:322–327, 1998; Sakurai et al. in Cell 92:573–585, 1998). Here is the account of the discovery of the hypocretin peptides, which was the result of a larger effort to characterize patterns of gene expression in the brain. The original hypothesis leading to the discovery of the peptides was that new mechanisms underlying the many functions of the hypothalamus (e.g. feeding, thermoregulation, circadian rhythmicity, sexual behavior and arousal) could be unraveled by identifying the expression patterns of hundreds of "cell-type" specific transcripts. This hypothesis ended up converging with a reverse pharmacology approach followed by Yanagisawa and colleagues, which is described in detail in other chapters. With the hypocretin story in the background, I also describe here the first implementation of optogenetic methods in a freely moving animal, which has led to a revolution in systems neuroscience.

1 Clones of Hypothalamus-Enriched mRNAs

The hypothalamus has long been recognized as a site for central regulation of homeostasis (Bloom 1987). In contrast to laminar cortical structures such as the cerebellum and hippocampus whose final functions rely on input from the thalamus and brain stem, the hypothalamus is organized as a collection of distinct, autonomously active nuclei with discrete functions (Swanson 1999). Ablation and electrical stimulation studies have implicated several of these nuclei as central regulatory centers for major autonomic and endocrine homeostatic systems mediating processes such as reproduction, lactation, fluid balance, blood pressure, thermoregulation,

L. de Lecea (✉)
Department of Psychiatry and Behavioral Sciences, Stanford University School of Medicine, Stanford, CA 94305, USA
e-mail: llecea@stanford.edu

© Springer International Publishing Switzerland 2015
T. Sakurai et al. (eds.), *Orexin and Sleep*,
DOI 10.1007/978-3-319-23078-8_2

metabolism, and aspects of behaviors, such as circadian rhythmicity, basic emotions, the sleep/wake cycle, feeding and drinking, mating activities, and responses to stress, as well as normal development of the immune system. Distinct hormones and releasing factors have been associated with some of these nuclei.

To illuminate additional molecules that contribute to the specialized functions of hypothalamic nuclei, Bloom and Sutcliffe embarked on a systematic analysis of the mRNAs whose expression is restricted to or enriched in the hypothalamus (Gautvik et al. 1996; Milner et al. 1987). The plan involved the following steps: Directional tag PCR subtractive hybridization (Usui et al. 1994) was used to enrich a cDNA library for clones of mRNA species selectively expressed in the hypothalamus. Candidate clones identified by their hybridization to a subtracted hypothalamus probe were validated in three stages. First, a high throughput cDNA library Southern blot was used to demonstrate that the candidate corresponded to a species enriched in the subtracted library. Second, candidate clones positive in the first assay were used as probes for Northern blots with RNA from several brain regions and peripheral tissues. Finally, candidate clones that were still positive were subjected to in situ hybridization analysis to detect the hypothalamic regions that express the corresponding mRNAs (Gautvik et al. 1996).

I joined the Sutcliffe lab as a postdoc in 1992. Hiroshi Usui, Mark Erlander and Ana Dopazo perfected the subtraction hybridization technique, and we used an in situ hybridization screening method I developed in Barcelona to generate a list of mRNAs enriched in the striatum (Usui et al. 1994). Using the subtraction and in situ screening methods, we reported the discovery of Cortistatin, a novel somatostatin-like gene expressed in the neocortex and hippocampus with neuronal depressant and sleep-inducing properties (de Lecea et al. 1996).

Kaare and Vigdis Gautvik of the University of Oslo visited Greg Sutcliffe's lab for a sabbatical year to learn the subtraction method. They were given the hypothalamus project and applied the directional tag PCR subtractive hybridization method to construct a rat hypothalamic cDNA library from which, in two consecutive steps, cerebellar and hippocampal sequences had been depleted, enriching 20-to-30 fold for sequences expressed selectively in the hypothalamus. They made a radioactive probe from the library inserts and used it to screen 648 clones from the subtracted library that they had spotted into grid arrays. Approximately 70 % of the colonies gave significant signals with the subtracted target probe compared to 50 % with an unsubtracted target probe. Only 10 % of the colonies gave signals with a mixed-driver probe. These data indicated that the enrichment by subtraction was substantial. Among the clones showing the highest degree of hypothalamus enrichment were cDNAs for oxytocin, vasopressin, CART, melanin concentrating hormone, POMC, VAT-1, and a novel species called clone 35.

As part of the screening effort, we conducted in situ hybridization on coronal sections of brain from adult male rats, using the inserts from clones representing many of the RNAs, including clone 35 (Fig. 1). The clone 35 mRNA displayed a striking pattern of bilaterally symmetric expression restricted to a few cells in the dorsolateral hypothalamic area, with no signals outside the hypothalamus (de Lecea et al. 1998). Other novel mRNAs showed substantial enrichment in basal

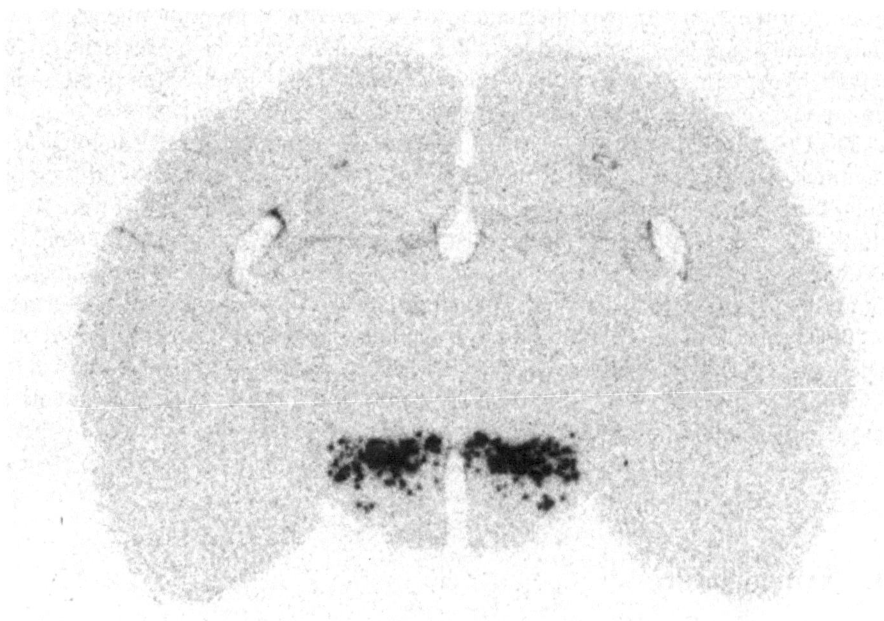

Fig. 1 The first glimpse of the hypocretin system. In situ hybridization of rat coronal section with cDNA isolated in subtractive hybridization study detecting a few thousand neurons in the dorsal-lateral hypothalamus [from Bloom (1987)]

diencephalic structures, particularly the hypothalamus, without restriction to single hypothalamic nuclei.

Among the novel species we identified, only clone 35 met our starting criterion in its strictest sense: the mRNA appeared to be restricted to a single nucleus in the dorsolateral area of the hypothalamus (the in situ hybridization image in reference 2 represents the first published description of the hypocretins).

2 Clone H35

We used the original rat cDNA clone 35 to isolate full-length cDNAs for both rat and mouse. The 569-nucleotide rat sequence (de Lecea et al. 1998) suggested that the corresponding mRNA encoded a 130-residue putative secretory protein with four pairs of tandem basic residues for potential proteolytic processing. We then isolated a cDNA clone from mouse libraries and its sequence revealed only that two of the four possible rat maturation products were conserved. Both of these terminated with glycine residues, which in proteolytically processed secretory peptides typically are substrates for peptidylglycine alpha-amidating monooxygenase, leaving a C-terminal amide in the mature peptide. These features suggested that the product of the clone 35 hypothalamic mRNA served as a preprohormone for two C-terminally amidated, secreted peptides. One of these, which was later to be named hypocretin 2 (hcrt2), was, on the basis of the putative preprohormone amino-acid sequence, predicted to contain precisely 28 residues.

The C-terminal 19 residues of these two putative peptides shared 13 amino acid identities. This region of one of the peptides contained a 7/7 match with secretin, suggesting that the preprohormone gave rise to two peptide products that were structurally related closely to each other and distantly to secretin. We initially commented on the secretin similarity. Sequence similarities with various members of the incretin family, especially secretin, suggested that the gene was formed from the secretin gene by three genetic rearrangements: first, a duplication of the secretin gene; second, deletions of the N-terminal portion of the 5′ duplicate and the C-terminal portion of the 3′ duplicate to yield a secretin with its N- and C- termini leap-frogged (circularly permuted); and third, a further duplication of the permuted gene, followed by modifications, to form a secretin derivative that encoded two related hypocretin peptides.

3 Nomenclature

As we began to write the paper describing our discovery of the peptides via cDNA cloning, their immunohistochemical detection, their presence in dense core vesicles at synapses, and their neuroexcitatory properties on hypothalamic neurons, we realized that we needed a name other than clone 35. There were several possible functions for the peptides, but direct evidence for none. Based on our previous discovery of Cortistatin (which we named after its predominantly *corti*cal expression and its similarity with somato*statin*), (de Lecea et al. 1996) we came up with several non-pejorative possibilities, most of which were variations on syllables abstracted from *hypo*thalamic member of the *incretin* family. The most straightforward of these possibilities was “hypocretin”. At the 1997 Society for Neuroscience Meeting, Greg Sutcliffe presented a poster describing the sequence of the new protein, the expression of its mRNA exclusively in a small number of neurons in the dorsolateral hypothalamus, the electron microscopic detection of immunoreactive vesicles in presynaptic boutons, and the neuroexcitatory properties of the amidated hcrt2 peptide. At the adjacent poster, Cristelle Peyron and Tom Kilduff presented the data detecting hypocretin immunoreactivity in the dorsolateral hypothalamic neurons and immunoreactive fibers through the CNS (Peyron et al. 1998).

At the poster, a ballot listing several of the names under consideration was listed and asked poster attendees to express their preference. The plurality of the votes were cast for hypocretin. Thus, this became the first democratically named neurotransmitter.

We sent a reformatted text using the revised name hypocretin to the *PNAS*, where it was accepted and published on January 6, 1998 (de Lecea et al. 1998). In the paper we noted that the existence of two hcrt peptides that differ in their amino acid sequences might indicate two Hcrt receptor subtypes (de Lecea et al. 1998).

4 Independent Discovery

The February 20, 1998 issue of *Cell* carried an article describing the identification of two endogenous peptides that stimulated calcium flux in cells transfected with an orphan G protein-coupled receptor (Sakurai et al. 1998). Sakurai and colleagues demonstrated that intracerebroventricular administration of either hcrt1 or hcrt2 increased food consumption in rats. Furthermore, rats fasted for 48 h increased the concentration of hypocretin mRNA and peptides. Based on these observations, they proposed the alternative name, orexins, for the hypocretin peptides (Sakurai et al. 1998).

This study demonstrated the actual presence of the two proteolytically processed hypocretin peptides within the brain, and elucidated by mass spectroscopy the exact structures of these endogenous peptides, which could not be deduced from nucleic acid sequences alone. The structure of orexin B was the same as that predicted from the hcrt2 cDNA sequence. The N-terminus of orexin A (hcrt1) was defined and found to correspond to a genetically encoded glutamine derivatized as pyroglutamate. The two intrachain disulfide bonds within hcrt1 were also defined.

A commentary about new hypothalamic factors accompanied the *Cell* report. It mentioned both the hypocretins and the orexins, without realizing that they were the same peptides. The omission was inadvertent but, unfortunately, this was the beginning of a dual nomenclature. The Yanagisawa group published an addendum in the March 6 *Cell* indicating that the orexins are the same as the hypocretins (Sakurai et al. 1998).

5 Hypocretin Functions

The papers describing the independent discovery and characterization of the hypocretins/orexins have catalyzed a great body of work. There are now already over 12,000 papers concerning aspects of these neurotransmitters, including their prominent role in arousal state transitions, brain reward, stress, panic and other neuropsychiatric conditions. The perifornical region of the hypothalamus has been associated with nutritional homeostasis, blood pressure and thermal regulation, neural control of endocrine secretion, brain reward and arousal. Thus, these activities ranked among those affected by the peptides (Sakurai 2014; Li et al. 2014).

The association of Hcrt deficiency with narcolepsy conducted in parallel in two different laboratories using, again, two independent approaches, clearly demonstrated that Hcrt system is a non-redundant peptidergic system whose critical function is to stabilize arousal states (de Lecea and Huerta 2014).

How do Hcrt neurons maintain wakefulness? In 2005, the groups of Siegel and Jones independently reported the recording of identified Hcrt neurons from awake animals (Mileykovskiy et al. 2005; Lee et al. 2005). These recordings revealed that Hcrt neurons are phasically active during active waking, and practically silent during quiet waking and sleep. Interestingly, the highest activity was observed during transitions between rapid eye movement sleep and wakefulness. This discovery also showed that the pharmacological approach to studying the role of Hcrt in sleep and wakefulness was not appropriate, as drugs take several minutes to diffuse into brain structures and therefore span several sleep/wake cycles in rodents. In order to determine precisely whether Hcrt release was instructive or permissive to wakefulness, other methodologies needed to be applied.

6 Optogenetics

To further explore the mechanism by which phasic Hcrt activity maintains wakefulness, we used a newly developed optogenetic approach to manipulate the activity of Hcrt neurons in vivo (Adamantidis et al. 2007). Optogenetics uses actuator opsin molecules (e.g., channelrhodopsin-2 (ChR2) or halorhodopsin—NpHR) to selectively activate or silence genetically-targeted cells, respectively, with flashes of light at specific wavelength (Boyden et al. 2005). Further information about optogenetic technology can be found in many other excellent reviews (Fenno et al. 2011).

In an effort to better understand the temporal dynamics of neural circuits of wakefulness, we applied optogenetics to reversibly and selectively manipulate the activity of hcrt neurons in freely-moving animals. To deliver light to the hcrt or LC field, we designed optical-neural interfaces in which optical fibers were chronically implanted on the mouse skull (Fig. 2). Using this strategy, we were able to control

Fig. 2 Light delivery into the brain. Picture of a mouse transduced with a virus expressing Channelrhodopsin in Hcrt neurons and cannulated with an optical fiber to stimulate Hcrt neurons with millisecond precision

hcrt neural activity both in vitro and in vivo with millisecond-precise optical stimulation (Adamantidis et al. 2007). The high temporal and spatial precision of stimulation allowed us to mimic the physiological range of hypocretin neu ron discharge rates (1–30 Hz) (Lee et al. 2005; Hassani et al. 2009).

Photostimulations of mice transduced with a Hcrt::ChR2 lentivirus at frequencies 5 Hz and higher dramatically decreased the latency to wakefulness compared with mice transduced with a control HCrt::mCherry lentivirus. An important control included stimulations of Hcrt neurons in Hcrt knockout animals, and demonstrated that the reduction on the latency to wakefulness was strictly dependent on Hcrt release. Interestingly, chronic stimulation studies revealed that persistent stimulation of Hcrt neurons cannot maintain wakefulness (Adamantidis et al. 2007).

Importantly, these results demonstrate a causal link between hcrt neuron activation and sleep-to-wake transitions, consistent with previous correlative studies. This was further supported by the fact that optical silencing of hcrt neurons promote SWS (Tsunematsu et al. 2011).

In a followup study (Carter et al. 2010) we genetically targeted noradrenergic neurons in the locus coeruleus, a major target of Hcrt cells, by stereotaxic injection of a Cre recombinase- dependent adeno-associated virus (rAAV) into knock-in mice selectively expressing Cre in tyrosine hydroxylase (TH) neurons. We found that both NpHR and ChR2 were functional and could inhibit and activate, respectively, LC-NE neurons both in vitro and in vivo. Optogenetic stimulation of LC-NE neurons at low frequencies (1–10 Hz) caused immediate (less than 5 s) sleep-to-wake transitions from both SWS and REM sleep. Approximately 20 pulses of light delivered over 5 s in the LC (presumably each leading to an action potential) were sufficient to induce an awakening deterministically (i.e. with a probability of 1). Stimulation of LC neurons during wakefulness increased locomotor activity and the total time spent awake, confirming the strong arousal effect. In contrast, NpHR-mediated silencing of LC-NE neurons decreased the duration of wake episodes but did not block sleep-to-wake transitions when animals were asleep. Taken together, this study showed that activation of LC-NE neurons is necessary for maintaining normal durations of wakefulness (NpHR experiment), and sufficient to induce immediate sleep-to-wake transitions, sustained wakefulness, and increased locomotor arousal. Thus, we proposed that the LC-NE neurons act as a fast tuning system to promote sleep-to-wake transitions and general arousal.

The question remained as to whether Hcrt and LC neurons, which had already been shown to be anatomically connected, were also functionally connected. We therefore used combinatorial optogenetics to determine whether LC neurons were necessary for Hcrt-induced awakenings. Simultaneous photostimulation of Hcrt neurons at 10 Hz and photoinhibition of LC neurons did not result in reduced latencies to wakefulness, strongly suggesting that noradrenergic transmission is necessary for at least one component of the awakenings elicited by Hcrt (Carter et al. (2012).

Since the first set of manuscripts using optogenetics to study behavioral state transitions, more than 5000 publications have used similar approaches in vivo. Optogenetics was selected as “Method of the Year” in 2010 and one of the

"Breakthroughs of the Decade" by the journal Nature. Clearly, the ability to manipulate genetically defined cell populations with millisecond precision is advancing our understanding of the mesoscale organization of the brain (Carter and de Lecea 2011; Deisseroth 2014). With regard to the Hcrt neurons, optogenetics has allowed us to functionally map connections with other neuronal structures associated with arousal and brain reward, two of the main functions of the hypocretinergic system (Carter et al. 2013).

Thus, the discovery of the hypocretins/orexins has led to many important advances in the neurosciences: providing a mechanism for narcolepsy and the regulation of sleep and wakefulness (Partinen et al. 2014); the first-in-class molecule for the treatment of primary insomnia (Osborne 2013), and is also leading the way of circuit functional mapping of neuromodulators (Giardino and de Lecea 2014). This is a lot to celebrate in only 15 years of existence.

Acknowledgments I thank Greg Sutcliffe for all of his generosity and support during my scientific career. I also thank Kaare Gautvik for his enthusiasm in the initial steps of the Hcrt journey.

References

Adamantidis A, Zhang F, Aravanis A, Deisseroth K, de Lecea L (2007) Neural substrates of awakening probed with optogenetic control of hypocretin neurons. Nature 450:420–424

Bloom FE (1987) Hypothalamus: past, present, future. Adv Biochem Psychopharmacol 43:1–11

Boyden ES, Zhang F, Bamberg E, Nagel G, Deisseroth K (2005) Millisecond-timescale, genetically targeted optical control of neural activity. Nat Neurosci 8:1263–1268

Carter ME, de Lecea L (2011) Optogenetic investigation of neural circuits in vivo. Trends Mol Med 17:197–206

Carter ME, Yizhar O, Chikahisa S, Nguyen H, Adamantidis A, Nishino S, Deisseroth K, de Lecea L (2010) Tuning arousal with optogenetic modulation of locus coeruleus neurons. Nat Neurosci 13: 1526–1533

Carter ME, Brill J, Bonnavion P, Huguenard JR, Huerta R, de Lecea L (2012) Mechanism for Hypocretin-mediated sleep-to-wake transitions. In: Proceedings of the National Academy of Sciences of the United States of America 2012

Carter ME, de Lecea L, Adamantidis A (2013) Functional wiring of hypocretin and LC-NE neurons: implications for arousal. Front Behav Neurosci 7:43

de Lecea L, Huerta R (2014) Hypocretin (orexin) regulation of sleep-to-wake transitions. Frontiers in pharmacol 5:16

de Lecea L, Criado JR, Prospero-Garcia O, Gautvik KM, Schweitzer P, Danielson PE, Dunlop CL, Siggins GR, Henriksen SJ, Sutcliffe JG (1996) A cortical neuropeptide with neuronal depressant and sleep-modulating properties. Nature 381:242–245

de Lecea L, Kilduff TS, Peyron C, Gao X, Foye PE, Danielson PE, Fukuhara C, Battenberg EL, Gautvik VT, Bartlett FS 2nd, Frankel WN, van den Pol AN, Bloom FE, Gautvik KM, Sutcliffe JG (1998) The hypocretins: hypothalamus-specific peptides with neuroexcitatory activity. Proc Natl Acad Sci USA 95:322–327

Deisseroth K (2014) Circuit dynamics of adaptive and maladaptive behaviour. Nature 505:309–317

Fenno L, Yizhar O, Deisseroth K (2011) The development and application of optogenetics. Ann Rev Neurosci 34:389–412

Gautvik KM, de Lecea L, Gautvik VT, Danielson PE, Tranque P, Dopazo A, Bloom FE, Sutcliffe JG (1996) Overview of the most prevalent hypothalamus-specific mRNAs, as identified by directional tag PCR subtraction. Proc Natl Acad Sci USA 93:8733–8738

Giardino WJ, de Lecea L (2014) Hypocretin (orexin) neuromodulation of stress and reward pathways. Curr Opin Neurobiol 29C:103–108

Hassani OK, Lee MG, Jones BE (2009) Melanin-concentrating hormone neurons discharge in a reciprocal manner to orexin neurons across the sleep-wake cycle. Proc Natl Acad Sci USA 106:2418–2422

Lee MG, Hassani OK, Jones BE (2005) Discharge of identified orexin/hypocretin neurons across the sleep-waking cycle. J Neurosci 25:6716–6720

Li J, Hu Z, de Lecea L (2014) The hypocretins/orexins: integrators of multiple physiological functions. Br J Pharmacol 171:332–350

Mileykovskiy BY, Kiyashchenko LI, Siegel JM (2005) Behavioral correlates of activity in identified hypocretin/orexin neurons. Neuron 46:787–798

Milner RJ, Bloom FE, Sutcliffe JG (1987) Brain-specific genes: strategies and issues. Curr Top Dev Biol 21:117–150

Osborne R (2013) First-in-class insomnia drug on the brink of approval nod. Nat Rev Drug Discov 12:492–493

Partinen M, Kornum BR, Plazzi G, Jennum P, Julkunen I, Vaarala O (2014) Narcolepsy as an autoimmune disease: the role of H1N1 infection and vaccination. Lancet Neurol 13:600–613

Peyron C, Tighe DK, van den Pol AN, de Lecea L, Heller HC, Sutcliffe JG, Kilduff TS (1998) Neurons containing hypocretin (Orexin) project to multiple neuronal systems. J Neurosci 18:9996–10015

Sakurai T (2014) The role of orexin in motivated behaviours. Nat Rev Neurosci 15:719–731

Sakurai T, Amemiya A, Ishii M, Matsuzaki I, Chemelli RM, Tanaka H, Williams SC, Richardson JA, Kozlowski GP, Wilson S, Arch JR, Buckingham RE, Haynes AC, Carr SA, Annan RS, McNulty DE, Liu WS, Terrett JA, Elshourbagy NA, Bergsma DJ, Yanagisawa M (1998a) Orexins and orexin receptors: a family of hypothalamic neuropeptides and G protein-coupled receptors that regulate feeding behavior. Cell 92:573–585

Sakurai T, Amemiya A, Ishii M, Matsuzaki I, Chemelli RM, Tanaka H, Williams SC, Richarson JA, Kozlowski GP, Wilson S, Arch JR, Buckingham RE, Haynes AC, Carr SA, Annan RS, McNulty DE, Liu WS, Terrett JA, Elshourbagy NA, Bergsma DJ, Yanagisawa M (1998) Orexins and orexin receptors: a family of hypothalamic neuropeptides and G protein-coupled receptors that regulate feeding behavior. Cell 92 (1 page following 696)

Swanson LW (1999) The neuroanatomy revolution of the 1970s and the hypothalamus. Brain Res Bull 50:397

Tsunematsu T, Kilduff TS, Boyden ES, Takahashi S, Tominaga M, Yamanaka A (2011) Acute optogenetic silencing of orexin/hypocretin neurons induces slow-wave sleep in mice. J Neurosci Official J Soc Neurosci 31:10529–10539

Usui H, Falk JD, Dopazo A, de Lecea L, Erlander MG, Sutcliffe JG (1994) Isolation of clones of rat striatum-specific mRNAs by directional tag PCR subtraction. J Neurosci 14:4915–4926

Input and Output Systems of Orexin Neurons

Takatoshi Mochizuki and Kyoko Yoshida-Court

Abstract Orexin/hypocretin neurons are localized in the lateral hypothalamic area (LH)/perifornical area (PF) and project their fibers widely to many regions in the brain. Two types of orexin receptors, OX1 and OX2, are expressed in various brain areas, including neurons controlling fundamental functions in the brain, such as sleep/wake behavior, feeding and drinking, homeostatic regulation, neuroendocrine and autonomic responses. In this chapter, we summarize the input and output pathways of orexin neurons in the rodent brain and discuss their physiological roles from an anatomical point of view. We then introduce recent publications of mouse studies, demonstrating functional evidence of these neural pathways from/to orexin neurons especially for controlling sleep/wake behavior. These new, integrated anatomical-physiological evidences well document the important circuitry of orexin neurons in the mammalian brain.

List of Abbreviations

AAD	Anterior amygdaloid area, dorsal part
ac	Anterior commissure
AcbSh	Nucleus accumbens
AHA	Anterior hypothalamic area
Arc	Arcuate nucleus
BST	Bed nucleus of the stria terminalis
CeM	Central amygdaloid nucleus
CM	Central medial nucleus of thalamus

T. Mochizuki (✉) · K. Yoshida-Court
Department of Neurology, Beth Israel Deaconess Medical Center, 3 Blackfan Circle, CLS-706, Boston, MA 02115, USA
e-mail: tmochizu@bidmc.harvard.edu

© Springer International Publishing Switzerland 2015

T. Sakurai et al. (eds.), *Orexin and Sleep*,
DOI 10.1007/978-3-319-23078-8_3

DMH	Dorsomedial hypothalamic nucleus
DR	Dorsal raphe nucleus
GiV	Gigantocellular reticular nucleus, ventral part
HDB	Horizontal limb of diagonal band of Broca
IL	Infralimbic cortex
LC	Locus coeruleus
LDTg	Laterodorsal tegmental nucleus
LH	Lateral hypothalamic area
LPB	Lateral parabrachial nucleus
LPO	Lateral preoptic area
LS	Lateral septal nucleus
MCPO	Magnocellular preoptic nucleus
MeA	Medial amygdaloid nucleus, anterior part
mfb	Medial forebrain bundle
MnR	Median raphe nucleus
MPA	Medial preoptic area
mPFC	Medial prefrontal cortex
MS	Medial septal nucleus
PAG	Periaqueductal gray
PB	Parabrachial nucleus
PF	Perifornical area
PH	Posterior hypothalamic area
PMnR	Paramedian raphe nucleus
POA	Preoptic area
PPTg	Pedunculopontine tegmental nucleus
PreC	Precoeruleus area
PVH	Paraventricular nucleus of hypothalamus
PVT	Paraventricular nucleus of thalamus
RMg	Raphe magnus nucleus
S	Subiculum
SCN	Suprachiasmatic nucleus
SI	Substantia innominata
SNC	Substantia nigra compact part
SNL	Substantia nigra lateral part
SNR	Substantia nigra reticular part
Sol	Nucleus of the solitary tract
SPZ	Subparaventricular zone
SuM	Supramammillary nucleus
TMN	Tuberomammillary nucleus
VDB	Ventral limb of diagonal band of Broca
VLPO	Ventrolateral preoptic area
VMH	Ventromedial hypothalamic nucleus
VTA	Ventral tegmental area

1 Projection of Orexin Neurons

The detailed distribution of orexin immunoreactive fibers are demonstrated in the rat brain and characterized well in the early work of Peyron, Date and Nambu (Peyron et al. 1998; Date et al. 1999; Nambu et al. 1999). The major projections of orexin neurons are outlined in Fig. 1. Orexin neurons yield two peptides, orexin A and orexin B (or hypocretin 1 and 2) from the single precursor peptide, prepro-orexin. Immunoreactivity for orexin A and B was found to be the same in these neurons and fibers (Date et al. 1999). Two types of orexin receptors (OX1 and OX2), both coupled with Gq protein-calcium signaling, were found in various target areas along the fiber projection (Sakurai et al. 1998; Marcus et al. 2001).

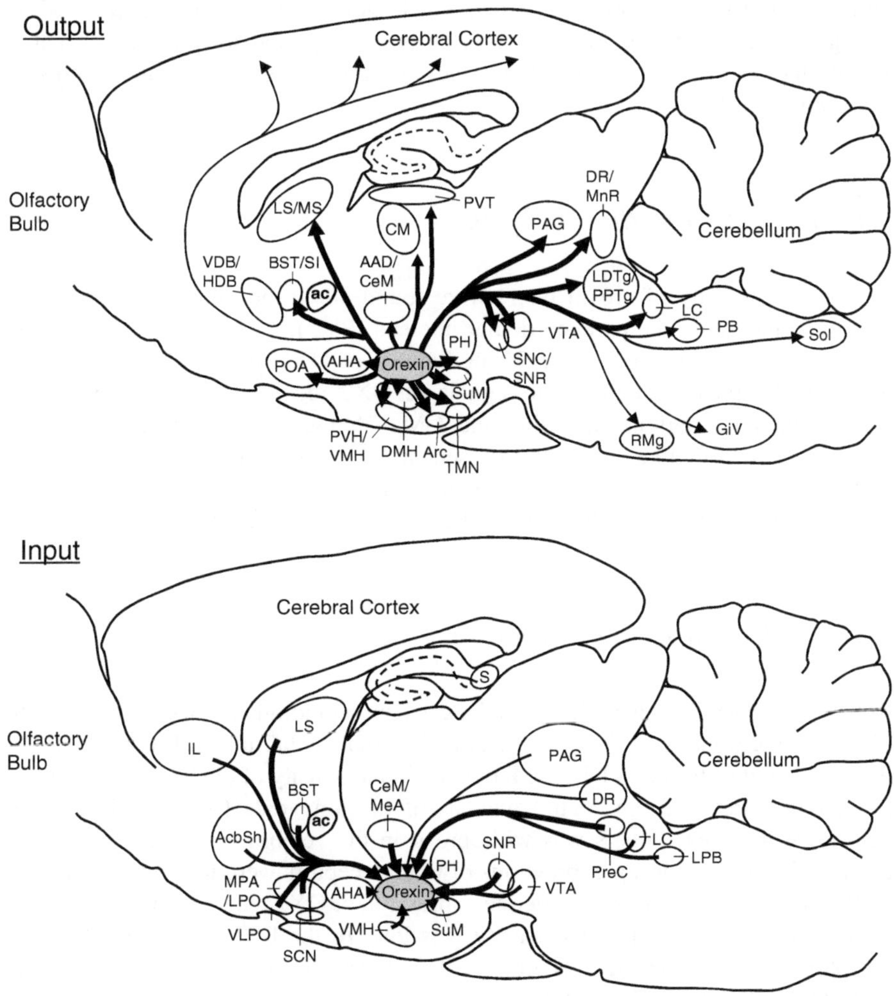

Fig. 1 Schematic summary showing fiber projections of orexin neurons (*top*) and afferent pathways to orexin neurons (*bottom*) in the rodent brain. Nomenclatures: see the list of abbreviations

1.1 Projection within the Hypothalamus

Orexin neurons innervate almost all regions of the hypothalamus. Dense fiber projections with varicose terminal structure were found widely in the LH, the dorsomedial hypothalamic nucleus (DMH), the supramammillary nucleus (SuM), the tuberomammillary nucleus (TMN), the posterior hypothalamic area (PH) and the arcuate nucleus (Arc). Relatively dense fibers were also found in the anterior hypothalamic area (AHA), the paraventricular nucleus (PVH), the ventromedial hypothalamic nucleus (VMH) and the preoptic area (POA), although no clear innervations were found in the suprachiasmatic nucleus (SCN). Both OX1 and OX2 receptors were widely expressed in the hypothalamus, but some nuclei showed uneven expression of the receptors; OX2 receptor is dominant in the Arc, LH, PVH, SuM and TMN, in contrast, OX1 receptor is dominant in VMH. These findings strongly suggest that orexin is involved in many hypothalamic functions, such as sleep/wake control, thermoregulation, feeding/drinking, neuroendocrine control and autonomic functions.

1.2 Projection to the Thalamus

A moderate density of orexin fibers were found along the midline regions of the thalamus, including the paraventricular nucleus (PVT), the central medial nucleus (CM), and the lateral habenula. Strong signals of OX1 and OX2 receptors were found in the PVT, and moderate signals of OX2 receptors were found in other medial thalamic regions including the CM. Though these thalamic nuclei are considered to function as a relay of sensory stimuli to the cortex and activate cortical arousal (Anderson et al. 2005; Fuller et al. 2011), the physiological importance of orexin signaling in the thalamus needs to be examined.

1.3 Projection to the Forebrain

Relatively dense fibers were found in the substantia innominata (SI), the bed nucleus of the stria terminalis (BST), the lateral and medial septal nuclei (LS, MS), and the diagonal band of Broca (VDB, HDB). A moderate density of the fibers was found in the anterior area and the central nucleus of the amygdala (AAD, CeM). A low density of the orexin fibers was widespread in the cerebral cortex and the hippocampus. Relatively strong expressions of OX2 receptors were found in the MS and VDB. In the hippocampus, a moderate expression of OX2 receptors were seen in the CA2-3 fields, though the fiber density was low. In the BST, both OX1 and OX2 receptors were expressed moderately. These results suggest that orexin plays important roles in the basal forebrain arousal mechanism, the septohippocampal functions and limbic systems (Arrigoni et al. 2010; Zhang et al. 2006). Interestingly, a strong expression of OX1 receptor was found in the tenia tecta with a moderate density of the fiber projection toward the prefrontal cortex (insular, infralimbic and prelimbic cortex) and the anterior olfactory nuclei, though the physiological importance of these pathways has not been examined yet.

1.4 Projection to Midbrain and Pons

Relatively dense fibers were found in the periaqueductal gray (PAG), the dorsal and median raphe nuclei (DR, MnR), the pedunculopontine tegmental nucleus (PPTg), the ventral tegmental area (VTA) and the substantia nigra compact/lateral part (SNC, SNL). The PAG, DR and PPTg express both OX1 and OX2 receptors, and are recognized as a component of sleep/wake control mechanism. The VTA and SNC also express both OX1 and OX2 receptors, and are associated with the motivation/reward, addiction and motor control mechanisms. Interestingly, a high level of OX2 receptor mRNA expression was seen in the pontine nuclei, however, the fiber projection was low level and the physiological importance of orexin in the pontine nuclei is unknown.

1.5 Projections to the Brainstem

A very dense fiber projection was found in the locus coeruleus (LC, A6). Other noradrenergic cell groups (A4, A5 and A7) were also innervated moderately, and these noradrenergic neurons dominantly express OX1 receptors. Moderately dense fibers were found in the parabrachial nucleus (PB) and the laterodorsal tegmental nucleus (LDTg) with the low level expressions of both OX1 and OX2 receptors. The LC, PB and LDTg are closely related to the control of sleep/wake behavior. Moderate dense fibers were also found in the nucleus of the solitary tract (Sol) and the dorsal motor nucleus of vagus, and a scattered fiber projection was found in broad areas of the reticular formation. These areas express both OX1 and OX2 receptors diffusely, and orexin may affect the autonomic functions through these nuclei.

2 Afferents to Orexin Neurons

The source of afferents to the orexin neurons have been characterized in detail using a combined retrograde (cholera toxin B, CTB) and anterograde (biotinylated dextrans, BD) tracing technique in the rat brain (Yoshida et al. 2006). This, when combined with the results of Sakurai, who generated a new line of transgenic mice expressing a retrograde tracing fusion protein (C-terminal fragment of tetanus toxin and green fluorescent protein, TTC::GFP) selectively in orexin neurons (Sakurai et al. 2005), provides conclusive results of the afferents in the rodent brain.

2.1 Hypothalamic Inputs to the Orexin Neurons

The Hypothalamus provides the heaviest innervation of the orexin neurons, with the densest inputs originating in the POA and PH. Most of hypothalamic regions innervate the medial part of the orexin field. Inputs to orexin neurons from the hypothalamic area are:

- *Medial preoptic area.* The medial preoptic area (MPA) heavily innervates the orexin neurons. The lateral part and caudal part of MPA are preferentially and heavily innervated neurons in the medial and perifornical parts of the orexin field, especially just dorsomedial to the fornix. Fibers from the MPA mainly follow the mediodorsal path into the orexin field.
- *Lateral preoptic area.* The orexin neurons receive heavy inputs from the center of the ventrolateral preoptic area (VLPO). The inputs are most abundant dorsal and medial to the fornix. Although the input patterns to the orexin neurons are similar, fewer appositions are seen from the lateral preoptic area (LPO, lateral and dorsal to the VLPO). Fibers from the LPO course caudally through the medial forebrain bundle (mfb), but the fibers from VLPO take the mediodorsal path.
- *Anterior hypothalamic area.* Moderate inputs to the perifornical area of the orexin field are seen from the anterior hypothalamic area (AHA), however, neurons of the suprachiasmatic nucleus (SCN) innervate the orexin neurons very sparsely in the rat brain and no GFP-positive cells was found in the SCN of orexin/TTC::GFP transgenic mice.
- *Ventromedial and dorsomedial hypothalamus.* The VMH neurons project moderately into the orexin neurons. Most of inputs are found medial to the fornix. The fibers from the VMH extend dorsally up into the medial part of the orexin field, DMH, and then into the dorsal hypothalamic area. The DMH neurons also show direct input to orexin neurons in orexin/TTC::GFP transgenic mice.
- *Posterior hypothalamus.* The neurons in the PH heavily innervate orexin neurons, mostly dorsal and medial to the fornix. These fibers from the PH enter the dorsomedial part of the orexin field and then course in a ventrolateral direction. Interestingly, the TMN neurons show no direct input to orexin neurons in orexin/TTC::GFP transgenic mice.
- *Supramammillary nucleus.* The SuM targeted the orexin neurons lightly. Fibers from the SuM innervate the perifornical and medial parts of the orexin field and also extend into the DMH, VMH, and Arc.

These afferents strongly suggest that orexin neurons are associated with many hypothalamic functions. In addition to above inputs, orexin neurons receive local excitatory input between themselves and/or from glutamatergic interneurons (Li et al. 2002; Sakurai et al. 2005; Yamanaka et al. 2010). This self-conditioning mechanism may help orexin neurons remain active for a long time when necessary, for example, to arouse animals during a variety of active behavior (España et al.

2003; Mochizuki et al. 2004, 2006; Alexandre et al. 2013). In contrast, the heaviest, presumably inhibitory input originates in the VLPO and MPA; possible sleep-active neurons that produce GABA and galanin (Sherin et al. 1998; McGinty et al. 2004). The VLPO mainly projects to the medial and perifornical parts of the orexin field, and orexin neurons in these regions show the greatest reduction in the activity during sleep (Estabrooke et al. 2001). Regions that control metabolism and feeding such as the Arc, DMH and VMH are likely to modulate the orexin neurons.

Sleep/wake behavior is closely regulated by the circadian biological rhythm, however, orexin neurons rarely receive the direct projection from the SCN, the main circadian oscillator in the mammalian brain (Reppert and Weaver 2002). An alternative multi-synaptic pathway was suggested by the neuroanatomical tracing and behavioral studies that the output of the SCN was first mediated by the neurons in the subparaventricular zone (SPZ), then to the DMH and finally to orexin neurons (Lu et al. 2001; Chou et al. 2003; Saper et al. 2005, 2010).

2.2 Forebrain Afferents to the Orexin Neurons

Most inputs from the forebrain regions innervate orexin neurons diffusely across the entire field, with slightly more projections in the perifornical region. Inputs to orexin neurons from the forebrain and limbic areas are:

- *Prefrontal cortex.* Although the infralimbic cortex (IL) projects dense fibers to the LH, these neurons innervate orexin neuron relatively lightly. Neurons of the cingular cortex project fibers to the zona incerta and subincertal nucleus, but rarely make direct contact with orexin neurons.
- *Septal nuclei.* The LS is one of the largest inputs to the orexin neurons, topographically innervating different parts of the orexin field. The intermediate part of the LS mainly targets the medial part of the field and the dorsal part of the LS projects to the lateral part. In all areas, the fibers course through the mfb and then radiated throughout the orexin field. In orexin/TTC::GFP transgenic mice, the retrogradely labeled neurons extended to the medial septal nucleus and the diagonal band of Broca, however, these area were largely negative by CTB/BD tracing in the rat brain.
- *Nucleus accumbens.* The rostral and caudal part of the shell of the nucleus accumbens (AcbSh) targets orexin neurons lightly. Fibers from the AcbSh run through the lateral part of the mfb and extended into the lateral part of the orexin field.
- *Bed nucleus of stria terminalis.* The BST is a major source of inputs to the orexin neurons. The ventral part of the lateral subdivision innervates orexin neurons, scattered across the field. The rostral, medial BST lightly innervates the orexin neurons. The posterior part of the medial division moderately innervates the orexin neurons, but the caudal edge of the medial division innervates fewer cells. These fibers pass through the medial part of the mfb to contact orexin neurons across the field, especially in the perifornical region.

- *Amygdala.* The lateral subdivision of the CeM innervates the orexin field, especially to the lateral half of the field. Fibers from the CeM course around the internal capsule to heavily innervate the perifornical region. The anterodorsal and posterodorsal parts of the medial amygdala innervate orexin neurons lightly in the lateral part of the field. In orexin/TTC::GFP transgenic mice, many retrogradely labeled neurons were also found in the basomedial nucleus of amygdala.
- *Subiculum.* The fibers from dorsal subiculum are scarce in the orexin field. The ventral subiculum project few orexin neurons, despite dense fibers in the fornix.

These forebrain inputs may drive orexin neurons in response to various emotional, stress and autonomic stimuli. The prefrontal cortex has an important role in emotional control, and suppression of these areas reduced food-elicited cataplexy in orexin knockout mice (Oishi et al. 2013). The amygdala is also one of the well-studied regions for processing emotional stimuli, and large neurotoxic lesion in the CeM and basolateral amygdala reduced emotion/motivation-triggered cataplexy in orexin knockout mice (Burgess et al. 2013). The cholinergic neurons in the basal forebrain, widely spread in the septal nuclei, the diagonal band of Broca, the magnocellular preoptic nucleus (MCPO) and the SI, are supposed to work for wake maintenance, and 20 % of orexin neurons were excited electrophysiologycally with a muscarinic agonist, carbachol (Yamanaka et al. 2003; Sakurai et al. 2005). However, more recent study suggested that the major afferents from the basal forebrain are non-cholinergic innervations (Agostinelli et al. 2013). Other inputs, such as the AcbSh and BST would influence stress, aggression, and anxiety, and all of these forebrain inputs may be important to activate orexin neurons when emotional stimuli drive arousal and autonomic responses in animals (Kayaba et al. 2003).

2.3 Brain Afferents to the Orexin Neurons

Projections from the SNR, DR, and VTA preferentially innervate orexin neurons in the lateral part of the orexin field. Many of these brainstem inputs project primarily to the ipsilateral orexin field, but fibers from the DR, ventrolateral PAG, and LC project bilaterally. Inputs to orexin neurons from brainstems are:

- *Substantia nigra.* Neurons in the reticular part of the substantia nigra (SNR) project to the orexin neurons. Numerous fibers from the SNR course through the LH to orexin neurons in the lateral part of the field, but actual contacts are sparse in the medial and perifornical parts of the orexin field.
- *Ventral tegmental area.* Neurons in the rostral and caudal VTA moderately innervate the orexin neurons with direct contacts, especially in the perifornical and lateral part of the field. In orexin/TTC::GFP transgenic mice, however, the retrogradely labeled neurons were limited to the most caudal part of the VTA and possibly non-tyrosine hydroxylase (TH) positive group.

- *Periaqueductal gray matter.* Neurons in the ventrolateral PAG make 1–2 appositions on orexin neurons, scattering throughout the orexin field. The neurons in the caudal lateral PAG project to the medial part of the orexin field. Both fibers ascend from the mfb into the dorsal hypothalamus.
- *Raphe nuclei.* The central portion of the anterior DR very lightly innervates the orexin neurons, mainly in the lateral orexin field. Fibers ascend bilaterally through the mfb and course through the orexin field. In contrast, in orexin/TTC:: GFP transgenic mice, many neurons in the MnR and paramedian raphe nucleus (PMnR) showed retrogradely-labeled signals, suggesting the major source of serotonergic input to orexin neurons instead.
- *Lateral parabrachial nucleus.* The rostral part of lateral parabrachial nucleus (LPB) lightly innervate the orexin neurons bilaterally. The cells with appositions are scattered throughout the orexin field.
- *Locus coeruleus.* The rostral LC projects a few fibers to the orexin field, whereas the caudal LC has no fiber projection in the field. In orexin/TTC::GFP transgenic mice, no retrogradely-labeled neurons was found in the LC, suggesting that LC is not the major source of noradrenergic input to orexin neurons. In contrast, the neurons in the precoeruleus area (PreC) mildly innervate the orexin field, mainly in the perifornical and medial parts.

Compared to the dense fiber projections from orexin neurons to wake-active nuclei in the brain stem (LC, DR, PB), afferent signals from these groups were considered minor. These neurons may work mostly as output pathway of orexin signaling. Many serotonergic neurons around the MnR innervate orexin neurons and may form a feedback circuit to modulate the activity of orexin neurons (Muraki et al. 2004). Orexin neurons in the lateral part of the field are activated when an animal anticipates a reward such as drugs or food (Harris et al. 2005), and these neurons may be more influenced by regions that regulate reward, emotion, and autonomic function such as the VTA, SN, or ascending visceral afferents.

3 Functional Mapping of Orexin Neurotransmission

Recent progress in molecular biology techniques, such as Cre-loxP gene recombination combined with gene-targeting elements, light-gated bacterial ion channels/pumps, drug-specific mutant receptors and transcription activators, enables us to study roles of focal brain regions or specific group of neurons in a variety of physiology and behavior in animals. Using these new genetic techniques, many new lines of transgenic mice, knock-out/in mice and recombinant viral vectors have been developed for detailed neuroanatomical and behavioral studies, including sleep/wake studies. Here, we summarize the recent mouse studies addressing the functional mapping of orexin neurotransmission with regard to the sleep/wake control circuitry (Fig. 2). These studies successfully demonstrate the integrated anatomical-physiological evidences of how orexin neurons regulate sleep/wake behavior in the brain.

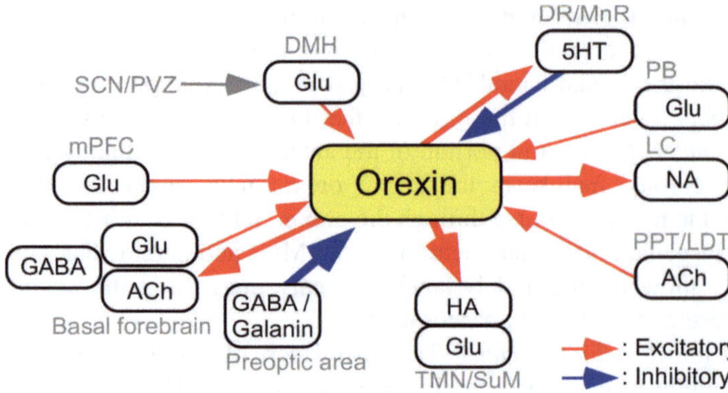

Fig. 2 Circuitry diagram of sleep/wake control by orexin neurons. Orexin neurons activate cholinergic and non-cholinergic neurons in the basal forebrain, histaminergic and other neurons in the TMN/SuM, and noradrenergic neurons in the LC to promote wakefulness. Serotonergic neurons in the DR/MnR may send negative feedback signals to orexin neurons. GABA/galanin neurons in the preoptic area suppress orexin neurons during sleep. Other wake (and REM) active neurons in the brainstem (PB, PPT, LDT) are supposed to modulate the activity of orexin neurons. The circadian clock signal is considered to input from the SCN via the PVZ and the DMH. The mPFC neurons may activate orexin neurons in response to emotional excitation

3.1 Orexin Receptor Rescue in the Posterior Hypothalamus

Loss of orexin signals in the brain causes narcolepsy syndrome in both humans and animals, showing chronic sleepiness, fragmented sleep/wake and cataplexy. One way to map the critical brain regions for orexin neurotransmission is to focally restore the orexin signaling and improve the behavioral deficits in narcoleptic animals. For this purpose, we produced a new line of mouse model in which a loxP-flanked gene-targeting cassette disrupts the transcription of OX2 receptor gene, but normal, eutopic expression of OX2 receptor can be restored by introducing Cre recombinase in the neurons (Mochizuki et al. 2011). These mice were born as OX2 receptor null mice, and had mildly fragmented wakefulness during the active period. We focally injected an adeno-associated viral vector (AAV) coding for Cre recombinase into the posterior hypothalamus, including the TMN (histamine neurons) and the SuM (probably glutamate neurons), and found that the restoration of OX2 in these neurons completely rescued the fragmented wakefulness in these mice. The results indicated that the posterior hypothalamic region works as an important relay pathway of orexin signaling to consolidate wake behavior. However, similar OX2 receptor rescue using OX2-coding viral vector was examined in constitutive OX1/OX2 double receptor knockout mice, showing severe sleep/wake fragmentation than the single receptor mutants, and the induction of OX2 receptor in the posterior hypothalamus was not effective to improve the fragmented wake (Hasegawa et al. 2014). The results suggest that arousal signal from the posterior hypothalamus may need further OX1-mediated downstream relays to fully recover the wake deficit in narcoleptic mice.

3.2 Cholinergic and Non-cholinergic Neurons in the BF

Cholinergic neurons in the BF project the fibers widely in the cortex and induce cortical activation associated with EEG desynchronization (Buzsaki et al. 1988; Cape and Jones 2000). Many of these neurons increase the firing rate over the transition period from REM sleep to wake, suggesting that these neurons function as an output pathway of cortical arousal (Lee et al. 2005; Takahashi et al. 2009). Because orexin activated the cholinergic neurons in vitro (Eggermann et al. 2001; Arrigoni et al. 2010), we examined the focal OX1/OX2 double receptor rescue in the BF using the OX1/OX2 double receptor transcription-disrupted mice (Alexandre et al. 2012). The restoration of OX1/OX2 receptors in the areas of the SI and the MCPO partially improved the fragmented wake behavior in these mice, indicating that the BF is another important relay site of orexin signaling. However, the BF also includes non-cholinergic neurons, such as glutamatergic and GABAergic neurons, and the state-dependent activity patterns of these neurons are not the same (Hassani et al. 2009; Takahashi et al. 2009). Thus, further studies are necessary to characterize the effects of orexin on different groups of the BF neurons to control sleep/wake behavior.

3.3 Serotonin Neurons in the DR/MnR

Serotonin neurons in the DR and MnR receive dense fiber projection from orexin neurons, and also send fibers to orexin neurons reciprocally. Because serotonin hyperpolarized orexin neurons in vitro, this pathway is considered as a negative feedback circuit to suppress orexin neurons when animals are actively awake (Muraki et al. 2004). This hypothesis was examined by Tabuchi et al. using an advanced transgenic mouse line overexpressing serotonin 5HT1A receptor under control of orexin-tetracycline-controlled transcriptional activator/operator system (Tabuchi et al. 2013). In these mice, orexin neurons overexpressed 5HT1A receptor and had enhanced serotonin signals under baseline conditions, but the expression of 5HT1A receptor returned to normal level when the mice were fed with doxycycline-containing food. These 5HT1A receptor overexpressing mice had more NREM and REM sleep with frequent wake to sleep transitions during the first 6 h of the active period. The results demonstrated that enhanced serotonergic tone in orexin neurons resulted in sleepiness, possibly due to the suppressed activity of orexin neurons. Interestingly, serotonin neurons reduced cataplexy occurrence in OX1/OX2 double receptor knockout mice by inducing OX2 receptor focally in the DR (Hasegawa et al. 2014). This mechanism is not fully understood yet, however, one possibility is that serotonin may suppress other wake-active neurons as well during the high arousal moment, and as a result, attenuate emotional changes/positive emotions which often trigger cataplexy in people and animal models of narcolepsy (Vetrugno et al. 2010; Clark et al. 2009; Oishi et al. 2013).

3.4 *Optogenetics/Pharmacogenitics*

Another way of characterizing the neuronal circuitry is to stimulate or inhibit the specific groups of neurons, and the recent genetic approach has provided us with better site-or neurochemical property-specificity of stimulation/inhibition. The first optogenetic study to examine sleep/wake circuitry was performed on orexin neurons themselves (Adamantidis et al. 2007). Adamantidis et al. introduced channelrhodopsin-2 (a light-gated cation channel)(Boyden et al. 2005) specifically in orexin neurons using a recombinant viral vector, then photo-stimulated orexin neurons when the mice were asleep. These mice had brief awakenings in response to the light stimuli with shorter latency than spontaneous awakenings (Adamantidis et al. 2007; Carter et al. 2009). An alternative genetic approach is to introduce mutant muscarinic acetylcholine receptors (M3 for excitation or M4 for inhibition) and stimulate/inhibit orexin neurons by injecting the ligand compound (Sasaki et al. 2011); this pharmacogenetic approach is known as DREADD (designer receptors exclusively activated by designer drugs) (Armbruster et al. 2007). All of these results strongly support the idea that orexin neurons play an important role to regulate wakefulness, as claimed previously from the transgenic/knockout mouse studies (Chemelli et al. 1999; Hara et al. 2001; Willie et al. 2003; Mochizuki et al. 2004). These optogenetic and pharmacogenetic approach have been further applied to identify the specific input/output pathways of orexin neurons.

3.5 *Noradrenaline Neurons in the LC*

The LC neurons have very dense innervation from orexin neurons with a strong expression of OX1 receptor. This pathway is considered an important output relay of orexin neurons to promote wakefulness, because the awakening response caused by the optogenetic stimulation of orexin neurons was suppressed by simultaneous optogenetic inhibition of LC noradrenergic neurons in TH-specific Cre-expressing mice (Carter et al. 2012). The receptor rescue study also showed similar results in narcoleptic mouse models (OX1/OX2 double receptor knockout mice and orexin-ataxin 3 transgenic mice) that the viral induction of OX1 receptor in the LC or pharmacogenetic activation of LC neurons improved the fragmented wake behavior in these mice during the active period (Hasegawa et al. 2014). Therefore, projections from orexin neurons to LC noradrenergic neurons form a strong excitatory pathway of arousal mechanism.

3.6 *GABA Neurons in the Preoptic Area*

Neurons in the preoptic area of the hypothalamus, including medial preoptic nucleus and the ventrolateral preoptic nucleus, innervate orexin neurons. Many of these preoptic neurons show sleep-active pattern of activity, and have GABA and galanin as an inhibitory neurotransmitter (Sherin et al. 1996; McGinty and Szymusiak 2001; Saper et al. 2001). Therefore, these preoptic neurons are considered to suppress orexin neurons during sleep. Indeed, Tsunematsu et al. (2011, 2013) demonstrated that the optogenetic silencing of orexin neurons themselves caused quick Wake to NREM sleep transitions in transgenic mice expressing halorhodopsin (a light-gated chloride pump) or archaerhodopsin (a light-gated proton pump) in orexin neurons. The functional link between GABAergic preoptic neurons and orexin neurons were examined by Saito et al. (2013), using DREADD and optogenetic technique to stimulate GABAergic neurons specifically in GAD67-Cre expressing mice. The activation of the preoptic GABAergic neurons in vivo caused mild increase of NREM sleep, and the activation of GABAergic nerve terminals decreased the firing rate of orexin neurons in vitro. Although these preoptic neurons innervate many other wake-active neurons, as nicely shown by the immunostaining of the channelrhodopsin-2 expressing fiber projections in this study (Saito et al. 2013), the GABAergic input to orexin neurons is one significant pathway to promote sleep in these mice.

4 Future Perspectives

Although optogenetic/pharmacogenetic approach has not been reported yet, some other areas are potentially important to study for regulating orexin neurons and the behavioral state control, such as the BST, the VTA and the PB. The integrated anatomical-behavioral analysis would encourage not only basic understanding of sleep/wake circuitry, but also pathogenetic mechanism underlying narcolepsy and other sleep disorders, as well as development of novel medications. In addition, it is fascinating to study plastic changes of synaptic connection from/to orexin neurons with a chronic treatment of sleeping pills or psychoactive stimulants.

Acknowledgement Writing of this paper was supported in part by the grant from the NIH/NHLBI (HL095491).

References

Adamantidis AR, Zhang F, Aravanis AM, Deisseroth K, de Lecea L (2007) Neural substrates of awakening probed with optogenetic control of hypocretin neurons. Nature 450(7168):420–424

Agostinelli LJ, Ferrari LL, Arrigoni E, Lowell BB, Scammell TE (2013) Basal forebrain efferents to the orexin neurons. Sleep 36 (Abstr Suppl):0061

Alexandre C, Mochizuki T, Arrigoni E, Yamamoto M, Clark EL, Scammell TE (2012) Orexin signaling in the basal forebrain promotes EEG activation and wakefulness. Sleep 35:A31

Alexandre C, Andermann ML, Scammell TE (2013) Control of arousal by the orexin neurons. Curr Opin Neurobiol 23(5):752–759

Anderson MP, Mochizuki T, Xie J, Fischler W, Manger JP, Talley EM, Scammell TE, Tonegawa S (2005) Thalamic Cav3.1 T-type Ca^{2+} channel plays a crucial role in stabilizing sleep. Proc Natl Acad Sci USA 102(5):1743–1748

Armbruster BN, Li X, Pausch MH, Herlitze S, Roth BL (2007) Evolving the lock to fit the key to create a family of G protein-coupled receptors potently activated by an inert ligand. Proc Natl Acad Sci USA 104(12):5163–5168

Arrigoni E, Mochizuki T, Scammell TE (2010) Activation of the basal forebrain by the orexin/hypocretin neurones. Acta Physiol (Oxf) 198(3):223–235

Boyden ES, Zhang F, Bamberg E, Nagel G, Deisseroth K (2005) Millisecond-timescale, genetically targeted optical control of neural activity. Nat Neurosci 8(9):1263–1268

Burgess CR, Oishi Y, Mochizuki T, Peever JH, Scammell TE (2013) Amygdala lesions reduce cataplexy in orexin knock-out mice. J Neurosci 33(23):9734–9742

Buzsaki G, Bickford RG, Ponomareff G, Thal LJ, Mandel R, Gage FH (1988) Nucleus basalis and thalamic control of neocortical activity in the freely moving rat. J Neurosci 8(11):4007–4026

Cape EG, Jones BE (2000) Effects of glutamate agonist versus procaine microinjections into the basal forebrain cholinergic cell area upon gamma and theta EEG activity and sleep-wake state. Eur J Neurosci 12(6):2166–2184

Carter ME, Adamantidis A, Ohtsu H, Deisseroth K, de Lecea L (2009) Sleep homeostasis modulates hypocretin-mediated sleep-to-wake transitions. J Neurosci 29(35):10939–10949

Carter ME, Brill J, Bonnavion P, Huguenard JR, Huerta R, de Lecea L (2012) Mechanism for hypocretin-mediated sleep-to-wake transitions. Proc Natl Acad Sci USA 109(39):E2635–E2644

Chemelli RM, Willie JT, Sinton CM, Elmquist JK, Scammell T, Lee C, Richardson JA, Williams SC, Xiong Y, Kisanuki Y, Fitch TE, Nakazato M, Hammer RE, Saper CB, Yanagisawa M (1999) Narcolepsy in orexin knockout mice: molecular genetics of sleep regulation. Cell 98:437–451

Chou TC, Scammell TE, Gooley JJ, Gaus SE, Saper CB, Lu J (2003) Critical role of dorsomedial hypothalamic nucleus in a wide range of behavioral circadian rhythms. J Neurosci 23 (33):10691–10702

Clark EL, Baumann CR, Cano G, Scammell TE, Mochizuki T (2009) Feeding-elicited cataplexy in orexin knockout mice. Neuroscience 161(4):970–977

Date Y, Ueta Y, Yamashita H, Yamaguchi H, Matsukura S, Kangawa K, Sakurai T, Yanagisawa M, Nakazato M (1999) Orexins, orexigenic hypothalamic peptides, interact with autonomic, neuroendocrine and neuroregulatory systems. Proc Natl Acad Sci USA 96(2):748–753

Eggermann E, Serafin M, Bayer L, Machard D, Saint-Mleux B, Jones BE, Mühlethaler M (2001) Orexins/hypocretins excite basal forebrain cholinergic neurones. Neuroscience 108(2):177–181

España RA, Valentino RJ, Berridge CW (2003) Fos immunoreactivity in hypocretin-synthesizing and hypocretin-1 receptor-expressing neurons: effects of diurnal and nocturnal spontaneous waking, stress and hypocretin-1 administration. Neuroscience 121(1):201–217

Estabrooke IV, McCarthy MT, Ko E, Chou TC, Chemelli RM, Yanagisawa M, Saper CB, Scammell TE (2001) Fos expression in orexin neurons varies with behavioral state. J Neurosci 21(5):1656–1662

Fuller PM, Sherman D, Pedersen NP, Saper CB, Lu J (2011) Reassessment of the structural basis of the ascending arousal system. J Comp Neurol 519(5):933–956

Hara J, Beuckmann CT, Nambu T, Willie JT, Chemelli RM, Sinton CM, Sugiyama F, Yagami K, Goto K, Yanagisawa M, Sakurai T (2001) Genetic ablation of orexin neurons in mice results in narcolepsy, hypophagia, and obesity. Neuron 30(2):345–354

Harris GC, Wimmer M, Aston-Jones G (2005) A role for lateral hypothalamic orexin neurons in reward seeking. Nature 437(7058):556–559

Hasegawa E, Yanagisawa M, Sakurai T, Mieda M (2014) Orexin neurons suppress narcolepsy via 2 distinct efferent pathways. J Clin Invest 124(2):604–616

Hassani OK, Lee MG, Henny P, Jones BE (2009) Discharge profiles of identified GABAergic in comparison to cholinergic and putative glutamatergic basal forebrain neurons across the sleep-wake cycle. J Neurosci 29(38):11828–11840

Kayaba Y, Nakamura A, Kasuya Y, Ohuchi T, Yanagisawa M, Komuro I, Fukuda Y, Kuwaki T (2003) Attenuated defense response and low basal blood pressure in orexin knockout mice. Am J Physiol Regul Integr Comp Physiol 285(3):R581–R593

Lee MG, Hassani OK, Alonso A, Jones BE (2005) Cholinergic basal forebrain neurons burst with theta during waking and paradoxical sleep. J Neurosci 25(17):4365–4369

Li Y, Gao XB, Sakurai T, van den Pol AN (2002) Hypocretin/orexin excites hypocretin neurons via a local glutamate neurons—a potential mechanism for orchestrating the hypothalamic arousal system. Neuron 36:1169–1181

Lu J, Zhang YH, Chou TC, Gaus SE, Elmquist JK, Shiromani P, Saper CB (2001) Contrasting effects of ibotenate lesions of the paraventricular nucleus and subparaventricular zone on sleep-wake cycle and temperature regulation. J Neurosci 21(13):4864–4874

Marcus JN, Aschkenasi CJ, Lee CE, Chemelli RM, Saper CB, Yanagisawa M, Elmquist JK (2001) Differential expression of orexin receptors 1 and 2 in the rat brain. J Comp Neurol 435(1):6–25

McGinty D, Szymusiak R (2001) Brain structures and mechanisms involved in the generation of NREM sleep: focus on the preoptic hypothalamus. Sleep Med Rev 5(4):323–342

McGinty D, Gong H, Suntsova N, Alam MN, Methippara M, Guzman-Marin R, Szymusiak R (2004) Sleep-promoting functions of the hypothalamic median preoptic nucleus: inhibition of arousal systems. Arch Ital Biol 142(4):501–509

Mochizuki T, Crocker A, McCormack S, Yanagisawa M, Sakurai T, Scammell TE (2004) Behavioral state instability in orexin knockout mice. J Neurosci 24(28):6291–6300

Mochizuki T, Klerman EB, Sakurai T, Scammell TE (2006) Elevated body temperature during sleep in orexin knockout mice. Am J Physiol Regul Integr Comp Physiol 291(3):R533–R540

Mochizuki T, Arrigoni E, Marcus JN, Clark EL, Yamamoto M, Honer M, Borroni E, Lowell BB, Elmquist JK, Scammell TE (2011) Orexin receptor 2 expression in the posterior hypothalamus rescues sleepiness in narcoleptic mice. Proc Natl Acad Sci USA 108(11):4471–4476

Muraki Y, Yamanaka A, Tsujino N, Kilduff TS, Goto K, Sakurai T (2004) Serotonergic regulation of the orexin/hypocretin neurons through the 5-HT_{1A} receptor. J Neurosci 24:7159–7166

Nambu T, Sakurai T, Mizukami K, Hosoya Y, Yanagisawa M, Goto K (1999) Distribution of orexin neurons in the adult rat brain. Brain Res 827(1–2):243–260

Oishi Y, Williams RH, Agostinelli L, Arrigoni E, Fuller PM, Mochizuki T, Saper CB, Scammell TE (2013) Role of the medial prefrontal cortex in cataplexy. J Neurosci 33 (23):9743–9751

Peyron C, Tighe DK, van den Pol AN, de Lecea L, Heller HC, Sutcliffe JG, Kilduff TS (1998) Neurons containing hypocretin (orexin) project to multiple neuronal systems. J Neurosci 18 (23):9996–10015

Reppert SM, Weaver DR (2002) Coordination of circadian timing in mammals. Nature 418 (6901):935–941

Saito YC, Tsujino N, Hasegawa E, Akashi K, Abe M, Mieda M, Sakimura K, Sakurai T (2013) GABAergic neurons in the preoptic area send direct inhibitory projections to orexin neurons. Front Neural Circuits 7:192. doi:10.110.3389/fncir.2013.00192

Sakurai T, Amemiya A, Ishii M, Matsuzaki I, Chemelli RM, Tanaka H, Williams SC, Richardson JA, Kozlowski GP, Wilson S, Arch JR, Buckingham RE, Haynes AC, Carr SA, Annan RS, McNulty DE, Liu WS, Terrett JA, Elshourbagy NA, Bergsma DJ, Yanagisawa M (1998) Orexins and orexin receptors: a family of hypothalamic neuropeptides and G protein-coupled receptors that regulate feeding behavior. Cell 92(4):573–585

Sakurai T, Nagata R, Yamanaka A, Kawamura H, Tsujino N, Muraki Y, Kageyama H, Kunita S, Takahashi S, Goto K, Koyama Y, Shioda S, Yanagisawa M (2005) Input of orexin/hypocretin neurons revealed by a genetically encoded tracer in mice. Neuron 46(2):297–308

Saper CB, Chou TC, Scammell TE (2001) The sleep switch: hypothalamic control of sleep and wakefulness. Trends Neurosci 24(12):726–731

Saper CB, Lu J, Chou TC, Gooley J (2005) The hypothalamic integrator for circadian rhythms. Trends Neurosci 28(3):152–157

Saper CB, Fuller PM, Pedersen NP, Lu J, Scammell TE (2010) Sleep state switching. Neuron 68 (6):1023–1042

Sasaki K, Suzuki M, Mieda M, Tsujino N, Roth B, Sakurai T (2011) Pharmacogenetic modulation of orexin neurons alters sleep/wakefulness states in mice. PLoS ONE 6(5):e20360. doi:10.20310.21371/journal.pone.0020360

Sherin JE, Shiromani PJ, McCarley RW, Saper CB (1996) Activation of ventrolateral preoptic neurons during sleep. Science 271(5246):216–219

Sherin JE, Elmquist JK, Torrealba F, Saper CB (1998) Innervation of histaminergic tuberomammillary neurons by GABAergic and galaninergic neurons in the ventrolateral preoptic nucleus of the rat. J Neurosci 18(12):4705–4721

Tabuchi S, Tsunematsu T, Kilduff TS, Sugio S, Xu M, Tanaka KF, Takahashi S, Tominaga M, Yamanaka A (2013) Influence of inhibitory serotonergic inputs to orexin/hypocretin neurons on the diurnal rhythm of sleep and wakefulness. Sleep 36(9):1391–1404

Takahashi K, Lin JS, Sakai K (2009) Characterization and mapping of sleep-waking specific neurons in the basal forebrain and preoptic hypothalamus in mice. Neuroscience 161(1):269–292

Tsunematsu T, Kilduff TS, Boyden ES, Takahashi S, Tominaga M, Yamanaka A (2011) Acute optogenetic silencing of orexin/hypocretin neurons induces slow-wave sleep in mice. J Neurosci 31(29):10529–10539

Tsunematsu T, Tabuchi S, Tanaka KF, Boyden ES, Tominaga M, Yamanaka A (2013) Long-lasting silencing of orexin/hypocretin neurons using archaerhodopsin induces slow-wave sleep in mice. Behav Brain Res 255:64–74

Vetrugno R, D'Angelo R, Moghadam KK, Vandi S, Franceschini C, Mignot E, Montagna P, Plazzi G (2010) Behavioural and neurophysiological correlates of human cataplexy: a video-polygraphic study. Clin Neurophysiol 121(2):153–162

Willie JT, Chemelli RM, Sinton CM, Tokita S, Williams SC, Kisanuki YY, Marcus JN, Lee C, Elmquist JK, Kohlmeier KA, Leonard CS, Richardson JA, Hammer RE, Yanagisawa M (2003) Distinct narcolepsy syndromes in Orexin receptor-2 and Orexin null mice: molecular genetic dissection of Non-REM and REM sleep regulatory processes. Neuron 38(5):715–730

Yamanaka A, Muraki Y, Tsujino N, Goto K, Sakurai T (2003) Regulation of orexin neurons by the monoaminergic and cholinergic systems. Biochem Biophys Res Commun 303:120–129

Yamanaka A, Tabuchi S, Tsunematsu T, Fukazawa Y, Tominaga M (2010) Orexin directly excites orexin neurons through orexin 2 receptor. J Neurosci 30(38):12642–12652

Yoshida K, McCormack S, España RA, Crocker A, Scammell TE (2006) Afferents to the orexin neurons of the rat brain. J Comp Neurol 494(5):845–861

Zhang W, Sakurai T, Fukuda Y, Kuwaki T (2006) Orexin neuron-mediated skeletal muscle vasodilation and shift of baroreflex during defense response in mice. Am J Physiol Regul Integr Comp Physiol 290(6):R1654–R1663

Physiological Roles of Orexin Receptors on Sleep/Wakefulness Regulation

Michihiro Mieda and Takeshi Sakurai

Abstract Hypothalamic neuropeptides orexin-A and orexin-B play critical roles in the regulation of sleep/wakefulness, as well as in a variety of physiological functions including emotion, reward and energy homeostasis. The effects of orexin peptides are mediated by two receptors, orexin 1 (OX1R) and orexin 2 (OX2R) receptors. These receptors show differential expression patterns depending on brain regions and neuron types, suggesting their differential roles. Here, we review the current understanding of the physiological roles of each orexin receptor subtype, focusing on the regulation of sleep/wakefulness.

1 Orexin Receptors

The actions of orexins are mediated by two G-protein-coupled receptors, orexin 1 (OX1R) and orexin 2 (OX2R) receptors (Sakurai et al. 1998). OX1R has a one-order higher affinity for orexin-A than for orexin-B, while OX2R binds orexin-A and orexin-B with similar affinities. Both receptors are coupled to the $G_{q/11}$ subclass of G-proteins and have caused strong excitatory effects on neurons examined thus far (Mieda et al. 2013), except in one study that reported the direct inhibitory action of orexin receptors on surprachiasmatic nucleus (SCN) neurons at night (Belle et al. 2014). When overexpressed, OX2R has also been reported to couple to $G_{i/o}$ in a neuronal cell line, suggesting that OX2R could exert inhibitory action in some neurons (Zhu et al. 2003).

In clear contrast to the restricted localization of *orexin* mRNA expression exclusively in neurons distributed within the lateral hypothalamus (LH) and perifornical area (PFA) (Date et al. 1999; de Lecea et al. 1998; Nambu et al. 1999; Peyron et al. 1998; Sakurai et al. 1998), *OX1R* and *OX2R* mRNA show wide

M. Mieda (✉) · T. Sakurai
Department of Molecular Neuroscience and Integrative Physiology, Faculty of Medicine, Kanazawa University, Ishikawa, Kanazawa 9208640, Japan
e-mail: mieda@med.kanazawa-u.ac.jp

© Springer International Publishing Switzerland 2015

T. Sakurai et al. (eds.), *Orexin and Sleep*,
DOI 10.1007/978-3-319-23078-8_4

distributions within the brain with partly overlapping but distinct and complementary distributions (Marcus et al. 2001; Mieda et al. 2011). This is consistent with the fact that orexin neurons project to almost all brain areas, with especially dense ones to monoaminergic and cholinergic nuclei of the brainstem and hypothalamus, which play important roles in the regulation of sleep/wakefulness states (Chemelli et al. 1999; Nambu et al. 1999; Peyron et al. 1998). These nuclei express OX1R and OX2R in differential manners. Histaminergic neurons in the tubelomammilary nucleus (TMN) exclusively express *OX2R*, while noradrenergic neurons in the locus coeruleus (LC) exclusively express *OX1R*. In the dorsal raphe (DR) and median raphe (MnR), serotonergic neurons express *OX1R* and/or *OX2R*. In the laterodorsal tegmental nucleus (LDT) and pedunculopontine tegmental nucleus (PPT), cholinergic neurons express *OX1R* but not *OX2R* mRNA (Marcus et al. 2001; Mieda et al. 2011). Intriguingly, there are many GABAergic neurons expressing orexin receptors which are intermingled with these monoaminergic and cholinergic neurons.

2 Disruption of the Orexin System Causes Narcolepsy

Degenerative loss of orexin neurons in humans is associated with narcolepsy (type I narcolepsy: narcolepsy with cataplexy), a debilitating neurological disorder, providing a unique perspective on the mechanisms of sleep/wakefulness control (Sakurai and Mieda 2011; Zeitzer et al. 2006). A series of discoveries leading to this conclusion started with findings by two animal studies: (i) functionally null mutations in the *OX2R* gene were found to be responsible for familial canine narcolepsy (Lin et al. 1999), and (ii) *orexin* knockout mice (*orexin*$^{-/-}$) were shown to exhibit a phenotype strikingly similar to human narcolepsy (Chemelli et al. 1999).

Human narcolepsy affects approximately 1 in 2000 individuals in the United States (Mignot 1998; Sakurai and Mieda 2011; Zeitzer et al. 2006). The syndrome consists of excessive daytime sleepiness that often results in sleep attacks (sudden onset of non-rapid eye movement [NREM] sleep), cataplexy (sudden bilateral skeletal muscle weakening triggered by emotions, without consciousness impairment), hypnagogic hallucinations, and sleep paralysis. These symptoms can be divided into two independent pathological phenomena (Dauvilliers et al. 2007). The first is dysregulation of NREM sleep onset: the inability to maintain a consolidated awake period, characterized by abrupt transitions from wakefulness to NREM sleep. This phenomenon manifests clinically as excessive daytime sleepiness or sleep attacks. The second pathological phenomenon is dysregulation of rapid eye movement (REM) sleep onset: the pathological intrusion of REM sleep or REM atonia into wakefulness or at sleep onset. It is during these periods that patients may experience cataplexy, hypnagogic hallucinations, and sleep paralysis.

Similarly, *orexin*$^{-/-}$ mice display a phenotype strikingly similar to human narcolepsy, with markedly decreased duration of wakefulness episodes during the dark phase (i.e., inability to maintain a long wakeful period, or sleepiness), abrupt behavioral arrests with muscle atonia (i.e., potential cataplexy), which manifest as direct transitions from wakefulness to REM sleep in electroencephalogram/electromyogram (EEG/EMG) recordings, decreased REM sleep latency, and increased REM sleep time during the dark phase (Chemelli et al. 1999).

3 Genetic Dissection of Sleep/Wakefulness Regulation by Orexin Receptors

The fact that functionally null mutations in the *OX2R* gene were responsible for two independent lines of familial narcoleptic canines indicates OX2R-mediated pathway as a critical signaling in the regulation of sleep and wakefulness (Lin et al. 1999). Studies of orexin receptor-deficient mice (*OX1R*$^{-/-}$ and *OX2R*$^{-/-}$ mice) have provided a deeper insight into the differential roles of OX1R and OX2R (Fig. 1) (Sakurai 2007). First, *OX1R*$^{-/-}$; *OX2R*$^{-/-}$ mice demonstrate a narcoleptic phenotype nearly similar to that in *orexin*$^{-/-}$ mice, confirming that orexinergic regulation of

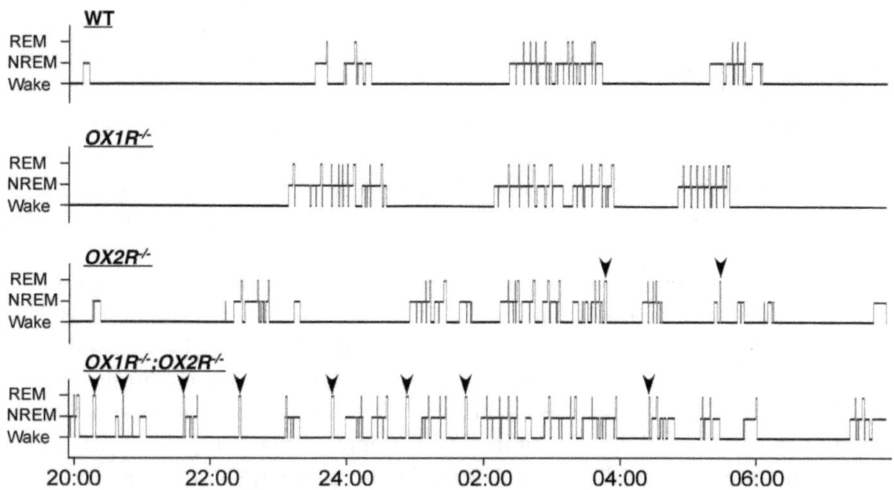

Fig. 1 Sleep state abnormalities in orexin receptor knockout mice. Representative 12-h dark-period (20:00–08:00) hypnograms for wild-type (*WT*), *OX1R*$^{-/-}$, *OX2R*$^{-/-}$, and *OX1R*$^{-/-}$; *OX2R*$^{-/-}$ mice, all with a C57BL/6J background, are shown. The different levels above the baseline indicate states of sleep and wakefulness (e.g., REM, NREM, and wakefulness) of mice at the time. Episodes of direct transition from wakefulness to REM sleep are shown by *arrows*. Note the greater awake and NREM sleep episode fragmentation and reduced duration of wakefulness in the hypnograms of *OX2R*$^{-/-}$ and *OX1R*$^{-/-}$; *OX2R*$^{-/-}$ mice compared with WT and *OX1R*$^{-/-}$ mice. Episodes of direct transition from wakefulness to REM sleep were not observed in *OX1R*$^{-/-}$ mice, and were hardly observed in *OX2R*$^{-/-}$ mice, though they were frequently observed in *OX1R*$^{-/-}$ and *OX2R*$^{-/-}$ mice. Modified from Sakurai (2007)

sleep/wakefulness is mediated by these two receptors. *OX1R*$^{-/-}$ mice show little abnormality in the states of sleep and wakefulness in the baseline condition (Hondo et al. 2010; Sakurai 2007). In contrast, *OX2R*$^{-/-}$ mice have clear narcoleptic symptoms, although their phenotype is milder as compared to that found in *orexin*$^{-/-}$ mice (Sakurai 2007; Willie et al. 2003). During the dark phase, *OX2R*$^{-/-}$ mice did show abrupt cataplexy-like behavioral arrests, which correlated with the occurrence of direct transitions from wakefulness to REM sleep in EEG/EMG recordings (Fig. 1). However, the frequency of such arrests was far less as compared to *orexin*$^{-/-}$ mice (31-fold lower frequency in *OX2R*$^{-/-}$ mice than in *orexin*$^{-/-}$ mice). Importantly, close observation of the behavior of *OX2R*$^{-/-}$ mice during the dark phase using infrared videophotography uncovered a distinct variety of behavioral arrests with more gradual onsets (gradual arrests). Moreover, such gradual arrests turned out to also manifest in *orexin*$^{-/-}$ mice with a frequency similar to *OX2R*$^{-/-}$ mice, in addition to plenty of abrupt arrests. Abrupt and gradual arrests have been characterized as the presumptive mouse correlates of cataplexy and sleep attacks in human narcolepsy, respectively, according to their behavioral, pharmacological, and electrophysiological features (Willie et al. 2003). Consistent with this observation, wakefulness episodes of *OX2R*$^{-/-}$ mice are fragmented to an extent similar to those of *orexin*$^{-/-}$ mice, indicating that these two strains of mouse suffer from comparable sleepiness (Fig. 1). In contrast, *OX2R*$^{-/-}$ mice show normal REM sleep latency and amount of REM sleep, which are profoundly shortened and increased, respectively, in both *orexin*$^{-/-}$ mice and *OX1R*$^{-/-}$; *OX2R*$^{-/-}$ mice (Willie et al. 2003).

Collectively, mouse reverse genetic studies suggest that the normal regulation of wakefulness and NREM sleep transitions depends critically on OX2R activation, whereas the profound dysregulation of REM sleep control unique to narcolepsy emerges from loss of signaling through both OX1R- and OX2R-dependent pathways.

This conclusion has been further confirmed by a complementary experiment, i.e., comparing the arousal effects of ICV orexin-A administration between wild-type, *OX1R*$^{-/-}$, and *OX2R*$^{-/-}$ mice (Fig. 2) (Mieda et al. 2011). The effects of orexin-A on wakefulness and NREM sleep were significantly attenuated in both *OX1R*$^{-/-}$ and *OX2R*$^{-/-}$ mice as compared to wild-type mice, with substantially larger attenuation in *OX2R*$^{-/-}$ than in *OX1R*$^{-/-}$ mice, suggesting the pivotal role of OX2R and the additional role of OX1R in the promotion of wakefulness. By contrast, the suppression of REM sleep via orexin-A administration was marginally and similarly attenuated in both *OX1R*$^{-/-}$ and *OX2R*$^{-/-}$ mice, suggesting a comparable contribution of the two receptors to REM sleep suppression (Mieda et al. 2011). The supplementary role of OX1R in the suppression of NREM sleep is consistent with the fact that *OX2R*$^{-/-}$ mice with a C57BL/6J, but not C57BL/6J-129/SvEv-mixed, genetic background show less fragmented wakefulness than *orexin*$^{-/-}$ mice and *OX1R*$^{-/-}$; *OX2R*$^{-/-}$ mice (Hasegawa et al. 2014; Mochizuki et al. 2011; Sakurai 2007; Willie et al. 2003), which suggests that OX1R is indispensable for the maintenance of wakefulness in the absence of OX2R.

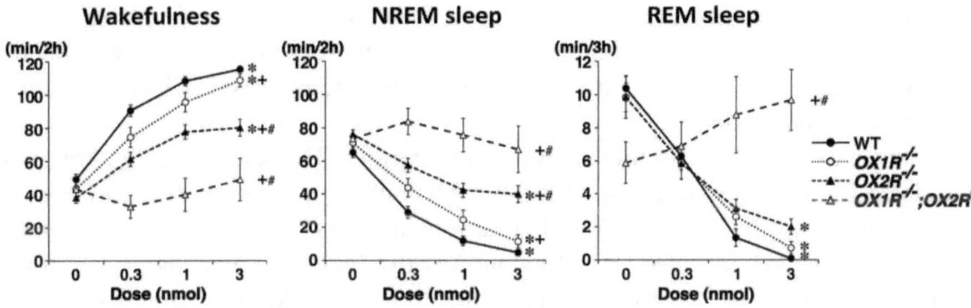

Fig. 2 Differential effects on sleep/wakefulness states following activation of OX1R and OX2R. Dose responses of the effect of ICV orexin-A in wild-type (*WT*), $OX1R^{-/-}$, $OX2R^{-/-}$, and $OX1R^{-/-}$; $OX2R^{-/-}$ mice on time spent in wakefulness, NREM sleep, and REM sleep following administration. Modified from Mieda et al. (2011)

The conclusion drawn from mouse genetics apparently contradicts the fact that an inherited canine model of narcolepsy, which demonstrates a frequent occurrence of cataplexy as well as excessive sleepiness, is attributable solely to mutations of the *OX2R* gene (Lin et al. 1999). Species differences (e.g., the precise expression patterns of the two orexin receptors) and/or selection bias may explain such an inconsistency. It should be noted that, even in canines, the absence of orexin peptides may cause severe narcoleptic symptoms as compared to *OX2R* mutation. Early studies reported that narcoleptic Dobermans and Labradors with *OX2R* mutations were much less severely affected with cataplexy than poodles with sporadic narcolepsy, which were supposed to lack orexin peptides (Baker et al. 1982).

4 Sites of Expression of Orexin Receptors Relevant to the Physiological Control of Sleep/Wakefulness

The application of exogenous orexins has been shown to excite many types of neurons. Monoaminergic and cholinergic nuclei of the hypothalamus and brainstem involved in the regulation of sleep and wakefulness especially receive dense projections of orexin neurons, express orexin receptors, and are activated by the application of orexin peptides in slice preparations (Bayer et al. 2001; Brown et al. 2001; Burlet et al. 2002; Eggermann et al. 2001; Eriksson et al. 2001; Horvath et al. 1999; Liu et al. 2002; van den Pol et al. 2002; Yamanaka et al. 2002; Mieda et al. 2013). Furthermore, the administration of orexin-A directly into the LC (Bourgin et al. 2000), TMN (Huang et al. 2001), BF cholinergic area (Espana et al. 2001; Thakkar et al. 2001), and LDT (Xi et al. 2001) has also been reported to increase wakefulness. However, neurons activated by the pharmacological application of exogenous orexin may not necessarily be essential to the endogenous mechanisms by which orexin neurons regulate sleep and wakefulness in a physiological condition. Thus, neurons directly downstream from orexin neurons in physiological conditions (i.e., the site and subtype of orexin receptors that mediate the wake-promoting and REM-gating effects by endogenous orexins) have remained incompletely understood.

Histaminergic neurons in the TMN, which express OX2R exclusively, are good candidates for such downstream neurons contributing the arousal effect of orexin. Wake-promoting effect of ICV orexin-A administration is both markedly attenuated by the histamine H1 receptor (H1R) antagonist pyrilamine (Yamanaka et al. 2002) and is absent in *H1R*$^{-/-}$ mice (Huang et al. 2001). In addition, Mochizuki et al. (2011) produced OX2R-deficient mice by inserting a *loxP*-flanked transcription-disrupter (TD) gene cassette into the *OX2R* gene, in which normal OX2R expression could be restored by Cre recombinase-mediated excision of TD cassette. Using such an elegant genetic model, they showed that focal restoration of OX2R expression in the TMN and adjacent regions completely reversed the fragmentation of wakefulness episodes observed in their OX2R-deficient mice.

However, this hypothesis remains controversial. Mice lacking both OX1R and H1R demonstrate no abnormality in sleep or wakefulness, which contradicts the idea that H1R-mediated histaminergic pathway is the principal downstream component of OX2R-mediated orexinergic signaling (Hondo et al. 2010). Moreover, a recent study showed that increased probability of sleep-to-wakefulness transitions by optogenetic activation of orexin neurons does not depend on histamine (Carter et al. 2009).

Recently, we searched for monoaminergic and cholinergic nuclei of the brainstem and hypothalamus in which the focal restoration of orexin receptor expression by recombinant AAV vectors ameliorates narcoleptic phenotype of *OX1R*$^{-/-}$; *OX2R*$^{-/-}$ mice (Fig. 3a) (Hasegawa et al. 2014). If the regional restoration of orexin receptors in a certain brain region suppresses narcoleptic symptoms in these mice, that particular region can at least be regarded as one of the important downstream targets of orexin neurons. The targeted restoration of orexin receptor expression in the DR and LC of these mice differentially inhibited cataplexy-like episodes and the fragmentation of wakefulness (i.e., sleepiness), respectively. The suppression of cataplexy-like episodes correlated with the number of serotonergic neurons restored with orexin receptor expression in the DR, while the consolidation of fragmented wakefulness correlated with the number of noradrenergic neurons restored in the LC. Furthermore, the pharmacogenetic activation of these neurons using DREADD (designer receptor exclusively activated by designer drug) technology ameliorated narcolepsy in mice that lacked orexin neurons (Fig. 3b). These results suggest that DR serotonergic and LC noradrenergic neurons play differential roles in the regulation of sleep and wakefulness by orexin neurons.

The suppression of cataplexy-like episodes by DR serotonergic neurons, but not by LC noradrenergic neurons, was an unexpected result (Hasegawa et al. 2014), because previous pharmacological and electrophysiological studies suggested that LC noradrenergic neurons are good candidates for downstream neurons to prevent cataplexy. For example, drugs that increase noradrenergic tone strongly suppress cataplexy in humans and canines, while blocking the noradrenergic signaling

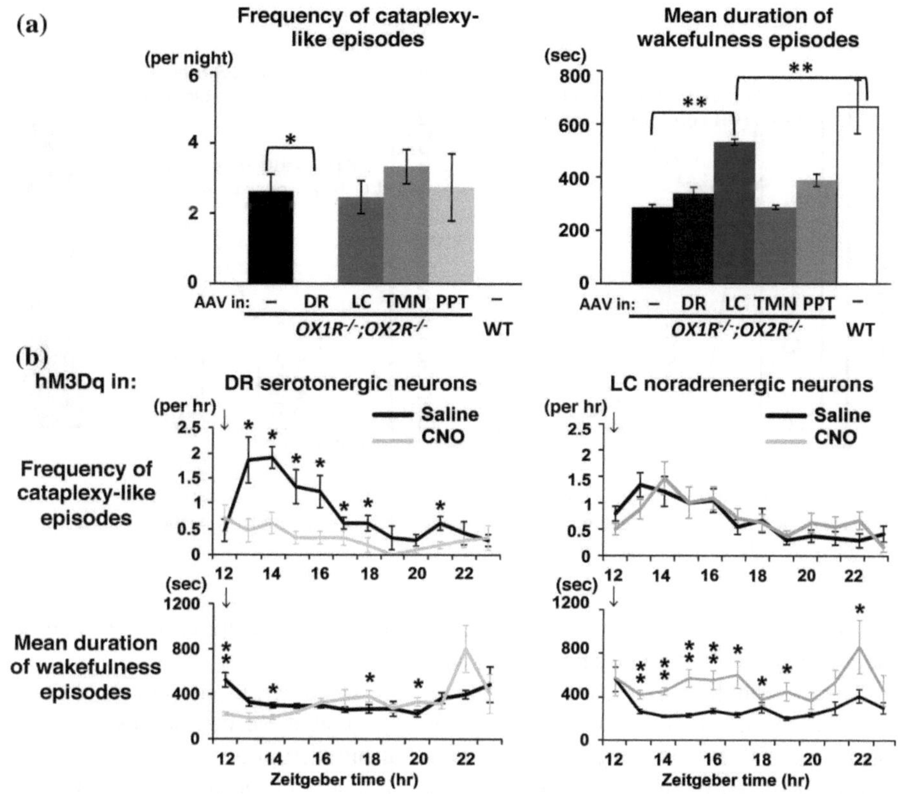

Fig. 3 Search for the downstream targets of orexin neurons to prevent narcolepsy. **a** Restoration of orexin receptor expression in the DR and LC inhibits cataplexy-like episodes and consolidates wakefulness, respectively, in narcoleptic $OX1R^{-/-}$; $OX2R^{-/-}$ mice. Orexin receptor expression was restored in the nuclei indicated in $OX1R^{-/-}$; $OX2R^{-/-}$ mice by recombinant AAV vectors. **b** Pharmacogenetic activation of DR serotonergic and LC noradrenergic neurons suppresses cataplexy-like episodes and consolidates wakefulness, respectively, in narcoleptic mice lacking orexin neurons (*orexin/ataxin-3* mice). *Orexin/ataxin-3* mice with DR serotonergic neuron–selective or LC noradrenergic neuron–selective expression of excitatory DREADD hM3Dq were injected with saline or CNO, the artificial ligand for DREADDs. Hourly plots of number of cataplexy-like episodes and mean duration of wakefulness episodes within 12 h after saline or CNO administration at ZT 12 (*arrows*) are shown. Modified from Hasegawa et al. (2014)

increases the frequency of cataplexy (Hirai and Nishino 2011; Nishino and Mignot 1997). In addition, LC neurons cease firing during cataplexy in canines (Wu et al. 1999). Nevertheless, our observations never deny the clinical importance of enhancing noradrenergic systems for preventing cataplexy, yet simply indicate that the sole regulation of LC noradrenergic neurons by endogenous orexins is not sufficient to suppress cataplexy in narcoleptic mice. Non-LC noradrenergic neurons may also play an important role in the suppression of cataplexy by the

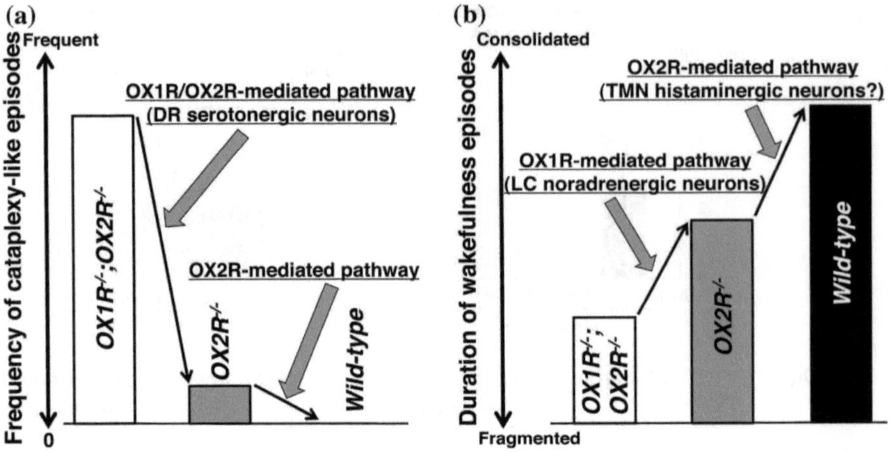

Fig. 4 Proposed model for OX1R- and OX2R-mediated pathways in suppressing narcoleptic symptoms. **a** Prevention of cataplexy-like episodes. **b** Consolidation of wakefulness episodes. Modified from Hasegawa et al. (2014)

pharmacological augmentation of systemic noradrenergic tone, which may be independent of the orexinergic regulation.

The contribution of orexin signaling in DR serotonergic neurons in the suppression of cataplexy fits with the observations that these neurons express both OX1R and OX2R (Mieda et al. 2011) and that the disruption of both OX1R- and OX2R-mediated pathways is required for the frequent occurrence of cataplexy, as described earlier (Sakurai 2007) (Fig. 4a). DR serotonergic neurons greatly reduce firing rates during cataplexy in canines (Wu et al. 2004). In addition, these neurons, as well as LC noradrenergic neurons, have been implicated in the suppression of REM sleep by inhibiting REM-on cholinergic neurons in the PPT/LDT and/or by activating REM-off GABAergic neurons in the ventrolateral periaqueductal gray (vlPAG) and adjacent lateral pontine tegmentum (LPT), also known as dorsal deep mesencephalic reticular nuclei (dDpMe) (Pace-Schott and Hobson 2002; Luppi et al. 2011). Indeed, we observed dense projections of DR serotonergic neurons to these brain areas, as well as to the amygdala (Hasegawa et al. 2014), which suggests that DR serotonergic neurons may coordinately control multiple brain regions involved in the regulation of REM sleep and emotion.

As described above, the fragmentation of wakefulness is less severe in *OX2R*$^{-/-}$ mice than in *orexin*$^{-/-}$ mice and *OX1R*$^{-/-}$; *OX2R*$^{-/-}$ mice with C57BL/6J genetic background (Mochizuki et al. 2011; Sakurai 2007), suggesting that OX1R plays an important role in the maintenance of wakefulness in the absence of OX2R (Mieda et al. 2011). Indeed, restoration of *OX1R* expression in the LC noradrenergic neurons of *OX1R*$^{-/-}$; *OX2R*$^{-/-}$ mice stabilized wakefulness episodes to an extent comparable to those in *OX2R*$^{-/-}$ mice (Hasegawa et al. 2014). Considering the fact that LC noradrenergic neurons exclusively express *OX1R* in wild-type mice (Mieda et al. 2011), these neurons may be responsible for the contribution of OX1R to the maintenance of wakefulness, while another OX2R-mediated mechanism, most likely mediated by TMN histaminergic neurons, is further required for fully maintained wakefulness as in normal mice (Fig. 4b). Recent optogenetic studies have provided supports for the importance of the orexinergic regulation of LC noradrenergic neurons in the consolidation of wakefulness. For instance, there is a causal relationship between the firing of LC noradrenergic neurons and transitions from sleep to wakefulness (Carter et al. 2010). Moreover, the optogenetic inactivation of these neurons prevents the arousal effects of the optogenetic stimulation of orexin neurons (Carter et al. 2012).

5 Pharmacological Dissection of Sleep/Wakefulness Regulation by Orexin Receptors

A series of non-selective (dual) antagonists for orexin receptors (DORA), as well as subtype-selective antagonists (SORA), have been developed. On the one hand, these drugs are drawing people's attention as novel medications for insomnia and other diseases (Scammell and Winrow 2011; Sakurai 2014). On the other hand, they are also useful for studying the roles of each subtype in the regulation of sleep/wakefulness.

To a large extent, results obtained by pharmacological studies utilizing DORAs and SORAs are consistent with those derived from genetic studies described in the previous sections. Selective blockade of OX2R efficiently increases NREM sleep and shortens NREM sleep latency (Dugovic et al. 2009, 2014; Betschart et al. 2013; Etori et al. 2014; Gozzi et al. 2011). Blockade of both OX1R and OX2R does increases NREM sleep, but also causes disproportionally large increase in REM sleep (Bonaventure et al. 2015; Dugovic et al. 2009, 2014; Etori et al. 2014; Hoyer et al. 2013). Except one study showing increases in REM and NREM sleep with SB-334867 (Morairty et al. 2012), selective blockade of OX1R alone does not cause any effects with statistical significance on baseline sleep/wakefulness (Bonaventure et al. 2015; Dugovic et al. 2009, 2014; Gozzi et al. 2011; Steiner et al. 2013). Thus, OX2R is the principal regulator of wakefulness/NREM sleep transition, while both OX1R- and OX2R-mediated pathways are critical for gating REM sleep.

Administration of DORAs seldom induces cataplexy in normal animals, although there is a report that less than half of rats treated with high doses of SB-649868 demonstrated direct transitions from wakefulness to REM sleep (Dugovic et al. 2014). Therefore, as compared to the induction of NREM and REM sleep, nearly complete and/or chronic absence of both OX1R- and OX2R-mediated pathways may be needed for cataplexy to occur. This notion is consistent with the observation that degeneration of more than 95 % of orexin neurons is required for the occurrence of cataplexy in mice, whose frequency subsequently increases along with further degeneration (Tabuchi et al. 2014).

6 Conclusions and Future Directions

Genetic and pharmacological studies have dissected the differential roles of orexin receptors in sleep/wakefulness regulation. These two approaches have both merits and demerits. Knockout mice provide clear animal models with specific and complete removal of one of two subtypes. However, chronic compensatory changes should always be taken into account. In contrast, SORAs can acutely block one of two, eliminating any compensatory effects. On the other hand, specificity, potency, efficacy, occupancy, and stability of the administered drugs should be considered.

Broad expression of orexin receptors throughout the brain complicates identifying neurons and orexin receptor subtypes directly regulated by endogenous orexins, mediating their wake-stabilizing effects. We and others have tackled this problem with a strategy of brain region/cell type-specific rescues of orexin receptors. In addition to such an approach, future studies utilizing brain region/cell type-specific genetic deletions of orexin receptors and pharmacological studies of focal administration of SORAs, as well as optogenetic and pharmacogenetic strategies, would further uncover the differential roles of orexin receptors and neuronal pathways downstream to orexin neurons in the regulation of sleep/wakefulness.

References

Baker TL, Foutz AS, McNerney V, Mitler MM, Dement WC (1982) Canine model of narcolepsy: genetic and developmental determinants. Exp Neurol 75(3):729–742

Bayer L, Eggermann E, Serafin M, Saint-Mleux B, Machard D, Jones B, Muhlethaler M (2001) Orexins (hypocretins) directly excite tuberomammillary neurons. Eur J Neurosci 14(9): 1571–1575

Belle MD, Hughes AT, Bechtold DA, Cunningham P, Pierucci M, Burdakov D, Piggins HD (2014) Acute suppressive and long-term phase modulation actions of orexin on the mammalian circadian clock. J Neurosci 34(10):3607–3621. doi:10.1523/JNEUROSCI.3388-13.2014

Betschart C, Hintermann S, Behnke D, Cotesta S, Fendt M, Gee CE, Jacobson LH, Laue G, Ofner S, Chaudhari V, Badiger S, Pandit C, Wagner J, Hoyer D (2013) Identification of a novel series of orexin receptor antagonists with a distinct effect on sleep architecture for the treatment of insomnia. J Med Chem 56(19):7590–7607. doi:10.1021/jm4007627

Bonaventure P, Yun S, Johnson PL, Shekhar A, Fitz SD, Shireman BT, Lebold TP, Nepomuceno D, Lord B, Wennerholm M, Shelton J, Carruthers N, Lovenberg T, Dugovic C (2015) A selective orexin-1 receptor antagonist attenuates stress-induced hyperarousal without hypnotic effects. J Pharmacol Exp Ther 352(3):590–601. doi:10.1124/jpet.114.220392

Bourgin P, Huitron-Resendiz S, Spier AD, Fabre V, Morte B, Criado JR, Sutcliffe JG, Henriksen SJ, de Lecea L (2000) Hypocretin-1 modulates rapid eye movement sleep through activation of locus coeruleus neurons. J Neurosci 20(20):7760–7765

Brown RE, Sergeeva O, Eriksson KS, Haas HL (2001) Orexin A excites serotonergic neurons in the dorsal raphe nucleus of the rat. Neuropharmacology 40(3):457–459

Burlet S, Tyler CJ, Leonard CS (2002) Direct and indirect excitation of laterodorsal tegmental neurons by hypocretin/orexin peptides: implications for wakefulness and narcolepsy. J Neurosci 22(7):2862–2872. doi:20026234

Carter ME, Adamantidis A, Ohtsu H, Deisseroth K, de Lecea L (2009) Sleep homeostasis modulates hypocretin-mediated sleep-to-wake transitions. J Neurosci 29(35):10939–10949. doi:10.1523/JNEUROSCI.1205-09.2009

Carter ME, Yizhar O, Chikahisa S, Nguyen H, Adamantidis A, Nishino S, Deisseroth K, de Lecea L (2010) Tuning arousal with optogenetic modulation of locus coeruleus neurons. Nat Neurosci 13(12):1526–1533. doi:10.1038/nn.2682

Carter ME, Brill J, Bonnavion P, Huguenard JR, Huerta R, de Lecea L (2012) Mechanism for hypocretin-mediated sleep-to-wake transitions. Proc Natl Acad Sci USA 109(39):E2635–E2644. doi:10.1073/pnas.1202526109

Chemelli RM, Willie JT, Sinton CM, Elmquist JK, Scammell T, Lee C, Richardson JA, Williams SC, Xiong Y, Kisanuki Y, Fitch TE, Nakazato M, Hammer RE, Saper CB, Yanagisawa M (1999) Narcolepsy in orexin knockout mice: molecular genetics of sleep regulation. Cell 98(4):437–451

Date Y, Ueta Y, Yamashita H, Yamaguchi H, Matsukura S, Kangawa K, Sakurai T, Yanagisawa M, Nakazato M (1999) Orexins, orexigenic hypothalamic peptides, interact with autonomic, neuroendocrine and neuroregulatory systems. Proc Natl Acad Sci USA 96(2): 748–753

Dauvilliers Y, Arnulf I, Mignot E (2007) Narcolepsy with cataplexy. Lancet 369(9560):499–511. doi:10.1016/S0140-6736(07)60237-2

de Lecea L, Kilduff TS, Peyron C, Gao X, Foye PE, Danielson PE, Fukuhara C, Battenberg EL, Gautvik VT, Bartlett FS 2nd, Frankel WN, van den Pol AN, Bloom FE, Gautvik KM, Sutcliffe JG (1998) The hypocretins: hypothalamus-specific peptides with neuroexcitatory activity. Proc Natl Acad Sci USA 95(1):322–327

Dugovic C, Shelton JE, Aluisio LE, Fraser IC, Jiang X, Sutton SW, Bonaventure P, Yun S, Li X, Lord B, Dvorak CA, Carruthers NI, Lovenberg TW (2009) Blockade of orexin-1 receptors attenuates orexin-2 receptor antagonism-induced sleep promotion in the rat. J Pharmacol Exp Ther 330(1):142–151. doi:10.1124/jpet.109.152009

Dugovic C, Shelton JE, Yun S, Bonaventure P, Shireman BT, Lovenberg TW (2014) Orexin-1 receptor blockade dysregulates REM sleep in the presence of orexin-2 receptor antagonism. Front Neurosci 8:28. doi:10.3389/fnins.2014.00028

Eggermann E, Serafin M, Bayer L, Machard D, Saint-Mleux B, Jones BE, Muhlethaler M (2001) Orexins/hypocretins excite basal forebrain cholinergic neurones. Neuroscience 108(2):177–181

Eriksson KS, Sergeeva O, Brown RE, Haas HL (2001) Orexin/hypocretin excites the histaminergic neurons of the tuberomammillary nucleus. J Neurosci 21(23):9273–9279

Espana RA, Baldo BA, Kelley AE, Berridge CW (2001) Wake-promoting and sleep-suppressing actions of hypocretin (orexin): basal forebrain sites of action. Neuroscience 106(4):699–715

Etori K, Saito YC, Tsujino N, Sakurai T (2014) Effects of a newly developed potent orexin-2 receptor-selective antagonist, compound 1 m, on sleep/wakefulness states in mice. Front Neurosci 8:8. doi:10.3389/fnins.2014.00008

Gozzi A, Turrini G, Piccoli L, Massagrande M, Amantini D, Antolini M, Martinelli P, Cesari N, Montanari D, Tessari M, Corsi M, Bifone A (2011) Functional magnetic resonance imaging reveals different neural substrates for the effects of orexin-1 and orexin-2 receptor antagonists. PLoS ONE 6(1):e16406. doi:10.1371/journal.pone.0016406

Hasegawa E, Yanagisawa M, Sakurai T, Mieda M (2014) Orexin neurons suppress narcolepsy via 2 distinct efferent pathways. J Clin Invest 124(2):604–616. doi:10.1172/JCI71017

Hirai N, Nishino S (2011) Recent advances in the treatment of narcolepsy. Curr Treat Options Neurol 13(5):437–457. doi:10.1007/s11940-011-0137-6

Hondo M, Nagai K, Ohno K, Kisanuki Y, Willie JT, Watanabe T, Yanagisawa M, Sakurai T (2010) Histamine-1 receptor is not required as a downstream effector of orexin-2 receptor in maintenance of basal sleep/wake states. Acta Physiol 198(3):287–294. doi:10.1111/j.1748-1716.2009.02032.x

Horvath TL, Peyron C, Diano S, Ivanov A, Aston-Jones G, Kilduff TS, van Den Pol AN (1999) Hypocretin (orexin) activation and synaptic innervation of the locus coeruleus noradrenergic system. J Comp Neurol 415(2):145–159

Hoyer D, Durst T, Fendt M, Jacobson LH, Betschart C, Hintermann S, Behnke D, Cotesta S, Laue G, Ofner S, Legangneux E, Gee CE (2013) Distinct effects of IPSU and suvorexant on mouse sleep architecture. Front Neurosci 7:235. doi:10.3389/fnins.2013.00235

Huang ZL, Qu WM, Li WD, Mochizuki T, Eguchi N, Watanabe T, Urade Y, Hayaishi O (2001) Arousal effect of orexin A depends on activation of the histaminergic system. Proc Natl Acad Sci USA 98(17):9965–9970 Epub 2001 Aug 9967

Lin L, Faraco J, Li R, Kadotani H, Rogers W, Lin X, Qiu X, de Jong PJ, Nishino S, Mignot E (1999) The sleep disorder canine narcolepsy is caused by a mutation in the hypocretin (orexin) receptor 2 gene. Cell 98(3):365–376

Liu RJ, van den Pol AN, Aghajanian GK (2002) Hypocretins (orexins) regulate serotonin neurons in the dorsal raphe nucleus by excitatory direct and inhibitory indirect actions. J Neurosci 22 (21):9453–9464

Luppi PH, Clement O, Sapin E, Gervasoni D, Peyron C, Leger L, Salvert D, Fort P (2011) The neuronal network responsible for paradoxical sleep and its dysfunctions causing narcolepsy and rapid eye movement (REM) behavior disorder. Sleep Med Rev 15(3):153–163. doi:10.1016/j.smrv.2010.08.002

Marcus JN, Aschkenasi CJ, Lee CE, Chemelli RM, Saper CB, Yanagisawa M, Elmquist JK (2001) Differential expression of orexin receptors 1 and 2 in the rat brain. J Comp Neurol 435(1):6–25

Mieda M, Hasegawa E, Kisanuki YY, Sinton CM, Yanagisawa M, Sakurai T (2011) Differential roles of orexin receptor-1 and -2 in the regulation of non-REM and REM sleep. J Neurosci 31 (17):6518–6526. doi:10.1523/JNEUROSCI.6506-10.2011

Mieda M, Tsujino N, Sakurai T (2013) Differential roles of orexin receptors in the regulation of sleep/wakefulness. Front Endocrinol (Lausanne) 4:57. doi:10.3389/fendo.2013.00057

Mignot E (1998) Genetic and familial aspects of narcolepsy. Neurology 50(2 Suppl 1):S16–S22

Mochizuki T, Arrigoni E, Marcus JN, Clark EL, Yamamoto M, Honer M, Borroni E, Lowell BB, Elmquist JK, Scammell TE (2011) Orexin receptor 2 expression in the posterior hypothalamus rescues sleepiness in narcoleptic mice. Proc Natl Acad Sci USA 108(11):4471–4476. doi:10.1073/pnas.1012456108

Morairty SR, Revel FG, Malherbe P, Moreau JL, Valladao D, Wettstein JG, Kilduff TS, Borroni E (2012) Dual hypocretin receptor antagonism is more effective for sleep promotion than antagonism of either receptor alone. PLoS ONE 7(7):e39131. doi:10.1371/journal.pone.0039131

Nambu T, Sakurai T, Mizukami K, Hosoya Y, Yanagisawa M, Goto K (1999) Distribution of orexin neurons in the adult rat brain. Brain Res 827(1–2):243–260

Nishino S, Mignot E (1997) Pharmacological aspects of human and canine narcolepsy. Prog Neurobiol 52(1):27–78

Pace-Schott EF, Hobson JA (2002) The neurobiology of sleep: genetics, cellular physiology and subcortical networks. Nat Rev Neurosci 3(8):591–605. doi:10.1038/nrn895

Peyron C, Tighe DK, van den Pol AN, de Lecea L, Heller HC, Sutcliffe JG, Kilduff TS (1998) Neurons containing hypocretin (orexin) project to multiple neuronal systems. J Neurosci 18 (23):9996–10015

Sakurai T (2007) The neural circuit of orexin (hypocretin): maintaining sleep and wakefulness. Nat Rev Neurosci 8(3):171–181. doi:10.1038/nrn2092

Sakurai T (2014) The role of orexin in motivated behaviours. Nat Rev Neurosci 15(11):719–731. doi:10.1038/nrn3837

Sakurai T, Mieda M (2011) Connectomics of orexin-producing neurons: interface of systems of emotion, energy homeostasis and arousal. Trends Pharmacol Sci 32(8):451–462. doi:10.1016/j.tips.2011.03.007

Sakurai T, Amemiya A, Ishii M, Matsuzaki I, Chemelli RM, Tanaka H, Williams SC, Richardson JA, Kozlowski GP, Wilson S, Arch JR, Buckingham RE, Haynes AC, Carr SA, Annan RS, McNulty DE, Liu WS, Terrett JA, Elshourbagy NA, Bergsma DJ, Yanagisawa M (1998) Orexins and orexin receptors: a family of hypothalamic neuropeptides and G protein-coupled receptors that regulate feeding behavior. Cell 92(4):573–585

Scammell TE, Winrow CJ (2011) Orexin receptors: pharmacology and therapeutic opportunities. Annu Rev Pharmacol Toxicol 51:243–266. doi:10.1146/annurev-pharmtox-010510-100528

Steiner MA, Gatfield J, Brisbare-Roch C, Dietrich H, Treiber A, Jenck F, Boss C (2013) Discovery and characterization of ACT-335827, an orally available, brain penetrant orexin receptor type 1 selective antagonist. Chem Med Chem 8(6):898–903. doi:10.1002/cmdc.201300003

Tabuchi S, Tsunematsu T, Black SW, Tominaga M, Maruyama M, Takagi K, Minokoshi Y, Sakurai T, Kilduff TS, Yamanaka A (2014) Conditional ablation of orexin/hypocretin neurons: a new mouse model for the study of narcolepsy and orexin system function. J Neurosci 34 (19):6495–6509. doi:10.1523/JNEUROSCI.0073-14.2014

Thakkar MM, Ramesh V, Strecker RE, McCarley RW (2001) Microdialysis perfusion of orexin-A in the basal forebrain increases wakefulness in freely behaving rats. Arch Ital Biol 139(3): 313–328

van den Pol AN, Ghosh PK, Liu RJ, Li Y, Aghajanian GK, Gao XB (2002) Hypocretin (orexin) enhances neuron activity and cell synchrony in developing mouse GFP-expressing locus coeruleus. J Physiol 541(Pt 1):169–185

Willie JT, Chemelli RM, Sinton CM, Tokita S, Williams SC, Kisanuki YY, Marcus JN, Lee C, Elmquist JK, Kohlmeier KA, Leonard CS, Richardson JA, Hammer RE, Yanagisawa M (2003) Distinct narcolepsy syndromes in Orexin receptor-2 and orexin null mice: molecular genetic dissection of non-REM and REM sleep regulatory processes. Neuron 38(5):715–730

Wu MF, Gulyani SA, Yau E, Mignot E, Phan B, Siegel JM (1999) Locus coeruleus neurons: cessation of activity during cataplexy. Neuroscience 91(4):1389–1399

Wu M, Zaborszky L, Hajszan T, van den Pol AN, Alreja M (2004) Hypocretin/orexin innervation and excitation of identified septohippocampal cholinergic neurons. J Neurosci: Official J Soc Neurosci 24(14):3527–3536. doi:10.1523/JNEUROSCI.5364-03.2004

Xi MC, Morales FR, Chase MH (2001) Effects on sleep and wakefulness of the injection of hypocretin-1 (orexin-A) into the laterodorsal tegmental nucleus of the cat. Brain Res 901 (1–2):259–264

Yamanaka A, Tsujino N, Funahashi H, Honda K, Guan JL, Wang QP, Tominaga M, Goto K, Shioda S, Sakurai T (2002) Orexins activate histaminergic neurons via the orexin 2 receptor. Biochem Biophys Res Commun 290(4):1237–1245

Zeitzer JM, Nishino S, Mignot E (2006) The neurobiology of hypocretins (orexins), narcolepsy and related therapeutic interventions. Trends Pharmacol Sci 27(7):368–374

Zhu Y, Miwa Y, Yamanaka A, Yada T, Shibahara M, Abe Y, Sakurai T, Goto K (2003) Orexin receptor type-1 couples exclusively to pertussis toxin-insensitive G-proteins, while orexin receptor type-2 couples to both pertussis toxin-sensitive and -insensitive G-proteins. J Pharmacol Sci 92(3):259–266

Orexin Receptor Functions in the Ascending Arousal System

Christopher S. Leonard and Masaru Ishibashi

Abstract Brainstem ascending arousal system neurons have long been thought to regulate the expression of waking and sleep. The finding that lateral hypothalamic orexin (hypocretin)-synthesizing neurons innervate these regions, including monoaminergic and cholinergic structures, where orexin receptors are strongly expressed, naturally suggested involvement of orexin peptides in regulating arousal. This general idea has been supported by multiple, compelling lines of investigation, but how orexins mediate these and their other functions is far from completely understood. Since the absence of orexin signaling results in Narcolepsy—a severe disruption in normal waking and sleep, it seems likely that disrupted orexin signaling at these monoaminergic and cholinergic targets play key roles in this disorder. In this review, we discuss the cellular actions of orexin at these targets and the roles played by each receptor based on results from constitutive receptor knockout mice. We then consider the implications of these actions for understanding the role played by orexin in arousal and narcolepsy.

1 Introduction

The orexins (orexin-A and -B), also called hypocretins (hypocretin-1 and -2), are hypothalamic neuropeptides (de Lecea et al. 1998; Sakurai et al. 1998) that influence multiple homeostatic systems including: feeding, metabolism, arousal, reward and stress via two G-protein coupled receptors (OX_1 and OX_2 receptors) (Sakurai et al. 1998). In a remarkable scientific convergence, results from forward genetics in dogs and reverse genetics in mice uncovered a surprising connection between orexin signaling and the regulation of waking and sleep. Indeed, the loss of orexin signaling produces a narcolepsy-like sleep disorder in animals, with unstable behavioral states, sleep attacks and cataplexy-like motor arrests (Lin et al. 1999;

C.S. Leonard (✉) · M. Ishibashi
Department of Physiology, New York Medical College, Valhalla, New York 10595, USA
e-mail: chris_leonard@nymc.edu

© Springer International Publishing Switzerland 2015

T. Sakurai et al. (eds.), *Orexin and Sleep*,
DOI 10.1007/978-3-319-23078-8_5

Chemelli et al. 1999; Hara et al. 2001; Willie et al. 2003; Beuckmann et al. 2004; Mochizuki et al. 2004; Kalogiannis et al. 2011). These early findings swiftly lead to the discovery that orexin-synthesizing neurons are lost in narcolepsy with cataplexy in humans (Peyron et al. 2000; Thannickal et al. 2000).

These discoveries have set the stage for major advances in understanding the etiology of narcolepsy (Scammell 2006), for developing new insomnia treatments using dual orexin receptor antagonists (DORAs; Winrow and Renger 2013) and for treating narcolepsy and other disorders through the development of small molecule orexin receptor agonists and selective orexin receptor antagonists (SORAs; Scammell and Winrow 2011). These seminal studies have also opened a new window into the brain circuitry underlying arousal, motivation and sleep. Essential to these advances is an understanding of the cellular and molecular consequences of orexin signaling and identification of the most relevant orexin targets for regulating orexin functions.

In addition to other chapters in this book, numerous reviews of the orexin system and its wide range of actions are available (e.g. Kukkonen and Leonard 2014; Leonard and Kukkonen 2014; Scammell and Winrow 2011; Tsujino and Sakurai 2009; de Lecea and Huerta 2014; Harris and Aston-Jones 2006). In this chapter we will first summarize the cellular actions of orexin in the CNS and then discuss orexin receptor actions on arousal system neurons, based on studies from orexin receptor knockout mice. Finally, we will consider the implication of these actions for understanding orexin's role in arousal, sleep and narcolepsy.

2 Actions of Native Orexin Receptors in the CNS

Orexins are generally recognized as excitatory neuropeptides since exogenous orexin-A or orexin-B produce a slow and long lasting post-synaptic depolarization which can be large enough to initiate or increase repetitive firing. This depolarisation results from three types of action: (1) closure of K^+ channels active at rest (Horvath et al. 1999; Ivanov and Aston-Jones 2000; Brown et al. 2001; Hwang et al. 2001; Bayer et al. 2002; Grabauskas and Moises 2003; van den Top et al. 2003; Yang and Ferguson 2003; Yang et al. 2003; Bayer et al. 2004; Hoang et al. 2004; Wu et al. 2004; Murai and Akaike 2005; Bisetti et al. 2006; Govindaiah and Cox 2006; Huang et al. 2006; Kolaj et al. 2008; Doroshenko and Renaud 2009; Zhang et al. 2010, 2011; Hoang et al. 2003; Ishibashi et al. 2005); (2) Activation of an electrogenic sodium-calcium exchanger (NCX) (Eriksson et al. 2001; Wu et al. 2002; Burdakov et al. 2003; Wu et al. 2004; Acuna-Goycolea and van den Pol 2009; Zhang et al. 2011); and (3) the activation of non-selective cation channels (NSCCs) (Eriksson et al. 2001; Hwang et al. 2001; Brown et al. 2002; Burlet et al. 2002; Liu et al. 2002; Wu et al. 2002; Yang and Ferguson 2002, 2003; Wu et al. 2004; Huang et al. 2006). Nevertheless, the specific molecular identities of these effectors have yet to be clarified.

In receptor expression systems, activation of OX_1 and OX_2 elevate intracellular Ca^{2+}, typically by release of Ca^{2+} from intracellular stores, although a necessary direct Ca^{2+} influx pathway has also been detected (for review Kukkonen and Leonard 2014). In central neurons, native orexin receptor activation also elevates intracellular Ca^{2+} (van den Pol et al. 1998; van den Pol 1999; van den Pol et al. 2001; Uramura et al. 2001; Lambe and Aghajanian 2003; Kohlmeier et al. 2004; Muroya et al. 2004; Ishibashi et al. 2005; Tsujino et al. 2005; Kohlmeier et al. 2008, 2013). However, there are only a few examples where this depends on intracellular Ca^{2+} stores (Korotkova et al. 2002; Burdakov et al. 2003; Muroya et al. 2004). Instead, Ca^{2+} elevation appears mediated by depolarisation and the activation of voltage-gated Ca^{2+} channels (VGCCs) (van den Pol et al. 1998; Uramura et al. 2001; Kohlmeier et al. 2004; Ishibashi et al. 2005; Kohlmeier et al. 2008), and the enhancement of Ca^{2+} influx mediated by VGCCs (Kohlmeier et al. 2008, 2013, see also Kukkonen and Leonard 2014). The basis of this difference between native and expressed receptors is not clear. Since Ca^{2+} imaging studies in central neurons have been limited to the somata and proximal dendrites, it is possible that orexin receptor-stimulated release of Ca^{2+} from intracellular stores occurs in compartments not yet examined, such as the dendrites or synaptic terminals.

It is also noteworthy that orexin receptors, particularly in expression systems, couple to multiple G-protein pathways, suggesting that native orexin receptor signaling may be more diverse than originally anticipated (Kukkonen and Leonard 2014) and that examples of native orexin receptor-mediated inhibition and modulation should be expected. Indeed, such an example of direct inhibition has been recently discovered in circadian clock neurons of the mouse suprachiasmatic nucleus (Belle et al. 2014). Intriguingly, during the day, indirect inhibition mediated by orexin-stimulated GABA inputs predominates, but in the night, this indirect action gives way to a large and long-lasting direct inhibition, proposed to be mediated by leak-K^+ channels. The prevalence of this action is not yet clear but since most slice studies are conducted during the day, when this mechanism is not engaged, the possibility of diurnal switching to post-synaptic inhibition should be further explored.

3 Effectors and Receptors Mediating Direct Actions on Arousal System Monoaminergic and Cholinergic Neurons

Orexin actions have been studied on arousal system monoaminergic and cholinergic neurons in brain slices, where bath application of orexin-A (30 nM–1 μM) produces a slow depolarization. Concentrations greater than about 100 nM produce depolarizations that are sufficient to drive repetitive firing in neurons that were previously quiescent. This action is mediated by different underlying receptors and effectors in each region.

Histaminergic neurons of the tuberomammilary nucleus (TMN) are directly excited by orexin and that the depolarization is mediated by activation of a Na^+/Ca^{2+}

exchanger (Eriksson et al. 2001) and a decrease in a GIRK current (Hoang et al. 2003). This depolarization leads to Ca^{2+} elevation which requires OX_2 receptors, based on slice measurements obtained from receptor knockouts mice (Willie et al. 2003).

Norepinephrine neurons of the locus coeruleus (LC) are directly depolarized by orexin and experiments from slices and dissociated cells suggest the depolarization is mediated by the closure of K^+ channels and the activation of a cation current (Ivanov and Aston-Jones 2000; Murai and Akaike 2005; Van Den Pol et al. 2002). Recordings in slices made from orexin receptor knockout mice indicate that this depolarization requires OX_1 receptors (Kohlmeier et al. 2013). This fits well with the high levels of OX_1 receptor mRNA identified in the LC by single and double in situ hybridization methods (Trivedi et al. 1998; Marcus et al. 2001; Mieda et al. 2011). Nevertheless, slice recordings also found that orexin-A (300 nM) augments depolarization-mediated Ca^{2+} transients mediated by L-type Ca^{2+} channels and that this effect did not require OX_1 receptors, but was absent in slices from double orexin receptor knockout (DKO) mice (Kohlmeier et al. 2013). This suggests that while direct excitation is mediated by the high density of OX_1 receptors, low levels of OX_2 receptors are present and are sufficient to modulate Ca^{2+} transients in LC neurons.

Serotonergic neurons of the dorsal raphe (DR) are strongly depolarized by bath application of orexin-A and voltage clamp experiments indicate that a major component of this depolarization is mediated by a noisy cation current (Brown et al. 2002; Liu et al. 2002; Kohlmeier et al. 2008). Initial studies using dose-response and receptor antagonists in rat brain slices suggested OX_2 was the primary receptor mediating this depolarization (Soffin et al. 2004). However, recent recordings from receptor knockout mice indicate OX_1 and OX_2 are each sufficient to support activation of this noisy cation current in a large fraction of serotonergic neurons (Kohlmeier et al. 2013). This is consistent with the distribution of mRNA for both receptors in the DR (Trivedi et al. 1998; Marcus et al. 2001) and the finding that both receptors are found in a large fraction of neurons also expressing TPH mRNA (Mieda et al. 2011). These findings indicate that most serotonergic neurons in the DR express both orexin receptors. DR slice recordings also indicate that Ca^{2+} transients mediated by L-type Ca^{2+} channels are augmented by orexin-A (300 nM) and that either receptor was competent to elicit this effect (Kohlmeier et al. 2004, 2008, 2013). Thus, both receptors mediate depolarization and both receptors augment Ca^{2+} transients in DR neurons.

Cholinergic neurons of the laterodorsal (LDT) and pedunculopontine tegmental nuclei (PPT) are depolarized by orexin and this effect is also mediated by a noisy cation current (Burlet et al. 2002; Kim et al. 2009). Evidence from radiographic in situ hybridization indicates higher levels of OX_1 than OX_2 mRNA but that both appear present (Marcus et al. 2001). Recordings in brain slices made from receptor knockout mice indicate that the orexin-induced inward current requires OX_1 receptors in LDT neurons (Kohlmeier et al. 2013). However, orexin also augments Ca^{2+}-transients mediated by L-type Ca^{2+} channels in these neurons (Kohlmeier et al. 2004, 2008) and studies in slices from receptor knockout mice indicate that

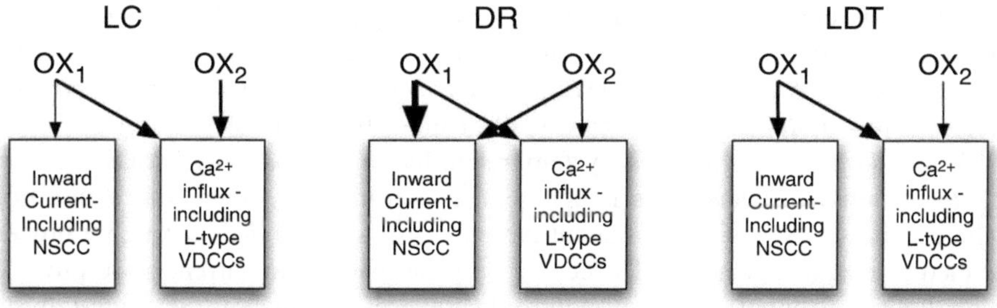

Fig. 1 Orexin receptor actions on brainstem arousal system neurons. Actions of orexin receptors on tyrosine hydroxylase positive (*TH*+) neurons from the locus coeruleus (*LC*), tryptophan hydroxylase positive (*TPH*+) dorsal raphe (*DR*) neurons and neuronal nitric oxide synthase positive (*nNOS*+) cholinergic laterodorsal tegmental (*LDT*) neurons. Arrow thickness indicates strength of functional coupling based on responses in slices from mice lacking one or the other OX receptor. Responses were determined using whole-cell recording and Ca^{2+} indicator fluorescence changes. See text for details

either receptor is competent to elicit this effect (Kohlmeier et al. 2013). Thus, like in the LC, even though a direct depolarization was absent in slices from OX_1 knockouts, remaining OX_2 receptors were sufficient to augment Ca^{2+} transients produced by prolonged depolarization. The strength of coupling from each orexin receptor to the depolarization and Ca^{2+} transient for LC, DR and LDT neurons is summarized in Fig. 1.

4 More Than Simple Excitation—Orexin Current Noise Provides High-Frequency Input During the Slow Depolarization

From the original electrophysiological description of orexins as excitatory peptides in primary cultures, it was clear that orexins could increase the rate of spontaneous transmitter release from terminals as manifest by increases in the miniature EPSC and IPSC (mEPSC/mIPSC) frequency (van den Pol et al. 1998). One or both of these presynaptic actions, have been observed in some, but not all central neurons studied in brain slices, suggesting orexin can play different roles in different circuits. These findings also suggest that orexin peptides do more than simply excite soma-dendritic targets (for a more general discussion, see Leonard and Kukkonen 2014). Emerging evidence indicates that in addition to modulating the strength of synaptic connections, orexins also influence the integrative properties of their target neurons.

In one example, we recently reported that in addition to providing a depolarizing input to serotonergic DR neurons and cholinergic LDT and PPT neurons, the noisy orexin cation current provides substantial high-frequency input at physiologically relevant frequencies (Ishibashi et al. 2015b). Measurement of the spectral properties of the orexin current revealed that the spectral amplitude significantly increased in theta, alpha, beta frequency bands and approximately doubled in the gamma frequency band for cholinergic LDT and PPT neurons, while in serotonergic DR neurons, this more than tripled across all of these frequencies.

While noise often reduces system reliability and sensitivity, noise can also confer beneficial effects. This is apparent in non-linear systems with a threshold where noise can selectively augment the effects of small signals through a process termed stochastic resonance. Hence, we propose that orexin current noise provides high frequency input to enhance the effectiveness of small EPSPs through stochastic resonance.

In this same study, we found that input of a virtual noisy orexin conductance to cholinergic LDT and PPT neurons, using a dynamic clamp activated an intrinsic Ca^{2+}-dependent resonance that peaked at theta and alpha frequencies. Neuronal resonances depend on the interplay between the passive and active membrane properties and function to boost the effectiveness of inputs occurring within the resonance bandwidth. Thus, we propose another function of the orexin noise is to engage the intrinsic resonance which could enhance inputs during the waking and REM states and help promote firing of these cholinergic LDT and PPT neurons during these states (Boucetta et al. 2014).

5 More Than Simple Excitation—Orexin Modulates the Late AHP and Alters Neuronal Firing Properties

In addition to producing a slow depolarization, orexins can slow spike repolarization and reduce the post-spike after hyperpolarization (AHP) (Yang and Ferguson 2003; Yang et al. 2003; Horvath et al. 1999; Murai and Akaike 2005) although the underlying channels have not been identified nor have the consequences of this modulation been studied. Recent findings in paraventricular thalamic neurons, indicate that orexins also can inhibit the long-lasting slow AHP (I_{sAHP}) that builds up following multiple spikes (Zhang et al. 2010). Although the underlying channel of the Ca^{2+}-dependent K^+ current are not identified, the I_{sAHP} is found in many neurons where it powerfully influences firing pattern. It is also a well known target for modulation and is suppressed by activation of both G_s and G_q coupled neurotransmitter receptors (for review Andrade et al. 2012). This current, produces strong spike-frequency adaptation and inhibition of this current by orexins enhances the excitability of these neurons to sustained depolarizing input (Zhang et al. 2009, 2010).

Recently, we found a novel effect of orexin-A to *enhance* a late after hyperpolarization following single and multiple spikes in serotonergic dorsal raphé neurons. These neurons have a large, medium AHP, mediated by small-conductance (SK) Ca^{2+}-activated K^+ channels, that promotes slow firing and rapid spike-frequency adaptation (Aghajanian and Vandermaelen 1982). We found that in addition to producing a noisy depolarization, orexin *enhances* this K^+ current and *induces* a longer-lasting, unidentified Ca^{2+}-dependent late AHP current that appears distinct from I_{sAHP}. Together, these enhanced outward currents increase spike-frequency adaptation and limit firing by slowing steady-state firing and reducing excitability (Ishibashi et al. 2015a).

Collectively, these diverse and potentially opposing modulatory actions imply that orexin receptors function to can powerfully tune how their target neurons respond to other inputs. Future studies aimed at understanding the diversity of mechanisms and impact of these emerging actions could provide important insights into how orexins normally regulate information flow in the brain and how this leads to perturbed circuit function in the absence of orexins in narcolepsy.

6 What Are the Behavioral Consequences of Orexin Receptor Signaling at Cholinergic and Monoaminergic Neurons of the Ascending Arousal System?

As described above, OX_1 receptors drive excitation of noradrenergic LC and cholinergic LDT neurons and both receptors drive orexin excitation of serotonergic DR neurons. To gain insight into the functional consequences of orexin receptor signaling at these neurons we compare the changes in orexin excitation to the behavioral impairment reported for these knockout mice.

Mice with constitutive knockouts of both orexin receptors (DKO mice) have a narcolepsy phenotype similar to prepro-orexin knockouts (Chemelli et al. 1999; Mochizuki et al. 2004) and show fragmented waking, sleep attacks and cataplexy (Kalogiannis et al. 2011; Hondo et al. 2010; Hasegawa et al. 2014). Since a large fraction of LDT, DR and LC neurons are stimulated by orexin, the absence of orexin signaling at these loci in DKO mice is consistent with a role in promoting wake and sleep consolidation, suppressing REM sleep, preventing sleep attacks and suppressing cataplexy. Indeed, selective excitation of orexin neurons promotes sleep-wake transitions (Adamantidis et al. 2007) and selective inhibition of orexin neurons promotes slow-wave sleep and reduced firing of DR neurons (Tsunematsu et al. 2011). Moreover, focal orexin injection into the LDT (Xi et al. 2001) or LC (Bourgin et al. 2000) prolongs waking bouts and suppresses REM sleep. Importantly, inhibition of TH + LC neurons using optogenetics blocks the ability of selective stimulation of orexin neurons to promote sleep-wake transitions (Carter et al. 2012), while stimulation of TH + LC neurons prolongs waking (Carter et al. 2010) and enhances the wake-promoting effect of orexin neuron stimulation (Carter et al. 2012). Thus, orexin–excitation of TH + LC neurons appears necessary for the acute ability of orexin neurons to generate transitions to arousal from sleep. This suggests OX_1 signaling in LC might also be critical for promoting consolidated waking bouts. A recent receptor rescue study, supports this idea since wake consolidation was increased in proportion to the number of TH + LC neurons that re-expressed OX_1 receptors in constitutive DKO mice (Hasegawa et al. 2014). This appeared quite specific since re-expression of OX_1 receptors in PPT or OX_2

receptors in either TMN or DR did not increase wake bout duration in DKO mice. Nevertheless, additional factors must play a role since OX_1−/− mice do not have fragmented sleep-wake states or sleep attacks (Mieda et al. 2011; Hondo et al. 2010), even though orexin excitation is abolished in LDT and LC neurons and is reduced in DR neurons (Kohlmeier et al. 2013). Thus, OX_1 excitation of LDT, DR and LC neurons is not necessary for consolidated wake bouts, at least when OX_2 receptor signaling is intact.

In contrast, OX_2 knockouts have sleep attacks at the same frequency as do prepro-orexin knockouts (Willie et al. 2003) and have shorter wake bouts even though orexin-mediated excitation of LDT, DR and LC neurons is intact in these mice. Thus, orexin excitation of LDT, DR and LC neurons is not sufficient for the expression of normally consolidated bouts of spontaneous waking in the absence of OX_2 signaling.

These above findings stress the importance of OX_2 receptor signaling in consolidating wakefulness, yet re-expression of OX_2 receptors in DR did not increase wake bout duration in DKO mice (Hasegawa et al. 2014) but re-expression of OX_2 in the tuberomammillary region did prolong wake bout duration in OX_2−/− mice (Mochizuki et al. 2011), but not in DKO mice (Hasegawa et al. 2014).

Collectively, these data are consistent with both OX_1 signaling in LC and OX_2 signaling in the TM region but not the DR in contributing to consolidation of wake activity. As noted by Hasegawa et al., it is important to interpret results from re-expression experiments with great cation since the effectiveness of the rescue will depend on many factors including re-expression efficacy and the degree of impairment to be overcome.

Cataplexy is another important impairment of narcolepsy, and cataplexy-like arrests are frequent in the dark phase of orexin ligand deficient mice (Chemelli et al. 1999; Mochizuki et al. 2004) and DKO mice (Kalogiannis et al. 2011). Since OX_2−/− mice have rare cataplexy compared to prepro-orexin null mice (Willie et al. 2003), the attenuated cataplexy in OX_2−/− mice likely results from residual OX_1 signaling. Orexin injections into the LDT and LC suppress REM sleep, and knockdown of OX_1 in LC increases REM during the dark phase (Chen et al. 2010). Hence, it is plausible that normal OX_1-mediated excitation in LDT, LC and DR, reduces cataplexy in OX_2−/− mice. However, neither re-expression of OX_1 in LC or PPT neurons or OX_2 in TMN rescued cataplexy-like arrests in DKO mice (Hasegawa et al. 2014). In contrast, re-expression of OX_2 in DR neurons completely rescued cataplexy in DKO mice (Hasegawa et al. 2014). Since signaling by OX_2 and OX_1 receptors converge onto similar, if not identical, effectors in serotonergic DR neurons (Kohlmeier et al. 2013), these findings strongly support the idea that residual OX_1 signaling in DR attenuates cataplexy in OX_2−/− mice. Moreover, since OX_1−/− mice do not have cataplexy, these findings suggest residual OX_2 signaling in DR neurons prevents cataplexy. Thus, orexin mediated DR activity appears to play a key role in coordinating muscle atonia with sleep states and preventing intrusion of muscle atonia into waking.

The previous discussion of orexin signaling in the ascending arousal system has focused on its role in regulating the expression of spontaneous bouts of sleep and waking. Of course, orexin signaling at these loci may also contribute to arousal levels in other circumstances such as during stress, anxiety, panic and/or food and drug seeking (Winskey-Sommerer et al. 2004; Johnson et al. 2012; Heydendael et al. 2013; Steiner et al. 2012; Piccoli et al. 2012; Nair et al. 2008; Boutrel et al. 2005; Harris et al. 2005). Moreover, such signaling likely also support other functions. For example, OX_1−/− mice show impaired acquisition and expression of fear conditioning and re-expression of OX_1 in TH + LC neurons rescues cued fear conditioning (Soya et al. 2013).

7 Conclusions

While orexins are recognized as excitatory neuropeptides, emerging evidence indicates orexins have actions that go beyond simple excitation, indicating they should be regarded as modulatory peptides that can also produce post-synaptic inhibition. In addition to producing a slow depolarization, orexin produces a set of modulatory actions at ascending arousal system targets. Evidence from whole-cell recordings from constitutive receptor knockout mice and receptor rescue experiments indicate that OX_1 signaling in the LC and both OX_1 and OX_2 receptors in the DR play important roles in consolidating bouts of waking and restricting muscle atonia to REM epochs, respectively. It remains to be determined how the modulatory actions of orexin contribute to the ability of these neuronal groups to organize arousal.

Acknowledgments This work was supported by grants NS027881 and HL064150 from the NIH of the USPHS

References

Acuna-Goycolea C, van den Pol AN (2009) Neuroendocrine proopiomelanocortin neurons are excited by hypocretin/orexin. J Neurosci 29:1503–1513

Adamantidis AR, Zhang F, Aravanis AM, Deisseroth K, de Lecea L (2007) Neural substrates of awakening probed with optogenetic control of hypocretin neurons. Nature 450:420–424

Aghajanian GK, Vandermaelen CP (1982) Intracellular recordings from serotonergic dorsal raphe neurons: pacemaker potentials and the effect of LSD. Brain Res 238:463–469

Andrade R, Foehring RC, Tzingounis AV (2012) The calcium-activated slow AHP: cutting through the Gordian knot. Front Cell Neurosci 6:47

Bayer L, Eggermann E, Saint-Mleux B, Machard D, Jones BE, Muhlethaler M, Serafin M (2002) Selective action of orexin (hypocretin) on nonspecific thalamocortical projection neurons. J Neurosci 22:7835–7839

Bayer L, Serafin M, Eggermann E, Saint-Mleux B, Machard D, Jones BE, Muhlethaler M (2004) Exclusive postsynaptic action of hypocretin-orexin on sublayer 6b cortical neurons. J Neurosci 24:6760–6764

Belle MD, Hughes AT, Bechtold DA, Cunningham P, Pierucci M, Burdakov D, Piggins HD (2014) Acute suppressive and long-term phase modulation actions of orexin on the mammalian circadian clock. J Neurosci 34:3607–3621

Beuckmann CT, Sinton CM, Williams SC, Richardson JA, Hammer RE, Sakurai T, Yanagisawa M (2004) Expression of a poly-glutamine-ataxin-3 transgene in orexin neurons induces narcolepsy-cataplexy in the rat. J Neurosci 24:4469–4477

Bisetti A, Cvetkovic V, Serafin M, Bayer L, Machard D, Jones BE, Muhlethaler M (2006) Excitatory action of hypocretin/orexin on neurons of the central medial amygdala. Neuroscience 142:999–1004

Boucetta S, Cissé Y, Mainville L, Morales M, Jones BE (2014) Discharge profiles across the sleep-waking cycle of identified cholinergic, GABAergic, and glutamatergic neurons in the pontomesencephalic tegmentum of the rat. J Neurosci 34:4708–4727

Bourgin P, Huitron-Resendiz S, Spier AD, Fabre V, Morte B, Criado JR, Sutcliffe JG, Henriksen SJ, de Lecea L (2000) Hypocretin-1 modulates rapid eye movement sleep through activation of locus coeruleus neurons. J Neurosci 20:7760–7765

Boutrel B, Kenny PJ, Specio SE, Martin-Fardon R, Markou A, Koob GF, de Lecea L (2005) Role for hypocretin in mediating stress-induced reinstatement of cocaine-seeking behavior. Proc Nat Acad Sci USA 102:19168–19173

Brown RE, Sergeeva O, Eriksson KS, Haas HL (2001) Orexin A excites serotonergic neurons in the dorsal raphe nucleus of the rat. Neuropharmacology 40:457–459

Brown RE, Sergeeva OA, Eriksson KS, Haas HL (2002) Convergent excitation of dorsal raphe serotonin neurons by multiple arousal systems (orexin/hypocretin, histamine and noradrenaline). J Neurosci 22:8850–8859

Burdakov D, Liss B, Ashcroft FM (2003) Orexin excites GABAergic neurons of the arcuate nucleus by activating the sodium–calcium exchanger. J Neurosci 23:4951–4957

Burlet S, Tyler CJ, Leonard CS (2002) Direct and indirect excitation of laterodorsal tegmental neurons by Hypocretin/Orexin peptides: implications for wakefulness and narcolepsy. J Neurosci 22:2862–2872

Carter ME, Brill J, Bonnavion P, Huguenard JR, Huerta R, de Lecea L (2012) Mechanism for Hypocretin-mediated sleep-to-wake transitions. Proc Natl Acad Sci U S A 109:E2635–E2644

Carter ME, Yizhar O, Chikahisa S, Nguyen H, Adamantidis A, Nishino S, Deisseroth K, de Lecea L (2010) Tuning arousal with optogenetic modulation of locus coeruleus neurons. Nat Neurosci 13:1526–1533

Chemelli RM, Willie JT, Sinton CM, Elmquist JK, Scammell T, Lee C, Richardson JA, Williams SC, Xiong Y, Kisanuki Y, Fitch TE, Nakazato M, Hammer RE, Saper CB, Yanagisawa M (1999) Narcolepsy in orexin knockout mice: molecular genetics of sleep regulation. Cell 98:437–451

Chen L, McKenna JT, Bolortuya Y, Winston S, Thakkar MM, Basheer R, Brown RE, McCarley RW (2010) Knockdown of orexin type 1 receptor in rat locus coeruleus increases REM sleep during the dark period. Eur J Neurosci 32:1528–1536

de Lecea L, Huerta R (2014) Hypocretin (orexin) regulation of sleep-to-wake transitions. Front Pharmacol 5:16

de Lecea L, Kilduff TS, Peyron C, Gao X, Foye PE, Danielson PE, Fukuhara C, Battenberg EL, Gautvik VT, Bartlett FS 2nd, Frankel WN, van den Pol AN, Bloom FE, Gautvik KM, Sutcliffe JG (1998) The hypocretins: hypothalamus-specific peptides with neuroexcitatory activity. Proc National Acad Sci (USA) 95:322–327

Doroshenko P, Renaud LP (2009) Acid-sensitive TASK-like K+ conductances contribute to resting membrane potential and to orexin-induced membrane depolarization in rat thalamic paraventricular nucleus neurons. Neuroscience 158:1560–1570

Eriksson KS, Sergeeva O, Brown RE, Haas HL (2001) Orexin/hypocretin excites the histaminergic neurons of the tuberomammillary nucleus. J Neurosci 21:9273–9279

Govindaiah G, Cox CL (2006) Modulation of thalamic neuron excitability by orexins. Neuropharmacology 51:414–425

Grabauskas G, Moises HC (2003) Gastrointestinal-projecting neurones in the dorsal motor nucleus of the vagus exhibit direct and viscerotopically organized sensitivity to orexin. J Physiol 549:37–56

Hara J, Beuckmann CT, Nambu T, Willie JT, Chemelli RM, Sinton CM, Sugiyama F, Yagami K, Goto K, Yanagisawa M, Sakurai T (2001) Genetic ablation of orexin neurons in mice results in narcolepsy, hypophagia, and obesity. Neuron 30:345–354

Harris GC, Aston-Jones G (2006) Arousal and reward: a dichotomy in orexin function. Trends Neurosci 29:571–577

Harris GC, Wimmer M, Aston-Jones G (2005) A role for lateral hypothalamic orexin neurons in reward seeking. Nature 437:556–559

Hasegawa E, Yanagisawa M, Sakurai T, Mieda M (2014) Orexin neurons suppress narcolepsy via 2 distinct efferent pathways. J Clin Invest 124:604–616

Heydendael W, Sengupta A, Beck S, Bhatnagar S (2013) Optogenetic examination identifies a context-specific role for orexins/hypocretins in anxiety-related behavior. Physiol Behav

Hoang QV, Bajic D, Yanagisawa M, Nakajima S, Nakajima Y (2003) Effects of orexin (hypocretin) on GIRK channels. J Neurophysiol 90:693–702

Hoang QV, Zhao P, Nakajima S, Nakajima Y (2004) Orexin (hypocretin) effects on constitutively active inward rectifier K+ channels in cultured nucleus basalis neurons. J Neurophysiol 92:3183–3191

Hondo M, Nagai K, Ohno K, Kisanuki Y, Willie JT, Watanabe T, Yanagisawa M, Sakurai T (2010) Histamine-1 receptor is not required as a downstream effector of orexin-2 receptor in maintenance of basal sleep/wake states. Acta Physiol (Oxf) 198:287–294

Horvath TL, Peyron C, Diano S, Ivanov A, Aston-Jones G, Kilduff TS, van Den Pol AN (1999) Hypocretin (orexin) activation and synaptic innervation of the locus coeruleus noradrenergic system. J Comp Neurol 415:145–159

Huang H, Ghosh P, van den Pol AN (2006) Prefrontal cortex-projecting glutamatergic thalamic paraventricular nucleus-excited by hypocretin: a feedforward circuit that may enhance cognitive arousal. J Neurophysiol 95:1656–1668

Hwang LL, Chen CT, Dun NJ (2001) Mechanisms of orexin-induced depolarizations in rat dorsal motor nucleus of vagus neurones in vitro. J Physiol 537:511–520

Ishibashi M, Gumenchuk I, Inoue T, Leonard CS (2015a) The arousal peptide orexin-A alters the firing properties of serotonergic dorsal raphe neurons by enhancing the late afterhyperpolarization and promoting slower, less variable steady-state firing. Submitted

Ishibashi M, Gumenchuk I, Kang B, Steger C, Lynn E, Molina NE, Eisenberg LM, Leonard CS (2015b) Orexin receptor activation generates gamma band input to cholinergic and serotonergic arousal system neurons and drives an intrinsic Ca2+-dependent resonance in LDT and PPT cholinergic neurons. Front Neurol 6:120

Ishibashi M, Takano S, Yanagida H, Takatsuna M, Nakajima K, Oomura Y, Wayner MJ, Sasaki K (2005) Effects of orexins/hypocretins on neuronal activity in the paraventricular nucleus of the thalamus in rats in vitro. Peptides 26:471–481

Ivanov A, Aston-Jones G (2000) Hypocretin/orexin depolarizes and decreases potassium conductance in locus coeruleus neurons. Neuroreport 11:1755–1758

Johnson PL, Samuels BC, Fitz SD, Lightman SL, Lowry CA, Shekhar A (2012) Activation of the orexin 1 receptor is a critical component of CO_2-mediated anxiety and hypertension but not bradycardia. Neuropsychopharmacology 37:1911–1922

Kalogiannis M, Hsu E, Willie JT, Chemelli RM, Kisanuki YY, Yanagisawa M, Leonard CS (2011) Cholinergic modulation of narcoleptic attacks in double orexin receptor knockout mice. PLoS ONE 6:e18697

Kim J, Nakajima K, Oomura Y, Wayner MJ, Sasaki K (2009) Electrophysiological effects of orexins/hypocretins on pedunculopontine tegmental neurons in rats: an in vitro study. Peptides 30:191–209

Kohlmeier KA, Inoue T, Leonard CS (2004) Hypocretin/orexin peptide signaling in the ascending arousal system: elevation of intracellular calcium in the mouse dorsal raphe and laterodorsal tegmentum. J Neurophysiol 92:221–235

Kohlmeier KA, Tyler CJ, Kalogiannis M, Ishibashi M, Kristensen MP, Gumenchuk I, Chemelli RM, Kisanuki YY, Yanagisawa M, Leonard CS (2013) Differential actions of orexin receptors in brainstem cholinergic and monoaminergic neurons revealed by receptor knockouts: implications for orexinergic signaling in arousal and narcolepsy. Front Neurosci 7:246

Kohlmeier KA, Watanabe S, Tyler CJ, Burlet S, Leonard CS (2008) Dual orexin actions on dorsal raphe and laterodorsal tegmentum neurons: noisy cation current activation and selective enhancement of Ca2+ transients mediated by L-type calcium channels. J Neurophysiol 100:2265–2281

Kolaj M, Coderre E, Renaud LP (2008) Orexin peptides enhance median preoptic nucleus neuronal excitability via postsynaptic membrane depolarization and enhancement of glutamatergic afferents. Neuroscience 155:1212–1220

Korotkova TM, Eriksson KS, Haas HL, Brown RE (2002) Selective excitation of GABAergic neurons in the substantia nigra of the rat by orexin/hypocretin in vitro. Regul Pept 104:83–89

Kukkonen JP, Leonard CS (2014) Orexin/hypocretin receptor signalling cascades. Br J Pharmacol 171:314–331

Lambe EK, Aghajanian GK (2003) Hypocretin (orexin) induces calcium transients in single spines postsynaptic to identified thalamocortical boutons in prefrontal slice. Neuron 40:139–150

Leonard CS, Kukkonen JP (2014) Orexin/hypocretin receptor signalling: A functional perspective. Br J Pharmacol 171:294–313

Lin L, Faraco J, Li R, Kadotani H, Rogers W, Lin X, Qiu X, de Jong PJ, Nishino S, Mignot E (1999) The sleep disorder canine narcolepsy is caused by a mutation in the hypocretin (orexin) receptor 2 gene. Cell 98:365–376

Liu RJ, van den Pol AN, Aghajanian GK (2002) Hypocretins (orexins) regulate serotonin neurons in the dorsal raphe nucleus by excitatory direct and inhibitory indirect actions. J Neurosci 22:9453–9464

Marcus JN, Aschkenasi CJ, Lee CE, Chemelli RM, Saper CB, Yanagisawa M, Elmquist JK (2001) Differential expression of orexin receptors 1 and 2 in the rat brain. J Comp Neurol 435:6–25

Mieda M, Hasegawa E, Kisanuki YY, Sinton CM, Yanagisawa M, Sakurai T (2011) Differential roles of orexin receptor-1 and -2 in the regulation of non-REM and REM sleep. J Neurosci 31:6518–6526

Mochizuki T, Arrigoni E, Marcus JN, Clark EL, Yamamoto M, Honer M, Borroni E, Lowell BB, Elmquist JK, Scammell TE (2011) Orexin receptor 2 expression in the posterior hypothalamus rescues sleepiness in narcoleptic mice. Proc Natl Acad Sci U S A 108:4471–4476

Mochizuki T, Crocker A, McCormack S, Yanagisawa M, Sakurai T, Scammell T (2004) Behavioral State Instability in Orexin Knockout Mice. J Neurosci 24:6291–6300

Murai Y, Akaike T (2005) Orexins cause depolarization via nonselective cationic and K+ channels in isolated locus coeruleus neurons. Neurosci Res 51:55–65

Muroya S, Funahashi H, Yamanaka A, Kohno D, Uramura K, Nambu T, Shibahara M, Kuramochi M, Takigawa M, Yanagisawa M, Sakurai T, Shioda S, Yada T (2004) Orexins (hypocretins) directly interact with neuropeptide Y, POMC and glucose-responsive neurons to regulate Ca2+ signaling in a reciprocal manner to leptin: orexigenic neuronal pathways in the mediobasal hypothalamus. Eur J Neurosci 19:1524–1534

Nair SG, Golden SA, Shaham Y (2008) Differential effects of the hypocretin 1 receptor antagonist SB 334867 on high-fat food self-administration and reinstatement of food seeking in rats. Br J Pharmacol 154:406–416

Peyron C, Faraco J, Rogers W, Ripley B, Overeem S, Charnay Y, Nevsimalova S, Aldrich M, Reynolds D, Albin R, Li R, Hungs M, Pedrazzoli M, Padigaru M, Kucherlapati M, Fan J, Maki R, Lammers GJ, Bouras C, Kucherlapati R, Nishino S, Mignot E (2000) A mutation in a case of early onset narcolepsy and a generalized absence of hypocretin peptides in human narcoleptic brains. Nat Med 6:991–997

Piccoli L, Micioni Di Bonaventura MV, Cifani C, Costantini VJ, Massagrande M, Montanari D, Martinelli P, Antolini M, Ciccocioppo R, Massi M, Merlo-Pich E, Di Fabio R, Corsi M (2012) Role of orexin-1 receptor mechanisms on compulsive food consumption in a model of binge eating in female rats. Neuropsychopharmacology 37:1999–2011

Sakurai T, Amemiya A, Ishii M, Matsuzaki I, Chemelli RM, Tanaka H, Williams SC, Richardson JA, Kozlowski GP, Wilson S, Arch JR, Buckingham RE, Haynes AC, Carr SA,

Annan RS, McNulty DE, Liu WS, Terrett JA, Elshourbagy NA, Bergsma DJ, Yanagisawa M (1998) Orexins and orexin receptors: a family of hypothalamic neuropeptides and G protein-coupled receptors that regulate feeding behavior. Cell 92:573–585

Scammell TE (2006) The frustrating and mostly fruitless search for an autoimmune cause of narcolepsy. Sleep 29:601–602

Scammell TE, Winrow CJ (2011) Orexin receptors: pharmacology and therapeutic opportunities. Annu Rev Pharmacol Toxicol 51:243–266

Soffin EM, Gill CH, Brough SJ, Jerman JC, Davies CH (2004) Pharmacological characterisation of the orexin receptor subtype mediating postsynaptic excitation in the rat dorsal raphe nucleus. Neuropharmacology 46:1168–1176

Soya S, Shoji H, Hasegawa E, Hondo M, Miyakawa T, Yanagisawa M, Mieda M, Sakurai T (2013) Orexin receptor-1 in the locus coeruleus plays an important role in cue-dependent fear memory consolidation. J Neurosci 33:14549–14557

Steiner MA, Lecourt H, Jenck F (2012) The brain orexin system and almorexant in fear-conditioned startle reactions in the rat. Psychopharmacology 223:465–475

Thannickal TC, Moore RY, Nienhuis R, Ramanathan L, Gulyani S, Aldrich M, Cornford M, Siegel JM (2000) Reduced number of hypocretin neurons in human narcolepsy. Neuron 27:469–474

Trivedi P, Yu H, MacNeil DJ, Van der Ploeg LHT, Guan XM (1998) Distribution of orexin receptor mRNA in the rat brain. FEBS Lett 438:71–75

Tsujino N, Sakurai T (2009) Orexin/hypocretin: a neuropeptide at the interface of sleep, energy homeostasis, and reward system. Pharmacol Rev 61:162–176

Tsujino N, Yamanaka A, Ichiki K, Muraki Y, Kilduff TS, Yagami K, Takahashi S, Goto K, Sakurai T (2005) Cholecystokinin activates orexin/hypocretin neurons through the cholecystokinin A receptor. J Neurosci 25:7459–7469

Tsunematsu T, Kilduff TS, Boyden ES, Takahashi S, Tominaga M, Yamanaka A (2011) Acute optogenetic silencing of orexin/hypocretin neurons induces slow-wave sleep in mice. J Neurosci 31:10529–10539

Uramura K, Funahashi H, Muroya S, Shioda S, Takigawa M, Yada T (2001) Orexin-a activates phospholipase C- and protein kinase C-mediated Ca2+ signaling in dopamine neurons of the ventral tegmental area. NeuroReport 12:1885–1889

van den Pol AN (1999) Hypothalamic hypocretin (orexin): robust innervation of the spinal cord. J Neurosci 19:3171–3182

van den Pol AN, Gao XB, Obrietan K, Kilduff TS, Belousov AB (1998) Presynaptic and postsynaptic actions and modulation of neuroendocrine neurons by a new hypothalamic peptide, hypocretin/orexin. J Neurosci 18:7962–7971

Van Den Pol AN, Ghosh PK, Liu RJ, Li Y, Aghajanian GK, Gao XB (2002) Hypocretin (orexin) enhances neuron activity and cell synchrony in developing mouse GFP-expressing locus coeruleus. J Physiol 541:169–185

van den Pol AN, Patrylo PR, Ghosh PK, Gao XB (2001) Lateral hypothalamus: early developmental expression and response to hypocretin (orexin). J Comp Neurol 433:349–363

van den Top M, Nolan MF, Lee K, Richardson PJ, Buijs RM, Davies CH, Spanswick D (2003) Orexins induce increased excitability and synchronisation of rat sympathetic preganglionic neurones. J Physiol 549:809–821

Willie JT, Chemelli RM, Sinton CM, Tokita S, Williams SC, Kisanuki YY, Marcus JN, Lee C, Elmquist JK, Kohlmeier KA, Leonard CS, Richardson JA, Hammer RE, Yanagisawa M (2003) Distinct narcolepsy syndromes in orexin receptor-2 and orexin null mice. molecular genetic dissection of non-REM and REM sleep regulatory processes. Neuron 38:715–730

Winrow CJ, Renger JJ (2013) Discovery and development of orexin receptor antagonists as therapeutics for insomnia. Br J Pharmacol 171(2):283–293

Winskey-Sommerer R, Yamanaka A, Diano S, Borok E, Roberts AJ, Sakurai T, Kilduff TS, Horvath TL, de Lecea L (2004) Interaction between corticotropin-releasing factor system and hypocretins (orexins): a novel circuit mediating stress response. J Neurosci 24:11439–11448

Wu M, Zaborszky L, Hajszan T, van den Pol AN, Alreja M (2004) Hypocretin/orexin innervation and excitation of identified septohippocampal cholinergic neurons. J Neurosci 24:3527–3536
Wu M, Zhang Z, Leranth C, Xu C, van den Pol AN, Alreja M (2002) Hypocretin increases impulse flow in the septohippocampal GABAergic pathway: implications for arousal via a mechanism of hippocampal disinhibition. J Neurosci 22:7754–7765
Xi M, Morales FR, Chase MH (2001) Effects on sleep and wakefulness of the injection of hypocretin-1 (orexin-A) into the laterodorsal tegmental nucleus of the cat. Brain Res 901:259–264
Yang B, Ferguson AV (2002) Orexin-A depolarizes dissociated rat area postrema neurons through activation of a nonselective cationic conductance. J Neurosci 22:6303–6308
Yang B, Ferguson AV (2003) Orexin-A depolarizes nucleus tractus solitarius neurons through effects on nonselective cationic and K+ conductances. J Neurophysiol 89:2167–2175
Yang B, Samson WK, Ferguson AV (2003) Excitatory effects of orexin-A on nucleus tractus solitarius neurons are mediated by phospholipase C and protein kinase C. J Neurosci 23:6215–6222
Zhang J, Li B, Yu L, He YC, Li HZ, Zhu JN, Wang JJ (2011) A role for orexin in central vestibular motor control. Neuron 69:793–804
Zhang L, Kolaj M, Renaud LP (2010) Ca2+-dependent and Na+-dependent K+ conductances contribute to a slow AHP in thalamic paraventricular nucleus neurons: a novel target for orexin receptors. J Neurophysiol 104:2052–2062
Zhang L, Renaud LP, Kolaj M (2009) Properties of a T-type Ca2+ channel-activated slow afterhyperpolarization in thalamic paraventricular nucleus and other thalamic midline neurons. J Neurophysiol 101:2741–2750

Elucidation of Neuronal Circuitry Involved in the Regulation of Sleep/Wakefulness Using Optogenetics

Akihiro Yamanaka and Tomomi Tsunematsu

Abstract The mechanisms of instinctive behaviors such as feeding, drinking, sexual activity and sleep/wakefulness can only be studied in the whole animal with intact neural networks. This has been difficult due to lack of techniques to control the activity of specific neurons in intact animals. Optogenetics is a recent technique, which enables such control using light and allows the study of regulatory mechanisms of instinctive behaviors. In this section, we introduce how optogenetics was applied to orexin/hypocretin neurons to reveal the regulatory mechanisms of sleep and wakefulness. Activity manipulation of orexin neurons controls the state changes among wakefulness, non-rapid eye movement (NREM) sleep and REM sleep state. Selective activation of orexin neurons using channelrhodopsin-2 (ChR2) or melanopsin (OPN4) induced transition from sleep to wakefulness. In contrast, suppression of these neurons using halorhodopsin (HaloR) or archaerhodopsin (ArchR) induced transition from wakefulness to NREM sleep and increased the time spent in NREM sleep. These studies help answer how orexin neurons contribute to regulate sleep/wakefulness.

Keywords Orexin · Locus coeruleus (LC) · Wakefulness · Non-rapid eye movement (NREM) sleep · Rapid eye movement (REM) sleep · Melanopsin/OPN4 · Channelrhodopsin-2 (ChR2) · Archaerhodopsin (ArchR) · Halorhodopsin (HaloR)

A. Yamanaka (✉) · T. Tsunematsu
Research Institute of Environmental Medicine, Nagoya University, Nagoya, Japan
e-mail: yamank@riem.nagoya-u.ac.jp

T. Tsunematsu
Strathclyde Institute of Pharmacy and Biomedical Sciences, University of Strathclyde, Glasgow, UK

© Springer International Publishing Switzerland 2015
T. Sakurai et al. (eds.), *Orexin and Sleep*,
DOI 10.1007/978-3-319-23078-8_6

1 Sleep/Wakefulness

Sleeping is a natural behavior and a universal phenomenon conserved in a wide range of species from fish to mammals. On average, we spend a third of our lifetime sleeping. Sleep deprivation reduces body weight while increasing food intake in animals like rats and dogs. Animals die within a few weeks of sleep deprivation, despite access to food and water (Rechtschaffen et al. 1983). These reports suggest that sleep is an absolutely essential physiological phenomenon that maintains normal health in animals.

There are two completely different states of sleep: rapid eye movement (REM) sleep and non-REM (NREM) sleep. To determine these distinct states (wakefulness, NREM sleep, and REM sleep), electroencephalogram (EEG) and electromyogram (EMG) are typically recorded. EEG waves are analyzed by fast Fourier transformation (FFT) to easily recognize frequency and amplitude. This analysis greatly helps us determine the vigilance state. During wakefulness, EEG shows a low amplitude with a relatively high frequency and EMG shows large-amplitude patterns due to muscle contraction. When NREM sleep starts, EEG shows slow-wave patterns with high amplitude and low frequency (delta wave), and EMG becomes low amplitude compared to patterns observed during wakefulness. After NREM sleep, REM sleep is initiated, with EEG showing a relatively low amplitude with a relatively high-frequency pattern that looks similar to the wakefulness EEG. However, the component of EEG frequency in this case is predominantly theta waves (6–10 Hz). EMG activity is almost nothing due to relaxation of voluntary muscles.

2 The Flip-Flop Hypothesis: Regulating the Timing of Sleep/Wakefulness—Sleep Center and Wake Center

What controls the sleep/wakefulness state in the brain, and how is it regulated? It has been generally believed that sleep is a passive phenomenon to recover from tiredness caused by continuous wakefulness. However, recent studies have revealed that sleep is actively initiated in a brain region, which is active during sleep, and is termed the "sleep center". This center is thought to be located in the anterior part of the hypothalamus, called the ventrolateral preoptic area (VLPO), which contains GABAergic neurons. The longer the time spent in wakefulness, the greater the increase in sleep pressure (sleepiness). As a result, sleep pressure increases activity of GABAergic neurons in the VLPO. In contrast with the sleep center, the "wake centers", which induce wakefulness, are thought to be located in the brain stem and the posterior part of the hypothalamus. Monoaminergic neurons, such as histaminergic neurons in the tuberomammillary nucleus (TMN), serotonergic neurons in the raphe nucleus, and noradrenergic neurons in the locus coeruleus (LC), have a major role in generating arousal as wake centers. The activity of monoaminergic neurons is strongly associated with sleep/wakefulness states. These neurons exhibit tonic firing during wakefulness, decrease firing during NREM sleep, and are almost quiescent during REM sleep. These monoaminergic neurons project axons broadly throughout the brain, not only in the brain stem and posterior part of the hypothalamus, but also into the cerebral cortex to activate entire brain regions. This activation results in the promotion of wakefulness.

Sleep and wake centers have mutually inhibitory circuitry; the sleep and wake centers function in a seesaw relationship, also known as the flip-flop hypothesis. Sleep is initiated if the sleep center activity overcomes wake center activity.

3 Orexin/Hypocretin and Narcolepsy

Orexin-producing neurons (orexin neurons) are exclusively and sparsely located in the lateral hypothalamic area, which is well known as the feeding center; orexin neurons project their axons throughout the brain. Orexin was initially recognized as a regulator of feeding behavior because intracerebroventricular injections of synthesized orexin peptide induced feeding behavior in rats and mice. Subsequently, *prepro-orexin* knockout mice (Chemelli et al. 1999), *OX2R* knockout mice (Willie et al. 2003) and orexin neuron-ablated mice (Hara et al. 2001) showed abnormality in the regulation of sleep/wakefulness. These mice displayed fragmentation of sleep/wakefulness (or a frequent repetition of sleep and wakefulness) and an inability to maintain wakefulness, showing sudden behavior arrest during wakefulness. These phenotypes are similar to the human sleep disorder of narcolepsy. Orexin concentrations in cerebrospinal fluid (CSF) are undetectable in narcolepsy patients (Nishino et al. 2000). Also, a post-mortem study of human narcolepsy patients revealed specific degeneration of orexin neurons in the hypothalamus (Peyron et al. 2000; Nishino et al. 2000). These facts support the idea that degeneration of orexin neurons for unknown reasons (possibly in an autoimmune manner) leads to narcolepsy. Narcolepsy typically begins during adolescence with the primary symptom of excessive daytime sleepiness. Patients tend to fall asleep in inappropriate times and places. Cataplexy is a specific symptom of narcolepsy, characterized by sudden weakness of voluntary muscles triggered by strong emotions such as surprise, anger, and humor. In addition to sleepiness and cataplexy, narcoleptic patients exhibit a direct transition from wakefulness to REM sleep, although normal subjects display a marked transition from wakefulness to REM sleep via NREM sleep. All these facts suggest that orexin has an important role in the regulation of sleep/wakefulness.

Orexin neurons project densely to monoaminergic neurons known as wake centers, such as the TMN, raphe nucleus, and LC. These monoaminergic neurons express orexin receptors, OX1R and/or OX2R, and these neurons are activated by orexin (Yamanaka et al. 2002). Orexin neurons are thought to be phasically active during wakefulness and almost silent during NREM and REM sleep (Mileykovskiy et al. 2005). Therefore, it seems that orexin neurons should contribute to stabilize the flip-flop circuit as an activator of wake centers, especially for the maintenance of wakefulness. In narcolepsy patients, the balance between sleep and wake centers is unstable due to specific degeneration of orexin neurons. This would cause the fragmentation of sleep/wakefulness states as a result of frequent transition between sleep and wakefulness. It is thus presumable that orexin neurons have a crucial role in the stabilization of arousal states. However, the question of how orexin neuron activity regulates sleep/wakefulness still remains. Optogenetics could provide a powerful tool to address this question.

4 Development and Basics of Optogenetics

Optogenetics is a recently developed technique to control the activity of specific neurons using light (Boyden et al. 2005). Optogenetics requires expression of a light activated protein, which senses a specific wavelength of light. This light-activated protein is capable of affecting membrane potential and cell function in neurons. Subsequently, the activity of these neurons can be manipulated with light, with very precise timing. Channelrhodopsin 2 (ChR2) and halorhodopsin (HaloR) are rhodopsin proteins, which were originally isolated from Green algae (Chlamydomonas reinhardtii) and Archaebacteria (Halobacteria). ChR2 is a blue light-gated ion channel, which pass through cations (Fig. 1). ChR2 opening induces depolarization, implying up-regulation of neural activity. On the other hand, HaloR is an orange light-activated chloride anion pump. It induces hyperpolarization of membrane potential, implying down-regulation of neural activity. Thus, neurons can instantly be switched from activation to inhibition or vice versa by changing the wavelength of light (blue to orange). This capability to activate certain neurons while inhibiting others at the same time is a prominent aspect of optogenetics. Optogenetics was first used to manipulate neural activity with light in primary cultured neurons. A viral vector was used to introduce the ChR2 gene in cultured neurons, and was illuminated with blue light while recording its electrical activity using the patch clamp technique. It resulted in an action potential with a frequency matching that of the pulse of blue light (Boyden et al. 2005). This is because ChR2 is a non-selective cation channel that opens upon sensing blue light. Cations (Na^+, H^+, K^+ and Ca^+)

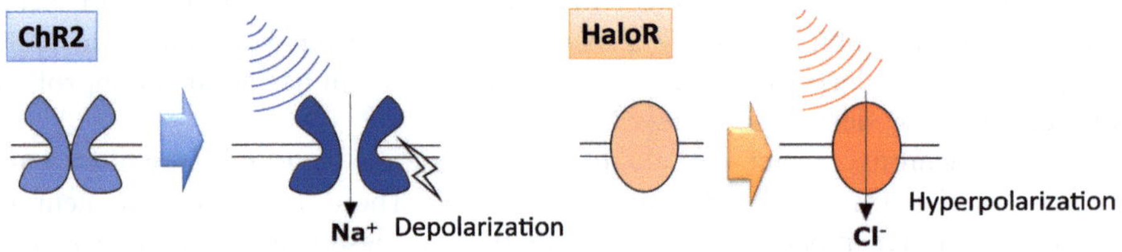

Fig. 1 ChR2 and HaloR ChR2 senses *blue light* and open non-selective cation channel to induce depolarization. HaloR senses *orange light* and pump into chloride ion to induce hyperpolarization (Color figure online)

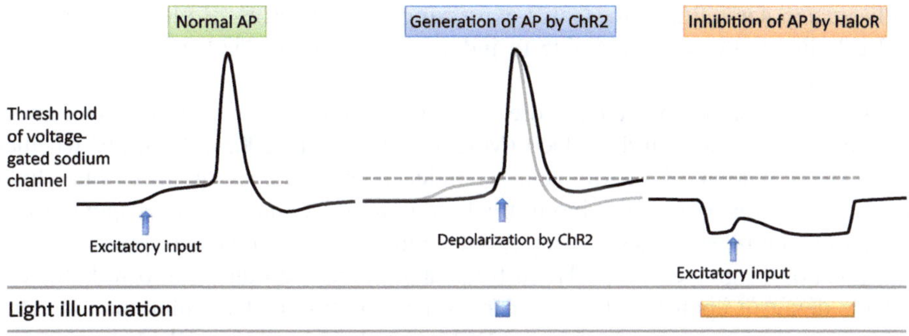

Fig. 2 Generation of action potential (*AP*) by ChR2 and inhibition of AP by HaloR. Depolarization by ChR2 reached to threshold of voltage-gated sodium channel generates AP. Hyperpolarization by HaloR prevents generation of AP

flow into a neuron from the extracellular space, causing the cell's membrane potential to be depolarized (Fig. 2 middle). Similarly, when neurons expressing HaloR are illuminated with orange light, their membrane potential is hyperpolarized, allowing the generation of an action potential to be inhibited. When Cl^- pumps start working, they pump Cl^- ions from the extracellular space into the neuron to hyperpolarize its membrane potential. Consequently, the membrane potential does not reach the threshold of voltage-gated sodium channel even in the presence of excitatory inputs (Fig. 2 right). ChR2 and HaloR were used to manipulate specific neural activity and control behavior in vivo in individual animals. This technique has allowed direct analysis of neural activity during animal behavior, and promotes major advances in elucidating mechanisms regulating physiological phenomena that are evident only in an animal as a whole. There are two factors necessary to be fulfilled for any research to succeed in manipulation of the neural activity and control of the behavior of individual animals: (1) an adequate number of molecules of light-activated proteins should be expressed in the cell membrane of the neurons of interest and (2) the optical system should illuminate the neurons with enough light intensity to activate the light-activated protein.

5 Activation of Orexin Neurons Using Optogenetics

In 2007, Dr. Deisseroth and Dr. de Lecea's group at Stanford University applied optogenetics to freely behaving animals and successfully controlled sleep/wakefulness by targeting orexin neurons in the hypothalamus (Adamantidis et al. 2007). ChR2 was exclusively expressed in orexin neurons using a lentiviral vector. They generated mice in which orexin neurons exclusively expressed ChR2 by injecting a lentiviral vector containing the 3.1-kb upstream region of the mice *prepro-orexin* gene as a promoter. To confirm the function of ChR2 expression in orexin neurons, slice patch clamp recordings were performed. Blue light illumination induced depolarization of the ChR2-expressing orexin neurons. Both continuous and pulse light illumination was able to induce depolarization and increases in firing frequency.

A previous study showed that orexin neurons phasically fire during wakefulness and are almost silent during sleep (Mileykovskiy et al. 2005). To study whether artificial activation of orexin neurons induces wakefulness in mice, blue light illumination was applied in vivo in freely moving mice that displayed spontaneous sleep or wakefulness states. During spontaneous sleep, continuous or pulsative light illumination of greater than 5 Hz in frequency induced a transition to wakefulness after a delay of approximately 30 s from the initiation of light illumination. This effect was observed not only during NREM sleep but also during REM sleep. These facts suggest that artificial activation of orexin neuronal activity is sufficient to promote the initiation of arousal (Adamantidis et al. 2007). This study was the first successful report linking activation of specific types of neurons in mice to behavior by using optogenetics. In addition, this study established that optogenetics is a powerful tool to regulate neuronal activity and to control animal behaviors. Since this report, optogenetic approaches have been applied to the study of many neuronal types and various behaviors.

Another independent study used transgenic mice with orexin neurons specifically expressing melanopsin (OPN4) under the tetracycline gene expression control system (Tet-off) (*Orexin-tTA*; *BitetO human OPN4/mCherry* bigenic mice) (Tsunematsu et al. 2012). OPN4 is a blue light-driven Gq-coupling G protein-coupled receptor (GPCR) normally expressed in a subpopulation of retinal ganglion cells. Blue light illumination depolarizes OPN4-expressing neurons via activation of the Gq signal transduction cascade. The advantage of using OPN4 is that long-lasting depolarization can be induced using relatively weaker and shorter light pulses, as compared with ChR2 and GPCR signaling, which are slower and long-lasting. Additionally, using a transgenic mouse strain leads to reliable and reproducible expression of OPN4 in the orexin neurons compared to using the viral vector injection method. Short light pulse illumination evoked long lasting depolarization even after blue light cessation. Similar to the activation of orexin neurons using ChR2, blue light pulses induced the transition from NREM sleep to wakefulness in these mice.

6 Inhibition of Orexin Neurons Using Optogenetics

To inhibit orexin neuronal activity, an orange light-driven chloride pump, HaloR, was exclusively expressed in orexin neurons using transgenic methods. To express HaloR in orexin neurons, the 3.2-kb upstream sequence of the human prepro-orexin gene was used as a promoter to generate transgenic mice (*orexin/HaloR* transgenic mice). Immunohistochemical analyses revealed that more than 90 % of orexin neurons expressed HaloR in the *orexin/HaloR* mice (Fig. 3). At first, to confirm the function of HaloR expressed in orexin neurons in the transgenic mouse brain, electrophysiological studies were performed using brain slice preparations from

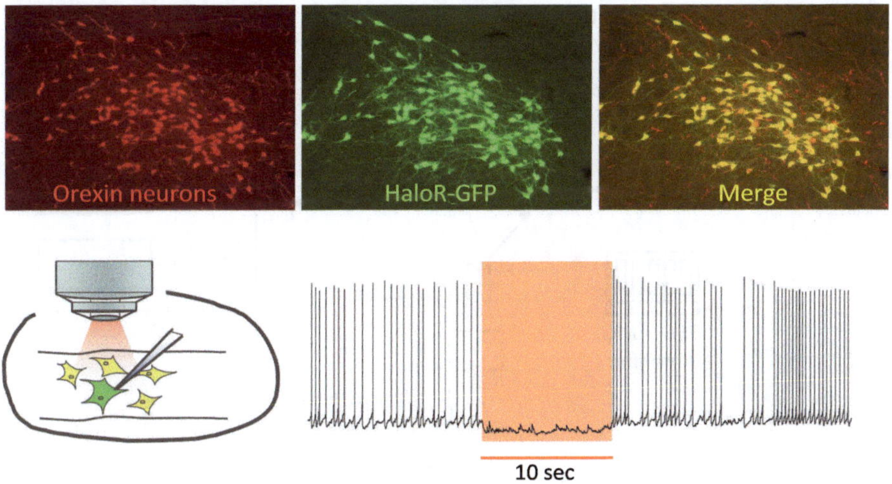

Fig. 3 Generation of transgenic mice in which orexin neurons express HaloR Immunohistochemical study confirmed that orexin-immunoreactive neurons (*red*) expressed HaloR-GFP (*green*). Slice patch clamp recording from orexin neurons confirmed the function of HaloR in orexin neurons. *Orange light* illumination from objective lens hyperpolarized membrane potential and inhibited generation of action potentials during illumination (Color figure online)

orexin/HaloR mice. Orange light illumination immediately hyperpolarized HaloR-expressing orexin neurons and completely inhibited the generation of action potentials. Firing was almost completely inhibited for approximately 10 s from the initiation of orange light illumination (Fig. 3). Inhibition was lasted for 1 min, however, firing gradually recovered suggesting that HaloR is not suitable to inhibit the activity of neurons for longer timescales. To obtain in vivo data, orexin neurons were acutely inhibited for 1 min in freely moving *orexin/HaloR* mice. To illuminate orange light into the hypothalamus, fiber optics were bilaterally inserted into the hypothalamus. To monitor sleep/wakefulness, EEG and EMG electrodes were implanted and recorded. Orange light was emitted by a strong light source such as laser or high-powered LED, and the light was guided through quartz fiber optics and connected to plastic fiber optics via an optical swivel (Fig. 4). Light illumination was applied when mice were in a state of arousal. Although mice maintained a wakefulness state for a few seconds even after the initiation of orange light illumination, EMG power gradually decreased and the slow-wave component of EEG increased. These facts indicate NREM sleep induction. The mice slept during orange light illumination, but they immediately returned to wakefulness when orange light was turned off. This result suggested that the acute inhibition of orexin neuronal activity is sufficient to promote the initiation of NREM sleep. However, phenotypes observed in narcolepsy, such as cataplexy and a direct transition from wakefulness to REM sleep were not observed in acute inhibition of orexin neurons for 1 min. Inhibition of orexin neurons for 1 min was not enough to reproduce all the phenotypes of narcolepsy. To further study the physiological significance of

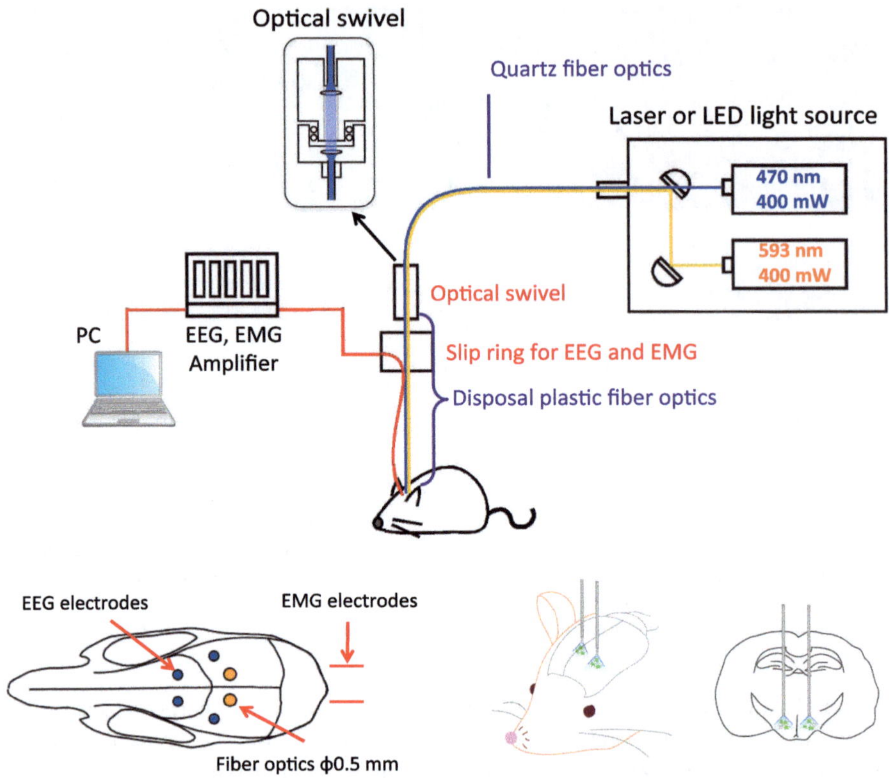

Fig. 4 In vivo light illumination in freely moving mice. *Upper panel* shows light illumination device. *Lower panel* shows schematic drawing of EEG and EMG electrode implantation and optical fiber implantation

orexin neuronal activity in sleep/wakefulness regulation in conjunction with wake centers, EEG and EMG activity and the activity of dorsal raphe (DR) serotonergic neurons were recorded simultaneously. In vivo single unit extracellular recording was performed using conscious, head-fixed *orexin/HaloR* mice (Fig. 5). The firing of serotonergic neurons in the DR was recorded when orexin neuronal activity was inhibited by light illumination via fiber optics located in the hypothalamus. Serotonergic neurons in the DR are densely innervated by orexin neurons and are activated by orexin either directly or indirectly (Liu et al. 2002; Brown et al. 2002). During orange light illumination, the firing frequency of serotonergic neurons in the DR gradually decreased in conjunction with the EEG delta power increase. These observations suggest that acute silencing of orexin neuronal activity by illuminating orange light into the hypothalamus decreased the firing frequency of serotonergic neurons in the DR and induced NREM sleep in mice (Fig. 5). This study provided the first example that selective inhibition of orexin neurons is sufficient to induce a transition from wakefulness to NREM sleep (Tsunematsu et al. 2011).

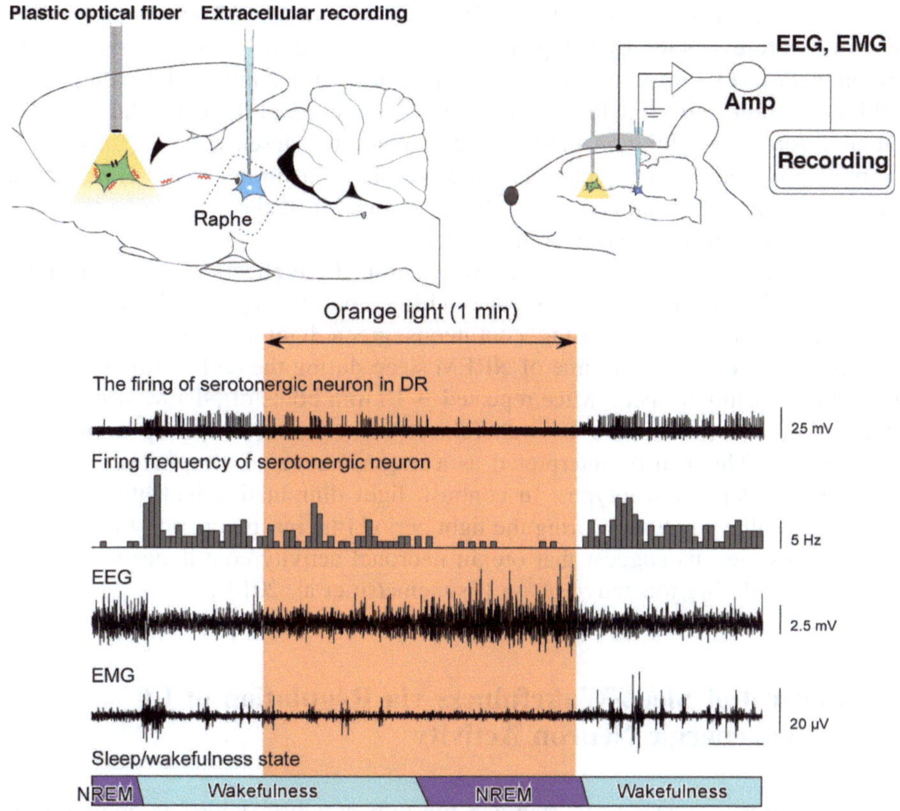

Fig. 5 Recording from serotonergic neurons in the raphe nucleus during inhibition of orexin neurons using *orange light*. *Upper panel* shows schematic drawing of extracellular recording from serotonergic neurons in the raphe nucleus and optogenetc inhibition of orexin neurons in the lateral hypothalamus. *Lower panel* shows activity of serotonergic neurons, EEG and EMG during optogenetic inhibition of orexin neurons (Color figure online)

7 Long-Lasting Inhibition of Orexin Neurons Using Archaerhodopsin

Archaerhodopsin-3 (ArchR) is a newly discovered light-activated protein (Chow et al. 2010). ArchR and Archaerhodopsin TP009 (ArchT) work as green light-driven proton pumps. Arch pumps out protons when it senses green light (Tsunematsu et al. 2013). Although this induces hyperpolarization similar to HaloR, photocurrent induced by Arch is three times larger than that of HaloR. To inhibit the activity of orexin neurons for a longer period, transgenic mice in which orexin neurons specifically express ArchT using the tet-off system in the case of ArchT (*orexin-tTA*; *TetO ArchT* mice). The expression rate of ArchT in orexin neurons in *orexin-tTA*; *TetO ArchT* mice was 72 %.

Light-induced responses of HaloR-expressing or ArchT-expressing orexin neurons were evaluated by using slice patch clamp. ArchT induced stronger inhibition in terms of light-induced current, hyperpolarization, and available duration of inhibition compared to HaloR (at least in orexin neurons). It seems that ArchT is suitable for longer duration inhibition due to limited desensitization. In fact, we were able to detect a significant inhibition of orexin neurons over at least 1 h not only by using slice patch clamp but also by immunohistochemical measurement of cFos, a marker of activated neurons.

Using these transgenic mice lines, we examined the effects of 1-h light illumination on the sleep/wakefulness state during both the light and dark periods. Inhibiting orexin neurons using continuous green light illumination induced a significant increase in total time of NREM sleep during the dark period (the active period for nocturnal mice). Mice repeated ~10 min bouts of NREM sleep several times, they could not maintain wakefulness but continued to sleep during light illumination. This can be interpreted as a fragmentation of sleep/wakefulness, as seen in narcolepsy phenotypes. In contrast, light illumination had little effect on sleep/wakefulness patterns during the light period (the inactive period for nocturnal mice). These results suggest that orexin neuronal activity contributes to maintain wakefulness during the active phase (Tsunematsu et al. 2013).

8 Control of Sleep/Wakefulness via Regulation of LC Noradrenergic Neuron Activity

Orexin neurons contribute to stabilize flip-flop circuitry, which is composed of mutual inhibitory circuits between sleep and wake centers in the brain. In addition to optogenetic control of orexin neurons, optogenetic strategies directly applied to neurons within sleep and wake centers would be expected to control sleep/ wakefulness states.

In 2010, Dr. de Lecea's group at Stanford University reported the optogenetic control of noradrenergic neurons of the LC, one of the primary arousal centers (Carter et al. 2010). Adeno-associated virus (AAV) was injected into transgenic mice in which tyrosine hydroxylase (TH)-expressing cells exclusively expressed Cre recombinase. Noradrenergic neurons in these mice specifically expressed ChR2 or HaloR. A 5 Hz frequency of blue light pulses over 5 s was applied to ChR2-expressing LC neurons in mice during the light period. Light illumination induced transitions from either NREM sleep or REM sleep to wakefulness within 5 s from the initiation of light illumination. This fact suggests that the activation of noradrenergic neurons in the LC acts as a transition switch from sleep to arousal state. For comparison, they also examined the effects of a 1-h light illumination on the sleep/wakefulness cycle using HaloR-expressing, freely moving mice. Yellow light illumination significantly increased total time in NREM sleep and decreased total time in wakefulness during the dark (active) period. However, light illumination had little effect on sleep/wakefulness states during the light (inactive) period.

This result is in agreement with inhibition of orexin neuronal activity. Furthermore, they also performed long-lasting activation via ChR2. A 3-Hz frequency of blue light pulses for 1 h induced significant increases in total time in wakefulness, decreases in total time in NREM sleep, and increases in locomotor activity. These results indicate that activation of orexin neurons induced wakefulness mainly via activation of noradrenergic neurons in the LC (Carter et al. 2010).

9 Perspective

The optical control of neuronal activity in the regulation of sleep/wakefulness indicates that such states can be controlled in vivo. These studies have helped answer how specific neuronal activity such as orexin neuron activity in the hypothalamus and noradrenergic neuron activity in the LC contributes to regulating and driving these behaviors.

The hope is not only to understand the neuronal circuits involved in sleep/wakefulness but also to determine the functions of sleep and precise definitions of sleep and wakefulness using optogenetics.

References

Adamantidis AR, Zhang F, Aravanis AM, Deisseroth K, de Lecea L (2007) Neural substrates of awakening probed with optogenetic control of hypocretin neurons. Nature 450(7168):420–424. doi:10.1038/nature06310

Boyden ES, Zhang F, Bamberg E, Nagel G, Deisseroth K (2005) Millisecond-timescale, genetically targeted optical control of neural activity. Nat Neurosci 8(9):1263–1268. doi:10.1038/nn1525

Brown RE, Sergeeva OA, Eriksson KS, Haas HL (2002) Convergent excitation of dorsal raphe serotonin neurons by multiple arousal systems (orexin/hypocretin, histamine and noradrenaline). J Neurosci: Official J Soc Neurosci 22(20):8850–8859

Carter ME, Yizhar O, Chikahisa S, Nguyen H, Adamantidis A, Nishino S, Deisseroth K, de Lecea L (2010) Tuning arousal with optogenetic modulation of locus coeruleus neurons. Nat Neurosci 13(12):1526–1533. doi:10.1038/nn.2682

Chemelli RM, Willie JT, Sinton CM, Elmquist JK, Scammell T, Lee C, Richardson JA, Williams SC, Xiong Y, Kisanuki Y, Fitch TE, Nakazato M, Hammer RE, Saper CB, Yanagisawa M (1999) Narcolepsy in orexin knockout mice: molecular genetics of sleep regulation. Cell 98(4):437–451

Chow BY, Han X, Dobry AS, Qian X, Chuong AS, Li M, Henninger MA, Belfort GM, Lin Y, Monahan PE, Boyden ES (2010) High-performance genetically targetable optical neural silencing by light-driven proton pumps. Nature 463(7277):98–102. doi:10.1038/nature08652

Hara J, Beuckmann CT, Nambu T, Willie JT, Chemelli RM, Sinton CM, Sugiyama F, Yagami K, Goto K, Yanagisawa M, Sakurai T (2001) Genetic ablation of orexin neurons in mice results in narcolepsy, hypophagia, and obesity. Neuron 30(2):345–354

Liu RJ, van den Pol AN, Aghajanian GK (2002) Hypocretins (orexins) regulate serotonin neurons in the dorsal raphe nucleus by excitatory direct and inhibitory indirect actions. J Neurosci: Official J Soc Neurosci 22(21):9453–9464

Mileykovskiy BY, Kiyashchenko LI, Siegel JM (2005) Behavioral correlates of activity in identified hypocretin/orexin neurons. Neuron 46(5):787–798. doi:10.1016/j.neuron.2005.04.035

Nishino S, Ripley B, Overeem S, Lammers GJ, Mignot E (2000) Hypocretin (orexin) deficiency in human narcolepsy. Lancet 355(9197):39–40. doi:10.1016/s0140-6736(99)05582-8

Peyron C, Faraco J, Rogers W, Ripley B, Overeem S, Charnay Y, Nevsimalova S, Aldrich M, Reynolds D, Albin R, Li R, Hungs M, Pedrazzoli M, Padigaru M, Kucherlapati M, Fan J, Maki R, Lammers GJ, Bouras C, Kucherlapati R, Nishino S, Mignot E (2000) A mutation in a case of early onset narcolepsy and a generalized absence of hypocretin peptides in human narcoleptic brains. Nat Med 6(9):991–997. doi:10.1038/79690

Rechtschaffen A, Gilliland MA, Bergmann BM, Winter JB (1983) Physiological correlates of prolonged sleep deprivation in rats. Science 221(4606):182–184

Tsunematsu T, Kilduff TS, Boyden ES, Takahashi S, Tominaga M, Yamanaka A (2011) Acute optogenetic silencing of orexin/hypocretin neurons induces slow-wave sleep in mice. J Neurosci: the official journal of the Society for Neuroscience 31(29):10529–10539. doi:10.1523/JNEUROSCI.0784-11.2011

Tsunematsu T, Tanaka KF, Yamanaka A, Koizumi A (2012) Ectopic expression of melanopsin in orexin/hypocretin neurons enables control of wakefulness of mice in vivo by blue light. Neurosci Res. doi:10.1016/j.neures.2012.07.005

Tsunematsu T, Tabuchi S, Tanaka KF, Boyden ES, Tominaga M, Yamanaka A (2013) Long-lasting silencing of orexin/hypocretin neurons using archaerhodopsin induces slow-wave sleep in mice. Behav Brain Res 255:64–74. doi:10.1016/j.bbr.2013.05.021

Willie JT, Chemelli RM, Sinton CM, Tokita S, Williams SC, Kisanuki YY, Marcus JN, Lee C, Elmquist JK, Kohlmeier KA, Leonard CS, Richardson JA, Hammer RE, Yanagisawa M (2003) Distinct narcolepsy syndromes in orexin receptor-2 and orexin null mice: molecular genetic dissection of Non-REM and REM sleep regulatory processes. Neuron 38(5):715–730

Yamanaka A, Tsujino N, Funahashi H, Honda K, Guan JL, Wang QP, Tominaga M, Goto K, Shioda S, Sakurai T (2002) Orexins activate histaminergic neurons via the orexin 2 receptor. Biochem Biophys Res Commun 290(4):1237–1245. doi:10.1006/bbrc.2001.6318

Optogenetic Dissection of Sleep-Wake Circuits in the Brain

Thomas C. Gent and Antoine R. Adamantidis

Abstract The use of optogenetics has increased markedly in recent years and is becoming a popular tool to determine the role of various circuits in behaviour, as well as identifying the neuronal components of those circuits. The lateral hypothalamus and its projections have been studied extensively with respect to sleep-wake behaviour using optogenetics. The hypocretin system has been found to be a powerful wake promoting region which recruits the noradrenergic locus coeruleus to evoke it's effects. A subset of neurones in the lateral hypothalamus producing melanin concentrating hormone (MCH) have been shown to promote REM sleep. The projections from the lateral hypothalamus are diverse, as are it's functions in other behaviours such as energy homeostasis and feeding. Future work is now needed to determine which part of this circuitry is required for regulation of the sleep-wake state and ultimately how consciousness is controlled.

1 Neural Substrates of the Sleep-Wake States

The sleep-wake cycle is a fundamental biological trait exhibited in all mammals and higher vertebrates (Saper et al. 2010). Even insects have been shown to display behavioural correlates of sleep (Gilestro et al. 2009; Raizen et al. 2008; Appelbaum et al. 2009). During wakefulness, individuals integrate a range of sensory inputs which allow a contextual interaction with their environment that involves a multitude of complex behaviours. However during sleep, meaningful sensory processing is dramatically reduced, with a corresponding relative behavioural inactivity. Both sleep and wakefulness have characteristic patterns of cortical

T.C. Gent · A.R. Adamantidis (✉)
University of Bern, Department of Neurology, Inselspital University Hospital, Freiburgstrasse 18, 3010 Bern, Switzerland
e-mail: antoine.adamantidis@dkf.unibe.ch

A.R. Adamantidis
McGill University, Department of Psychiatry, Douglas Institute, Montreal, Canada

© Springer International Publishing Switzerland 2015

T. Sakurai et al. (eds.), *Orexin and Sleep*,
DOI 10.1007/978-3-319-23078-8_7

activity on an electroencephalogram (EEG) and muscle activity on an electromyogram (EMG). The wakeful EEG displays fast low amplitude rhythms due to an abundance of sensory processing in the neocortex, whereas during sleep, the EEG is slow with a large amplitude indicative of large ensembles of cortical neurones firing in synchrony during low activity. Psychiatric and metabolic disorders are often associated with perturbations of the sleep-wake cycle architecture such as fragmented sleep or sleep during inappropriate periods of the circadian cycle. Sleep perturbations, and changes in arousal threshold, are now considered as pre-symptomatic phases of many psychiatric disorders including, schizophrenia, addiction, attention deficit hyperactivity disorder (ADHD), and generalized anxiety disorder (GAD). Elucidating the neurobiological substrates of consciousness control is an important step towards developing novel therapeutics for sleep disorders, which are frequently impaired in psychiatric conditions.

Despite recent progress in identifying neural populations that regulate wakefulness and arousal, it is still unclear how these circuits function to control behavioural states in different environmental contexts.

Wakefulness is regulated by distinct neural populations in the brain. Activity in these nuclei is correlated with arousal. Activity is increased in wakefulness over asleep, and importantly is increased further during states of enhanced arousal, such as moments of high alertness or stress. These arousal systems include the hypocretin (Hcrt)-expressing neurons in the lateral hypothalamus, the noradrenergic locus coeruleus (LC)-expressing neurons in the brainstem, the neuropeptide S neurons (NPS) in the brainstem, the serotoninergic dorsal raphe nuclei (DRN) in the brainstem, the histaminergic tuberomammillary nucleus (TMN) in the posterior hypothalamus, the cholinergic pedunculopontine (PPT) and laterodorsal tegmental (LDT) nuclei in the midbrain, as well as cholinergic neurons in the basal forebrain. Many of these nuclei have recently also been demonstrated to contain glutamatergic and GABAergic neurons which are important regulators of arousal.

Activation of these systems not only promotes wakefulness, but also engages other arousal related behaviours such as reward-seeking, sexual activity and flight-or-fight responses. Interestingly, specific behavioural outputs differ from arousal system to another. For example, activation of the LC increases arousal and can cause anxiety-like behaviour, whereas the NPS system increases arousal but decreases anxiety (de Lecea et al. 2012). Therefore such arousal areas must have highly specific as well as common projections to affective brain regions in order to regulate and array of behaviours.

Although neural circuits promoting the onset and maintenance of wakefulness are known to exhibit a high level of redundancy, the current understanding of the neural mechanisms underlying slow-wave sleep remains somewhat limited and only few brain areas have been shown to be active during this state. Following studies of patients with post-influenza encephalitis, von Economo reported that inflammatory lesions of the preoptic area (POA) were often associated with insomnia and therefore proposed that the this region was critical for the regulation of normal sleep (von Economo 1926). This observation was confirmed experimentally in monkeys (Ranson 1939), rats (Nauta 1946), and cats (McGinty and

Sterman 1968) (McGinty) where POA lesions reliably induced a profound and persistent insomnia (Lu et al. 2000). Consistent with these results, POA electrical stimulation induces EEG slow wave activity and sleep (SWS) (Sterman and Clemente 1962). Furthermore, neurons recorded from feely moving cats in the horizontal limb of the diagonal bands of Broca and the lateral preoptic area-substantia innominata were found to exhibit higher firing rates during slow wave sleep compared to wakefulness (Szymusiak and McGinty 1986). Together, these results support the notion that the POA encompasses neurons that promote slow wave sleep, however its exact role in the initiation of natural sleep remains to be determined.

It is thought that the initiation of sleep-to-wake transitions involves a coordinated inhibition the multiple arousal nuclei which make up the ascending reticular activating system (ARAS) by discrete sleep promoting areas (Jego et al. 2013).

2 Hypocretin System and Sleep-Wake Circuits in the Mammalian Brain

Regulation of this cycle involves several regions of the brain, mostly discrete monoaminergic wake promoting nuclei in the hypothalamus and an inhibitory sleep promoting nucleus. However the manner in which these areas orchestrate transitions from wakefulness to sleep, and back again, are yet to be fully appreciated, as do the downstream mechanisms that lead to the archetypal synchronisation states in the neocortex. One such arousal nucleus that has been demonstrated to play a pivotal role in the sleep-wake cycle is the hypocretin/orexin producing lateral hypothalamus (LH). The two hypocretin peptides (de Lecea et al. 1998; Sakurai et al. 1998) are released exclusively from a subset of neurones in the LH and was one of the first arousal systems to be investigated using optogenetics (Carter et al. 2009, 2012; Adamantidis et al. 2007). Indeed the orexinergic system is often thought to stabilize the behaviour of vigilance states (de Lecea et al. 2006).

In order to start considering how hypocretin-LH neurons have a stabilizing effect on the sleep-wake cycle, it is useful to consider the anatomical connectivity of this nucleus. Immunohistochemistry and in situ hybridisation studies have demonstrated hypocretin projections to the adrenergic locus coeruleus (LC), histaminergic tuberomammillary nucleus (TMN) dopaminergic ventral tegmental area (VTA) and serotonergic dorsal raphe (DR) (Peyron et al. 1998), all of which have been shown to have strong arousal properties. Hypocretin projections have recently been demonstrated to be vital for the sustained firing patterns of TMN neurones (Schöne et al. 2014a) which are highly active during wakefulness. Furthermore, optogenetic stimulation of hypocretin neurons increases the activity of both LC and TMN neurons (Carter et al. 2009).

The most compelling evidence for the stabilising effect of hypocretin is that dysfunction of the orexinergic system leads to the clinical condition of narcolepsy in humans and Doberman dogs (Peyron et al. 2000; Lin et al. 1999). Despite a differing molecular mechanism for the disease between the two species, the behavioural phenotype of both sets of patients is very similar. Narcolepsy is characterised by increased daytime sleeping with wake to REM transitions and fragmented night-time sleep (Nishino et al. 2000). Mice with a double knockout of the hypocretin 1 and 2 receptors or those that lack the precursor preproorexin also exhibit cataplexy, a behavioural signature of narcolepsy, during active wakeful periods and therefore parallel the clinical condition (Chemelli et al. 1999).

3 Optogenetics to the Rescue of Sleep-Wake Circuit Dissection

In the last few years there has been a surge of investigators turning to optogenetic tools to investigate sleep-wake control, and the majority of these have focused on the various hypothalamic nuclei. The hypothalamus consists of an intricate network of excitatory and inhibitory neuronal populations, each of which has a specific chemical nature and distinct roles in homeostatic function. Much of the classical evidence supporting a functional role for these circuits in sleep regulation have limited spatial (e.g. pharmacological and electrical stimulation) and temporal (e.g. pharmacological and lesioning) resolution. Many of these hypothalamic nuclei have been demonstrated to consist of more than one cell type with different neurotransmitter profiles, each having a different role in sleep regulation. Previously used techniques cannot discriminate between these cell types. Furthermore, genetic modification of animals is likely to lead to compensatory mechanisms, which shroud the native function of the mutated allele (Adamantidis et al. 2010).

Optogenetic technology ("*opto*" for optical stimulation and "*genetics*" for genetically targeted cell types) is a new-generation tool that allows remote control of specific neural circuits with physiologically relevant spatial and temporal resolution (Boyden et al. 2005; Zhang et al. 2007; Yizhar et al. 2011).

Optogenetics is a particularly attractive modality for this since the sleep-wake nuclei of the hypothalamus are discrete regions of the brain composing of small numbers of neurones (typically less than 6000 cells in the mouse brain) and can therefore all be modulated using a single small beam of light. Additionally, the increasing number of available transgenic mouse lines expressing Cre-recombinase allows in vivo viral transfection of light sensitive opsins (channelrhodopsin, halorhodopsin, archaerhodopsin, etc.) (Mattis et al. 2012) to be performed in many experimental models.

Channelrhodopsins are a subfamily of membrane-bound proteins that originate from algae to allow navigation in response to light and related to the protein rhodopsin, found in the mammalian retina (Hegemann and Nagel 2013). They function as light gated ion channels, which have a non-specific cation conductance. Activation with blue light therefore serves to depolarise cell membranes leading to excitability. The ability to code channelrhodopsin in viral vectors has facilitated expression of the protein in a wide variety of brain areas allowing investigations of neural circuitry and behaviour without many of the complications associated with pharmacological manipulation of brain regions.

Since the initial demonstration of the ChR2 in hippocampal neurons in 2005 (Boyden et al., Nature Neuroscience), the availability and divergence of the channelrhodopsins has increased. These include mutations of ChR2 such as ChETA (Gunaydin et al. 2010), which has a faster decay time allowing quicker repolarization of target cells. Functional silencing of neurons is also possible via the use of halorhodopsin and archaerhodopsin, which absorb yellow light to produce a chloride or proton conductance respectively (Chow et al. 2010). Together these offer a powerful toolbox for interrogating the function of neural circuits and have been used together to investigate sleep pathways (Carter et al. 2009).

4 Hcrt Modulation of Arousal

We first used optogenetics to manipulate the hypocretin system in 2007 (Adamantidis et al. 2007). We used a lentivirus to stably express ChR2 in orexin LH neurones only. The LH was then optically stimulated with blue light, but due to the specificity of the viral transfection, stimulation occurs only in targeted hypocretin neurones. We found that optical stimulation of the LH at certain frequencies reduced the latency to waking from REM and NREM sleep by up to 50 % (Carter et al. 2009; Adamantidis et al. 2007). Importantly we found that successful arousal required a pulsatile stimulation (1–30 Hz) rather than continuous stimulation. This would mimic the natural tendency of LH neurones to burst fire just before natural wakefulness (Hassani et al. 2009; Lee et al. 2005). Furthermore, we showed that the expression of ChR2 without optical stimulation did not alter the sleep-wake structure when compared to virus injection alone. The mechanism was shown to be due to hypocretin transmission by the use of the OR-1A antagonist SB334867 which antagonised the arousal effect in a dose dependant manner. This work was furthered by the observation that hypocretin neuron activation is insufficient to promote wakefulness after short term sleep deprival (Carter et al. 2009), suggesting that rather than being an omnipotent arousal centre, the hypocretin system is responsible for maintaining the sleep-wake cycle ultra-structure.

More recently, a reversal of these experiments has been used to confirm these previous findings. Tsunematsu et al. (2011), expressed halorhodopsin in hypocretin neurones which when illuminated with yellow (593 nm) light are silenced. They found that such silencing in vivo resulted in the onset of NREM sleep, however interestingly the mice did not take on a narcoleptic phenotype that would be expected from hypocretin silencing. This is proposed to be due to the difference in chronicity of hypocretin inhibition since in the clinical condition of narcolepsy, plasticity is likely to result in downstream target reorganisation. This is backed up by pharmacological experiments showing that the dual hypocretin receptor antagonist (Brisbare-Roch et al. 2007) increases night-time sleepiness without causing cataplexy. A follow up study (Tsunematsu et al. 2013) demonstrated that sustained orexinergic silencing only increased SWS during the active period, further augmenting the view that the orexin system primarily acts as a stabiliser of wakefulness.

5 Functional Targets of Hcrt System

Since the hypocretin system has regulatory controls over hypothalamic arousal pathways (Sakurai 2007), optogenetics may also be useful in determining downstream events that lead to wakefulness. Two such candidates that have been investigated are the TMN and LC. Carter et al. (2009), also showed that hypocretin induced wakefulness was not dependent on mice lacking histidine decarboxylase, the enzyme required for the production of histamine. The TMN provides histamine to large areas of the brain and has been implicated in arousal (Parmentier et al. 2002; Anaclet et al. 2012) from both natural sleep and anaesthesia (Zecharia and et al. 2012). However, animals lacking the ability to produce histamine have a largely undisturbed sleep-wake cycle (Parmentier et al. 2002). This finding may have a number of implications, namely that either the TMN facilitates arousal states by an alternative pathway, or that a compensatory mechanism occurs. To tackle this question, more recent studies have shown a robust connectivity between hypocretin neurones and the TMN and found that hypocretin cells also co-release the fast neurotransmitter glutamate (Schöne et al. 2014a, b). Optical stimulation of hypocretin LH neurones for a duration that mimics their in vivo behaviour shortly prior to arousal, showed that the resulting activity in TMN neurones has both a fast (glutamatergic) and slow (hypocretin) component. TMN activity could not be sustained by glutamate during blockade of the OX-R2. TMN activity resulting from orexinergic input increased during the stimulation period, suggesting a mechanism for the stabilising effect of orexins on arousal. Furthermore, orexinergic output occurred only at higher stimulation frequencies, whereas glutamatergic activity occurred across a range of frequencies.

The second hypocretin system target is the LC which is a strong arousal centre (Aston-Jones and Bloom 1981; Vazey and Aston-Jones 2014) which when optogenetically activated leads to increased wakefulness from NREM and REM sleep (Carter et al. 2010) and also receives stimulatory inputs from the LH. The authors of this study showed that stimulation of the LC caused rapid wakefulness, rather than a gradual emergence as seen with LH stimulation. Furthermore, studies which simultaneously optogenetically stimulated the LH and silenced the LC found that the LC activity was required to mediate orexinergic induced wakefulness (18). However mice lacking dopamine beta-hydroxylase, required for adrenaline synthesis, do not have a disrupted sleep-wake cycle (Hunsley and Palmiter 2003), nor do animals with chemical ablation of the LC (Blanco-Centurion et al. 2007), suggesting that LC activity alone is not required to maintain wakefulness, whereas that of the orexin system is. Indeed prolonged stimulation of the LC causes behavioural arrest (Carter et al. 2012), possibly because of depletion of cortical noradrenaline, suggesting that other ascending arousal systems must work in concerto to maintain arousal. Further experiments investigating wakefulness by optogenetically silencing the LC would be informative on this and whether or not this hypothalamic co-operation is entirely orchestrated by orexinergic neurones remains to be demonstrated.

In addition to the direct synaptic control of histaminergic TMN and norepinephrine LC cells by Hcrt neurons, others circuits participate to the onset and maintenance of wakefulness. Exactly how these regions orchestrate onset and maintenance of different states of wakefulness (attention, stress, anxiety, etc.) or their transitions into sleep states await further investigations.

6 The MCH System: A Functional Counterpart to the Hcrt System?

Like Hcrt neurons, MCH cells are exclusively found in the lateral hypothalamus (LH) with the exception of a subpopulation located in the zona incerta (ZI) (Bittencourt et al. 1992). A subpopulation of these neurons also express the glutamate decarboxylase GAD67/65 suggesting that MCH neurons are GABAergic (Jego et al. 2013; Elias et al. 2001; Sapin et al. 2009) and thus inhibitory. MCH peptide acts through specific G Protein-Coupled Receptors including MCH-R1 and MCH-R2 (MCH-R2 is only functional in human, dog, ferret, but not in rodents) (Pissios and Maratos-Flier 2003). The pattern of expression of MCH-R1 perfectly matches the distribution of MCH-containing terminals (Saito et al. 2001; Chee et al. 2013). In addition, MCH receptors may be coupled to different G proteins (G_i, G_α et G_q) or other transmembrane proteins (Pissios and Maratos-Flier 2003), suggesting possible multimodal fine-tuning of the signalling cascade activated by MCH binding to its receptors.

MCH neurons are the target of arousal systems and, in turn, are known to project back to these nuclei (Bittencourt et al. 1992), which is consistent with their role in sleep-wake cycle regulation. Indeed, neurotransmitters from extra-hypothalamic arousal systems, including noradrenaline, serotonin, acetylcholine (muscarinic agonists), all inhibit MCH neurons (van den Pol et al. 2004). In addition, Glutamate, ATP and Hcrt-1 & 2 effectively increase the activity of MCH neurons. Accordingly, Some MCH neurons express leptin receptor Ob-R, glutamate and GABA receptors, adrenoreceptor alpha2, muscarinic and serotoninergic receptors and $HcrtR_{1,2}$ (Figs. 1 and 2).

Recent studies suggest a sleep-promoting role for the MCH system. A large number of c-Fos^+ cells (marker of neuronal activity) during a sleep rebound induced by total or a selective REM sleep deprivation were found to be immunoreactive for MCH (Modirrousta et al. 2005; Verret et al. 2003). While 60 %

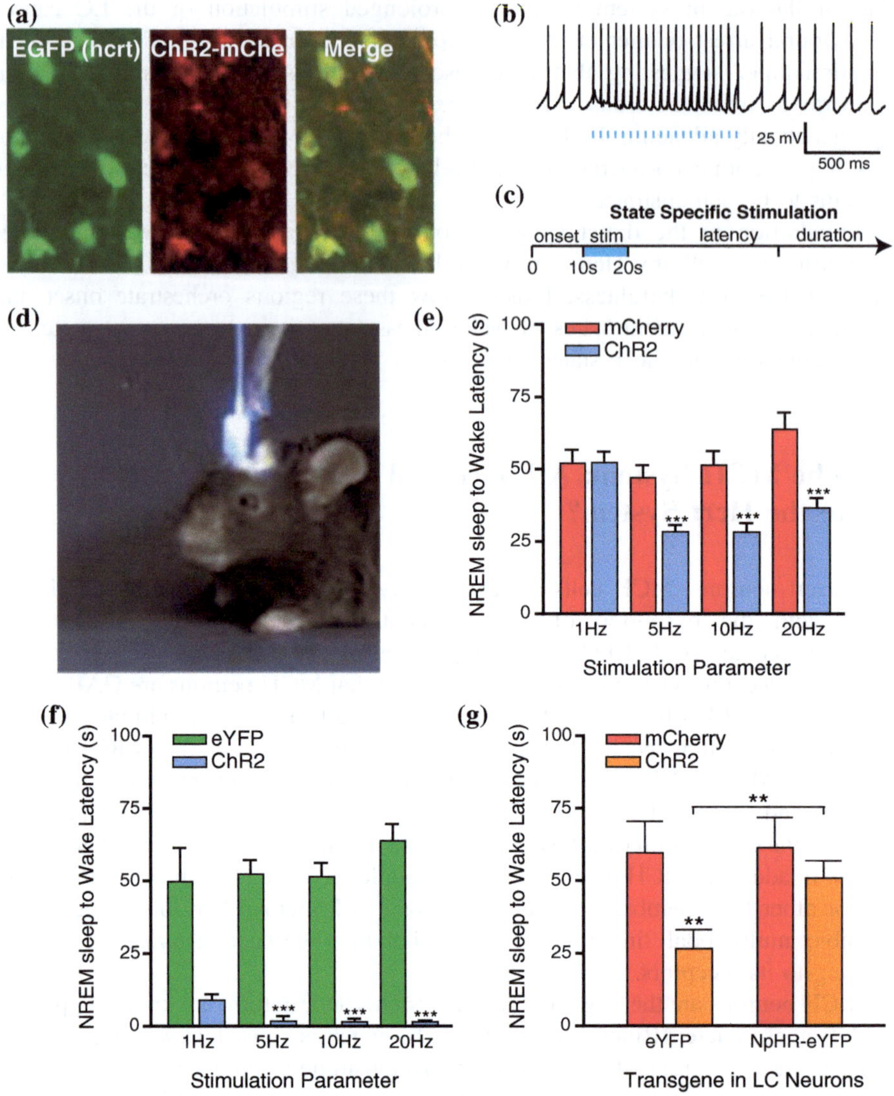

Fig. 1 Optogenetic stimulation of Hcrt neurons and their downstream targets during NREM result in wakefulness. **a** Fluorescence micrographs of neurones in the lateral hypothalamus expressing EGFP (*green*), ChR2-mCherry (*red*) and merge showing co expression. **b** *Blue-light* pulse trains (15 ms per pulse, 20 Hz) evoked reliable firing of ChR2 transfected Hcrt neurons in vitro. Evoked action potentials show high temporal precision to stimulation, despite the presence of basal spontaneous activity. **c** Time line for the behavioural state dependant simulation of Hcrt neurons in vivo. Stimulation is performed 10 s after the onset of behavioural state and lasts for 10 s. The latency to state transition is recorded. **d** Schematic showing mouse implanted with EEG/EMG electrodes as well as fibre optics for optical stimulation. **e** Latencies to wake transitions following optogenetic stimulation of Hcrt neurons during NREM. Data analysis is based on an average of 15 stimulations per frequency per mouse. **f** Optogenetic stimulation of LC neurons at any frequency reduces the time to wakening from NREM. **g** Stimulation of the LC increases membrane excitability and enhances Hcrt-mediated sleep-to-wake transitions. [Figure **b** and **e** adapted from Adamantidis et al. (2007). Figure **f** adapted from Carter et al. (2010), Figure **g** from Carter et al. (2012)] (Color figure online)

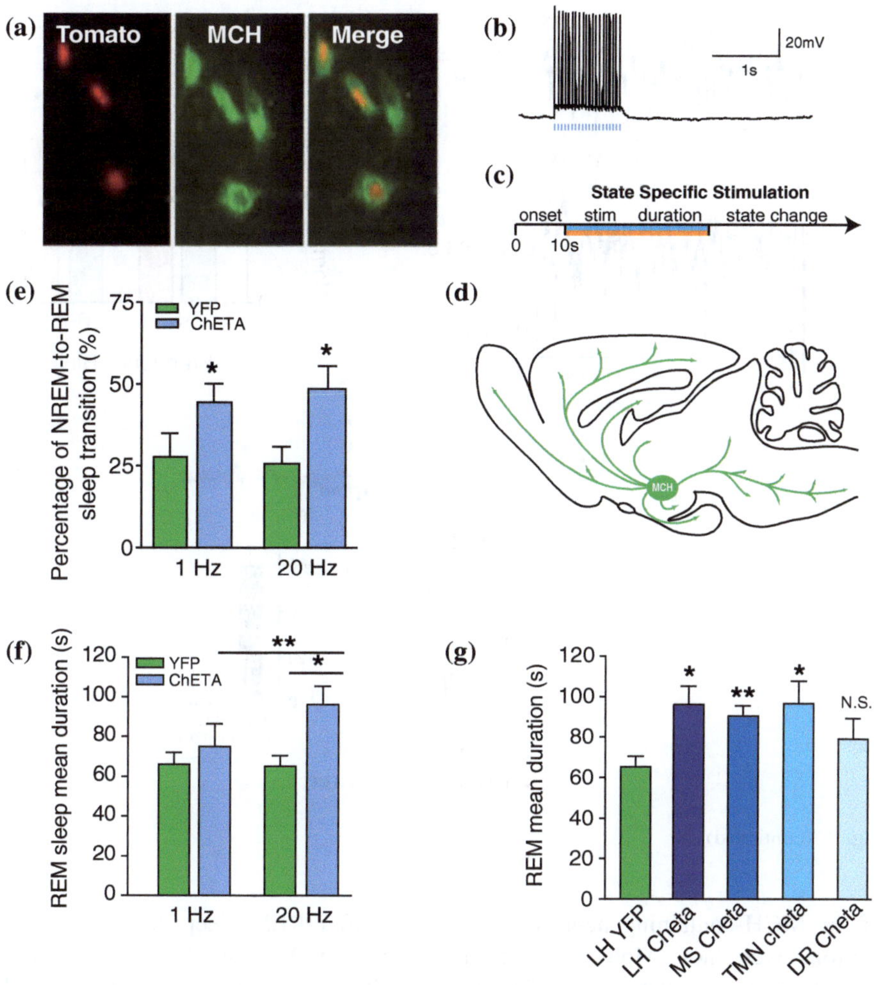

Fig. 2 Optogenetic activation of MCH neurons augments REM sleep. **a** Fluorescence micrograph of MCH neurones in the lateral hypothalamus showing expression of td-Tomato (*red*), MCH (*green*) and merge to show co-expression. **b** *Blue-light* pulse trains (5 ms per pulse, 20 Hz) evoked reliable firing of ChETA transfected MCH neurons in vitro. Evoked action potentials show high temporal precision to stimulation, despite the presence of basal spontaneous activity. **c** Time line for the behavioural state dependant simulation of MCH neurons in vivo. Neurones are either stimulated (*blue*) or inhibited (*orange*) until change of behavioural state occurs. **d** Schematic of projections of MCH neurones throughout the brain. **e** Percentage of successful NREM-to-REM sleep transitions following optogenetic stimulation during NREM. Data are shown as a percentage of the total number of NREM-to-REM sleep transitions on the total number of stimulations during NREM sleep. **f** Mean duration of REM sleep episodes following optogenetic stimulation at 1–20 Hz of control and ChETA-EYFP mice. Data analysis is based on an average of at least 15 stimulations per frequency and per mouse during REM sleep episodes. **g** Mean REM sleep duration of mice stimulated at 20 Hz during REM sleep in the TMN, dorsal raphe or medial septum. The results of lateral hypothalamus stimulation are reported for comparison. **h** Representative EEG traces during optogenetic silencing of MCH neurones with eNpHr3.0 and eYFP. **i** Quantification of slow theta oscillations during optogenetic silencing of EYFP and eNpHR3.0-YFP transfected MCH neurons in vivo. **j** Evoked IPSCs (*black*) in TMN neurons expressing ChETA-EYFP are blocked with bicuculline. The amplitudes of the currents are shown in the *inset*. [Figures adapted from Jego et al. (2013)] (Color figure online)

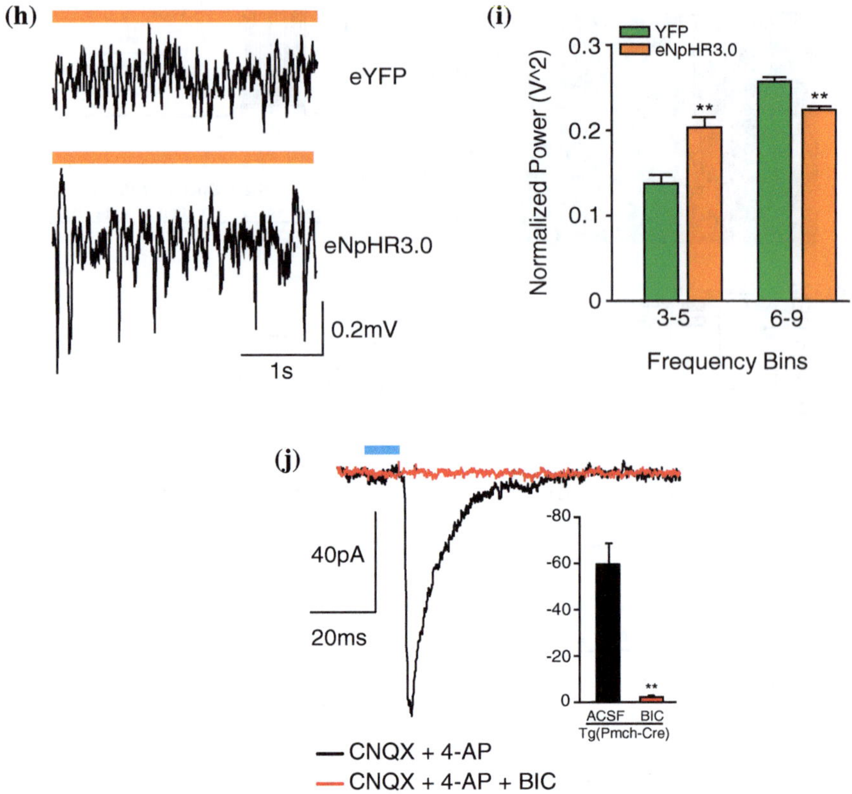

Fig. 2 (continued)

of the MCH-containing neurons were c-Fos$^+$ after REM sleep deprivation, Hcrt neurons were not (Modirrousta et al. 2005; Verret et al. 2003). In addition, intracerebroventricular (icv) infusion of MCH in rats causes hypersomnia by dose-dependent increases in SWS (+70 %) and REM sleep (+200 %) (Verret et al. 2003), whereas a MCH-R1 antagonist has the opposite effect. Collectively, these functional data suggested that activation of MCH neurons may promote sleep by inhibiting arousal centers of the brain, whereas the Hcrt system induces arousal by activating them. However, the precise role the MCH system in sleep remains unclear (Adamantidis et al. 2008; Jego and Adamantidis 2013).

In agreement with this hypothesis, we recently demonstrated that acute optogenetic activation of MCH neurons (ChETA, SSFO) at the onset of REM sleep extended the duration of REM, but not non-REM sleep episodes (Jego et al. 2013). In contrast, their acute silencing (eNpHR3.0, ArchT) reduced the frequency and amplitude of hippocampal theta rhythm, without affecting REM sleep duration. In vitro activation of MCH neuron terminals induced $GABA_A$-mediated inhibitory post-synaptic currents (IPSCs) in wake-promoting histaminergic neurons of the tuberomammillary nucleus (TMN), whilst in vivo activation of MCH neuron terminals in TMN or medial septum also extend the duration of REM sleep episodes.

Collectively, these results demonstrated that selective activation of LH_{MCH} neurons promotes and maintains REM sleep, possibly through inhibition of arousal circuits in the mammalian brain.

7 Perspectives

Optogenetics is proving to be a valuable tool in dissecting the components of the sleep-cycle at the hypothalamic level. It has in some instances confirmed previous less specific ablation, pharmacological and genetic studies and also given new insights into the neural circuits involved in the control of brain states during sleep and wakefulness. Different sleep-wake states are characterised by markedly different EEG signatures due to changes in cortical synchronicity. Since long-range cortical networks are regulated by the thalamus, future optogenetic experiments should focus in elucidating circuitry downstream from the hypothalamus, in order to form a unifying theory of brain control of sleep, as well as consciousness.

Acknowledgments We thank the Tidis Lab members for their technical help and comments. A. A. is supported by the Human Frontier Science Program, Canadian Institute for Health Research (CIHR), the University of Bern, the Inselspital and the Swiss National Funds.

References

Adamantidis AR, Zhang F, Aravanis AM, Deisseroth K, de Lecea L (2007) Neural substrates of awakening probed with optogenetic control of hypocretin neurons. Nature 450:420–424

Adamantidis A et al (2008) Sleep architecture of the melanin-concentrating hormone receptor 1-knockout mice. Eur J Neurosci 27:1793–1800

Adamantidis A, Carter MC, de Lecea L (2010) Optogenetic deconstruction of sleep-wake circuitry in the brain. Front Mol Neurosci 2:31

Anaclet C et al (2012) Identification and characterization of a sleep-active cell group in the rostral medullary brainstem. J Neurosci 32:17970–17976

Appelbaum L et al (2009) Sleep-wake regulation and hypocretin-melatonin interaction in zebrafish. Proc Natl Acad Sci USA 106:21942–21947

Aston-Jones G, Bloom FE (1981) Activity of norepinephrine-containing locus coeruleus neurons in behaving rats anticipates fluctuations in the sleep-waking cycle. J Neurosci 1:876–886

Bittencourt JC et al (1992) The melanin-concentrating hormone system of the rat brain: an immuno- and hybridization histochemical characterization. J Comp Neurol 319:218–245

Blanco-Centurion C, Gerashchenko D, Shiromani PJ (2007) Effects of saporin-induced lesions of three arousal populations on daily levels of sleep and wake. J Neurosci 27:14041–14048

Boyden ES, Zhang F, Bamberg E, Nagel G, Deisseroth K (2005) Millisecond-timescale, genetically targeted optical control of neural activity. Nat Neurosci 8:1263–1268

Brisbare-Roch C et al (2007) Promotion of sleep by targeting the orexin system in rats, dogs and humans. Nat Med 13:150–155

Carter ME, Adamantidis A, Ohtsu H, Deisseroth K, de Lecea L (2009) Sleep homeostasis modulates hypocretin-mediated sleep-to-wake transitions. J Neurosci 29:10939–10949
Carter ME et al (2010) Tuning arousal with optogenetic modulation of locus coeruleus neurons. Nat Neurosci 13:1526–1533
Carter ME et al (2012) Mechanism for Hypocretin-mediated sleep-to-wake transitions. Proc Natl Acad Sci USA 109:E2635–E2644
Chee MJ, Pissios P, Maratos-Flier E (2013) Neurochemical characterization of neurons expressing melanin-concentrating hormone receptor 1 in the mouse hypothalamus. J Comp Neurol 521:2208–2234
Chemelli RM et al (1999) Narcolepsy in orexin knockout mice: molecular genetics of sleep regulation. Cell 98:437–451
Chow BY et al (2010) High-performance genetically targetable optical neural silencing by light-driven proton pumps. Nature 463:98–102
de Lecea L et al (1998) The hypocretins: hypothalamus-specific peptides with neuroexcitatory activity. Proc Natl Acad Sci USA 95:322–327
de Lecea L et al (2006) Addiction and arousal: alternative roles of hypothalamic peptides. J Neurosci 26:10372–10375
de Lecea L, Carter ME, Adamantidis A (2012) Shining light on wakefulness and arousal. Biol Psychiatry 71:1046–1052
Elias CF et al (2001) Characterization of CART neurons in the rat and human hypothalamus. J Comp Neurol 432:1–19
Gilestro GF, Tononi G, Cirelli C (2009) Widespread changes in synaptic markers as a function of sleep and wakefulness in Drosophila. Science 324:109–112
Gunaydin LA et al (2010) Ultrafast optogenetic control. Nat Neurosci 13:387–392
Hassani OK, Lee MG, Jones BE (2009) Melanin-concentrating hormone neurons discharge in a reciprocal manner to orexin neurons across the sleep-wake cycle. Proc Natl Acad Sci USA 106:2418–2422
Hegemann P, Nagel G (2013) From channelrhodopsins to optogenetics. EMBO Mol Med 5:173–176
Hunsley MS, Palmiter RD (2003) Norepinephrine-deficient mice exhibit normal sleep-wake states but have shorter sleep latency after mild stress and low doses of amphetamine. Sleep 26:521–526
Jego S, Adamantidis A (2013) MCH neurons: vigilant workers in the night. Sleep 36:1783–1786
Jego S et al (2013) Optogenetic identification of a rapid eye movement sleep modulatory circuit in the hypothalamus. Nat Neurosci 16:1637–1643
Lee MG, Hassani OK, Jones BE (2005) Discharge of identified orexin/hypocretin neurons across the sleep-waking cycle. J Neurosci 25:6716–6720
Lin L et al (1999) The sleep disorder canine narcolepsy is caused by a mutation in the hypocretin (orexin) receptor 2 gene. Cell 98:365–376
Lu J, Greco MA, Shiromani P, Saper CB (2000) Effect of lesions of the ventrolateral preoptic nucleus on NREM and REM sleep. J Neurosci 20:3830–3842
Mattis J et al (2012) Principles for applying optogenetic tools derived from direct comparative analysis of microbial opsins. Nat Methods 9:159–172
McGinty DJ, Sterman MB (1968) Sleep suppression after basal forebrain lesions in the cat. Science 160:1253–1255
Modirrousta M, Mainville L, Jones BE (2005) Orexin and MCH neurons express c-Fos differently after sleep deprivation vs. recovery and bear different adrenergic receptors. Eur J Neurosci 21:2807–2816
Nauta WJ (1946) Hypothalamic regulation of sleep in rats; an experimental study. J Neurophysiol 9:285–316
Nishino S, Ripley B, Overeem S, Lammers GJ, Mignot E (2000) Hypocretin (orexin) deficiency in human narcolepsy. Lancet 355:39–40
Parmentier R et al (2002) Anatomical, physiological, and pharmacological characteristics of histidine decarboxylase knock-out mice: evidence for the role of brain histamine in behavioral and sleep-wake control. J Neurosci 22:7695–7711

Peyron C et al (1998) Neurons containing hypocretin (orexin) project to multiple neuronal systems. J Neurosci 18:9996–10015
Peyron C et al (2000) A mutation in a case of early onset narcolepsy and a generalized absence of hypocretin peptides in human narcoleptic brains. Nat Med 6:991–997
Pissios P, Maratos-Flier E (2003) Melanin-concentrating hormone: from fish skin to skinny mammals. Trends Endocrinol Metab 14:243–248
Raizen DM et al (2008) Lethargus is a Caenorhabditis elegans sleep-like state. Nature 451:569–572
Ranson SW (1939) Somnolence caused by hypothalamic lesions in the monkey. Arch Neurol Psychiat 41:1–23
Saito Y, Cheng M, Leslie FM, Civelli O (2001) Expression of the melanin-concentrating hormone (MCH) receptor mRNA in the rat brain. J Comp Neurol 435:26–40
Sakurai T (2007) The neural circuit of orexin (hypocretin): maintaining sleep and wakefulness. Nat Rev Neurosci 8:171–181
Sakurai T et al (1998) Orexins and orexin receptors: a family of hypothalamic neuropeptides and G protein-coupled receptors that regulate feeding behavior. Cell 92, 1 page following 696
Saper CB, Fuller PM, Pedersen NP, Lu J, Scammell TE (2010) Sleep state switching. Neuron 68:1023–1042
Sapin E et al (2009) Localization of the brainstem GABAergic neurons controlling paradoxical (REM) sleep. PLoS ONE 4:e4272
Schöne C, Apergis-Schoute J, Sakurai T, Adamantidis A, Burdakov D (2014a) Coreleased orexin and glutamate evoke nonredundant spike outputs and computations in histamine neurons. Cell Rep 7:697–704
Schöne C et al (2014b) Optogenetic probing of fast glutamatergic transmission from hypocretin/orexin to histamine neurons in situ. J Neurosci 32:12437–12443
Sterman MB, Clemente CD (1962) Forebrain inhibitory mechanisms: cortical synchronization induced by basal forebrain stimulation. Exp Neurol 6:91–102
Szymusiak R, McGinty D (1986) Sleep-related neuronal discharge in the basal forebrain of cats. Brain Res 370:82–92
Tsunematsu T et al (2011) Acute optogenetic silencing of orexin/hypocretin neurons induces slow-wave sleep in mice. J Neurosci 31:10529–10539
Tsunematsu T et al (2013) Long-lasting silencing of orexin/hypocretin neurons using archaerhodopsin induces slow-wave sleep in mice. Behav Brain Res 255:64–74
van den Pol AN, Acuna-Goycolea C, Clark KR, Ghosh PK (2004) Physiological properties of hypothalamic MCH neurons identified with selective expression of reporter gene after recombinant virus infection. Neuron 42:635–652
Vazey EM, Aston-Jones G (2014) Designer receptor manipulations reveal a role of the locus coeruleus noradrenergic system in isoflurane general anesthesia. Proc Natl Acad Sci USA 111:3859–3864
Verret L et al (2003) A role of melanin-concentrating hormone producing neurons in the central regulation of paradoxical sleep. BMC Neurosci 4:19
von Economo C (1926). In: Von Bethe A, Bergman GV, Embden G, Ellinger UA (eds) Handbuch des Normalen und Pathologischen Physiologie, Berlin, pp 591–610
Yizhar O, Fenno LE, Davidson TJ, Mogri M, Deisseroth K (2011) Optogenetics in neural systems. Neuron 71:9–34
Zecharia AY et al (2012) GABAergic inhibition of histaminergic neurons regulates active waking but not the sleep-wake switch or propofol-induced loss of consciousness. J Neurosci 32:13062–13075
Zhang F et al (2007) Multimodal fast optical interrogation of neural circuitry. Nature 446:633–639

Modulation of Thalamocortical Pathways by Orexins

Y. Audrey Hay

Modern research on sleep-wake cycle and arousal is based on the ascending reticular arousal system (ARAS), described in the 1940s by Moruzzi and Magoun. This system is located in the upper brainstem and separates in two branches, one innervating the thalamus and the other the orexin and cholinergic nuclei of the basal forebrain (Moruzzi and Magoun 1949; Ropert and Steriade 1981). The thalamus is organized in two groups of nuclei: the specific nuclei, that relay sensory information to the sensory cortices, and the non-specific nuclei that have been thought to promote cortical arousal via a diffuse system of projections throughout the neocortex (Morison et al. 1941; Macchi et al. 1977; Herkenham 1980). The non-specific thalamic nuclei are the main recipient of upper brainstem terminals but they are also innervated by the orexin nuclei of the basal forebrain, which together with the cholinergic nuclei are also known to have a prominent influence on cortical attention and arousal. Therefore, understanding the regulations of cortical states of vigilance by the thalamic and orexin systems requires a deep comprehension of their interplay.

To decipher how orexin modulates thalamocortical dynamics, I will first briefly describe the two main thalamocortical pathways as well as their role in the modulation of cortical states of arousal. Then, I will continue with the orexinergic system, its projections to the cortical areas and thalamic nuclei and electrical responses to orexin. Finally, I will investigate how orexin could modulate cortical arousal via its influence on thalamocortical dynamics.

Y.A. Hay (✉)
Department of Physiology, Development and Neuroscience, University of Cambridge, Physiological Laboratory, Downing Street, Cambridge CB2 3EG, UK
e-mail: audrey.hay@normalesup.org

© Springer International Publishing Switzerland 2015

T. Sakurai et al. (eds.), *Orexin and Sleep*,
DOI 10.1007/978-3-319-23078-8_8

1 Thalamocortical Loops and Sleep-Wake Cycles

1.1 Specific Thalamic Nuclei: Relays of Sensory Information and Involvement in Cortical Oscillations

Specific thalamic nuclei are the major gateways for sensory information to flow toward the neocortex. These nuclei are composed of glutamatergic neurons called relay cells, which transfer relatively unprocessed sensory information to sensory cortices. Each specific nucleus relays sensory signal of a particular modality, for instance visual inputs are relayed by the lateral geniculate nucleus, somatosensory inputs are relayed by the ventrobasal nucleus, etc.

Thalamic neuron outputs are shaped by the resting membrane potential that varies across sleep-wake cycle. During sleep, thalamic neurons are hyperpolarized and likely to discharge in bursts while during awake state they are depolarized and display a tonic discharge of action potential. As a consequence, during sleep, the transfer of sensory inputs to cortical areas often fails because of the hyperpolarization of the cell while in awake states, information are reliably transmitted to cortical areas.

Furthermore, sensory nuclei play a major role in the generation and maintenance of cortical oscillations, especially the slow oscillations characteristic of deep sleep. These cortical oscillations are shaped by the strong reciprocal connections between thalamic nuclei and cortical areas, known as the corticothalamocortical loop (Fig. 1).

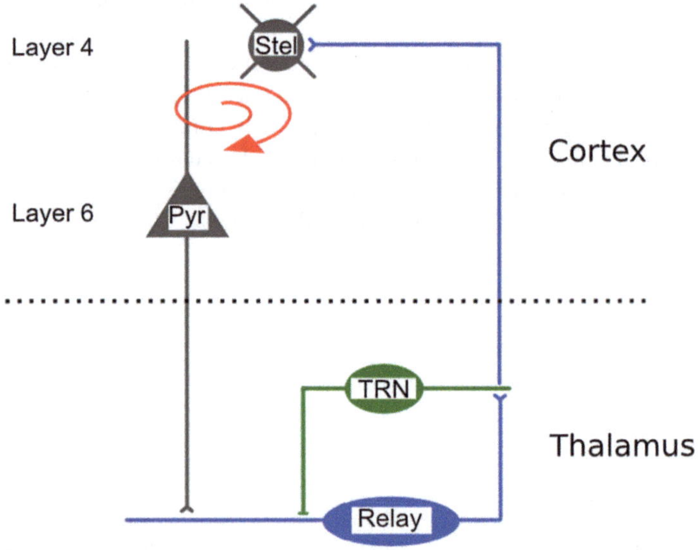

Fig. 1 The thalamocortical loop. The *red curved arrow* symbolizes the intra-cortical processing of sensory information. Note that layer 6 pyramidal cell projects on the distal part of the thalamic dendrite (Color figure online)

1.1.1 The Corticothalamocortical Loop

The organization of the thalamocortical loop is roughly similar for all sensory thalamic nuclei and their cortical counterparts (Fig. 1). Relay cells of the sensory thalamus send glutamatergic projections to the neocortex and to the thalamic reticular nucleus (TRN). In rodent species, TRN cells provide the only source of inhibitory inputs to the sensory nuclei (Jones 2007). In the cortex, relay cells project mainly on layer 4 neurons, while corticothalamic feedback is provided by a sub-population of layer 6a pyramidal neurons. In contrast to sensory inputs, which arrive on the proximal part of the thalamic dendrite, layer 6a neurons target the distal part of the relay cells. These corticothalamic returns contribute to almost 50 % of the distal synapses on thalamic relay cells and provide a powerful tool for the cortex to regulate thalamic activity (Sherman and Guillery 1996; Erisir and Sherman 1997; Varela 2014). However, this dense synaptic network is balanced by the small and long-lasting responses elicited by cortical inputs which are due to the weak activation of ionotropic receptors and to the reliable activation of metabotropic receptors (Miyata and Imoto 2006; Turner and Salt 1998). In contrast, thalamic inputs trigger a strong time-scaled response in the cortex, consistent with their role in the transfer of sensory information (Swadlow 2003; Cruikshank et al. 2007). These reciprocal interactions play a central role in shaping thalamocortical oscillations during deep sleep, and may be involved in the sleep-wake transition.

1.1.2 Thalamocortical Dynamics Across Sleep-Wake Cycle

Oscillations measured by electroencephalogram (EEG) at different stages of a sleep-wake cycle reflect thalamocortical loop dynamics. Indeed, deep sleep is characterized by slow and ample oscillations of the EEG resulting from the synchronization of thalamic and cortical networks while EEG of awake animals exhibits rapid and short oscillations characteristic of desynchronized networks.

Slow and ample oscillations are the main EEG hallmarks of deep sleep. These oscillations are quasi-synchronous across cortical areas and involve virtually all the cortical neurons. Although both cortical and thalamic networks in isolation are able to trigger slow-wave like oscillations in certain conditions, the constant communication between both networks through the thalamocortical loop is essential to produce physiological slow oscillations (Steriade et al. 1993; David et al. 2013). For instance, thalamic neurons require the activation of distal mGluRs receptors by corticothalamic inputs to generate oscillations, whereas cortical oscillations are fine-tuned by thalamic oscillations to reach physiological frequencies (Crunelli and Hughes 2010; Gutierrez et al. 2013; David et al. 2013).

The frequency of slow cortical oscillations varies across sleep stages according to the bursting frequency of thalamic neurons (Crunelli and Hughes 2010). This bursting frequency is mainly shaped by the hyperpolarization of the resting membrane potential. Indeed, the duration of a thalamic burst depends on the aperture of T-type channels, which trigger a low-threshold Ca^{2+} current crowned by

Na^+ action potentials (Contreras and Steriade 1995). These T-type channels are inactivated for depolarized membrane potentials and deinactivated by membrane hyperpolarization, meaning that the percentage of T-type Ca^{2+} channel putatively activated is correlated to the membrane hyperpolarization. Hyperpolarization is due to the aperture of a K^+ leak conductance, sustained by several types of channels, among them the TWISK-related acid-sensitive K^+-channel (Task1 and Task3) (Meuth et al. 2006). K^+ conductances are of primary importance since orexin affects membrane dynamics primarily by closure of K^+ channels, meaning that orexin may directly impact the waveform of the slow oscillations of the thalamus.

1.2 Non-specific Thalamic Nuclei: Relays of Ascending Reticular Activating System and Dedicated Roles

1.2.1 Non-specific Thalamic Nuclei as Broad Cortical Activators

Whereas the role and organization of specific thalamic nuclei have been well characterized, non-specific thalamic nuclei have long been neglected and only few studies aim at dissecting the precise role of these nuclei in the maintenance of arousal states. Because of their widespread projections to the neocortex, midline and intrathalamic nuclei, the two main sets of non-specific thalamic nuclei, have long been considered as global cortical activators (Lorente de No 1938; Herkenham 1980; Steriade and Glenn 1982). This hypothesis is supported by the pioneering experiments of Dempsey and Morison showing that electrical stimulation of midline and intralaminar thalamic nuclei is able to induce a broad activation of the cortical network (Morison et al. 1941; Dempsey et al. 1941; Morison and Dempsey 1942; Dempsey and Morison 1942). Subsequent experiments by Moruzzi and Magoun demonstrate that non-specific thalamic nuclei relay information from the ascending brainstem arousal system to the cortical areas and is thus involved in the promotion of cortical arousal (Moruzzi and Magoun 1949; Steriade and Glenn 1982).

Clinical evidence supports this hypothesis. Large lesions of the non-specific thalamic nuclei induced by strokes or by the alcoholic Korsakoff syndrome have been shown to affect the consciousness of patients (Schmahmann 2003; Carrera and Bogousslavsky 2006). Moreover, injuries of the midline thalamic nuclei, which are the nuclei that receive the densest projections from orexin neurons, lead to major deficits in arousal and memory (Schmahmann 2003). In contrast, lesions disrupting other thalamic nuclei affect primarily mood and lead to disinhibited behaviors (Schmahmann 2003). Brain imaging indicates that the spatial extent of the thalamic lesion is correlated to the severity of the syndromes. Partial lesions impact attention, which often trigger a deficit in working and long-term memory, while entire lesions of the median thalamic nuclei are associated with a state close to coma (Façon et al. 1958; Castaigne et al. 1962; Tinuper et al. 1989; Llinas et al. 1998; Carrera and

Bogousslavsky 2006). Moreover, stimulation of non-specific thalamic nuclei in patient suffering from minimal conscious state has been shown to improve their symptoms, indicating that these nuclei can promote cortical arousal (Schiff et al. 2007).

Evidence from physiological experiments explicits the activity of non-specific thalamic nuclei across sleep-wake cycle. In rodents, electrical activity and neuronal excitability increase during the dark period, which corresponds to their active period (Novak and Nunez 1998; Steriade and Glenn 1982). Electrical stimulation of these non-specific thalamic nuclei produces a broad activation of cortical networks, that is likely to drive a tonic state of readiness in cortical networks, allowing performing high-demanding tasks (Glenn et al. 1982; Steriade and Glenn 1982). Moreover, high frequency stimulation of non-specific thalamic nuclei is sufficient to shift the cortical dynamics from slow and ample oscillations to fast awake-like activity (Steriade and Llinas 1988; Steriade 1997). Recordings of human brain activity using positron emission topography indicate that besides being involved in the sleep to wake transition, non-specific nuclei are also implicated in the transition between relaxed wake and high general attention (Kinomura et al. 1996). This tonic activation of cortical networks is driven by the widespread projections of non-specific thalamic nuclei to layers 1, 5 and 6, which contain the majority of neurons activated in attentive behaviors (Hyvarinen et al. 1980; Herkenham 1980; Glenn et al. 1982; Berendse and Groenewegen 1991; Clasca et al. 2012; Cruikshank et al. 2012).

1.2.2 Non-specific Thalamic Nuclei and Their Dedicated Roles

Refinement of anatomical tracing as well as behavioral and functional studies however indicate that non-specific thalamic nuclei take part in more dedicated tasks. Each nucleus sends projections to a small number of associative brain regions, among them the striatum, hippocampus, amygdala and several cortical areas, to provide information about the environmental and behavioral context (Jones and Leavitt 1974; Macchi et al. 1975, 1977; Berendse and Groenewegen 1991; Groenewegen and Berendse 1994; Giménez-Amaya et al. 1995; Van der Werf et al. 2002; Vertes et al. 2006). Furthermore, functional studies indicate a preferential activation of certain non-specific thalamic nuclei according to the task performed by the animal (Deutch et al. 1995; Vann et al. 2000; Loureiro et al. 2012; Mitchell and Chakraborty 2013; Xu and Südhof 2013).

Among them, the rhomboid and reuniens thalamic nuclei, two nuclei belonging to the midline thalamic nuclei, are activated during goal-directed tasks that require a high level of attention (Hembrook and Mair 2011; Hembrook et al. 2012; Loureiro et al. 2012; Cholvin et al. 2013). Due to the privileged position of rhomboid and reuniens nuclei at the interface between prefrontal cortex and hippocampus, they are thought to orchestrate the communication between these regions during high-demanding tasks (Cholvin et al. 2013; Cassel et al. 2013).

As for the other non-specific thalamic nuclei, they have been involved in tasks as diverse as stress response, spatial navigation, drug addiction or head direction coding. Therefore, thalamic nuclei have dedicated roles that all require increased attention and contribute all together to the maintenance of states of arousal.

In conclusion, though acting at different steps of the sleep-wake cycle, both thalamic pathways are involved in the regulation of cortical states of arousal. Specific thalamic nuclei take part in the maintenance and shaping of slow-wave oscillations occurring during deep sleep, while activation of non-specific thalamic nuclei is associated with sleep-wake transitions and enhancement of cortical arousal. Thalamic pathways are differentially impacted by upper brainstem inputs, and interestingly they are differentially targeted by orexin neurons.

2 Projections of Orexin Neurons on Thalamic Nuclei and Cortical Areas

Orexin neurons of the lateral hypothalamus project widely throughout the brain with densest projections in the regions involved in arousal and attention.

2.1 Projections on Non-specific Thalamic Nuclei

Labeling of orexin fibers using antibodies against the two orexin proteins, orexin-A and orexin-B, or their common precursor, the prepro-orexin, reveal a heterogenous pattern of projection throughout the thalamus (Peyron et al. 1998; Chen et al. 1999; Nambu et al. 1999; Date et al. 1999; Cutler et al. 1999; Mintz et al. 2001; Kirouac et al. 2005). Orexin fibers project widely to non-specific thalamic nuclei while they poorly target specific nuclei. Orexin neurons projecting to the thalamus are distributed within the hypothalamus with a highest density in the perifornical nucleus and in the ventral part of the lateral hypothalamus (Kirouac et al. 2005; Lee 2005). Axons exit the hypothalamus by the ascending pathway and project throughout the majority of the intralaminar and median thalamic nuclei following a dorsolateral to ventromedial orientation (Peyron et al. 1998). They enter the thalamus through the zona incerta, which provides GABAergic inputs to higher order thalamic nuclei, and ramify in this structure, indicating that inhibitory neurons of the thalamus are strongly modulated by orexin (Peyron et al. 1998; McGranaghan and Piggins 2001; Jones 2007). Orexin fibers target primarily the paraventricular nucleus and to a lesser extent the central medial, rhomboid, reuniens, intramediodorsal, centrolateral and lateral habenular nuclei (Peyron et al. 1998; Nambu et al. 1999; Date et al. 1999; Mintz et al. 2001; Kirouac et al. 2005; Hsu and Price 2009). In these nuclei, orexin axons can exhibit a smooth or a varicose morphology, suggesting the coexistence of synaptic and volume transmission. The colocalization of orexin and synaptophysin, a protein present in synaptic buttons, supports the hypothesis of a synaptic release of orexin in non-specific nuclei, in agreement with electron microscopy analysis in other brain areas (Peyron et al. 1998; Kirouac et al. 2005). In addition to this point-to-point release in the thalamus, some orexin fibers crossing the thalamic nuclei are thought to release orexin in the dorsal third ventricle. These fibers have been reported in the ventricular wall at the boundary of the paraventricular nucleus and may contribute to a global effect of orexin in the brains (Mintz et al. 2001).

In contrast, orexin fibers are very sparse in sensory nuclei, except in the intergeniculate leaflet (IGL). The IGL belongs to the lateral geniculate nucleus, which primarily relays visual information from the retina to visual cortices (Peyron et al. 1998; Mintz et al. 2001). In contrast to most subnuclei of the lateral geniculate nucleus, the IGL is not only a relay of primary visual information but is also an important component of the circadian regulator system. Indeed, it is involved in transferring photic and non-photic cues from the retina to the suprachiasmatic nucleus, which in mammals is the main circadian oscillator (Reppert and Weaver 2002). The IGL has also been involved in the transfer of non-specific information to the suprachiasmatic nucleus, especially information from structures involved in the sleep-wake cycle. Thus, the specific innervation of the IGL by orexin fibers is consistent with their common involvement in sleep-wake cycle regulation.

Interestingly, alongside a high homology of orexin sequences, the differential targeting of thalamic nuclei is highly conserved among vertebrates (Alvarez and Sutcliffe 2002). Orexin fibers are restricted to thalamic areas that project to the limbic system in both diurnal and nocturnal mammals (Mintz et al. 2001; McGranaghan and Piggins 2001), birds (Singletary et al. 2006), amphibians (Shibahara et al. 1999; Lopez et al. 2009a; Suzuki et al. 2007) and fish (Huesa et al. 2005; Lopez et al. 2009b). The conservation of the projection pattern of orexin neurons contrasts with the divergence observed for other neuropeptides (Mintz et al. 2001), suggesting that the innervation of non-specific thalamic nuclei by orexin neurons plays a conserved role in animal physiology.

Measure of orexin concentration in brain tissues reveals an equivalent release of orexin in the paraventricular thalamic nucleus, the lateral hypothalamus and the locus coeruleus (Mondal et al. 1999b). Accordingly, the density of post-synaptic buttons expressing orexin receptors is comparable in the locus coeruleus and in the paraventricular nucleus (Huang et al. 2006). The strong modulation locus coeruleus dynamics by orexin has been assessed in the literature at the cellular and system levels, as reported in other chapters of this book (Carter et al. 2009, 2012). This first confirms that non-specific thalamic nuclei are one of the major target for orexin neurons and second suggests that orexin play a strong and under-estimated role on the modulation of thalamic dynamics. Orexin-A and orexin-B positive fibers largely overlap in the thalamic nuclei, suggesting that most orexin fibers release both peptides (Date et al. 1999; Zhang et al. 2002, 2004). However, local disparities have been reported throughout the brains which point out the differential involvement of these peptides in the modulation of physiological modalities. In the thalamus, orexin-B antibody labels thinner and sparser axons in most nuclei and only orexin-A positive fibers are found to reach the ventricular wall at the border of the paraventricular nucleus (Cutler et al. 1999; Zhang et al. 2002, 2004). In contrast to these labelings, measures of orexin concentration reveal a three times higher concentration of orexin-B compared to orexin-A in thalamic tissues (Mondal et al. 1999a). Conversely, in the locus coeruleus, labelings and measure of orexin concentrations are consistent and indicate a predominant release of orexin-A. These contrasting results indicate first a differential release of orexin-A and -B according to the cerebral structure and second suggest that orexin-B plays a primary role in the modulation of orexin-dependent processes in thalamic neurons.

The sensibility to orexin is conferred by post-synaptic expression of G-coupled receptors for orexin, which present two isoforms called OX1R and OX2R (Sakurai et al. 1998). OX1R exhibits a preferential affinity for orexin-A while OX2R affinity is equivalent for both orexin peptides. Distribution of OX1R and OX2R throughout the thalamus has been extensively studied using in situ *hybridization*. It shows that orexin receptors are mainly expressed by midline and intralaminar thalamic nuclei, in agreement with the projection profile of orexin fibers (Trivedi et al. 1998; Lu et al. 2000; Marcus et al. 2001). In contrast, orexin receptors are barely detected in the specific thalamic nuclei. Within non-specific thalamic nuclei, OX1R and OX2R are distributed according to a precise pattern. Expression of mRNAs coding for OX1R is restricted to paraventricular neurons and is present to a lesser extent in the reuniens and parafascicular nuclei (Trivedi et al. 1998; Lu et al. 2000). In contrast, mRNAs coding for OX2R are detected in most intralaminar and midline thalamic nuclei, the highest density being noticed in the central median, rhomboid, paraventricular and centrolateral nuclei (Trivedi et al. 1998; Lu et al. 2000; Marcus et al. 2001). At the protein level, OXRs are observed widely throughout the brain and especially in the thalamus, suggesting that OXR expression was underestimated by in situ *hybridization* experiments (Hervieu et al. 2001; Cluderay et al. 2002). However, the strength of the labeling varies between regions and consistently with mRNA detection, the strongest labelings are detected in the central median, rhomboid, paraventricular, parafascicular, reuniens and centrolateral nuclei.

Altogether, these results indicate that non-specific thalamic nuclei are privileged targets for orexin modulation in contrast to specific nuclei that are poorly targeted by orexin fibers. Labelings show an overlap of the OXR expression and orexin fibers in most non-specific thalamic nuclei, with a highest density of fibers and receptors measured in the paraventricular, centromedian, rhomboid and reuniens nuclei. Furthermore, orexin-B is primarily involved in non-specific thalamic modulation, in agreement with the preferential expression of OX2R in these nuclei.

2.2 Projections on Cortical Areas

Orexin neurons send sparse projections to the neocortical areas through the dorsal ascending pathway (Peyron et al. 1998). These afferences, especially those projecting to the medial prefrontal cortex, come from a small subset of neurons distributed within the entire lateral hypothalamus (Fadel et al. 2002). Labelings using either anti-prepro-orexin, anti-orexin-A or anti-orexin-B antibodies reveal a substantial targeting of associative cortical areas whereas sensory cortices receive only sparse projections. Among associative areas, the retrosplenial and prefrontal cortices, especially the insular, prelimbic and infralimbic cortices and to a lesser extent the perirhinal and piriform cortices exhibit the highest density of fibers (Peyron et al. 1998; Date et al. 1999; Nambu et al. 1999; Cutler et al. 1999). Interestingly, neocortical regions exhibiting the highest density of orexin fibers belong to the associative areas, that processes higher-order information, highly dependent of arousal states. These labelings indicate that orexin differentially modulates cortical areas according to the nature of information processed.

Across cortical layers, orexin fibers are primarily located in layer 6 and to a lesser extent in layer 1 (Peyron et al. 1998; Nambu et al. 1999). Orexin fibers are oriented radially from the white matter to the pial surface, except in layer 6 where they run along the white matter towards the corpus callosum (Peyron et al. 1998; Nambu et al. 1999). Cortically projecting fibers exhibit both smooth and varicose morphology, as observed in the thalamus (Cutler et al. 1999). The concentration of orexins measured in cortical tissues is roughly equivalent for orexin-A and -B, in agreement with the global overlapping of orexin-A and -B positive fibers in the cortex (Date et al. 1999; Mondal et al. 1999a). However, this concentration is ten times lower than those measured in the thalamus, probably due to the sparse projections of orexin fibers in the neocortex (Mondal et al. 1999a). It is noteworthy that this concentration has been measured within the entire cortical column and may not reflect the real concentration in the layer 6, which receives the majority of the orexin fibers. Refinement of this measure would thus be important to better understand the physiological impact of orexin in the deep layers of cortical areas.

Expression of orexin receptors is weaker in the neocortex than in the non-specific thalamus and displays high heterogeneity (Trivedi et al. 1998; Hervieu et al. 2001; Cluderay et al. 2002). OX2R is strongly expressed in the deep layers whereas a diffuse expression of OX1R and OX2R mRNAs is detected throughout the rest of the cortical layers (Trivedi et al. 1998; Lu et al. 2000). Refinement of this observation using single-cell RT-PCR indicates that the strong expression of OX2R in the deep layers is restricted to layer 6b in the somatosensory cortices (Hay et al. 2014). In the infralimbic, cingulate, dorsal peduncular and granular insular regions of the prefrontal cortex, layers 2, 5 and 6 neurons express a relatively high level of OX1R mRNA compared to the very weak detection of OX1R in other cortical areas (Marcus et al. 2001). Accordingly, protein level of OX1R is higher in the frontal cortex than in somatosensory cortices (Aracri et al. 2013). Furthermore, layer 5 neurons of the somatosensory cortex exhibit a low labeling, although in some cases a clear pyramidal morphology can be observed. In contrast, layer 5 cells of the frontal cortex exhibit a dense labeling of the soma and some punctiform labeling of the neuropil (Hervieu et al. 2001; Aracri et al. 2013). Labeling of OX2R protein confirms the strong expression of this receptor in layer 6b neurons but also indicates a substantial expression by pyramidal cells of the prefrontal cortex (Cluderay et al. 2002).

Taken together, these anatomical data indicate a heterogenous distribution of orexin fibers and orexin receptors throughout cortical areas. This suggests that orexin impacts strongly and globally the dynamics of the associative cortical areas whereas sensory cortical networks are precisely modulated through the activation of layer 6b neurons. Furthermore, in associative cortices activation of both OX1R and OX2R participate to the modulation of cortical dynamics whereas in sensory cortices OX2R only is implicated.

3 Orexin Regulates Thalamocortical Pathway by Pre- and Post-synaptic Mechanisms

3.1 *Somatodendritic Activation of Thalamic and Cortical Neurons by Orexin*

Electrophysiological recordings in rodent brains slices allow describing precisely the electrical responses to orexin application. In thalamic and cortical neurons, orexin application induces a long-lasting depolarization of the cells peaking 2–4 min after the onset of the application. Suprathreshold depolarizations are crowned by a long train of tonic action potentials or by a sustained train of bursts (Bayer et al. 2002, 2004; Kolaj et al. 2007; Hay et al. 2014; Govindaiah and Cox 2006; Ishibashi et al. 2005). Upon washing, the response decreases slowly and gradually, until returning to the initial state after 10–15 min (Kolaj et al. 2007). Similar profile of responses are obtained upon application of orexin-A or -B although the amplitude of response varies across regions according to the relative expression of each OXR.

Proportion of orexin-sensitive neurons varies across areas according to the density of orexin fibers while the response magnitude to either orexin-A or -B in each area is correlated to the local expression of OXRs. In the thalamus, virtually all paraventricular, central median and rhomboid neurons are depolarized by orexins, whereas responses are detected in only 70 % neurons of the centrolateral nucleus and even less in the mediodorsal and reticularis nuclei (Bayer et al. 2002; Govindaiah and Cox 2006; Kolaj et al. 2007; Ishibashi et al. 2005; Huang et al. 2006). Orexin-A and -B trigger similar responses in most non-specific thalamic neurons, indicating that the sensitivity to orexin is likely to be mediated by OX2R. In contrast, in the reticularis neurons, orexin-A drives a stronger depolarization than orexin-B, indicating that the depolarization is likely to be mediated by OX1R activation (Govindaiah and Cox 2006). We mentioned previously that specific thalamic nuclei are free of orexin fibers and display few or no expression of OXRs. Consistently application of orexin has no effect on the resting membrane potential or on the membrane resistance of these neurons (Bayer et al. 2002; Govindaiah and Cox 2006).

In most cortical areas, neurons sensitive to orexin are located in layer 6b, in agreement with the exclusive expression of OX2R in deep layers and with the sparse density of orexin fibers throughout the cortical layers except in deep layers (Trivedi et al. 1998; Peyron et al. 1998; Bayer et al. 2004; Hay et al. 2014). Interestingly, both excitatory and inhibitory neurons of layer 6b are activated by orexin (Hay et al. 2014). This restricted sensitivity of cortical neurons to orexin is consistent with the low occurrence of orexin-induced response in primary culture of cortical cells (Pol et al. 1998). In prefrontal cortices, which exhibit the highest density of orexin fibers in the neocortex, the situation remains to be clarified as orexin sensitivity has been reported in layer 5 pyramids and to a lesser extend in layer 2 pyramids by some groups (Xia et al. 2005; Song et al. 2005; Li et al. 2010), whereas others mention restricted responses to layer 6b neurons at least in the cingulate cortex (Bayer et al. 2004). One may suggest that this difference is due to the fact that Bayer and colleague. have used preferentially orexin-B for their experiments, which would have made the OX1R-mediated response undetectable. Furthermore, the highest density of orexin fibers in the prefrontal cortex is consistent with a broader responsivity across cortical layers, and OX1R labeling has been observed in layer 2 and 5 pyramidal cells (Li et al. 2010; Aracri et al. 2013).

3.2 *Mechanisms of Orexin-Induced Somatodendritic Activation*

The amplitude of orexin-mediated depolarization in both thalamic and cortical neurons depends on orexin concentration. For neurons expressing mainly OX2R, i.e. neocortical layer 6b and non-specific thalamic neurons, orexin-B is slightly more potent for low doses of orexin while higher doses lead to similar amplitude of depolarization in response to orexin-A and -B. Indeed, a subset of these neurons exhibit responses to orexin-B from 10 nM while no response to orexin-A is observed at this concentration. Furthermore, EC50 is reached at a concentration of 20 nM for orexin-B and 32 nM for orexin-A. At higher concentration (1000 nM), both peptides trigger a large depolarization crowned with a massive discharge of action potentials (Bayer et al. 2002, 2004; Ishibashi et al. 2005; Govindaiah and Cox 2006; Huang et al. 2006). While the EC50 for orexin-A is consistent with previous report using human isoform of the OXRs, the EC50 for rodent orexin-B is at least two times lower than those previously described (Sakurai et al. 1998; Beuckmann and Yanagisawa 2002). That suggests that the EC50 measured for orexin-B in heterologous system may have been underestimated and as a consequence that OX2R would be more sensitive for orexin-B than for orexin-A.

Direct somatodendritic activation of thalamic and cortical neurons is assessed by the persistence of orexin responses in presence of TTX (Bayer et al. 2002, 2004; Ishibashi et al. 2005; Huang et al. 2006; Govindaiah and Cox 2006; Kolaj et al. 2007; Li et al. 2010). Indeed, TTX, by acting on Na^+ voltage-dependent channels, prevents propagation of action potentials, blocks synaptic transmission and as a consequence isolates neurons from each other. Furthermore, orexin effects persist in the presence of glutamatergic and GABAergic inhibitors confirming a direct action of orexin on the somatodendritic compartment instead of an indirect network-mediated effect (Huang et al. 2006). A final argument to assess the post-synaptic activation of non-specific thalamic and layer 6b cortical neurons is provided by the enhancement of the orexin response in low-Ca^{2+} and high-Mg^{2+} concentration conditions. Indeed, in these conditions, the release probably decreases dramatically, which blocks the synaptic transmission (Bayer et al. 2002, 2004; Ishibashi et al. 2005; Huang et al. 2006). Thus, several pieces of evidence indicate that, in non-specific thalamic nuclei as well as in cortical layer 6b, orexin acts on neurons by a direct somatodendritic activation.

Most orexin-mediated depolarizations are coupled with an increase of the membrane resistance, signaling channel closure. This increase of membrane resistance reaches a maximum of 150 % in the paraventricular neurons and is milder in other non-specific thalamic nuclei and in layer 6b neurons (Ishibashi et al. 2005; Bayer et al. 2002, 2004; Govindaiah and Cox 2006; Huang et al. 2006; Doroshenko and Renaud 2009; Kolaj et al. 2007). Information about the ion involved in this conductance is provided by plotting the current-voltage relationship (I–V), which reveals that the orexin-induced current reverses at the equilibrium potential of K^+. Moreover the I–V curve is right-shifted in the presence of an increased concentration of external K^+, pointing out that orexin-mediated current is due to the closure of a K^+ conductance (Govindaiah and Cox 2006; Bayer et al. 2002, 2004; Huang et al. 2006). Several K^+-channels are expressed by thalamic and cortical cells. In the thalamus, it is noteworthy that Task1 and Task3 channels, which are involved in the hyperpolarization of thalamic cells during sleep, are also implicated in orexin-induced depolarization (Doroshenko and Renaud 2009).

Several other mechanisms have been implicated in orexin-induced depolarization, such as the aperture of a cationic non-specific conductance or the inactivation of a HCN channel (Huang et al. 2006; Kolaj et al. 2007; Li et al. 2009). Aperture of a conductance is confirmed by the increase of the background noise in the presence of orexin (Huang et al. 2006). This mechanism has been observed in some cells in layer 2 and 5 of the prefrontal cortex and in few non-specific thalamic neurons (Huang et al. 2006; Kolaj et al. 2007), whereas inactivation of HCN channels have been reported only in pyramidal cells of the prefrontal cortex upon activation of OX1R (Li et al. 2009).

3.3 Pre-synaptic Activation of Non-specific Thalamocortical Terminals in the Prefrontal Cortex

In addition to the somatodendritic activation of cortical neurons, orexin has been shown to act directly on a subset of pre-synaptic terminals in the prefrontal cortex

(Lambe and Aghajanian 2003). Using a combination of patch-clamp recording and calcium imaging, Lambe and colleague have demonstrated that the application of orexin-B does not affect the resting membrane potential of layer 5 cells but selectively depolarize a limited number of dendritic spines. These local depolarizations are due the indirect activation of glutamatergic terminals, as assessed by the large increase of EPSC frequency in response to orexin-B application. A subsequent report shows that the increase of EPSC frequency persists in the presence of TTX, meaning that the orexin-induced EPSCs are due to the local depolarization of pre-synaptic glutamatergic terminals (Aracri et al. 2013). Non-specific thalamic fibers exhibit a close vicinity with the dendritic buttons depolarized during orexin application in the prefrontal cortex, suggesting that orexin activates specifically these thalamic terminals (Lambe and Aghajanian 2003). This hypothesis is supported by the absence of EPSC frequency increase when thalamic nuclei are lesioned (Lambe and Aghajanian 2003). On layer 5 pyramidal cells, orexin-activated buttons are primarily located on the apical tuft of the dendritic tree and to a lesser extent on the basilar dendrites, which is consistent with the projection pattern of non-specific thalamic nuclei (Berendse and Groenewegen 1991; Lambe and Aghajanian 2003). Thus, in addition to a direct post-synaptic activation of both cortical and thalamic neurons, non-specific thalamocortical pathway is also potentiated by orexin through the specific activation of pre-synaptic terminals.

However, the authors indicate that roughly 10 % of dendritic buttons are not apposed to non-specific thalamocortical fibers (Lambe and Aghajanian 2003). The discovery in 2004 of the specific sensitivity of layer 6b cells to orexin may provide a cue to explain the origin of these remaining buttons (Bayer et al. 2004; Lambe et al. 2007).

The presence of orexin receptors in pre-synaptic terminals was not investigated in other cortical areas, probably due to the weak effect of orexin on EPSC frequency compared to the strong one measured in prefrontal neurons (Lambe and Aghajanian 2003). Recently, the development of optogenetics tools that allow a specific photo-stimulation of non-specific thalamocortical pathways has permitted to address this question in the somatosensory cortex (Hay et al. 2014). Measure of the paired pulse ratio between two successive stimulations of the presynaptic terminals allow evaluating the pre-synaptic impact of a drug. Orexin does not impact this ratio at the synapse between non-specific thalamocortical fibers and cortical dendrites, indicating an absence of pre-synaptic modulation in the somatosensory cortex (Hay et al. 2014). Moreover, thalamocortical inputs recorded in layer 6a pyramidal cells are followed by a disynaptic current in the presence of orexin, which is prevented by the physical separation of layer 6b from the rest of the cortical layers (Hay et al. 2014). This indicates that in the somatosensory cortex, orexin acts selectively by post-synaptic activation of layer 6b neurons, as suggested by Lambe and Aghajanian (2003). Finally, analysis of the paired-pulse ratio in paraventricular thalamic neurons suggests as well an absence of pre-synaptic modulation by orexin within the thalamic nuclei (Kolaj et al. 2007).

As a consequence, in most non-specific thalamocortical pathways, orexin acts primarily through a post-synaptic activation of cortical and thalamic neurons. However, in certain cortical areas, such as the prefrontal cortex, orexin can also trigger presynaptic activation of non-specific thalamic terminals. It is noteworthy that the prefrontal cortex receives the strongest orexin inputs, acting via pre- and post-synaptic mechanisms, since prefrontal activity is highly correlated to the state of vigilance of the animal.

4 Orexin Modulates Vigilance States by Acting on the Thalamocortical System

We have shown that deep sleep is characterized by slow oscillations of the thalamocortical system while arousal is characterized by depolarization of the cells and desynchronization of the network. It is also known that sustained attention involves an increase of neuronal excitability and a synchronization of fast oscillations across cortical areas. Finally it has been reported that the induction of attention can be triggered by thalamic bursting. In this chapter, I will describe how orexin acts on these different parameters linked to the states of vigilance.

4.1 Modulation of Cortical Dynamics and Integration of Extra-Cortical Inputs by Orexin

4.1.1 Increase of Neuronal Excitability Throughout Cortical Layers in Primary Sensory Cortices

In the somatosensory cortex, activation of layer 6b neurons by orexin results in increased EPSC frequency primarily in infragranular layers and secondarily in layer 1 interneurons (Lambe and Aghajanian 2003; Bayer et al. 2004; Hay et al. 2014). This pattern is consistent with the projections of layer 6b in the cortical column and indicates that activation of layer 6b by orexin promotes an increase of cortical excitability (Clancy and Cauller 1999; Marx and Feldmeyer 2013). Besides this direct activation of the cortical network, orexin triggers action potential discharge in non-specific thalamic neurons by a direct somatodendritic mechanism, which may result in an increase of glutamatergic inputs in the cortex. Interestingly, layer 6b and non-specific thalamic neurons exhibit an overlapping projection pattern in the neocortex, since thalamic neurons also target primarily layers 1 and 5–6 (Herkenham 1980). This suggests that in most cortical areas, release of orexin activates two pathways running in parallel, leading to an increase of neuronal excitability specifically in layers 1 and 5–6. In layer 1, thalamocortical fibers target both layer 1 interneurons and the dendritic tuft of the pyramidal neurons (Krettek and Price 1977; Cruikshank et al. 2012; Clasca et al. 2012). This activation of apical dendrites has been hypothesized to promote a tonic state of readiness in cortical networks, typical of attentive states (Steriade and Morin 1981). Furthermore, this specific activation of cortical layers 1 and 6 is consistent with the observation that these layers contain the majority of neurons activated during attentive states, indicating a prominent role of these two cortical layers in the maintenance of attentive states (Hyvarinen et al. 1980).

In contrast, layer 4 cells are neither depolarized nor indirectly excited by orexin (Hay et al. 2014). This layer is the main target of specific thalamic nuclei and as a consequence receives mostly somatosensory inputs, whereas non-specific thalamic fibers only sparsely project to this layer. This suggests that in sensory cortices, sensory inputs are not directly influenced by cortical states of vigilance at their entrance to the neocortex (Fig. 2).

The functional role of layer 6b is still poorly understood. Layer 6b neurons express Nurr1, a marker of intracortical long-range projections, suggesting that this layer might be involved in long-range transfer of information or maybe in the synchronization of distant cortical areas (Arimatsu et al. 2003; Andjelic et al. 2009). However, its specific sensitivity to arousal-related modulators such as acetylcholine and orexin suggests that this layer might be involved in the regulation of cortical arousal. In agreement with this hypothesis, layer 6b has been shown to act as an

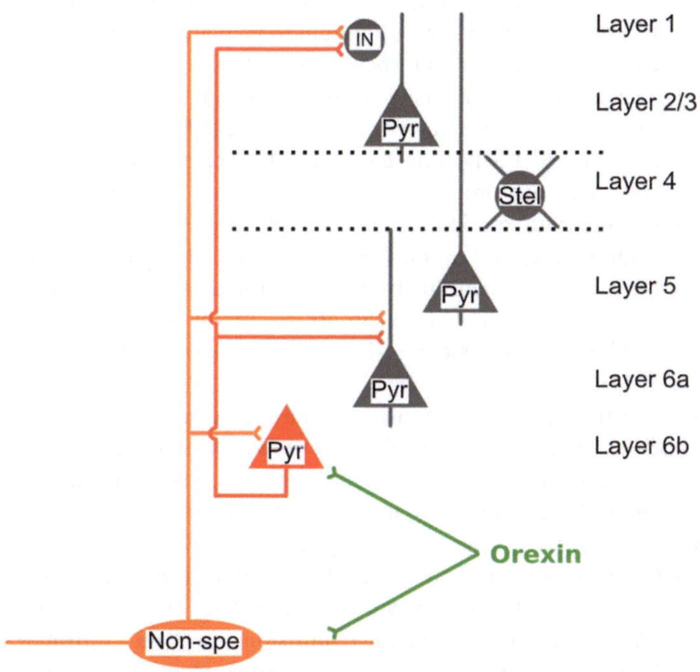

Fig. 2 The two parallel pathways for orexin-mediated excitation of sensory cortices. Note that non-specific thalamic and layer 6b pathways target primarily layer 1 and 5–6 but avoid layer 4

orexin-dependent feedforward excitatory loop that potentiates layer 6a responses to non-specific thalamic inputs (Hay et al. 2014).

In conclusion, orexin activates primarily the cortical layers involved in the intracortical regulation of the state of arousal and promotes a tonic state of readiness in cortical circuits. Furthermore, by activating layer 6b neurons which send long-range projections, orexin may be involved in the synchronization of cortical areas.

4.1.2 Increase of Attention by the Complex Activation of the Prefrontal Cortex

Besides the direct activation of layer 6b neurons, orexin acts on several other pathways to modulate cortical dynamics in prefrontal cortices. Indeed, as we mentioned before, prefrontal cortices exhibit a wider expression of orexin receptors and receive denser projections from orexin and non-specific thalamic neurons. Orexin-B reliably depolarizes layer 6b neurons and activates non-specific thalamocortical terminals in prefrontal cortices (Lambe and Aghajanian 2003; Bayer et al. 2004; Lambe et al. 2007). Both mechanisms induce a much stronger increase of EPSC frequency in layer 5 pyramidal cells of the prefrontal cortices compared to other cortical areas (Lambe and Aghajanian 2003). Moreover, application of orexin-A induces a direct somatodendritic activation of layer 2 and 5 pyramidal neurons of the prelimbic cortex, and is thought to activate presynaptic glutamatergic terminals located on corticocortical fibers (Song et al. 2005; Li et al. 2010; Aracri et al. 2013). These two mechanisms lead to a strong activation of the cortical network in the presence of orexin-A (Aracri et al. 2013).

In addition to its direct action on the prefrontal network, orexin activates many structures involved in the promotion and maintenance of arousal states. Among these structures, the paraventricular thalamic (glutamate), raphe (serotonin), locus coeruleus (noradrenaline) and ventral tegmental area (dopamine) nuclei project to the neocortex. Interestingly neurons from these structures that specifically target the prefrontal cortex exhibit a higher sensitivity to orexin than other neurons of the same structure (Huang et al. 2006; Cid-Pellitero and Garzon 2011; Del Cid-Pellitero and Garzon 2011a, b).

Prefrontal cortical areas are known to play a central role in high cognitive processes such as attention (Funahashi et al. 1989; Kinomura et al. 1996; Euston et al. 2012; Bloem et al. 2014). This region is strongly activated during high-demanding tasks and lesioning the prefrontal cortex has been shown to impair dramatically the ability to realize correctly a given task (Muir et al. 1996; Kahn et al. 2012). On the other hand, orexin levels as well as firing rate of orexin neurons are closely correlated with states of vigilance (Yoshida et al. 2001; Kiyashchenko et al. 2002). As a consequence, orexin-B infusion in rodent prefrontal cortex has been shown to increase performance under sustained attention (Lambe et al. 2005). This mechanism relies primarily on the orexin-triggered activation of non-specific thalamocortical terminals (Lambe et al. 2005). However, although direct evidence are still lacking, the somatodendritic activation of non-specific thalamic neurons by orexin may also be involved in the regulation of cortical arousal and attention (Huang et al. 2006).

Evidence from narcoleptic patients assess the major role of non-specific thalamic—prefrontal cortex pathway in the maintenance of arousal states. Indeed, narcoleptic patients report concentration difficulties which are linked to an altered working of the prefrontal cortex (Rieger et al. 2003; Naumann et al. 2001; Saletu et al. 2009). Attention deficit in narcoleptic patients are correlated with a decrease of orexin concentration in the prefrontal cortex (Mishima et al. 2008). According to this observation, modafinil, a drug used to reduce sleepiness in narcoleptic patients by activating orexin neurons, is shown to improve the speed of information processing in the prefrontal cortex (Saletu et al. 2009). Thus, the modulation of prefrontal cortical excitability by orexin may play an important role in the regulation of attention.

In conclusion, orexin induces a direct increase of neuronal excitability primarily in layers known to be activated during high-demanding tasks. In the prefrontal cortex, orexin induces a stronger and more complex activation throughout the cortical layers, consistent with the central role played by this area in high cognitive processes. Moreover, cortical areas are targeted by orexin-sensitive neurons from ascending modulatory structures, which are involved in the regulation of states of vigilance.

4.2 Induction and Maintenance of Attention by Thalamic Burst Firing

4.2.1 Modulation of Firing Mode in Non-specific Thalamic Nuclei

Initial pharmacological studies aiming at characterizing the sensitivity to orexin throughout the thalamus have shown that application of orexin on silent cells triggers a tonic discharge of action potentials (Bayer et al. 2002; Ishibashi et al. 2005). Subsequent refined studies indicate that the mode of action potential discharge in response to orexin depends greatly on the resting state of the cell (Kolaj et al. 2007). Indeed, in silent hyperpolarized cells ($Vm < -80$ mV), orexin application induces a depolarization either sub-threshold or crown by a volley of bursts due to the activation of T-type channels, whereas application of orexin on silent depolarized cells ($Vm > -70$ mV) triggers a tonic action potential discharge (Kolaj et al. 2007).

However, in physiological conditions, thalamic neurons are barely silent and generally display either a regular bursting discharge or an irregular discharge of action potentials mixing tonic and bursting modes (Glenn et al. 1982; Ramcharan et al. 2005). As we mentioned previously the mode of discharge of thalamic neurons is correlated with the state of arousal, burst being associated to deep sleep and tonic discharge to arousal. If this contrasted description is relevant in the case of the specific thalamic nuclei, it is not as clear cut in non-specific thalamus neurons as they exhibit roughly 15 % of burst discharge during wakefulness (Ramcharan et al. 2005). In drowsy animals, the discharge of a burst of action potential by specific relay cells has been shown to powerfully activate the cortical network (Swadlow and Gusev 2001; Sherman 2001). However, relay cells do not discharge bursts during attentive states in contrary to non-specific thalamic neurons (Glenn et al. 1982; Ramcharan et al. 2005). As a consequence, burst firing may be used by non-specific thalamic neurons to influence cortical arousal. Deciphering the influence of orexin on the discharge of burst of action potentials in non-specific thalamic cells is thus central to understand how non-specific neurons maintain cortical arousal.

Slice electrophysiology provides evidence on the impact of orexin on non-specific thalamic burst firing. In depolarized neurons exhibiting a tonic and regular firing, orexin-induced depolarization promotes T-type channels inactivation, that triggers an increase of action potential frequency but never allows switching to burst firing (Govindaiah and Cox 2006; Kolaj et al. 2007). In bursting cells, orexin either increases the length of the burst and the number of action potentials in each burst but decreases the instantaneous frequency of action potential during the burst or allow switching from burst to tonic discharge (Kolaj et al. 2007). It is important to note that these results were obtained in artificial conditions where neurons are bursting regularly and not in thalamic cells firing irregularly burst and tonic action potentials as observed during wakefulness. This indicates nonetheless that orexin triggers a large variety of responses, which is consistent with the combination of burst and tonic discharge observed in awake animals (Ramcharan et al. 2005). Moreover, orexin is likely to promote tonic action potential discharge in several conditions. Although burst discharge of relay cells has been shown to potentially activate the neocortex, tonic firing is known to promote the desynchronization of the cerebral network. In the case of non-specific thalamic nuclei, tonic discharge has been hypothesized to promote a tonic state of readiness in cortical networks, which is triggered by the depolarization of pyramidal dendritic trees (Steriade and Morin 1981). Indeed, because of their strong projections in layer 1, non-specific inputs impact primarily distal apical dendrite located in this layer. We may hypothesize that the diversity of responses triggered by orexin may account for its role in promoting thalamocortical awakens. However, although we know that non-specific thalamic nuclei are activated by in vivo injection of orexin or Modafinil, an activator of orexin neurons used to treat sleepiness, more investigations are needed to understand the impact of orexin on non-specific thalamic neuron discharge (Date et al. 1999; Scammell et al. 2000).

4.2.2 Antidromic Burst and Synchronization of Cortical Areas

High general attention is sustained by the synchronization of cortical areas that oscillate at high frequency, primarily in the gamma band (~40 Hz, Gray et al.

1989). This process is of considerable importance since a synchronized dialog between areas is necessary to generate an adapted behavioral response to an external stimulus (Saalmann 2014). Synchronization of cortical areas has been well described in visual tasks, where salient or informative clues have to be extracted from the visual background. In these high-demanding tasks, the degree of neuronal population synchrony is correlated to the relevance of the behavioral response (Saalmann 2014). The synchronization of two distinct cortical areas may be achieved either by a direct corticocortical communication or by the involvement of an extracortical structure connected to both of them. Non-specific thalamic nuclei are ideally positioned to synchronize cortical areas since interconnected cortical areas are also indirectly connected through non-specific thalamic nuclei (Berendse and Groenewegen 1991; Groenewegen and Berendse 1994; Van der Werf et al. 2002; Vertes et al. 2006; Jones 2007; Saalmann 2014). Furthermore, stimulation of the reticular formation, which provides the main source of excitation to non-specific thalamic nuclei, promotes the synchronization of cortical areas, suggesting that non-specific thalamic nuclei may also be involved in this process (Munk et al. 1996). Non-specific cells can fire at high frequency, which may trigger coherent cortical oscillations in distant areas and thus facilitate long-range corticocortical interactions. During a complex visual task, neurons of the pulvinar thalamic nucleus, which is involved in high-level processing of visual information, are synchronized and oscillate in the alpha band (8–15 Hz). These thalamic oscillations are shown to drive two cortical areas, which are also involved in processing visual information and in elaborating behavioral responses, thus confirming that thalamic oscillations are sufficient to trigger cortical oscillations and as a consequence to promote their synchronization (Saalmann et al. 2012). Consistently, correlated activity between non-specific thalamic nuclei and their cortical targets have been reported in several other experimental paradigms, suggesting that non-specific thalamic nuclei may act as a conductor of cortical activity (Di Prisco and Vertes 2006; Zhang et al. 2012; Cholvin et al. 2013; Xu and Südhof 2013).

Although there is no direct evidence indicating that orexin promotes the synchronization of cortical areas, infusion of orexin in the prefrontal cortex has been shown to enhance the performance in attentional tasks (Lambe et al. 2005). It has thus been hypothesized that orexin would induce synchronization of cortical areas by the antidromic activation of non-specific thalamic nuclei. Indeed, I mentioned that orexin depolarizes presynaptic terminals of non-specific thalamic fibers in the prefrontal cortex (Lambe and Aghajanian 2003; Lambe et al. 2005). Such presynaptic depolarizations by neuromodulators are sufficient to trigger antidromic action potential and bursts in thalamic relay cells (Gutnick and Prince 1972; Pinault 1995). This suggests that orexin release in the prefrontal cortex may drive antidromic action potential toward the thalamic soma (Lambe and Aghajanian 2003; Lambe et al. 2005). Antidromic bursts may synchronize numerous adjacent thalamic neurons, leading to a synchronized activation of cortical areas. As a consequence, presynaptic activation of thalamocortical neurons by orexin may participate to the synchronization of cortical areas observed during high-demanding processing and may thus explain how orexin perfusion in the prefrontal cortex enhance attention performance.

4.3 Sleep-Wake Transitions: Role of the Non-specific Thalamic Nuclei and Their Modulators

The sleep-wake transition is associated with decreased slow wave amplitude and disappearance of deep sleep markers: K-complex, spindles, δ-oscillations, etc. The progressive disynchronization of the cortical network correlates with the depolarization of specific relay cells and TRN GABAergic neurons (Hirsch et al. 1983; Steriade et al. 1986). In the cortex, the situation is considerably more complex because of the high diversity of cell types and their complex inter-cellular connectivity, but on average cortical neurons are more hyperpolarized during sleep than during active periods (McCormick and von Krosigk 1992). Thalamic cell depolarization is mediated by disinhibition and direct depolarization. Both mechanisms can be mediated by the ascending reticular formation inputs which trigger depolarization via slow activation of the adrenergic receptor $\alpha 2$ and disinhibition via activation of the muscarinic receptor M2 (Moruzzi and Magoun 1949; McCormick and von Krosigk 1992; Pinault and Deschênes 1992). Conversely, the onset of sleep is characterized by a progressive synchronization of the thalamocortical loop and hyperpolarization of thalamic cells, which are both mediated partly by the reduced inputs from the reticular formation (Saper 2010).

However, several lines of evidence indicate that alterations of orexin and non-specific thalamic nuclei activity precedes sleep-wake transitions, suggesting that both structures may promote changes in the thalamocortical loop dynamics and thus may trigger these transitions. Activation of orexin neurons slightly precedes the transition from sleep to wakefulness, measured by the return of muscular tone (Lee et al. 2005). Similarly, in both natural and anesthetics-induced wake-to-sleep transition, deactivation of non-specific thalamic nuclei occurs before any change can be measured in cortical areas or in specific thalamic nuclei (Magnin et al. 2010; Baker et al. 2014). At the onset of sleep, the local field potential power measured in midline thalamic nuclei exhibits an abrupt change at δ-frequency, which is small and insignificant in the other cortical and thalamic areas recorded (Baker et al. 2014). This early transition of non-specific thalamic nuclei activity is likely to be due to dense projections from the brainstem reticular formation and from orexin neurons (Moruzzi and Magoun 1949; Ropert and Steriade 1981; Peyron et al. 1998; Saper et al. 2010). In Humans, the deafferation of the cortical network due to thalamus deactivation seems to be a prerequisite to the fading of consciousness (Magnin et al. 2010). Besides cortical network deafferation, which produces a decrease of cortical excitability, I mentioned previously that nonspecific thalamic nuclei are involved in the synchronization of cortical areas through dense thalamocortical connections. This synchronization of cortical areas is a fundamental process to maintain consciousness (Guillery and Sherman 2002). As a consequence, alteration of non-specific thalamic activity could impact the ability of cortical areas to be synchronized, thus driving transition between one stage to another (Guillery and Sherman 2002).

Other clues are provided by the analysis of brain activity during the onset of anesthesia. Anesthesia can be induced by diverse drugs, which are all aiming at deactivating the arousal system to induce loss of consciousness (Alkire et al. 2008). Similarly to deep sleep, anesthesia triggers slow wave oscillations at low frequency (generally less than 1 Hz). A striking effect produced by anesthetics at the transition between wakefulness and unconsciousness is the reduction of non-specific thalamic metabolism and blood flow, indicating that these nuclei are powerfully impacted by anesthetics (Alkire et al. 2008). Furthermore, the transition of non-specific thalamic nuclei activity in response to anesthetics precedes those of cortical and specific thalamic activity, suggesting that non-specific thalamic nuclei may promote loss of consciousness (Baker et al. 2014). According to this hypothesis, the injection of GABA in the intralaminar nuclei triggers rapid sleep onset whereas injection of nicotine at the same location promotes wakefulness (Miller and Ferrendelli 1990; Alkire et al. 2007). The sensitivity of non-specific thalamic nuclei to anesthetics, and especially to isofluorane is conferred amongst others by the expression of Tasks channels. I already mentioned that Tasks channels play a fundamental role in the regulation of slow oscillation during sleep and that these channels are one of the main targets of orexin-induced depolarization (Doroshenko and Renaud 2009). Thus, interestingly, anesthetics impact the same molecular pathways as orexin and natural sleep.

In conclusion, several mechanisms may describe how non-specific thalamic nuclei, modulated by orexin, could affect the thalamocortical dynamics at the transitions between wakefulness and sleep. A first hypothesis is that the transition of non-specific thalamic activity at sleep onset would impair the interactions between cortical areas, whose desynchronization would drive loss of consciousness. Secondarily, I report that thalamic oscillations are strongly modulated by cortico-thalamic inputs. Indeed, thalamic oscillations are driven by the activation of mGluR receptors which are activated by the layer 6a corticothalamic pyramids. By promoting the synchronized activation of layer 6a cells, which are main targets of non-specific thalamic nuclei, these nuclei may promote the switch to a bursting mode in specific thalamic neurons.

5 Conclusion

We have seen that orexin acts differentially on the two main thalamocortical pathways. Indeed, orexin potentiates the non-specific thalamocortical pathway while it does not directly interfere with the transfer of sensory information. Modulation of non-specific thalamic nuclei impacts both sleep-wake transitions and attention, confirming the important role of these nuclei in the regulation of vigilance states. Furthermore, I have pointed out the strong impact of non-specific thalamic and orexin inputs in the prefrontal cortices, a region which processes are highly dependent on arousal. The sum of these observations indicates first that non-specific thalamic nuclei act in an orexin-dependent manner on the broad cortical activation, which may indirectly affect the specific thalamocortical oscillations and thus trigger sleep-wake transitions. Second, it suggests a restricted stronger influence of orexin-dependent non-specific inputs to the prefrontal cortex that may be involved in the precise modulation of attention. In conclusion, although numerous other structures are involved in the regulation of arousal states, the non-specific thalamic nuclei thanks to their widespread projections to cortical areas and to the strong orexin inputs they receive, are ideally positioned to control cortical arousal states.

Acknowledgements I thank Bénédicte Babayan, Piotr Tourkine, Esther Klingler and Agnès Bonnot for their valuable help. This work was supported by the Herchel Smith Found of Cambridge, UK.

References

Alkire MT, McReynolds JR, Hahn EL, Trivedi AN (2007) Thalamic microinjection of nicotine reverses sevoflurane-induced loss of righting reflex in the rat. Anesthesiology 107:264–272

Alkire MT, Hudetz AG, Tononi G (2008) Consciousness and anesthesia. Science 322:876–880

Alvarez CE, Sutcliffe JG (2002) Hypocretin is an early member of the incretin gene family. Neurosci Lett 324:169–172

Andjelic S, Gallopin T, Cauli B, Hill EL, Roux L, Badr S, Hu E, Tamàs G, Lambolez B (2009) Glutamatergic nonpyramidal neurons from neocortical layer VI and their comparison with pyramidal and spiny stellate neurons. J Neurophysiol 101:641–654

Aracri P, Banfi D, Pasini ME, Amadeo A, Becchetti A (2013) Hypocretin (orexin) regulates glutamate input to fast-spiking interneurons in layer v of the fr2 region of the murine prefrontal cortex. Cereb Cortex 25:1330–1347

Arimatsu Y, Ishida M, Kaneko T, Ichinose S, Omori A (2003) Organization and development of corticocortical associative neurons expressing the orphan nuclear receptor. J Comp Neurol 466:180–196

Baker R, Gent TC, Yang Q, Parker S, Vyssotski AL, Wisden W, Brickley SG, Franks NP (2014) Altered activity in the central medial thalamus precedes changes in the neocortex during transitions into both sleep and propofol anesthesia. J Neurosci 34:13326–13335

Bayer L, Eggermann E, Saint-Mleux B, Machard D, Jones BE, Muhlethaler M, Serafin M (2002) Selective action of orexin (hypocretin) on nonspecific thalamocortical projection neurons. J Neurosci 22:7835–7839

Bayer L, Serafin M, Eggermann E, Saint-Mleux B, Machard D, Jones BE, Muhlethaler M (2004) Exclusive postsynaptic action of hypocretin-orexin on sublayer 6b cortical neurons. J Neurosci 24:6760–6764

Berendse HW, Groenewegen HJ (1991) Restricted cortical termination fields of the midline and intralaminar thalamic nuclei in the rat. Neuroscience 42:73–102

Beuckmann CT, Yanagisawa M (2002) Orexins: from neuropeptides to energy homeostasis and sleep/wake regulation. J Mol Med 80:329–342

Bloem B, Poorthuis RB, Mansvelder HD (2014) Cholinergic modulation of the medial prefrontal cortex: the role of nicotinic receptors in attention and regulation of neuronal activity. Front Neural Circuits 8:17

Carrera E, Bogousslavsky J (2006) The thalamus and behavior: effects of anatomically distinct strokes. Neurology 66:1817–1823

Carter ME, Adamantidis A, Ohtsu H, Deisseroth K, de Lecea L (2009) Sleep homeostasis modulates hypocretin-mediated sleep-to-wake transitions. J Neurosci 29:10939–10949

Carter ME, Brill J, Bonnavion P, Huguenard JR, Huerta R, de Lecea L (2012) Mechanism for hypocretin-mediated sleep-to-wake transitions. Proc Natl Acad Sci USA 109:E2635–E2644

Cassel JC, Pereira de Vasconcelos A, Loureiro M, Cholvin T, Dalrymple-Alford JC, Vertes RP (2013) The reuniens and rhomboid nuclei: neuroanatomy, electrophysiological characteristics and behavioral implications. Prog Neurobiol 111:34–52

Castaigne P, Buge A, Escourolle R, Masson M (1962) Ramollissement pédunculaire médian, tegmento-thalamique avec ophtalmoplégie et hypersomnie. Rev Neurol (Paris) 106:357–367

Chen CT, Dun SL, Kwok EH, Dun NJ, Chang JK (1999) Orexin a-like immunoreactivity in the rat brain. Neurosci Lett 260:161–164

Cholvin T, Loureiro M, Cassel R, Cosquer B, Geiger K, De Sa Nogueira D, Raingard H, Robelin L, Kelche C, Pereira de Vasconcelos A, Cassel JC (2013) The ventral midline thalamus contributes to strategy shifting in a memory task requiring both prefrontal cortical and hippocampal functions. J Neurosci 33:8772–8783

Cid-Pellitero Ed, Garzon M (2011) Hypocretin1/OrexinA axon targeting of laterodorsal tegmental nucleus neurons projecting to the rat medial prefrontal cortex. Cereb Cortex 21:2762–2773

Clancy B, Cauller LJ (1999) Widespread projections from subgriseal neurons (layer VII) to layer i in adult rat cortex. J Comp Neurol 407:275–286

Clasca F, Rubio-Garrido P, Jabaudon D (2012) Unveiling the diversity of thalamocortical neuron subtypes. Eur J Neurosci 35:1524–1532

Cluderay JE, Harrison DC, Hervieu GJ (2002) Protein distribution of the orexin-2 receptor in the rat central nervous system. Regul Pept 104:131–144

Contreras D, Steriade M (1995) Cellular basis of EEG slow rhythms: a study of dynamic corticothalamic relationships. J Neurosci 15:604–622

Cruikshank SJ, Lewis TJ, Connors BW (2007) Synaptic basis for intense thalamocortical activation of feedforward inhibitory cells in neocortex. Nat Neurosci 10:462–468

Cruikshank SJ, Ahmed OJ, Stevens TR, Patrick SL, Gonzalez AN, Elmaleh M, Connors BW (2012) Thalamic control of layer 1 circuits in prefrontal cortex. J Neurosci 32:17813–17823

Crunelli V, Hughes SW (2010) The slow (<1 hz) rhythm of non-REM sleep: a dialogue between three cardinal oscillators. Nat Neurosci 13:9–17

Cutler DJ, Morris R, Sheridhar V, Wattam TAK, Holmes S, Patel S, Arch JRS, Wilson S, Buckingham RE, Evans ML, Leslie RA, Williams G (1999) Differential distribution of orexin-a and orexin-b immunoreactivity in the rat brain and spinal cord. Peptides 20:1455–1470

Date Y, Ueta Y, Yamashita H, Yamaguchi H, Matsukura S, Kangawa K, Sakurai T, Yanagisawa M, Nakazato M (1999) Orexins, orexigenic hypothalamic peptides, interact with autonomic, neuroendocrine and neuroregulatory systems. Proc Natl Acad Sci USA 96:748–753

David FC, Schmiedt JT, Taylor HL, Orban G, Giovanni GD, Uebele VN, Renger JJ, Lambert RC, Leresche N, Crunelli V (2013) Essential thalamic contribution to slow waves of natural sleep. J Neurosci 33:19599–19610

Del Cid-Pellitero E, Garzon M (2011a) Hypocretin1/OrexinA-containing axons innervate locus coeruleus neurons that project to the rat medial prefrontal cortex. Implication in the sleep-wakefulness cycle and cortical activation. Synapse 65:843–857

Del Cid-Pellitero E, Garzon M (2011b) Medial prefrontal cortex receives input from dorsal raphe nucleus neurons targeted by hypocretin1/orexinA-containing axons. Neuroscience 172:30–43

Dempsey EW, Morison RS (1942) The production of rhythmically recurrent cortical potentials after localized thalamic stimulation. Amer J Physiol 135:293–300

Dempsey EW, Morison RS, Morison BR (1941) Some afferent diencephalic pathways related to cortical potentials in the cat. Amer J Physiol 131:718–732

Deutch AY, Ongür D, Duman RS (1995) Antipsychotic drugs induce fos protein in the thalamic paraventricular nucleus: a novel locus of antipsychotic drug action. Neuroscience 66:337–346

Di Prisco GV, Vertes RP (2006) Excitatory actions of the ventral midline thalamus (rhomboid/reuniens) on the medial prefrontal cortex in the rat. Synapse 60:45–55
Doroshenko P, Renaud LP (2009) Acid-sensitive TASK-like k+ conductances contribute to resting membrane potential and to orexin-induced membrane depolarization in rat thalamic paraventricular nucleus neurons. Neuroscience 158:1560–1570
Erisir A, Van Horn SC, Sherman SM (1997) Relative numbers of cortical and brainstem inputs to the lateral geniculate nucleus. Proc Natl Acad Sci USA 94:1517–1520
Euston DR, Gruber AJ, McNaughton BL (2012) The role of medial prefrontal cortex in memory and decision making. Neuron 76:1057–1070
Façon E, Steriade M, Wertheim N (1958) Prolonged hypersomnia caused by bilateral lesions of the medial activator system; thrombotic syndrome of the bifurcation of the basilar trunk. Revue Neurologique 98:117–133
Fadel J, Bubser M, Deutch AY (2002) Differential activation of orexin neurons by antipsychotic drugs associated with weight gain. J Neurosci 22:6742–6746
Funahashi S, Bruce CJ, Goldman-Rakic PS (1989) Mnemonic coding of visual space in the monkey's dorsolateral prefrontal cortex. J Neurophysiol 61:331–349
Giménez-Amaya JM, McFarland NR, de las Heras S, Haber SN (1995) Organization of thalamic projections to the ventral striatum in the primate. J Comp Neurol 354:127–149
Glenn LL, Hada J, Roy JP, Descheènes M, Steriade M (1982) Anterograde tracer and field potential analysis of the neocortical layer I projection from nucleus ventralis medialis of the thalamus in cat. Neuroscience 7:1861–1877
Govindaiah G, Cox CL (2006) Modulation of thalamic neuron excitability by orexins. Neuropharmacology 51:414–425
Gray CM, König P, Engel AK, Singer W (1989) Oscillatory responses in cat visual cortex exhibit inter-columnar synchronization which reflects global stimulus properties. Nature 338:334–337
Groenewegen HJ, Berendse HW (1994) The specificity of the 'nonspecific' midline and intralaminar thalamic nuclei. Trends Neurosci 17:52–57
Guillery RW, Sherman SM (2002) Thalamic relay functions and their role in corticocortical communication: generalizations from the visual system. Neuron 33:163–175
Gutierrez GJ, O'Leary T, Marder E (2013) Multiple mechanisms switch an electrically coupled, synaptically inhibited neuron between competing rhythmic oscillators. Neuron 77:845–858
Gutnick MJ, Prince DA (1972) Thalamocortical relay neurons: antidromic invasion of spikes from a cortical epileptogenic focus. Science 176:424–426
Hay YA, Andjelic S, Badr S, Lambolez B (2014) Orexin-dependent activation of layer VIb enhances cortical network activity and integration of non-specific thalamocortical inputs. Brain Struct Funct
Hembrook JR, Mair RG (2011) Lesions of reuniens and rhomboid thalamic nuclei impair radial maze win-shift performance. Hippocampus 21:815–826
Hembrook JR, Onos KD, Mair RG (2012) Inactivation of ventral midline thalamus produces selective spatial delayed conditional discrimination impairment in the rat. Hippocampus 22:853–860
Herkenham M (1980) Laminar organization of thalamic projections to the rat neocortex. Science 207:532–535
Hervieu GJ, Cluderay JE, Harrison DC, Roberts JC, Leslie RA (2001) Gene expression and protein distribution of the orexin-1 receptor in the rat brain and spinal cord. Neuroscience 103:777–797
Hirsch JC, Fourment A, Marc ME (1983) Sleep-related variations of membrane potential in the lateral geniculate body relay neurons of the cat. Brain Res 259:308–312
Hsu DT, Price JL (2009) Paraventricular thalamic nucleus: subcortical connections and innervation by serotonin, orexin, and corticotropin-releasing hormone in macaque monkeys. J Comp Neurol 512:825–848
Huang H, Ghosh P, Pol ANvd (2006) Prefrontal cortex-projecting glutamatergic thalamic paraventricular nucleus-excited by hypocretin: a feedforward circuit that may enhance cognitive arousal. J Neurophysiol 95:1656–1668

Huesa G, van den Pol AN, Finger TE (2005) Differential distribution of hypocretin (orexin) and melanin-concentrating hormone in the goldfish brain. J Comp Neurol 488:476–491

Hyvarinen J, Poranen A, Jokinen Y (1980) Influence of attentive behavior on neuronal responses to vibration in primary somatosensory cortex of the monkey. J Neurophysiol 43:870–882

Ishibashi M, Takano S, Yanagida H, Takatsuna M, Nakajima K, Oomura Y, Wayner MJ, Sasaki K (2005) Effects of orexins/hypocretins on neuronal activity in the paraventricular nucleus of the thalamus in rats in vitro. Peptides 26:471–481

Jones E (2007) Principles of thalamic organization. In: The thalamus, 2nd edn. Cambridge University Press, Cambridge, pp 87–170

Jones EG, Leavitt RY (1974) Retrograde axonal transport and the demonstration of non-specific projections to the cerebral cortex and striatum from thalamic intralaminar nuclei in the rat, cat and monkey. J Comp Neurol 154:349–377

Kahn JB, Ward RD, Kahn LW, Rudy NM, Kandel ER, Balsam PD, Simpson EH (2012) Medial prefrontal lesions in mice impair sustained attention but spare maintenance of information in working memory. Learn Mem 19:513–517

Kinomura S, Larsson J, Gulyas B, Roland PE (1996) Activation by attention of the human reticular formation and thalamic intralaminar nuclei. Science 271:512–515

Kirouac GJ, Parsons MP, Li S (2005) Orexin (hypocretin) innervation of the paraventricular nucleus of the thalamus. Brain Res 1059:179–188

Kiyashchenko LI, Mileykovskiy BY, Maidment N, Lam HA, Wu MF, John J, Peever J, Siegel JM (2002) Release of hypocretin (orexin) during waking and sleep states. J Neurosci 22:5282–5286

Kolaj M, Doroshenko P, Yan Cao X, Coderre E, Renaud LP (2007) Orexin-induced modulation of state-dependent intrinsic properties in thalamic paraventricular nucleus neurons attenuates action potential patterning and frequency. Neuroscience 147:1066–1075

Krettek JE, Price JL (1977) The cortical projections of the mediodorsal nucleus and adjacent thalamic nuclei in the rat. J Comp Neurol 171:157–191

Lambe EK, Aghajanian GK (2003) Hypocretin (orexin) induces calcium transients in single spines postsynaptic to identified thalamocortical boutons in prefrontal slice. Neuron 40:139–150

Lambe EK, Olausson P, Horst NK, Taylor JR, Aghajanian GK (2005) Hypocretin and nicotine excite the same thalamocortical synapses in prefrontal cortex: correlation with improved attention in rat. J Neurosci 25:5225–5229

Lambe EK, Liu RJ, Aghajanian GK (2007) Schizophrenia, hypocretin (orexin), and the thalamocortical activating system. Schizophr Bull 33:1284–1290

Lee HS (2005) Hypothalamic orexin projections to midline thalamic nuclei in the rat. Integr Biosci 9:145–152

Lee MG, Hassani OK, Jones BE (2005) Discharge of identified orexin/hypocretin neurons across the sleep-waking cycle. J Neurosci 25:6716–6720

Li Y, Li S, Sui N, Kirouac GJ (2009) Orexin-a acts on the paraventricular nucleus of the midline thalamus to inhibit locomotor activity in rats. Pharmacol Biochem Behav 93:506–514

Li B, Chen F, Ye J, Chen X, Yan J, Li Y, Xiong Y, Zhou Z, Xia J, Hu Z (2010) The modulation of orexin a on HCN currents of pyramidal neurons in mouse prelimbic cortex. Cereb Cortex 20:1756–1767

Llinas R, Ribary U, Contreras D, Pedroarena C (1998) The neuronal basis for consciousness. Philos Transc R Soc Lond B, Bio Sci 353:1841–1849

Lopez JM, Dominguez L, Moreno N, Gonzalez A (2009a) Comparative immunohistochemical analysis of the distribution of orexins (hypocretins) in the brain of amphibians. Peptides 30:873–887

Lopez JM, Dominguez L, Moreno N, Morona R, Joven A, Gonzalez A (2009b) Distribution of orexin/hypocretin immunoreactivity in the brain of the lungfishes protopterus dolloi and neoceratodus forsteri. Brain Behav Evol 74:302–322

Lorente de No R (1938) Cerebral cortex: architecture. In: Physiologie of the nervous system. Oxford University Press, Oxford, pp 274–313

Loureiro M, Cholvin T, Lopez J, Merienne N, Latreche A, Cosquer B, Geiger K, Kelche C, Cassel JC, Pereira de Vasconcelos A (2012) The ventral midline thalamus (reuniens and rhomboid nuclei) contributes to the persistence of spatial memory in rats. J Neurosci 32:9947–9959

Lu XY, Bagnol D, Burke S, Akil H, Watson SJ (2000) Differential distribution and regulation of OX1 and OX2 orexin/hypocretin receptor messenger RNA in the brain upon fasting. Horm Behav 37:335–344

Macchi G, Quattrini A, Chinzari P, Marchesi G, Capocchi G (1975) Quantitative data on cell loss and cellular atrophy of intralaminar nuclei following cortical and subcortical lesions. Brain Res 89:43–59

Macchi G, Bentivoglio M, D'Atena C, Rossini P, Tempesta E (1977) The cortical projections of the thalamic intralaminar nuclei restudied by means of the HRP retrograde axonal transport. Neurosci Lett 4:121–126

Magnin M, Rey M, Bastuji H, Guillemant P, Fc Mauguière, Garcia-Larrea L (2010) Thalamic deactivation at sleep onset precedes that of the cerebral cortex in humans. Proc Natl Acad Sci USA 107:3829–3833

Marcus JN, Aschkenasi CJ, Lee CE, Chemelli RM, Saper CB, Yanagisawa M, Elmquist JK (2001) Differential expression of orexin receptors 1 and 2 in the rat brain. J Comp Neurol 435:6–25

Marx M, Feldmeyer D (2013) Morphology and physiology of excitatory neurons in layer 6b of the somatosensory rat barrel cortex. Cereb Cortex 23:2803–2817

McCormick DA, von Krosigk M (1992) Corticothalamic activation modulates thalamic firing through glutamate "metabotropic" receptors. Proc Natl Acad Sci USA 89:2774–2778

McGranaghan PA, Piggins HD (2001) Orexin a-like immunoreactivity in the hypothalamus and thalamus of the syrian hamster (*Mesocricetus auratus*) and siberian hamster (*Phodopus sungorus*), with special reference to circadian structures. Brain Res 904:234–244

Meuth SG, Kanyshkova T, Meuth P, Landgraf P, Munsch T, Ludwig A, Hofmann F, Pape HC, Budde T (2006) Membrane resting potential of thalamocortical relay neurons is shaped by the interaction among TASK3 and HCN2 channels. J Neurophysiol 96:1517–1529

Miller JW, Ferrendelli JA (1990) Characterization of GABAergic seizure regulation in the midline thalamus. Neuropharmacology 29:649–655

Mintz EM, van den Pol AN, Casano AA, Albers HE (2001) Distribution of hypocretin-(orexin) immunoreactivity in the central nervous system of syrian hamsters (*Mesocricetus auratus*). J Chem Neuroanat 21:225–238

Mishima K, Fujiki N, Yoshida Y, Sakurai T, Honda M, Mignot E, Nishino S (2008) Hypocretin receptor expression in canine and murine narcolepsy models and in hypocretin-ligand deficient human narcolepsy. Sleep 31:1119–1126

Mitchell AS, Chakraborty S (2013) What does the mediodorsal thalamus do? Front Syst Neurosci 7:37

Miyata M, Imoto K (2006) Different composition of glutamate receptors in corticothalamic and lemniscal synaptic responses and their roles in the firing responses of ventrobasal thalamic neurons in juvenile mice. J Physiol 575:161–174

Mondal MS, Nakazato M, Date Y, Murakami N, Hanada R, Sakata T, Matsukura S (1999a) Characterization of orexin-a and orexin-b in the microdissected rat brain nuclei and their contents in two obese rat models. Neurosci Lett 273:45–48

Mondal MS, Nakazato M, Date Y, Murakami N, Yanagisawa M, Matsukura S (1999b) Widespread distribution of orexin in rat brain and its regulation upon fasting. Biochem Biophys Res Commun 256:495–499

Morison RS, Dempsey EW (1942) A study of thalamo-cortical relations. Amer J Physiol 135:281–292

Morison RS, Dempsey EW, Morison BR (1941) Cortical responses from electrical stimulation of the brain stem. Amer J Physiol 131:732–743

Moruzzi G, Magoun HW (1949) Brain stem reticular formation and activation of the EEG. Electroencephalogr Clin Neurophysiol 1:455–473

Muir JL, Everitt BJ, Robbins TW (1996) The cerebral cortex of the rat and visual attentional function: dissociable effects of mediofrontal, cingulate, anterior dorsolateral, and parietal cortex lesions on a five-choice serial reaction time task. Cereb Cortex 6:470–481

Munk MH, Roelfsema PR, König P, Engel AK, Singer W (1996) Role of reticular activation in the modulation of intracortical synchronization. Science 272:271–274

Nambu T, Sakurai T, Mizukami K, Hosoya Y, Yanagisawa M, Goto K (1999) Distribution of orexin neurons in the adult rat brain. Brain Res 827:243–260

Naumann A, Bierbrauer J, Przuntek H, Daum I (2001) Attentive and preattentive processing in narcolepsy as revealed by event-related potentials (ERPs). NeuroReport 12:2807–2811

Novak CM, Nunez AA (1998) Daily rhythms in fos activity in the rat ventrolateral preoptic area and midline thalamic nuclei. Am J Physiol 275:R1620–R1626

Peyron C, Tighe DK, van den Pol AN, de Lecea L, Heller HC, Sutcliffe JG, Kilduff TS (1998) Neurons containing hypocretin (orexin) project to multiple neuronal systems. J Neurosci 18:9996–10015

Pinault D (1995) Backpropagation of action potentials generated at ectopic axonal loci: hypothesis that axon terminals integrate local environmental signals. Brain Res Rev 21:42–92

Pinault D, Deschênes M (1992) Muscarinic inhibition of reticular thalamic cells by basal forebrain neurones. NeuroReport 3:1101–1104

Pol ANvd, Gao XB, Obrietan K, Kilduff TS, Belousov AB (1998) Presynaptic and postsynaptic actions and modulation of neuroendocrine neurons by a new hypothalamic peptide, hypocretin/orexin. J Neurosci 18:7962–7971

Ramcharan EJ, Gnadt JW, Sherman SM (2005) Higher-order thalamic relays burst more than first-order relays. PNAS 102:12236–12241

Reppert SM, Weaver DR (2002) Coordination of circadian timing in mammals. Nature 418:935–941

Rieger M, Mayer G, Gauggel S (2003) Attention deficits in patients with narcolepsy. Sleep 26:36–43

Ropert N, Steriade M (1981) Input-output organization of midbrain reticular core. J Neurophysiol 46:17–31

Saalmann YB (2014) Intralaminar and medial thalamic influence on cortical synchrony, information transmission and cognition. Front Syst Neurosci 8:83

Saalmann YB, Pinsk MA, Wang L, Li X, Kastner S (2012) The pulvinar regulates information transmission between cortical areas based on attention demands. Science 337:753–756

Sakurai T, Amemiya A, Ishii M, Matsuzaki I, Chemelli RM, Tanaka H, Williams SC, Richarson JA, Kozlowski GP, Wilson S, Arch JR, Buckingham RE, Haynes AC, Carr SA, Annan RS, McNulty DE, Liu WS, Terrett JA, Elshourbagy NA, Bergsma DJ, Yanagisawa M, (1998) Orexins and orexin receptors: a family of hypothalamic neuropeptides and g protein-coupled receptors that regulate feeding behavior. Cell 92, 1 page following 696

Saletu M, Anderer P, Saletu-Zyhlarz GM, Mandl M, Saletu B, Zeitlhofer J (2009) Modafinil improves information processing speed and increases energetic resources for orientation of attention in narcoleptics: double-blind, placebo-controlled ERP studies with low-resolution brain electromagnetic tomography (LORETA). Sleep Med 10:850–858

Saper CB, Fuller PM, Pedersen NP, Lu J, Scammell TE (2010) Sleep state switching. Neuron 68:1023–1042

Scammell TE, Estabrooke IV, McCarthy MT, Chemelli RM, Yanagisawa M, Miller MS, Saper CB (2000) Hypothalamic arousal regions are activated during modafinil-induced wakefulness. J Neurosci 20:8620–8628

Schiff ND, Giacino JT, Kalmar K, Victor JD, Baker K, Gerber M, Fritz B, Eisenberg B, Biondi T, O'Connor J, Kobylarz EJ, Farris S, Machado A, McCagg C, Plum F, Fins JJ, Rezai AR (2007) Behavioural improvements with thalamic stimulation after severe traumatic brain injury. Nature 448:600–603

Schmahmann JD (2003) Vascular syndromes of the thalamus. Stroke 34:2264–2278

Sherman SM (2001) Tonic and burst firing: dual modes of thalamocortical relay. Trends Neurosci 24:122–126

Sherman SM, Guillery RW (1996) Functional organization of thalamocortical relays. J Neurophy 76:1367–1395

Shibahara M, Sakurai T, Nambu T, Takenouchi T, Iwaasa H, Egashira SI, Ihara M, Goto K (1999) Structure, tissue distribution, and pharmacological characterization of xenopus orexins. Peptides 20:1169–1176

Singletary KG, Deviche P, Strand C, Delville Y (2006) Distribution of orexin/hypocretin immunoreactivity in the brain of a male songbird, the house finch, *Carpodacus mexicanus*. J Chem Neuroanat 32:81–89

Song CH, Xia JX, Ye JN, Chen XW, Zhang CQ, Zhang QC, Gao EQ, Hu ZA (2005) Signaling pathways of hypocretin-1 actions on pyramidal neurons in the rat prefrontal cortex. NeuroReport 16:1529–1533

Steriade M (1997) Synchronized activities of coupled oscillators in the cerebral cortex and thalamus at different levels of vigilance. Cereb Cortex 7:583–604

Steriade M, Glenn LL (1982) Neocortical and caudate projections of intralaminar thalamic neurons and their synaptic excitation from midbrain reticular core. J Neurophysiol 48:352–371

Steriade M, Llinas RR (1988) The functional states of the thalamus and the associated neuronal interplay. Physiol Rev 68:649–742

Steriade M, Morin D (1981) Reticular influences on primary and augmenting responses in the somatosensory cortex. Brain Res 205:67–80

Steriade M, Domich L, Oakson G (1986) Reticularis thalami neurons revisited: activity changes during shifts in states of vigilance. J Neurosci 6:68–81

Steriade M, McCormick DA, Sejnowski TJ (1993) Thalamocortical oscillations in the sleeping and aroused brain. Science 262:679–685

Suzuki H, Takemoto Y, Yamamoto T (2007) Differential distribution of orexin-a-like and orexin receptor 1 (OX1r)-like immunoreactivities in the xenopus pituitary. Tissue Cell 39:423–430

Swadlow HA (2003) Fast-spike interneurons and feedforward inhibition in awake sensory neocortex. Cereb Cortex 13:25–32

Swadlow HA, Gusev AG (2001) The impact of 'bursting' thalamic impulses at a neocortical synapse. Nat Neurosci 4:402–408

Tinuper P, Montagna P, Medori R, Cortelli P, Zucconi M, Baruzzi A, Lugaresi E (1989) The thalamus participates in the regulation of the sleep-waking cycle. A clinico-pathological study in fatal familial thalamic degeneration. Electroencephalogr Clin Neurophysiol 73:117–123

Trivedi P, Yu H, MacNeil DJ, Ploeg LHVd, Guan XM (1998) Distribution of orexin receptor mRNA in the rat brain. FEBS Lett 438:71–75

Turner JP, Salt TE (1998) Characterization of sensory and corticothalamic excitatory inputs to rat thalamocortical neurones in vitro. J Physiol 510(Pt 3):829–843

Van der Werf YD, Witter MP, Groenewegen HJ (2002) The intralaminar and midline nuclei of the thalamus. Anatomical and functional evidence for participation in processes of arousal and awareness. Brain Res Rev 39:107–140

Vann SD, Brown MW, Aggleton JP (2000) Fos expression in the rostral thalamic nuclei and associated cortical regions in response to different spatial memory tests. Neuroscience 101:983–991

Varela C (2014) Thalamic neuromodulation and its implications for executive networks. Front Neural Circuits 8:69

Vertes RP, Hoover WB, Do Valle AC, Sherman A, Rodriguez JJ (2006) Efferent projections of reuniens and rhomboid nuclei of the thalamus in the rat. J Comp Neurol 499:768–796

Xia J, Chen X, Song C, Ye J, Yu Z, Hu Z (2005) Postsynaptic excitation of prefrontal cortical pyramidal neurons by hypocretin-1/orexin a through the inhibition of potassium currents. J Neurosci Res 82:729–736

Xu W, Südhof TC (2013) A neural circuit for memory specificity and generalization. Science 339:1290–1295

Yoshida Y, Fujiki N, Nakajima T, Ripley B, Matsumura H, Yoneda H, Mignot E, Nishino S (2001) Fluctuation of extracellular hypocretin-1 (orexin a) levels in the rat in relation to the light-dark cycle and sleep-wake activities. Eur J Neurosci 14:1075–1081

Zhang JH, Sampogna S, Morales FR, Chase MH (2002) Co-localization of hypocretin-1 and hypocretin-2 in the cat hypothalamus and brainstem. Peptides 23:1479–1483
Zhang JH, Sampogna S, Morales FR, Chase MH (2004) Distribution of hypocretin (orexin) immunoreactivity in the feline pons and medulla. Brain Res 995:205–217
Zhang Y, Yoshida T, Katz DB, Lisman JE (2012) NMDAR antagonist action in thalamus imposes delta oscillations on the hippocampus. J Neurophysiol 107:3181–3189

Orexin, Alcohol and Sleep Homeostasis

Rishi Sharma, Pradeep Sahota and Mahesh M. Thakkar

Abstract Alcoholism is a third leading preventable cause of death in the United States. One of the most significant and severe consequences of alcohol use is its effects on sleep. Strong and convincing evidences suggest that alcohol-induced sleep disruptions may have a major role towards the development of alcoholism. However, very little is known about how and where alcohol acts to affect sleep. Adenosine (AD) is strongly implicated in sleep promotion as well as in mediating alcohol-induced sleep. Since orexinergic system plays a pivotal role in the initiation and maintenance of wakefulness, we hypothesized that alcohol may act via AD to inhibit the orexinergic system and promote sleep. In this chapter, we have described a series of experiments performed on Sprague-Dawley rats to test our hypothesis. Our first group of experiments suggests that adenosinergic mechanism in the perifornical hypothalamus is involved in spontaneous as well as recovery sleep after prolonged wakefulness. Our second group of experiments suggests that alcohol increases extracellular levels of AD in the orexin-rich perifornical hypothalamus, increased AD activates A1 receptors on orexinergic neurons and inhibits them, and inhibition of orexinergic neurons may result in sleep promotion. In our third group of experiments, we have shown that alcohol withdrawal-induced sleep disruptions are associated with downregulation of orexin gene expression suggesting a crucial link between orexinergic system and alcohol induced sleep disruptions. In conclusion, the results of our studies suggest that interaction between adenosinergic mechanism and the orexinergic system in the perifornical hypothalamus is crucial for alcohol's effects on sleep.

R. Sharma · P. Sahota · M.M. Thakkar (✉)
Harry S. Truman Memorial Veterans Hospital and Department of Neurology, University of Missouri, Research, Room A023, 800 Hospital Drive, Columbia, MO 65201, USA
e-mail: thakkarmm@health.missouri.edu

© Springer International Publishing Switzerland 2015

T. Sakurai et al. (eds.), *Orexin and Sleep*,
DOI 10.1007/978-3-319-23078-8_9

1 Introduction

The use of alcoholic beverages has been an integral part of social culture since the dawn of human civilization (Keller 1979). However, it was only in the 19th century when alcohol addiction or compulsive usage of alcohol was considered as a disease and the term "alcoholism" was coined (Levine 1978; Gunzerath et al. 2011). While the Diagnostic and Statistical Manual-IV (DSM-IV) has classified alcoholism into two distinct disorders: alcohol abuse and alcohol dependence, recently issued DSM-5 integrates the two DSM–IV disorders into a single disorder called as "alcohol use disorder" with mild, moderate, and severe sub-classifications (National Institute on Alcohol Abuse and Alcoholism 2013). Since alcoholism and alcohol use disorder are synonymous and alcoholism is widely used, we will use "alcoholism" in this chapter.

Alcoholism is highly prevalent in our society with high lifetime prevalence of more than 30 % (Hasin et al. 2007). According to the CDC, alcoholism is the third leading preventable cause of death in the United States, with an average of 88,000 deaths each year (CDC 2006, 2012). The societal cost of alcoholism in 2006 was estimated to be $223.5 billion (Bouchery et al. 2011; Stahre et al. 2014). In addition, alcoholism is also associated with a myriad of health and social problems including automobile accidents, violence, new HIV infections, accidents and unintended pregnancies (Esser et al. 2014). Thus, alcoholism is a severe debilitating disease that has profound socio-economic impact.

Strong and convincing evidence exist to suggest that alcohol induced sleep disruptions may have a major role towards the development of alcoholism.

Acute alcohol intake in healthy non-alcoholics promotes sleep by reducing sleep onset latency (SOL, time to fall asleep) and increasing the quality (increased delta power) and quantity of non-rapid eye movement (NREM) sleep (Ebrahim et al. 2013; Roehrs and Roth 2001). It is this "deceptive" NREM sleep promoting effects of alcohol which makes it an extensively used "over the counter" sleep aid (Roehrs and Roth 2012). Deceptive, because alcohol's NREM sleep promoting effect is short-lived and fades away rapidly resulting in sleep disruptions during the latter half of nocturnal sleep time (Roehrs and Roth 2001). Alcoholics, both during drinking period and during abstinences suffer from multitude of sleep disruptions (Brower 2001). During abstinence, recovering alcoholics suffer from severe and protracted sleep disruptions manifested by profound insomnia, increased REM sleep coupled with excessive daytime sleepiness (Brower 2001). Furthermore, subjective and objective indicators of sleep disturbances are predictors of relapse (Brower and Perron 2009). Finally within USA, it is estimated that the cost of alcohol-related sleep problems exceeds $22 billion (Bouchery et al. 2011; Stoller 1994). Thus, although sleep problems associated with alcohol have significant economic and clinical consequences, very little is known about how and where alcohol acts to affect sleep.

In this chapter, we have described our attempts to understand how and where alcohol acts to disrupt sleep. We will begin this chapter by providing a brief description about sleep regulation and the role of orexin in the sleep followed by a brief review of human and animal studies describing the effects of acute and chronic alcohol on sleep-wakefulness. Finally, we will describe our results focused on describing the role of orexins in alcohol induced sleep-disruptions.

2 Fundamentals of Sleep-Wakefulness

Sleep has always fascinated mankind. There is a myriad of treatises and reviews, scientific and non-scientific, trying to explain the phenomenon of sleep, yet, none have been comprehensive enough to gain general acceptance. It is now well established that sleep is neither a unitary nor a passive process rather; intricate neuronal systems via complex mechanism are responsible for the initiation and maintenance of sleep. In this section, we have highlighted some aspects of sleep regulation pertaining to this review. An interested reader is recommended to recent reviews (Brown et al. 2012; Datta and Maclean 2007).

Sleep is defined as a rapidly reversible state of immobility and greatly reduced sensory responsiveness. An important further criterion is that sleep is homeostatically regulated, i.e. lost sleep is made up with an increased 'drive' for sleep and a consequent 'sleep rebound'. Sleep is not a homogenous state, rather a continuum of mixed states, broadly divided into: non-rapid eye movement (NREM) sleep and rapid eye movement (REM) sleep. A combination of electrophysiological measurements including (1) the electroencephalogram (EEG), which traces the electrical activity of the brain, (2) the electro-oculogram (EOG), which measures eye movements, and, (3) the electromyogram (EMG), which measures the electrical activity of muscles are used to objectively identify different states of sleep and wakefulness (Datta and Maclean 2007).

2.1 Physiology of Sleep-Wakefulness

The state of wakefulness is characterized by the presence of low voltage, high frequency (>15 Hz) waves in the EEG, REMs in the EOG and high amplitude activity in the EMG. The onset of sleep is heralded by slowing down of the brain activity. In humans, NREM sleep is divided into three stages: *Stage I* is characterized by relatively low voltage, mixed frequency activity (3–7 Hz) and vertex sharp waves in the EEG. *Stage II* is characterized by slow (<1 Hz) oscillation with distinctive sleep spindles (waxing and waning of 12–14 Hz waves lasting between 0.5 and 1.0 s) and K-complex (a negative sharp wave followed immediately by slower positive component) waveforms. *Stage III* (also termed as slow wave sleep) is the deepest stage of NREM sleep, characterized by the dominance of high amplitude, low frequency (<4 Hz) EEG.

In laboratory animals (cats, rats and mice) NREM sleep is generally not subdivided into stages. It is identified by the presence of high amplitude, low frequency (delta) waves (0.1–4.0 Hz) in the EEG (Datta and Maclean 2007).

REM sleep is characterized by an ensemble of concomitant events including: (1) Low-amplitude and high frequency waves in the EEG (also called activated EEG); (2) Very low or complete absence of activity in the EMG (muscle atonia) and (3) Singlets and clusters of rapid eye movements (REMs) in the EOG. Supplemental to these polysomnographic signs, other REM sleep-specific physiological signs in mammals are: tonic hippocampal theta activity, myoclonic twitches, most apparent in the distal limb musculature; pronounced fluctuations in cardio-respiratory rhythms and core body temperature; penile erection and clitoral tumescence.

The sleep cycle is structured. In adult humans and non-human primates, circadian distribution of sleep period is mainly monophasic and the sleep architecture consists of 4–5 NREM—REM sleep cycle, with each cycle of approximately 90 min duration. While NREM sleep occupies majority of time during the first half of sleep time, REM sleep is predominant in the second half.

In rodents, circadian distribution of sleep period is polyphasic and NREM-REM sleep cycles are shorter and continue throughout sleep periods during day and night.

2.2 *Cellular Substrates of Sleep-Wakefulness*

Wakefulness or behavioral arousal is the result of cortical activation. Cortical activation is caused by a concerted increase in the activity of multiple neuronal aggregates (wake-promoting systems), localized in various brain regions, and utilizing multiple neurotransmitters (Thakkar 2011). Each wake-promoting system is distinct and has a special role in the maintenance of wakefulness. For example, the glutamatergic neurons of brainstem reticular formation are believed to be the core of the ascending reticular activating system (ARAS; Jones 2003), responsible for cortical activation. While the cholinergic neurons of the basal forebrain represent the terminal node of the extra-thalamic relay of the ARAS and are responsible for the promotion of cortical desynchronization, the hypothalamic orexins/hypocretin system is essential for the maintenance of wakefulness with muscle tone. In contrast, the monoaminergic systems namely, the norepinephrine containing locus coeruleus, serotonin containing raphe nuclei and the histamine containing tuberomammillary nucleus (TMN) act in concert with other arousal systems to maintain wakefulness and inhibit REM sleep (Brown et al. 2012; Burgess and Peever 2013; Chemelli et al. 1999; Lin et al. 1999; Thakkar et al. 1999; Thakkar 2011).

Transitions from waking to NREM sleep involve coordinated inhibition of multiple arousal systems. A major source of inhibition involves homeostasis mediated regulation of sleep (discussed separately). Another important source of sleep-related inhibition of arousal systems arises from the GABAergic neurons located in the ventrolateral and median preoptic nuclei (Brown et al. 2012). These neurons are activated during NREM sleep and exhibit state-dependent discharge patterns that are reciprocal to that observed in arousal systems (sleep-on/wake-off). Recently, melanin concentrating hormone containing neurons of zona-incerto-hypothalamic region have also been implicated in sleep regulation (Konadhode et al. 2013; Monti et al. 2013).

The perpetration of REM sleep occurs through an interplay between mesopontine cholinergic REM-promoting (On) and monoaminergic and orexinergic REM-permitting (Off) neuronal systems, resulting in the concurrent promotion of forebrain activation and peripheral muscle inhibition/atonia (Datta and Maclean 2007; McCarley 2007).

2.3 Regulation of Sleep-Wakefulness

The physiological regulation of mammalian sleep is controlled by: (1) homeostatic sleep process (Process S) and (2) circadian alerting process (Process C), that is independent of sleep and wakefulness (Borbely 1982). During normal, spontaneous wakefulness, increasing circadian alerting signal counteracts with rising homeostatic sleep pressure to maintain the wake state. Conversely, during the night, reduction in circadian alerting signal coupled with residual homeostatic sleep drive ensures the maintenance of sleep (Edgar et al. 1993). The changes in sleep-wakefulness across 24-h, observed in humans and animals is the result of combined influences of Process S and Process C and the assessment of relative contribution of sleep homeostasis and circadian rhythmicity is complex (Thakkar et al. 2014).

Circadian alerting signal (Process C) is controlled by the intrinsic body clock, the suprachiasmatic nucleus (SCN). Although neural mechanisms by which the circadian clock influences sleep–wake system remain to be fully specified, SCN is known to communicate with several sleep–wakefulness centers including the preoptic area and the BF. Although complete ablation of the SCN eliminates sleep-wake rhythms, sleep loss results in compensatory increase in the duration and intensity of sleep suggesting that Process S is independent of Process C (Tobler et al. 1983).

2.4 Sleep Homeostasis

Sleep homeostasis defines an innate regulatory mechanism that maintain the “constancy” of sleep. Thus, loss of sleep will results in compensatory increase in sleep whereas, excess sleep will result in compensatory reduction in sleep. The core of sleep homeostasis is sleep propensity or the urge to fall asleep, also described as sleep pressure. Sleep pressure starts to accumulate as soon as one is awake, and continues to build-up until sleep is initiated. Once sleep is initiated, sleep pressure starts to dissipate. Sleep pressure is the manifestation of a build-up or an accumulation of sleep homeostatic factor/s (Thakkar et al. 2014).

Although several sleep factors have been implicated, only few satisfy the established criteria to be considered as homeostatic sleep factors (Brown et al. 2012). These are: adenosine (AD), nitric oxide (NO), prostaglandin D_2 (PGD_2) and cytokines. Out of these four neuromodulators only AD has gained utmost attention because the remaining three: NO, PGD_2 and the cytokine, interleukin-1, mediate their sleep promoting effects via AD (Brown et al. 2012; Hayaishi et al. 2004; Luk et al. 1999; Mizoguchi et al. 2001; Rosenberg et al. 2000). In addition, AD links sleep with energy metabolism and neuronal activity. For example, during wakefulness, energy (ATP) usage is high in wake-promoting systems, due to increased neuronal firing, synaptic activity, and synaptic potentiation. This increased energy usage during wakefulness is reflected in increased accumulation of extracellular AD, a breakdown product of ATP metabolism, which corresponds to increased accumulation of sleep pressure. The longer the period of wakefulness, greater the accumulation of sleep pressure and/or AD and the longer it takes for sleep pressure to dissipate during sleep (Porkka-Heiskanen 2013). Increased AD acts via A1 receptors (A1R) to inhibit wake-promoting systems and to promote sleep (Thakkar et al. 2003). Homeostatic response or dissipation of sleep pressure is measured by sleep latency along with the duration and intensity of recovery sleep that follows after sleep loss (Borbely 1982; Porkka-Heiskanen 2013).

Now that we have described the fundamentals of sleep, let us describe how sleep is affected after acute and chronic alcohol consumption.

3 Alcohol and Sleep: Clinical Studies

3.1 Effects of Acute Alcohol on Sleep in Healthy Non-alcoholics

The effects of alcohol on human sleep were first described by Kleitman in late 1930s (Kleitman 1963). However, it was only in the late 1960s-early 1970s, that scientists began conducting laboratory studies to examine the effects of alcohol on sleep in healthy non-alcoholics as well as in alcoholics. The impetus behind alcohol and sleep research was the development of sleep medicine as a distinct field.

The standard protocol used to investigate the effects of acute alcohol intake on sleep in healthy non-alcoholics was to administer alcohol 30–60 min before bedtime so as to allow alcohol concentrations in the breath or blood to peak at "lights-out", as soon as the subject went to sleep. The doses ranged from 0.16 to 1.0 g of alcohol/kg body weight (g/kg) were used to yield breath alcohol concentrations (BrAC) ranging from 0.025 to 0.105 % (equivalent to blood alcohol concentration of 0.25–105 mg/dL) which approximately corresponds to the consumption of one to six standard drinks (see reviews Ebrahim et al. 2013; Roehrs and Roth 2001). The most common sleep variables examined were (1) sleep onset latency (SOL; time to fall asleep); (2) wake after sleep onset (WASO); (3) Slow wave sleep (SWS; Stage 3 sleep), and (4) REM sleep.

Irrespective of the gender, age and amount, majority of studies have reported that acute alcohol consumption reduces SOL and WASO along with an increase in SWS during the first half of the night (sleep period). However, sleep is severely disrupted in the second half of the night as evident by increased wakefulness along with increased light (Stage 1) sleep during the second half. Since the peak BrAC (0.025–0.105 %) is achieved before sleep onset, increase in SWS during the first half of the night is observed in the descending limb of the blood alcohol concentration curve, whereas sleep disruption are observed only after alcohol is completely metabolized and cleared from the body.

In contrast, the effects of acute alcohol consumption on REM sleep are dose dependent. While low dose of alcohol have minimal effects on REM sleep, majority of studies suggest that moderate to high consumption of alcohol reduces REM sleep especially in the first half of the sleep period. Some studies that have reported increase in REM sleep ("REM rebound") during the second half of the sleep period (reviewed in Ebrahim et al. 2013; Roehrs and Roth 2001).

Thus, convincing literature from humans studies suggest that acute alcohol intake reduces the time to fall asleep and consolidates NREM sleep especially during the first half of the night. During the second half, sleep is severely disrupted as evident by increase in wakefulness, light (stage 1 sleep) and REM sleep.

3.2 Effects of Alcohol on Sleep in Alcoholics

Several earlier studies investigated the effects of continued alcohol consumption in alcoholics [Gross et al. 1973; Gross and Hastey 1975; Wagman and Allen 1975 reviewed by Brower (2001)]. Alcoholics undergoing inpatient treatment before withdrawal were provided with alcohol for 1 or more days and sleep was monitored. These studies suggest that alcohol consumption in alcoholics increased sleep latency, decreased total sleep time and decreased REM sleep. This experiment design raises ethical issues and is no longer used (Brower 2001).

3.3 Sleep in Alcoholics During Abstinence

Several polysomnographic studies have been performed to investigate sleep-wake behavior in alcoholics who have been abstinent for days, weeks and months (Adamson and Burdick 1973; Colrain et al. 2009; Drummond et al. 1998; Irwin et al. 2000; Johnson et al. 1970; Snyder and Karacan 1985). These studies suggest that acute withdrawal resulted in insomnia like symptoms including difficulty in falling and staying asleep, reduced sleep efficiency (ratio of time spent asleep to the amount of time spent in bed) and increased wakefulness (Brower 2001). Sustained withdrawal produced sleep fragmentation, excessive daytime sleepiness, reduced NREM sleep during the night and increased REM sleep that lasted for more than 3 years during abstinence (Drummond et al. 1998; Imatoh et al. 1986; Ishibashi et al. 1987). Delirium tremens (DTs), most severe manifestation of alcohol withdrawal, is characterized by presence of tremors, agitation, hallucinations and severe sleep disruption (reviewed in Brower 2001).

3.4 Alcohol and Sleep: Other Effects

As described above alcohol is a potent somnogen. It is this sleep promoting characteristic of alcohol which makes it one of the most commonly used "over the counter" sleep aid (Johnson et al. 1998; Roehrs and Roth 2001). It has been estimated that approximately 12 % of general population and 20 % of insomniac have used alcohol to promote sleep while 50 % of alcoholics use alcohol to self-medicate sleep disorders (Brower et al. 2001; Roehrs and Roth 2012).

Frequent use of alcohol, to medicate sleep disorders, results in tolerance development. Alcohol tolerance, defined as a need to progressively consume higher amounts of alcohol to achieve the same effect, is a major contributor towards the development of alcoholism. Functional tolerance to alcohol is the result of the brain adapting to alcohol-induced disruption and is classified as: (1) acute tolerance, which is observed during a single session of alcohol consumption; (2) rapid tolerance, which is observed between 8 and 24 h after the effects of first alcohol administration has disappeared; and (3) chronic tolerance, which is detected after long term (days) alcohol exposure. Of these three, rapid tolerance development has significant importance because it is an index of chronic tolerance (Kalant 1998; Khanna et al. 1991).

Administration of the same dose alcohol for 2 days, in healthy human subjects, resulted in the development of rapid tolerance to the REM sleep suppressing effects of alcohol (Yules et al. 1966).

Finally, subjective and objective indicators of insomnia including difficulty in falling asleep, decreased total sleep time, decreased sleep efficiency and increased REM sleep pressure are predictors of relapse to alcoholism (Brower et al. 1998, 2001; Gillin et al. 1994).

4 Alcohol and Sleep: Animal Studies

Results obtained from animal studies corroborate human studies. Irrespective of the time or the route (systemic or central) of administration, rodents exposed to acute alcohol treatment display reduction in SOL and increases NREM sleep

(Ghosh et al. 1991; Hattan and Eacho 1978; Hill and Reyes 1978a; Kubota et al. 2002; Papale et al. 2008; Sharma et al. 2014c; Thakkar et al. 2010; Ticho et al. 1992).

In contrast, chronic alcohol administration produces sleep disruptions. Ehlers and Slawecki (2000) exposed rats to 6 weeks of ethanol vapor and reported a reduction in the spectral power of the EEG in both low frequency and high frequency bands. Mukherjee and Simasko (2008, 2009) performed 6 weeks of chronic exposure to 6 and 12 % of alcohol in water and 3 % of alcohol in liquid diet and observed increased NREM sleep only in the dark period. In contrast, 6 % alcohol in liquid diet increased wakefulness and reduced sleep (NREM and REM) during the light period, but decrease wakefulness and increased sleep during the dark period.

Alcohol withdrawal also causes sleep disruptions in animals. While two studies reported decreased REM during withdrawal (Gitlow et al. 1973; Rouhani et al. 1998), one study reported increased REM sleep during withdrawal (Mendelson et al. 1978). Kubota et al. (2002) observed increased NREM sleep along with significant circadian changes in REM sleep during alcohol withdrawal in rats. Veatch (2006) found increased REM with a concomitant reduction in NREM sleep during initial withdrawal period. While REM remained increased, NREM sleep normalized within 4 days. The inconsistencies among animal studies are likely due to methodological differences for alcohol administration, duration of EtOH exposure, etc. (Veatch 2006).

Similar to humans, mice exposed to voluntary alcohol consumption (self-administration) in their home cage (stress free environment) display rapid tolerance development to NREM sleep promoting effects (Sharma et al. 2014a) whereas, increased REM pressure induced by selective REM sleep deprivation increases alcohol consumption (Aalto and Kiianmaa 1984).

In summary, alcoholism is a dreadful disease with profound socio-economic cost. It is imperative that we find and use efficacious treatment strategy. In order to treat alcoholism, it is critical to manage withdrawal symptoms and prevent relapse. Strong and consistent literature suggests that alcohol consumption causes sleep disruptions and sleep disruptions are among the most severe and protracted symptoms, observed both during drinking period, and during abstinence in recovering alcoholics. In addition, sleep disturbance during withdrawal are predictor of relapse. Thus, based on the review described above, sleep may have a causal role in all three stages of alcoholism including development, maintenance and relapse. Therefore, in order for develop efficacious strategies to treat alcoholism and prevent relapse, it is imperative that we understand as to how and where alcohol act to affect sleep.

5 Alcohol and Sleep: Anatomical and Cellular Mediators

"Intoxication" following alcohol intake describes an ensemble of behavioral changes, one of which is a profound sleepiness. While significant advances have been made into our understanding of motor in-coordination following alcohol intake (Dar 2001; Dar and Meng 2004), very little is known about cellular substrates mediating sleepiness after alcohol intake.

Initial studies suggested that serotonin may have a role in mediating the sedative effects of alcohol (Hill and Reyes 1978b) however, serotonin in NREM sleep is yet unclear. Norepinephrine (NE), another monoamine, has also been implicated in alcohol induced sleep promotion (Brower 2001) however, the role of NE in sleep-wakefulness appears to be limited; mice lacking NE have almost normal sleep-wake cycle (Hunsley and Palmiter 2003).

It has been suggested that alcohol induced reduction in glutamate may be responsible for REM suppressive effects of alcohol (Prospero-Garcia et al. 1994) although, direct evidence supporting this hypothesis is lacking.

Since GABA is an inhibitory neurotransmitter which is implicated in sleep promotion and in mediating behavioral effects of alcohol, we asked: Does GABA mediate sleepiness after alcohol intake?

GABA and alcohol may act via common cellular mechanisms (reviewed Wallner et al. 2006). In vivo and vitro studies suggest that alcohol may enhance GABA mediated actions via $GABA_A$ receptor, similar to barbiturates or benzodiazepines, known hypnotics that promote sleep. However, the effects of alcohol on $GABA_A$ receptor are complex and not clearly understood (reviewed in Aguayo et al. 2002; Siggins et al. 2005; Wallner et al. 2006). While there is evidence to suggest that physiological concentration of alcohol alters $GABA_A$ receptor function, whether alcohol directly affects $GABA_A$ receptors remains controversial. In fact, a growing body of evidence suggests that alcohol may indirectly potentiate $GABA_A$ receptor function in a GABA-mimetic manner (Breese et al. 2006; Wallner et al. 2006). Multiple mechanisms may influence the sensitivity of $GABA_A$ receptors to alcohol. For example, activation of β-adrenergic receptor is necessary to detect alcohol's effect on GABA (Freund and Palmer 1996; Wang et al. 1999). Similarly endogenous neurosteroids, potent positive allosteric modulators of the $GABA_A$ receptor, are known to modulate various effects of alcohol (Kumar et al. 2004). Finally recent reports suggest that the intoxicating effects of alcohol may be mediated via extra-synaptic δ-subunit containing $GABA_A$ receptors (Breese et al. 2006; Wallner et al. 2006).

With regards to sleep, although, almost all hypnotics are believed to activate $GABA_A$ receptor, the role of GABA in sleep is yet unclear. For example: (1) GABA release in pontine reticular formation is highest during wakefulness and lowest during REM sleep (Thakkar et al. 2004; Vanini et al. 2011). (2) Activation of $GABA_A$ receptors in the pons promotes wakefulness (Xi et al. 2001). (3) Blockade of $GABA_AR$ in the pontine reticular formation promotes REM sleep (Boissard et al. 2002; Pollock and Mistlberger 2003). (4) Finally, while, activation of extra-synaptic GABA receptors in the hypothalamus promotes sleep, systemic administration of extra-synaptic GABA receptors agonist, THIP, has no effect on sleep-wakefulness (Thakkar et al. 2008; Winsky-Sommerer et al. 2007).

Thus, based on the review of literature described above, it appears that GABA may not be the mediator of sleep following alcohol intake.

6 Alcohol and Sleep: Adenosine Is the Mediator

On one side, there is considerable evidence suggesting that AD is a key mediator of neuronal responses to alcohol. On the other side, convincing evidence exist suggesting that adenosine via its action on A1 receptors promotes NREM sleep and enhances EEG power in delta frequency, remarkably similar to the effects of alcohol intake on sleep (Basheer et al. 2004; Dunwiddie and Masino 2001; Hack and Christie 2003; Newton and Messing 2006; Radulovacki 1985). Does AD mediate the sleep promoting effects of alcohol? Let us review the literature in detail.

6.1 AD and Sleep

Strong and compelling evidence suggest that AD is a modulator of sleep-wakefulness (Alam et al. 1999; Basheer et al. 2000, 2001a, b; Benington and Heller 1995; Murillo-Rodriguez et al. 2004; Porkka-Heiskanen et al. 1997, 2000; Portas et al. 1997; Radulovacki et al. 1984; Strecker et al. 2000; Thakkar et al. 2003). AD, a byproduct of ATP metabolism, build up selectively in the wake-promoting centers especially the BF and inhibits them resulting in the transition from wakefulness to sleep (Basheer et al. 2004). While AD and its agonists reduces cortical and behavioral arousal and promotes sleep, caffeine and theophylline, powerful blockers of AD receptors, promote cortical and behavioral arousal while suppressing sleep (Fredholm 1995). In fact, of the known biochemical actions of caffeine, only the blockade of AD receptors occurs at concentrations achieved during normal human consumption (Fredholm 1995). The prominent actions of AD in the brain is its ability to inhibit neuronal activity (Dunwiddie and Masino 2001) including the discharge activity of wake-promoting neurons (Alam et al. 1999; Arrigoni et al. 2006; Thakkar et al. 2003) and inhibit the release of a neurotransmitters by acting on presynaptic terminals (Dunwiddie 1985).

A1 receptor is an important AD receptor mediating AD's sleep inducing actions (Basheer et al. 2004). *There are* few reports that suggest that prostaglandin–D2 promotes sleep via AD acting on *A2a receptor* on sleep promoting ventrolateral preoptic area neurons (Basheer et al. 2004; Huang et al. 2007; Scammell et al. 2001). The role of A3R in sleep is yet unknown.

Recently studies conducted in humans also support the role of AD in sleep regulation. Caffeine, a potent antagonist at AD A_1 and A_{2A} receptors, selectively interferes with NREM sleep homeostasis with minimal effects on circadian facet of sleep–wake regulation (Landolt 2008). Caffeine attenuates the build-up of theta activity associated with prolonged wakefulness (Landolt et al. 2004). Low dose of caffeine (100 mg), administered immediately prior to sleep, prolongs sleep latency, reduces slow-wave sleep in the first NREM/REM sleep cycle, and impairs sleep efficiency. The EEG spectral power in the low delta range is decreased, whereas power in the spindle frequency range is increased (Landolt et al. 1995a). Early morning caffeine intake (200 mg) produces similar effects as observed during nocturnal sleep at a time when the caffeine concentration in saliva is approaching zero. These caffeine-induced EEG changes in NREM sleep are comparable in rested and sleep deprived subjects and mimic the changes in sleep EEG activity associated with a physiological reduction in nocturnal NREM sleep propensity (Landolt et al. 1995b). Humans with a functional variation in adenosine deaminase (enzyme involved in AD degradation) gene have increased NREM sleep and slow wave activity in the EEG during sleep (Landolt 2008; Retey et al. 2005). Finally, sleep deprivation in humans increases A_1 receptor occupancy throughout cortical and subcortical regions (Elmenhorst et al. 2007).

6.2 AD and Alcohol

It was almost 30 years ago when Dar and colleagues first demonstrated a functional relationship between alcohol and adenosine (Dar et al. 1983). Since then, strong and consistent evidence has been gathered, suggesting that adenosine is responsible for mediating several cellular and behavioral effects of alcohol.

In vitro studies have revealed that acute alcohol exposure inhibits AD reuptake via equilibrative nucleoside transporter 1 (ENT1), whereas chronic alcohol exposure down-regulates ENT1 expression (Krauss et al. 1993; Nagy et al. 1990). The ENT1-null mice show decreased AD tone resulting in increased consumption coupled with reduced hypnotic (loss of righting reflex) and ataxic responses to alcohol. Treatment with an A1 receptor agonist decreases alcohol consumption in ENT1-null mice (Choi et al. 2004).

There is strong and convincing evidence demonstrating the importance of A1R in mediating alcohol effects on ataxia (Dar 1993, 1997, 2001; Phan et al. 1997; Saeed 2006) and other behaviors including anxiety, tremors and seizures, during alcohol withdrawal (Batista et al. 2005; Concas et al. 1996; Dar 2001; Kaplan et al. 1999; Prediger et al. 2006). Although, there is evidence implicating A2A receptor in alcohol induced behavioral changes (Barwick and Dar 1998; Jarvis and Becker 1998; Kaplan et al. 1999; Newton and Messing 2006), the role of A3R is unclear.

Several clinical studies also support the role of AD in alcohol induced behaviors including sleep. For example, caffeine, a non-specific AD receptor antagonist, offsets debilitating effects of alcohol and improve alertness (Franks et al. 1975; Hasenfratz et al. 1993; Kerr et al. 1991; Liguori and Robinson 2001). Caffeine has also been reported to prevent sleepiness and performance impairment associated with alcohol (Drake et al. 2003). Finally, the combined administration of caffeine and alcohol increases the development of alcohol tolerance in humans (Fillmore 2003). Alcoholics display an impaired response to sleep deprivation suggesting an impaired adenosinergic system (Irwin et al. 2002).

Thus, based on the review of literature describe above, there is a strong rationale to implicate AD as a mediator of sleep following alcohol intake. Since orexinergic system is pivotal for initiation and maintenance of wakefulness, we hypothesized that alcohol

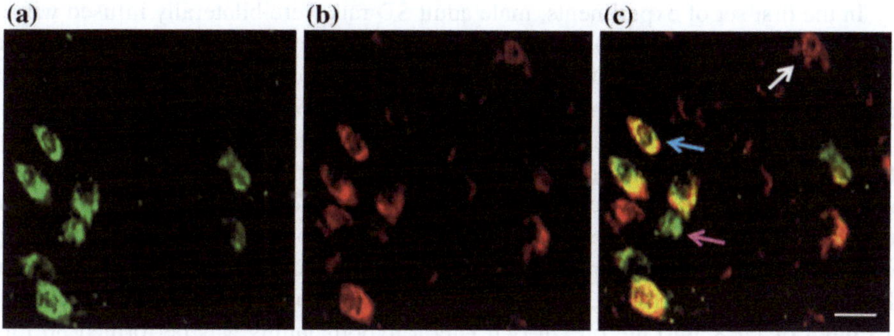

Fig. 1 Orexin neurons expresses adenosine A1 receptors as evident by double labelled fluorescence. Panel **a** Orexin-A containing neurons in the perifornical region (*green* fluorescence). Panel **b** Same visual field and section as Panel **a** showing adenosine A1 receptors (*red* fluorescence). Panel **c** Double labeled orexin-A and the A1 receptor (*green arrow*, indicative of A1 receptors on orexin neurons) along with single labeled orexin neurons (*purple arrow*) and A1 receptor-labeled neuron are also observed in the same visual field. Calibration bar = 25 μm. (Reproduced from Thakkar et al. 2002) (color figure online)

acts via AD to inhibit the orexinergic system and affect sleep. In the subsequent sections, we have described a series of experiments designed to test our hypothesis and to examine the role of AD as a mediator of sleep following alcohol intake.

6.3 *Orexin Neurons Express A1 Receptors*

If AD, acting on orexin neurons is the mediator of sleep promoting effects of alcohol, then orexin neurons should express A1 receptors. Thus, our first study was designed to examine whether orexin neurons express A1 receptors.

Male Sprague-Dawley (SD) rats were euthanized, brains removed, blocked and the region containing perifornical hypothalamus was sectioned to examine the expression of A1 receptors on orexin neurons. A dual color, double-labeling, fluorescence immunohistochemistry was used to examine the orexin neurons expressing A1 receptors. Our results a significant proportion of orexin neurons (>30 %) expressed A1 receptors (Fig. 1). This study was the first to demonstrate the presence of A1 receptors on orexin neuron (Thakkar et al. 2002).

7 A1 Receptors Expressed on Orexin Neurons Promote Sleep

The next question we asked was: Does A1 receptor expressed on orexin neuron have a physiological role in sleep regulation. We performed two experiment to address this question.

In the first set of experiments, male adult SD rats were bilaterally infused with a selective AD A1R antagonist 1, 3-Dipropyl-8-phenylxanthine (DPX) into the orexinergic hypothalamus at sleep (light) onset and its effects on spontaneous bouts of sleep–wakefulness was monitored (Daly et al. 1985).

Since AD is implicated in the homeostatic regulation of sleep and recovery sleep following sleep deprivation is an essential paradigm to study the homeostatic mechanisms regulating sleep–wakefulness, our second set of experiments monitored the effects of A1 receptor antagonist on recovery sleep following sleep deprivation. Rats were sleep deprived by gentle handling during the last 6 h of the light cycle (Thakkar et al. 2003). Bilateral microinjections of either saline (0.9 %) or 15 pmol of DPX into the orexin-rich lateral hypothalamus was performed 30 min before the animals were allowed to enter recovery sleep and recovery sleep was monitored for 3 h.

The results obtained are as follow: Bilateral microinjections of the selective A1 receptor antagonist, DPX into the orexin-rich perifornical hypothalamus, dose dependently, increased the amount of time spent in wakefulness and decreased the amount of time spent in sleep (NREM and REM phases) during 3 h of spontaneous bouts of sleep–wakefulness during the normal sleep (light) period (Fig. 2).

Similarly, bilateral microinjections of DPX into the orexin-rich perifornical hypothalamus significantly increased the latency to the NREM phase of recovery sleep while significantly reducing both NREM and REM phases of sleep. Wakefulness was increased (Fig. 3). Thus, the results of our study confirm our hypothesis and suggest that A1 receptors on orexin neurons may have a physiological role in sleep-wakefulness.

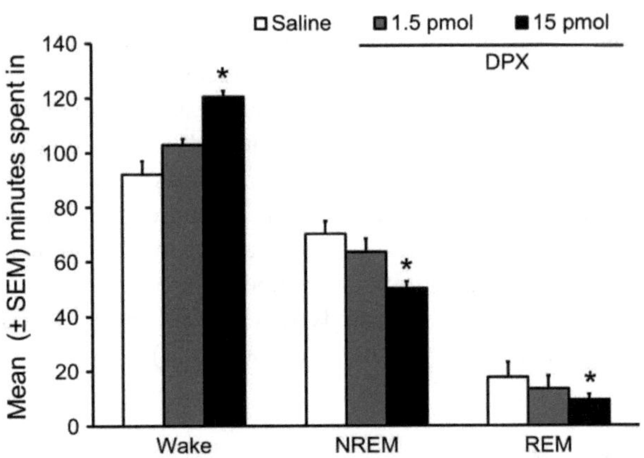

Fig. 2 Bilateral administration of DPX, adenosine A1 receptor antagonist, in the orexin-rich perifornical hypothalamus significantly and dose-dependently increased wakefulness with a concomitant reduction in NREM and REM sleep during first three hours of light period. $*p < 0.05$. (Reproduced from Thakkar et al. 2008)

Fig. 3 Bilateral local microinjections of A1 receptor antagonist, DPX, into the orexin-rich perifornical hypothalamus significantly attenuated the recovery sleep following 6 h of sleep deprivation as evident by increased NREM sleep latency (**a**) and reduced total time spent in NREM sleep during first three hours of recovery sleep with a concomitant increase in the wakefulness (**b**). *$p < 0.05$. (Reproduced from Thakkar et al. 2008)

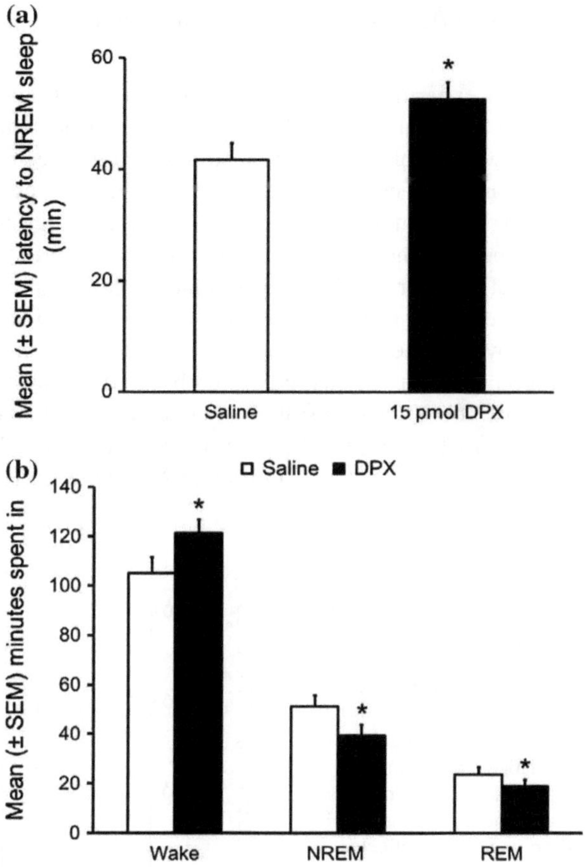

Thus, these results clearly demonstrate that AD via A1 receptors expressed on orexin neurons may play a significant physiological role in regulating sleep. Indeed, our results were confirmed by several subsequent studies. For example, in vitro electrophysiological studies have confirmed the role of AD via A1 receptor in the inhibition of orexin neurons (Liu and Gao 2007). In vivo studies in rats suggested that activation of A1 receptor in the orexin-rich perifornical hypothalamus suppressed the discharge activity of wake-active neurons and reduced the number of orexinergic neurons exhibiting c-Fos. Behaviorally, activation of A1 receptor in the orexin-rich perifornical hypothalamus increased sleep and reduced wakefulness. In contrast, blockade of A1 receptor resulted in increased wakefulness coupled with reduction in sleep (NREM and REM phase) along with an increase in the numbers of orexinergic neurons exhibiting c-Fos (Alam et al. 2009; Rai et al. 2010).

Now that we have established that AD via A1 receptor interacts with orexinergic system. Our next series of experiments were designed to examine the role of AD and orexin in alcohol's effects on sleep.

7.1 Alcohol Is a Potent Somnogen

Our first experiment verified the effects of acute alcohol administration on spontaneous bouts of sleep-wakefulness in freely behaving SD rats. The experimental design was simple. Rats were instrumented for electrographic recording to examine and quantify sleep-wake states. To examine the potency of alcohol in sleep promotion, we administered alcohol (3 g/kg; intragastric) at the onset of circadian active period when rats are maximally active and all wake-promoting systems are very active. Our results suggest that alcohol administration at dark onset significantly reduced sleep onset latency (Fig. 4a) and increased NREM sleep (Fig. 4b). While wakefulness was reduced, REM sleep remained unaffected. Maximal sleep promoting effects of alcohol were observed during the first six hours after alcohol administration. These results suggest that alcohol is a potent somnogen and is able to counteract a strong wake-promoting drive, initiated by combined activation of all wake-promoting centers, to promote NREM sleep (Thakkar et al. 2010).

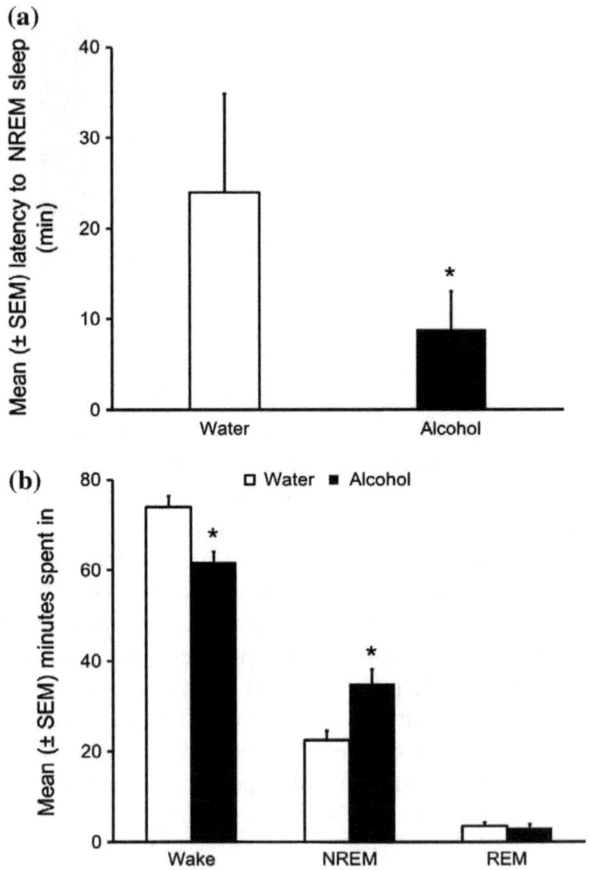

Fig. 4 Systemic alcohol administration (3 g/kg) at dark onset promote sleep as evident by reduction in NREM sleep latency (**a**) and increase in time spent in NREM sleep during 12 h of the dark period with a concomitant reduction in wakefulness whereas REM sleep remained unaffected (**b**). *$p < 0.05$. (Reproduced from Thakkar et al. 2010)

7.2 Alcohol Inhibits Orexin Neurons

Since the orexinergic system is among the most potent wake-promoting systems, we asked: Does alcohol mediates its sleep promoting effects by inhibiting orexin neurons? Our next experiment was designed to address this. Adult male SD rats were used as our animal model. Expression of c-Fos in the nucleus of orexin neurons was used as a marker of orexinergic activation. Double-label, orexin and c-Fos, immunohistochemistry (IH) revealed that alcohol (3 g/kg; intragastric) administration at dark onset significantly reduced the activation of orexin neurons as evident by a significant reduction in orexin neurons with c-Fos immunoreactivity (Fig. 5). Thus, based on these results we suggest that alcohol inhibits orexin neurons to promote sleep. The next question we asked was: What mediates alcohol's inhibition of orexin neurons?

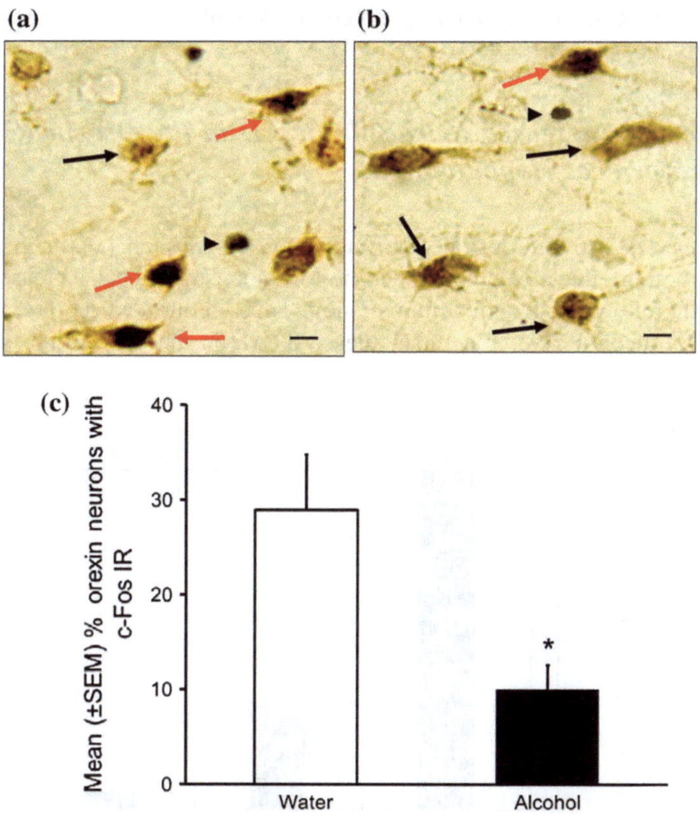

Fig. 5 A representative photomicrograph describing orexin neurons with c-Fos immunoreactivity (*IR*) in controls (**a**) and alcohol group (**b**). *Red arrow* double labeled orexin and c-Fos +ve neuons; *black arrow* single labeled orexin neurons; *black arrowhead* single labeled c-Fos +ve cells. Calibration bar = 10 μm. The percentage of orexin neurons with c-Fos IR was significantly reduced after acute alcohol administration (**c**). *$p < 0.05$ versus water (color figure online)

8 Alcohol Induced Inhibition of Orexin Neurons Is Mediated by Adenosinergic A1 Receptor

Since A1 receptors, expressed on orexin neurons, regulate sleep-wakefulness (see Figs. 1, 2 and 3), we hypothesized that alcohol may act via AD to activate A1 receptor resulting in the inhibition of orexin neurons and sleep promotion. Thus, we examined the effects of bilateral infusion of A1 receptor antagonist DPX on alcohol induced sleep. We performed the infusion of DPX at dark onset because the animals are maximally awake and spent >80 % time in wakefulness. DPX infusion at dark onset will have minimal effects on spontaneous wake-promotion due to "ceiling" effects (Thakkar et al. 2014). However, DPX will suppress alcohol induced sleep and promote wakefulness. The results of our studies are described in Fig. 6. Bilateral infusion of DPX into orexin-rich perifornical hypothalamus resulted in an attenuation of alcohol induced sleep promotion implicating an interaction of orexin, AD and its A1R in sleep promoting effects of alcohol.

8.1 Alcohol Increases Extracellular AD in the Orexin-Rich Perifornical Hypothalamus

The "litmus test" to demonstrate the interaction of AD and orexin in mediating sleep promoting effects of alcohol will be to demonstrate local increase in extracellular AD following alcohol administration. Thus, in our next set of experiments, the effect of alcohol on extracellular release of AD in the orexin-rich perifornical

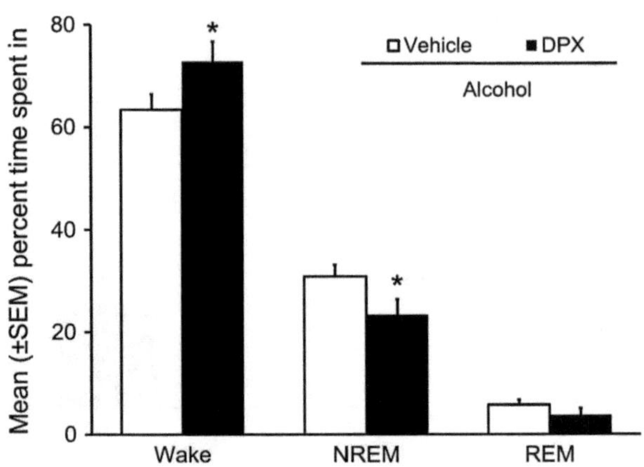

Fig. 6 Bilateral infusion of A1R antagonist, DPX, into the orexin-rich perifornical hypothalamus attenuated alcohol-induced NREM sleep promotion and promoted wakefulness during 12 h of dark period post-alcohol. Rapid eye movement (*REM*) sleep remained unchanged. *$p < 0.05$ versus vehicle

hypothalamus was examined. Microdialysis was used to measure AD release. The specificity of alcohol's action on orexin neurons was provided by reverse microdialysis, through the same microdialysis probe, delivery of pharmacological relevant alcohol doses into orexin-rich perifornical region.

As predicted, our results suggest that alcohol produced a dose-dependent increase in extracellular AD in the orexin-rich perifornical hypothalamus with maximal increase observed with 300 mM dose of alcohol perfused (Fig. 7). Reverse microdialysis perfusion of 300 mM dose of alcohol provides a concentration of $\sim$25 mM directly outside the probe. This concentration of alcohol is in the range of alcohol concentration observed in the brain after a systemic administration of sedative dose ($\sim$3 g/kg) of alcohol (Sharma et al. 2014c). This study provides confirmatory evidence support the role AD and orexins in sleep promoting effects of alcohol.

To summarize, based on the results of our studies described above, we suggest that alcohol increase extracellular AD in the orexin-rich perifornical hypothalamus. Increased AD acts on A1 receptors expressed on orexin neurons to inhibit them. Inhibition of orexin neurons will result in sleep promotion.

8.2 Alcohol Withdrawal Disrupts Sleep and Downregulates Orexin Expression

The next question we asked was: Are sleep disruptions observed during alcohol withdrawal due to impaired orexinergic system?

Our first experiment examined sleep-wakefulness during alcohol withdrawal. We used the extensively used (>300 citation of the original article in last 20 years) Majchrowicz's chronic binge alcohol method to induce alcohol withdrawal in SD rats (Faingold 2008; Majchrowicz 1975). This method offer other advantages including (1) mimicking high blood alcohol levels and heavy alcohol consumption in alcoholics, (2) rapid induction of alcohol dependency and relatively high and sustained blood alcohol concentration achieved within a short duration, (3) high incidence of overt signs of withdrawal (Sharma et al. 2010).

A brief description of Majchrowicz protocol used is as follows: Two hour before the light onset, experimental animals are primed with an intragastric administration of 5 g/kg dose of alcohol. Subsequent doses are adjusted depending on the level of intoxication (assessed by observing the behavior of the animal (see Table 1 in Sharma et al. 2014b) and administered every 8 h for 4 days (treatment days). As shown in Fig. 8a, during alcohol withdrawal, rats displayed significant increase in wakefulness, both during their sleep (light) and active (dark) period with a concomitant reduction in sleep (NREM and REM) suggesting insomnia-like symptoms observed during alcohol withdrawal in humans.

(a) (b)

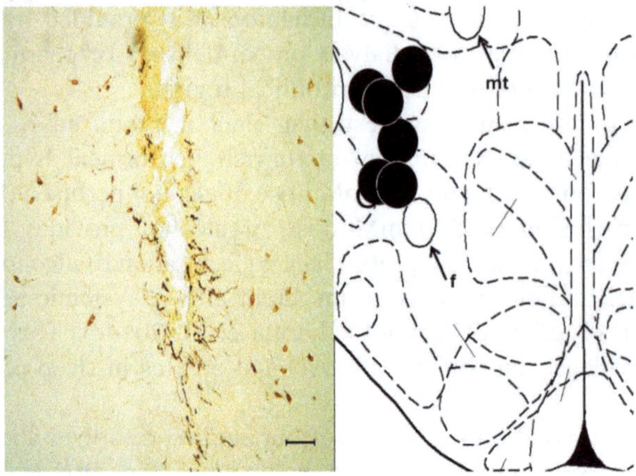

(c)

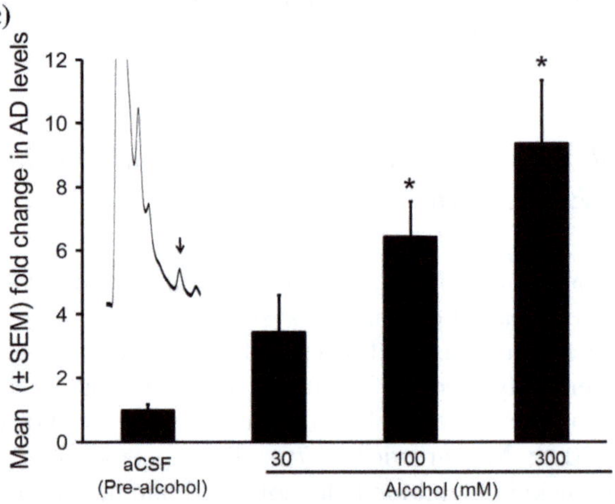

Fig. 7 Representative photomicrograph depicting lesions caused by microdialysis guide cannula (*black arrow*) in the midst of orexin-A containing neurons (**a**). All probe sites (*black circles*) were located between AP −3.2 to −3.6 mm from the bregma and are mapped onto one side of the coronal schematic brain section at the level of AP −3.3 (**b**). *f = fornix*; *mt = mammillothalamic tract.* Calibration bar = 100 μm. Local alcohol perfusion significantly and dose-dependently increased adenosine (*AD*) release in the orexin-rich perifornical region (**c**). The chromatogram in panel (**c**) depicts the AD peak from a sample (10 μl) collected from the orexinergic perifornical hypothalamus during 100 mM alcohol perfusion (**c**). *$p < 0.05$ versus aCSF. *aCSF* Artificial cerebrospinal fluid

Our next experiment examined the expression of orexin gene during acute alcohol withdrawal. The same Majchrowicz protocol (described above) was used to induce alcohol dependency in SD rats. The animals were euthanized 12 h after the

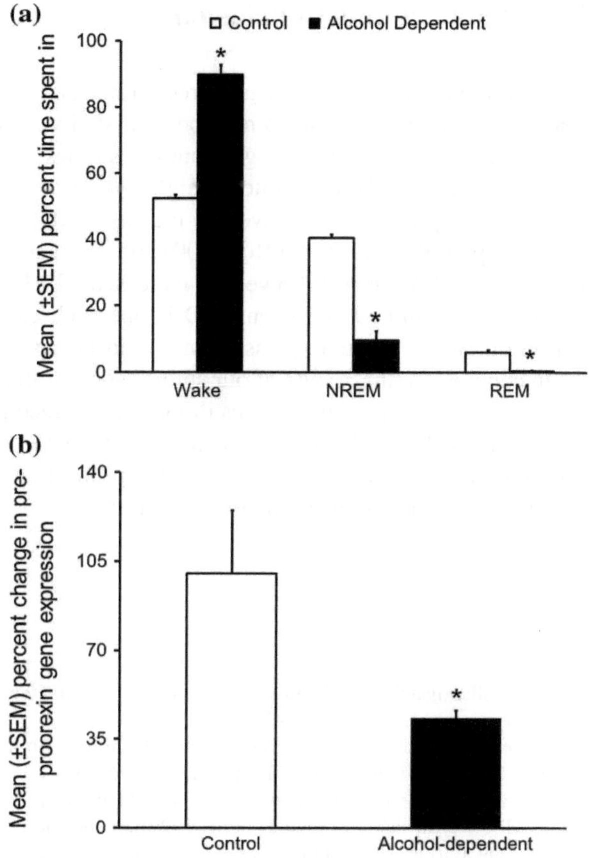

Fig. 8 Alcohol withdrawal following chronic binge alcohol exposure disrupts sleep and downregulates orexin expression. **a** During 24 h of withdrawal, alcohol dependent rats displayed significant increase in the amount of time spent in the wakefulness along with a concomitant reduction in NREM and REM sleep as compared to the controls. $^{*}p < 0.05$ versus control. (Reproduced from Sharma et al. 2010). **b** As compared to controls, animals exposed to chronic binge alcohol followed by withdrawal displayed a significant reduction in the pre-proorexin gene expression in the hypothalamus. $^{*}p < 0.05$ versus control. (Unpublished results)

last administration. At this point, blood alcohol level is almost zero and the withdrawal symptoms are at peak (Faingold 2008; Morris et al. 2010; Penland et al. 2001). The hypothalamus was rapidly dissected out and processed for quantitative Real-Time PCR to examine the expression of pre-proorexin gene (Sharma et al. 2010).

Interestingly, our preliminary/unpublished results suggested that as compared to controls, the rat undergoing alcohol withdrawal showed a significant reduction in pre-proorexin gene expression (Fig. 8b). While we are in the process of understanding the mechanisms responsible for the downregulation of orexin gene expression, we believe that chronic alcohol exposure induced perturbations in the epigenome may be responsible for reduction in orexin gene expression (Lodhi et al. 2011; Pandey et al. 2008). Downregulation of orexin gene expression is likely to result in increased REM sleep which is a common phenomenon observed in recovering alcoholics (Gillin et al. 1994).

9 Summary and Conclusion

As described above, we have gathered strong and convincing evidence suggesting that alcohol's effects on sleep may be mediated via an interaction between AD and the orexin neurons. Interestingly, human studies suggest that acute alcohol consumption, (especially moderate to high doses) in non-alcoholics results in robust NREM sleep promotion followed by increased REM sleep during the second half of the sleep period (Roehrs and Roth 2001; Ebrahim et al. 2013). We believe that both, increased NREM sleep followed by increased REM sleep may occur only if orexin neurons are inhibited and silent. AD is one such candidate that will inhibit via its action on A1 receptors expressed on orexin neurons.

In conclusion, the results of our studies suggest that alcohol interacts and inhibits ENT1 on orexin neurons resulting in extracellular accumulation of AD. Extracellular AD will act on A1 receptors expressed on orexin neurons resulting in the inhibition of orexin neurons and sleep promotion. Thus, alcohol promotes sleep via AD induced inhibition of orexin neurons.

References

Aalto J, Kiianmaa K (1984) Increased voluntary alcohol drinking concurrent with REM-sleep deprivation. Alcohol 1(1):77–79

Adamson J, Burdick JA (1973) Sleep of dry alcoholics. Arch Gen Psychiatry 28(1):146–149

Aguayo LG, Peoples RW, Yeh HH et al (2002) GABA(A) receptors as molecular sites of ethanol action. Direct or indirect actions? Curr Top Med Chem 2(8):869–885

Alam MN, Szymusiak R, Gong H et al (1999) Adenosinergic modulation of rat basal forebrain neurons during sleep and waking: neuronal recording with microdialysis. J Physiol 521 (Pt 3):679–690

Alam MN, Kumar S, Rai S et al (2009) Role of adenosine A(1) receptor in the perifornical-lateral hypothalamic area in sleep-wake regulation in rats. Brain Res 1304:96–104

Arrigoni E, Chamberlin NL, Saper CB et al (2006) Adenosine inhibits basal forebrain cholinergic and noncholinergic neurons in vitro. Neuroscience 140(2):403–413

Barwick VS, Dar MS (1998) Adenosinergic modulation of ethanol-induced motor incoordination in the rat motor cortex. Prog Neuropsychopharmacol Biol Psychiatry 22(4):587–607

Basheer R, Porkka-Heiskanen T, Strecker RE et al (2000) Adenosine as a biological signal mediating sleepiness following prolonged wakefulness. Biol Signals Recept 9(6):319–327

Basheer R, Halldner L, Alanko L et al (2001a) Opposite changes in adenosine A1 and A2A receptor mRNA in the rat following sleep deprivation. NeuroReport 12(8):1577–1580

Basheer R, Rainnie DG, Porkka-Heiskanen T et al (2001b) Adenosine, prolonged wakefulness, and A1-activated NF-kappaB DNA binding in the basal forebrain of the rat. Neuroscience 104(3):731–739

Basheer R, Strecker RE, Thakkar MM et al (2004) Adenosine and sleep-wake regulation. Prog Neurobiol 73(6):379–396

Batista LC, Prediger RD, Morato GS et al (2005) Blockade of adenosine and dopamine receptors inhibits the development of rapid tolerance to ethanol in mice. Psychopharmacology 181(4):714–721

Benington JH, Heller HC (1995) Restoration of brain energy metabolism as the function of sleep. Prog Neurobiol 45(4):347–360

Boissard R, Gervasoni D, Schmidt MH et al (2002) The rat ponto-medullary network responsible for paradoxical sleep onset and maintenance: a combined microinjection and functional neuroanatomical study. Eur J Neurosci 16(10):1959–1973

Borbely AA (1982) A two process model of sleep regulation. Human Neurobiol 1:195–204

Bouchery EE, Harwood HJ, Sacks JJ et al (2011) Economic costs of excessive alcohol consumption in the U.S., 2006. Am J Prev Med 41(5):516–524

Breese GR, Criswell HE, Carta M et al (2006) Basis of the gabamimetic profile of ethanol. Alcohol Clin Exp Res 30(4):731–744

Brower KJ (2001) Alcohol's effects on sleep in alcoholics. Alcohol Res Health 25(2):110–125

Brower KJ, Perron BE (2009) Sleep disturbance as a universal risk factor for relapse in addictions to psychoactive substances. Med Hypotheses 74(5):928–933

Brower KJ, Aldrich MS, Hall JM (1998) Polysomnographic and subjective sleep predictors of alcoholic relapse. Alcohol Clin Exp Res 22(8):1864–1871

Brower KJ, Aldrich MS, Robinson EA et al (2001) Insomnia, self-medication, and relapse to alcoholism. Am J Psychiatry 158(3):399–404

Brown RE, Basheer R, McKenna JT et al (2012) Control of sleep and wakefulness. Physiol Rev 92(3):1087–1187

Burgess CR, Peever JH (2013) A noradrenergic mechanism functions to couple motor behavior with arousal state. Curr Biol 23(18):1719–1725

CDC (2006) Alcohol-related disease impact (ARDI)

CDC (2012) Vital signs: binge drinking prevalence, frequency, and intensity among adults—United States, 2010. MMWR Morb Mortal Wkly Rep 61(1):12–19

Chemelli RM, Willie JT, Sinton CM et al (1999) Narcolepsy in orexin knockout mice: molecular genetics of sleep regulation. Cell 98(4):437–451

Choi DS, Cascini MG, Mailliard W et al (2004) The type 1 equilibrative nucleoside transporter regulates ethanol intoxication and preference. Nat Neurosci 7(8):855–861

Colrain IM, Turlington S, Baker FC (2009) Impact of alcoholism on sleep architecture and EEG power spectra in men and women. Sleep 32(10):1341–1352

Concas A, Mascia MP, Cuccheddu T et al (1996) Chronic ethanol intoxication enhances [3H] CCPA binding and does not reduce A1 adenosine receptor function in rat cerebellum. Pharmacol Biochem Behav 53(2):249–255

Daly JW, Padgett W, Shamim MT et al (1985) 1, 3-dialkyl-8-(p-sulfophenyl)xanthines: potent water-soluble antagonists for A1- and A2-adenosine receptors. J Med Chem 28(4):487–492

Dar MS (1993) Brain adenosinergic modulation of acute ethanol-induced motor impairment. Alcohol Alcohol Suppl 2:425–429

Dar MS (1997) Mouse cerebellar adenosinergic modulation of ethanol-induced motor incoordination: possible involvement of cAMP. Brain Res 749(2):263–274

Dar MS (2001) Modulation of ethanol-induced motor incoordination by mouse striatal A(1) adenosinergic receptor. Brain Res Bull 55(4):513–520

Dar MS, Meng ZH (2004) Acute ethanol-induced adenosine diphosphate ribosylation regulates the functional activity of rat striatal pertussis toxin-sensitive g proteins. Alcohol Clin Exp Res 28 (9):1299–1307

Dar MS, Mustafa SJ, Wooles WR (1983) Possible role of adenosine in the CNS effects of ethanol. Life Sci 33(14):1363–1374

Datta S, Maclean RR (2007) Neurobiological mechanisms for the regulation of mammalian sleep-wake behavior: reinterpretation of historical evidence and inclusion of contemporary cellular and molecular evidence. Neurosci Biobehav Rev 31(5):775–824

Drake CL, Roehrs T, Turner L et al (2003) Caffeine reversal of ethanol effects on the multiple sleep latency test, memory, and psychomotor performance. Neuropsychopharmacology 28 (2):371–378

Drummond SP, Gillin JC, Smith TL et al (1998) The sleep of abstinent pure primary alcoholic patients: natural course and relationship to relapse. Alcohol Clin Exp Res 22(8):1796–1802

Dunwiddie TV (1985) The physiological role of adenosine in the central nervous system (review). Int Rev Neurobiol 27:63–139

Dunwiddie TV, Masino SA (2001) The role and regulation of adenosine in the central nervous system. Annu Rev Neurosci 24:31–55
Ebrahim IO, Shapiro CM, Williams AJ et al (2013) Alcohol and sleep I: effects on normal sleep. Alcohol Clin Exp Res 37(4):539–549
Edgar DM, Dement WC, Fuller CA (1993) Effect of SCN lesions on sleep in squirrel monkeys: evidence for opponent processes in sleep-wake regulation. J Neurosci 13(3):1065–1079
Ehlers CL, Slawecki CJ (2000) Effects of chronic ethanol exposure on sleep in rats. Alcohol 20(2):173–179
Elmenhorst D, Meyer PT, Winz OH et al (2007) Sleep deprivation increases A1 adenosine receptor binding in the human brain: a positron emission tomography study. J Neurosci 27(9):2410–2415
Esser MB, Hedden SL, Kanny D et al (2014) Prevalence of alcohol dependence among US adult drinkers, 2009–2011. Prev Chronic Dis 11:E206
Faingold CL (2008) The Majchrowicz binge alcohol protocol: an intubation technique to study alcohol dependence in rats. Curr Protoc Neurosci 9–28 (chapter 9 unit)
Fillmore MT (2003) Alcohol tolerance in humans is enhanced by prior caffeine antagonism of alcohol-induced impairment. Exp Clin Psychopharmacol 11(1):9–17
Franks HM, Hagedorn H, Hensley VR et al (1975) The effect of caffeine on human performance, alone and in combination with ethanol. Psychopharmacologia 45(2):177–181
Fredholm BB (1995) Adenosine, adenosine receptors and the action of caffeine. Pharmacol Toxicol 76(2):93–101
Freund RK, Palmer MR (1996) 8-Bromo-cAMP mimics beta-adrenergic sensitization of GABA responses to ethanol in cerebellar Purkinje neurons in vivo. Alcohol Clin Exp Res 20(2):408–412
Ghosh TK, Copeland J, Pradhan SN (1991) Effects of ethanol inhalation on EEG in rats. Pharmacol Biochem Behav 38(2):293–297
Gillin JC, Smith TL, Irwin M et al (1994) Increased pressure for rapid eye movement sleep at time of hospital admission predicts relapse in nondepressed patients with primary alcoholism at 3-month follow-up. Arch Gen Psychiatry 51(3):189–197
Gitlow SE, Bentkover SH, Dziedzic SW et al (1973) Persistence of abnormal REM sleep response to ethanol as a result of previous ethanol ingestion. Psychopharmacologia 33(2):135–140
Gross MM, Hastey JM (1975) The relation between baseline slow wave sleep and the slow wave sleep response to alcohol in alcoholics. Adv Exp Med Biol 59:467–475
Gross MM, Goodenough DR, Hastey J et al (1973) Experimental study of sleep in chronic alcoholics before, during, and after four days of heavy drinking with a nondrinking comparison. Ann N Y Acad Sci 215:254–265
Gunzerath L, Hewitt BG, Li TK, Warren KR (2011) Alcohol research: past, present, and future. Ann N Y Acad Sci 1216:1–23
Hack SP, Christie MJ (2003) Adaptations in adenosine signaling in drug dependence: therapeutic implications. Crit Rev Neurobiol 15(3–4):235–274
Hasenfratz M, Bunge A, Dal PG et al (1993) Antagonistic effects of caffeine and alcohol on mental performance parameters. Pharmacol Biochem Behav 46(2):463–465
Hasin DS, Stinson FS, Ogburn E et al (2007) Prevalence, correlates, disability, and comorbidity of DSM-IV alcohol abuse and dependence in the United States: results from the National Epidemiologic Survey on alcohol and related conditions. Arch Gen Psychiatry 64(7):830–842
Hattan DG, Eacho PI (1978) Relationship of ethanol blood level to REM and non-REM sleep time and distribution in the rat. Life Sci 22(10):839–846
Hayaishi O, Urade Y, Eguchi N et al (2004) Genes for prostaglandin d synthase and receptor as well as adenosine A2A receptor are involved in the homeostatic regulation of nrem sleep. Arch Ital Biol 142(4):533–539
Hill SY, Reyes RB (1978a) Effects of chronic and acute ethanol administration on sleep in laboratory rats. J Stud Alcohol 39(1):47–55
Hill SY, Reyes RB (1978b) Effects of L-tryptophan and ethanol on sleep parameters in the rat. Psychopharmacology 58(3):229–233

Huang ZL, Urade Y, Hayaishi O (2007) Prostaglandins and adenosine in the regulation of sleep and wakefulness. Curr Opin Pharmacol 7(1):33–38

Hunsley MS, Palmiter RD (2003) Norepinephrine-deficient mice exhibit normal sleep-wake states but have shorter sleep latency after mild stress and low doses of amphetamine. Sleep 26 (5):521–526

Imatoh N, Nakazawa Y, Ohshima H et al (1986) Circadian rhythm of REM sleep of chronic alcoholics during alcohol withdrawal. Drug Alcohol Depend 18(1):77–85

Irwin M, Gillin JC, Dang J, Weissman J, Phillips E, Ehlers CL (2002) Sleep deprivation as a probe of homeostatic sleep regulation in primary alcoholics. Biol Psychiatry 51(8):632–641

Irwin M, Miller C, Gillin JC et al (2000) Polysomnographic and spectral sleep EEG in primary alcoholics: an interaction between alcohol dependence and African-American ethnicity. Alcohol Clin Exp Res 24(9):1376–1384

Ishibashi M, Nakazawa Y, Yokoyama T et al (1987) Cerebral atrophy and slow wave sleep of abstinent chronic alcoholics. Drug Alcohol Depend 19(4):325–332

Jarvis MF, Becker HC (1998) Single and repeated episodes of ethanol withdrawal increase adenosine A1, but not A2A, receptor density in mouse brain. Brain Res 786(1–2):80–88

Johnson LC, Burdick JA, Smith J (1970) Sleep during alcohol intake and withdrawal in the chronic alcoholic. Arch Gen Psychiatry 22(5):406–418

Johnson EO, Roehrs T, Roth T et al (1998) Epidemiology of alcohol and medication as aids to sleep in early adulthood. Sleep 21(2):178–186

Jones BE (2003) Arousal systems. Front Biosci 8(s438–51):s438–s451

Kalant H (1998) Research on tolerance: what can we learn from history? Alcohol Clin Exp Res 22(1):67–76

Kaplan GB, Bharmal NH, Leite-Morris KA et al (1999) Role of adenosine A1 and A2A receptors in the alcohol withdrawal syndrome. Alcohol 19(2):157–162

Keller M (1979) A historical overview of alcohol and alcoholism. Cancer Res 39(7 Pt 2):2822–2829.

Kerr JS, Sherwood N, Hindmarch I (1991) Separate and combined effects of the social drugs on psychomotor performance. Psychopharmacology 104(1):113–119

Khanna JM, Kalant H, Shah G et al (1991) Rapid tolerance as an index of chronic tolerance. Pharmacol Biochem Behav 38(2):427–432

Kleitman N (1963) Sleep and wakefulness. University of Chicago Press, Chicago, pp 1–552

Konadhode RR, Pelluru D, Blanco-Centurion C, Zayachkivsky A, Liu M, Uhde T, Glen WB, Jr., Van den Pol AN, Mulholland PJ, Shiromani PJ (2013) Optogenetic stimulation of MCH neurons increases sleep. J Neurosci 33(25):10257–10263

Krauss SW, Ghirnikar RB, Diamond I et al (1993) Inhibition of adenosine uptake by ethanol is specific for one class of nucleoside transporters. Mol Pharmacol 44(5):1021–1026

Kubota T, De A, Brown RA et al (2002) Diurnal effects of acute and chronic administration of ethanol on sleep in rats. Alcohol Clin Exp Res 26(8):1153–1161

Kumar S, Fleming RL, Morrow AL (2004) Ethanol regulation of gamma-aminobutyric acid A receptors: genomic and nongenomic mechanisms. Pharmacol Ther 101(3):211–226

Landolt HP (2008) Sleep homeostasis: a role for adenosine in humans? Biochem Pharmacol 75(11):2070–2079

Landolt HP, Dijk DJ, Gaus SE et al (1995a) Caffeine reduces low-frequency delta activity in the human sleep EEG. Neuropsychopharmacology 12(3):229–238

Landolt HP, Werth E, Borbely AA et al (1995b) Caffeine intake (200 mg) in the morning affects human sleep and EEG power spectra at night. Brain Res 675(1–2):67–74

Landolt HP, Retey JV, Tonz K et al (2004) Caffeine attenuates waking and sleep electroencephalographic markers of sleep homeostasis in humans. Neuropsychopharmacology 29(10):1933–1939

Levine HG (1978) The discovery of addiction. Changing conceptions of habitual drunkenness in America. J Stud Alcohol 39(1):143–174

Liguori A, Robinson JH (2001) Caffeine antagonism of alcohol-induced driving impairment. Drug Alcohol Depend 63(2):123–129

Lin L, Faraco J, Li R et al (1999) The sleep disorder canine narcolepsy is caused by a mutation in the hypocretin (orexin) receptor 2 gene. Cell 98(3):365–376
Liu ZW, Gao XB (2007) Adenosine inhibits activity of hypocretin/orexin neurons by the A1 receptor in the lateral hypothalamus: a possible sleep-promoting effect. J Neurophysiol 97(1):837–848
Lodhi S, Sharma R, deRoode D et al (2011) Epigentic modifications in the wake-promoting basal forebrain may contribute to insomnia during alcohol withdrawal. In: XIXth world congress of psychiatric genetics, 10–14 Sept 2011, Washington DC (PP157), p 149
Luk WP, Zhang Y, White TD et al (1999) Adenosine: a mediator of interleukin-1beta-induced hippocampal synaptic inhibition. J Neurosci 19(11):4238–4244
Majchrowicz E (1975) Induction of physical dependence upon ethanol and the associated behavioral changes in rats. Psychopharmacologia 43(3):245–254
McCarley RW (2007) Neurobiology of REM and NREM sleep. Sleep Med 8(4):302–330
Mendelson WB, Majchrowicz E, Mirmirani N et al (1978) Sleep during chronic ethanol administration and withdrawal in rats. J Stud Alcohol 39(7):1213–1223
Mizoguchi A, Eguchi N, Kimura K et al (2001) Dominant localization of prostaglandin D receptors on arachnoid trabecular cells in mouse basal forebrain and their involvement in the regulation of non-rapid eye movement sleep. Proc Natl Acad Sci U S A 98(20):11674–11679
Monti JM, Torterolo P, Lagos P (2013) Melanin-concentrating hormone control of sleep-wake behavior. Sleep Med Rev 17(4):293–298
Morris SA, Kelso ML, Liput DJ et al (2010) Similar withdrawal severity in adolescents and adults in a rat model of alcohol dependence. Alcohol 44(1):89–98
Mukherjee S, Simasko SM (2009) Chronic alcohol treatment in rats alters sleep by fragmenting periods of vigilance cycling in the light period with extended wakenings. Behav Brain Res 198 (1):113–124
Mukherjee S, Kazerooni M, Simasko SM (2008) Dose-response study of chronic alcohol induced changes in sleep patterns in rats. Brain Res 1208:120–127
Murillo-Rodriguez E, Blanco-Centurion C, Gerashchenko D et al (2004) The diurnal rhythm of adenosine levels in the basal forebrain of young and old rats. Neuroscience 123(2):361–370
Nagy LE, Diamond I, Casso DJ et al (1990) Ethanol increases extracellular adenosine by inhibiting adenosine uptake via the nucleoside transporter. J Biol Chem 265(4):1946–1951
National Institute on Alcohol Abuse and Alcoholism (2013) Alcohol use disorder: a comparison between DSM–IV and DSM–5
Newton PM, Messing RO (2006) Intracellular signaling pathways that regulate behavioral responses to ethanol. Pharmacol Ther 109(1–2):227–237
Pandey SC, Ugale R, Zhang H et al (2008) Brain chromatin remodeling: a novel mechanism of alcoholism. J Neurosci 28(14):3729–3737
Papale LA, Andersen ML, Perry JC et al (2008) Participation of the cholinergic system in the ethanol-induced suppression of paradoxical sleep in rats. Braz J Med Biol Res 41(9):782–788
Penland S, Hoplight B, Obernier J et al (2001) Effects of nicotine on ethanol dependence and brain damage. Alcohol 24(1):45–54
Phan TA, Gray AM, Nyce JW (1997) Intrastriatal adenosine A1 receptor antisense oligodeoxy-nucleotide blocks ethanol-induced motor incoordination. Eur J Pharmacol 323(2–3):R5–R7
Pollock MS, Mistlberger RE (2003) Rapid eye movement sleep induction by microinjection of the GABA-A antagonist bicuculline into the dorsal subcoeruleus area of the rat. Brain Res 962 (1–2):68–77
Porkka-Heiskanen T (2013) Sleep homeostasis. Curr Opin Neurobiol 23(5):799–805
Porkka-Heiskanen T, Strecker RE, Thakkar M et al (1997) Adenosine: a mediator of the sleep-inducing effects of prolonged wakefulness. Science 276(5316):1265–1268
Porkka-Heiskanen T, Strecker RE, McCarley RW (2000) Brain site-specificity of extracellular adenosine concentration changes during sleep deprivation and spontaneous sleep: an in vivo microdialysis study. Neuroscience 99(3):507–517
Portas CM, Thakkar M, Rainnie DG et al (1997) Role of adenosine in behavioral state modulation: a microdialysis study in the freely moving cat. Neuroscience 79(1):225–235

Prediger RD, da Silva GE, Batista LC et al (2006) Activation of adenosine A1 receptors reduces anxiety-like behavior during acute ethanol withdrawal (hangover) in mice. Neuropsychopharmacology 31(10):2210–2220

Prospero-Garcia O, Criado JR, Henriksen SJ (1994) Pharmacology of ethanol and glutamate antagonists on rodent sleep: a comparative study. Pharmacol Biochem Behav 49(2):413–416

Radulovacki M (1985) Role of adenosine in sleep in rats. Rev Clin Basic Pharm 5(3–4):327–339

Radulovacki M, Virus RM, Djuricic-Nedelson M et al (1984) Adenosine analogs and sleep in rats. J Pharmacol Exp Ther 228(2):268–274

Rai S, Kumar S, Alam MA et al (2010) A1 receptor mediated adenosinergic regulation of perifornical-lateral hypothalamic area neurons in freely behaving rats. Neuroscience 167 (1):40–48

Retey JV, Adam M, Honegger E et al (2005) A functional genetic variation of adenosine deaminase affects the duration and intensity of deep sleep in humans. Proc Natl Acad Sci U S A 102(43):15676–15681

Roehrs T, Roth T (2001) Sleep, sleepiness, and alcohol use. Alcohol Res Health 25(2):101–109

Roehrs T, Roth T (2012) Insomnia pharmacotherapy. Neurotherapeutics 9(4):728–738

Rosenberg PA, Li Y, Le M et al (2000) Nitric oxide-stimulated increase in extracellular adenosine accumulation in rat forebrain neurons in culture is associated with ATP hydrolysis and inhibition of adenosine kinase activity. J Neurosci 20(16):6294–6301

Rouhani S, Dall'Ava-Santucci J, Bajenaru O et al (1998) Effects of muscimol or homotaurine on sleep-wake states in alcohol-dependent rats during withdrawal. Pharmacol Biochem Behav 59(4):955–960

Saeed DM (2006) Co-modulation of acute ethanol-induced motor impairment by mouse cerebellar adenosinergic A1 and GABA(A) receptor systems. Brain Res Bull 71(1–3):287–295

Scammell TE, Gerashchenko DY, Mochizuki T et al (2001) An adenosine A2a agonist increases sleep and induces Fos in ventrolateral preoptic neurons. Neuroscience 107(4):653–663

Sharma R, Engemann S, Sahota P et al (2010) Role of adenosine and wake-promoting basal forebrain in insomnia and associated sleep disruptions caused by ethanol dependence. J Neurochem 115(3):782–794

Sharma R, Sahota P, Thakkar M (2014a) Rapid tolerance development to the NREM sleep promoting effect of alcohol. Sleep 37(4):821–824

Sharma R, Bradshaw K, Sahota P et al (2014b) Acute binge alcohol administration reverses sleep-wake cycle in sprague dawley rats. Alcohol Clin Exp Res 38(7):1941–1946

Sharma R, Sahota P, Thakkar MM (2014c) Role of adenosine and the orexinergic perifornical hypothalamus in sleep-promoting effects of ethanol. Sleep 37(3):525–533

Siggins GR, Roberto M, Nie Z (2005) The tipsy terminal: presynaptic effects of ethanol. Pharmacol Ther 107(1):80–98

Snyder S, Karacan I (1985) Sleep patterns of sober chronic alcoholics. Neuropsychobiology 13(1–2):97–100

Stahre M, Roeber J, Kanny D et al (2014) Contribution of excessive alcohol consumption to deaths and years of potential life lost in the United States. Prev Chronic Dis 11:E109

Stoller MK (1994) Economic effects of insomnia. Clin Ther 16(5):873–897

Strecker RE, Morairty S, Thakkar MM et al (2000) Adenosinergic modulation of basal forebrain and preoptic/anterior hypothalamic neuronal activity in the control of behavioral state. Behav Brain Res 115(2):183–204

Thakkar MM (2011) Histamine in the regulation of wakefulness. Sleep Med Rev 15(1):65–74

Thakkar MM, Ramesh V, Cape EG et al (1999) REM sleep enhancement and behavioral cataplexy following orexin (hypocretin)-II receptor antisense perfusion in the pontine reticular formation. Sleep Res Online 2(4):112–120

Thakkar MM, Winston S, McCarley RW (2002) Orexin neurons of the hypothalamus express adenosine A1 receptors. Brain Res 944(1–2):190–194

Thakkar MM, Delgiacco RA, Strecker RE et al (2003) Adenosinergic inhibition of basal forebrain wakefulness—active neurons: a simultaneous unit recording and microdialysis study in freely behaving cats. Neuroscience 22(4):1107–1113

Thakkar MM, Tao R, Yunren B et al (2004) GABA release in the pontine reticular formation is lowest during REM sleep. Soc Neurosci Abs 895–5490

Thakkar MM, Winston S, McCarley RW (2008) Effect of microdialysis perfusion of 4, 5, 6, 7-tetrahydroisoxazolo-[5, 4-c]pyridine-3-ol in the perifornical hypothalamus on sleep-wakefulness: role of delta-subunit containing extrasynaptic GABAA receptors. Neuroscience 153(3):551–555

Thakkar MM, Engemann SC, Sharma R et al (2010) Role of wake-promoting basal forebrain and adenosinergic mechanisms in sleep-promoting effects of ethanol. Alcohol Clin Exp Res 34(6):997–1005

Thakkar MM, Sharma R, Sahota P (2014) Alcohol disrupts sleep homeostasis. Alcohol 49 (4):299–310

Ticho SR, Stojanovic M, Lekovic G et al (1992) Effects of ethanol injection to the preoptic area on sleep and temperature in rats. Alcohol 9(3):275–278

Tobler I, Borbely AA, Groos G (1983) The effect of sleep deprivation on sleep in rats with suprachiasmatic lesions. Neurosci Lett 42(1):49–54

Vanini G, Wathen BL, Lydic R et al (2011) Endogenous GABA levels in the pontine reticular formation are greater during wakefulness than during rapid eye movement sleep. J Neurosci 31 (7):2649–2656

Veatch LM (2006) Disruptions in sleep time and sleep architecture in a mouse model of repeated ethanol withdrawal. Alcohol Clin Exp Res 30(7):1214–1222

Wagman AM, Allen RP (1975) Effects of alcohol ingestion and abstinence on slow wave sleep of alcoholics. Adv Exp Med Biol 59:453–466

Wallner M, Hanchar HJ, Olsen RW (2006) Low dose acute alcohol effects on GABA A receptor subtypes. Pharmacol Ther 112(2):513–528

Wang Y, Freund RK, Palmer MR (1999) Potentiation of ethanol effects in cerebellum by activation of endogenous noradrenergic inputs. J Pharmacol Exp Ther 288(1):211–220

Winsky-Sommerer R, Vyazovskiy VV, Homanics GE et al (2007) The EEG effects of THIP (Gaboxadol) on sleep and waking are mediated by the GABA(A)delta-subunit-containing receptors. Eur J Neurosci 25(6):1893–1899

Xi MC, Morales FR, Chase MH (2001) Induction of wakefulness and inhibition of active (REM) sleep by GABAergic processes in the nucleus pontis oralis. Arch Ital Biol 139 (1–2):125–145

Yules RB, Freedman DX, Chandler KA (1966) The effect of ethyl alcohol on man's electroencephalographic sleep cycle. Electroencephalogr Clin Neurophysiol 20(2):109–111

Orexin Induced Modulation of REM Sleep and Its Loss Associated Patho-Physiological Changes Are Mediated Through Locus Coeruleus

Birendra Nath Mallick, Mudasir Ahmad Khanday and Abhishek Singh

Abstract Orexinergic neurons are located in the perifornical (PeF) area and their projections have been reported in many areas in the brain including the hypothalamus and locus coeruleus (LC). Orexin is known to influence many patho-physiological processes, including REM sleep (REMS) and associated processes in health and diseases. Based on the findings/reports from this lab and that of others, we conclude that orexin-induced modulation (loss) of REMS and associated pathophysiological changes are mediated to a large extent at least by influencing the noradrenalin (NA)-ergic neurons in the LC, which possesses REM-OFF neurons.

Keywords Brainstem · Hypothalamus · Narcolepsy · Neural circuitry · Noradrenalin · Orexin-1 receptor

Abbreviations

EEG	Electroencephalogram
EMG	Electromyogram
EOG	Electrooculogram
LC	Locus coeruleus
LDT	Latero-dorsal tegementum
NA	Noradrenalin
NREMS	Non-REMS
Orx	Orexin
OX1R	Orexin-1 receptor
PeF	Perifornical area
PPT	Pedunclo-pontine tegmentum
REMS	Rapid eye movement sleep
REMSD	REMS deprivation
TMN	Tuberomammillary nucleus

B.N. Mallick (✉) · M.A. Khanday · A. Singh
School of Life Sciences, Jawaharlal Nehru University, New Delhi 110067, India
e-mail: remsbnm@yahoo.com

© Springer International Publishing Switzerland 2015

T. Sakurai et al. (eds.), *Orexin and Sleep*,
DOI 10.1007/978-3-319-23078-8_10

1 Introduction

Sleep and wakefulness are instinct behaviors expressed in living species higher in the evolutionary ladder. These states are believed to have been evolved form of the *basic rest and activity cycle* expressed even in the living species lower in evolution. Initially wakefulness was considered to be an *active process*, while sleep as a state left over due to withdrawal of wakefulness and therefore, a *passive phenomenon*. By the end of the second decade of the twentieth century, based on the electrophysiological signals recorded from the brain, the electroencephalogram (EEG), the muscles, the electromyogram (EMG) and the eye movements, the electrooculogram (EOG), the waking and the sleep were objectively defined, identified and classified into multiple sub-stages. These characteristics led to the identification of specific sites in the brain, which actively induce wakefulness, the wake inducing area(s) and sleep, the sleep inducing area(s). Such findings confirmed that sleep is also an *active process* and had put to rest that it is not a *passive phenomenon* as was believed before (reviewed by Moruzzi 1972).

Subsequently, based on the characteristic electrophysiological features (EEG, EMG and EOG, mentioned above) the sleep has been objectively divided into rapid eye movement sleep (REMS) and non-REMS (NREMS) (Aserinsky and Kleitman 1953; Dement 1958; Jouvet 1999). Classically REMS is characterized by the simultaneous presence of desynchronized EEG, rapid eye movements and muscle atonia during behavioral sleep. The REMS with all or some of its characteristic features has been identified in most of the species higher in evolution so far studied starting from marsupial mammals, the platypus (Siegel et al. 1998); its presence in lower species is debatable. The primary limitation of identification of REMS, if any, in lower species is the lack of an identifiable characteristic marker or feature. At present identification of REMS relies heavily on expression of EEG, which is present only in species having well-developed and evolved brain. Thus, whether REMS exists in lower species is still an open question and more importantly, there is certainly a need for identification of a possible more fundamental characteristic feature or a marker for defining REMS.

REMS is an integral component of sleep. It is present throughout the life of an individual; it is controlled by the neurons located in the central portion of the brainstem. The latter regulates most of the other fundamental processes necessary for the maintenance of normal life processes e.g. cardio-vascular and respiratory systems, sleep-waking, etc. REMS is important for maintenance of homeostatic processes in the body, e.g. thermoregulation, food intake, memory, brain excitability and so on (Siegel 1975; Mallick and Singh 2011). REMS has been reported to be affected in most of the altered psycho-patho-physiological disorders, and under experimental conditions its loss has been reported to affect several physiological processes and bio-molecules (Vogel 1975; Stickgold and Walker 2005; Gagnon et al. 2008). It has been proposed that REMS serves house-keeping function of the brain (Mallick and Singh 2011). Therefore, it is important to understand the regulation of REMS in its totality.

Based on the characteristic electrophysiological parameters the REMS was identified by Aserinsky and Kleitman (1953). After the discovery of REMS in the mid-twentieth century, by the end of the last century, it was clear that neurons located in the central core of the brainstem control this state, however, other regions of the higher brain structure e.g. hypothalamus and sub-cortical structures have modulatory influence (Moruzzi 1972). It may be noted that in older studies the EEG desynchronization in chronic cerveau isole preparation (Moruzzi 1972) could have been associated to REMS; however, as in those days REMS had not been identified, it was interpreted as waking process (Jouvet 1999). In the brainstem the neurons in the dorso-lateral pontine regions, the locus coeruleus (LC) and the latero-dorsaltegmentum (LDT)/pedunculo-pontine tegmentum (PPT) areas were found to possess REM-OFF and REM-ON neurons, respectively (Chu and Bloom 1973; Hobson et al. 1975). Findings from series of systematic studies have led us to propose that cessation of the LC-REM-OFF neurons is a pre-requisite for the generation of REMS (Pal and Mallick 2007; Mallick et al. 2012). The other aspect to be highlighted for REMS generation is that normally it does not appear during waking and it is expressed only after a period of NREMS has been set in. The findings from these physiological in vivo studies have been reinforced by the results of a mathematical modeling and simulation study of the neural circuit (Kumar et al. 2012).

Observations from clinical and autopsy reports, as well as findings from experimental transection, lesion and stimulation studies, it was concluded that the rostral brainstem and posterior hypothalamic regions are responsible for waking inducing, while the caudal brainstem and anterior hypothalamic regions are responsible for NREMS (Moruzzi 1972). Reciprocal influence and interaction among the neurons of some of the NREMS and wakefulness regulating brain areas for the control of sleep-waking has been reported (McGinty and Szymusiak 1971; Mallick et al. 1984, 1986, 1998, 2004; Thankachan et al. 2001). Subsequently, it was observed that the waking and NREMS inducing areas have excitatory and inhibitory influences (opposite effects) on the REM-OFF and REM-ON neurons, respectively (Mallick and Inoue 1999; Thankachan et al. 2001; Mallick et al. 2004). *Finally, it has been summarized that for REMS generation the inputs from waking area(s) on the REM-OFF neurons have to be withdrawn, while that from the NREMS area(s) on the REM-ON neurons have to be activated* (Mallick et al. 2012).

2 Orexinergic Neurons and REMS Modulation

As mentioned above based on lesion, transection and stimulation studies the posterior hypothalamus as well as the postero-lateral histaminergic neurons were identified as a wake promoting areas (Nauta 1946; Moruzzi 1972). In more recent years a small perifornical (PeF) region in the postero-lateral hypothalamus, orexin (Orx, also known as hypocretin) containing neurons have been identified; these neurons are the exclusive source of Orx in the brain (de Lecea et al. 1998; Sakurai et al. 1998). The Orx-ergic neurons in the PeF receive afferent inputs from many brain regions and send efferent output throughout the brain (Peyron et al. 1998; Nambu et al. 1999). Such neural connections enable these neurons to influence many biological processes including sleep-waking (Sutcliffe and de Lecea 1999; Nishino and Sakurai 2006; Prober et al. 2006). The critical role of Orx-ergic neurons in REMS regulation will be discussed below.

2.1 Support from Phylogenetic and Ontogenetic Comparison

We performed a search using STRING (www.string-db.org) to identify a gene regulatory network of proteins involved in Orx-ergic and NA-ergic metabolism. As shown in Fig. 1, Orx is present in all mammals except in *Dasypus noveminctus* (Nine Band Armadillo) and in *Ornithorhyncus anatinus* (Platypus) (monotremes).

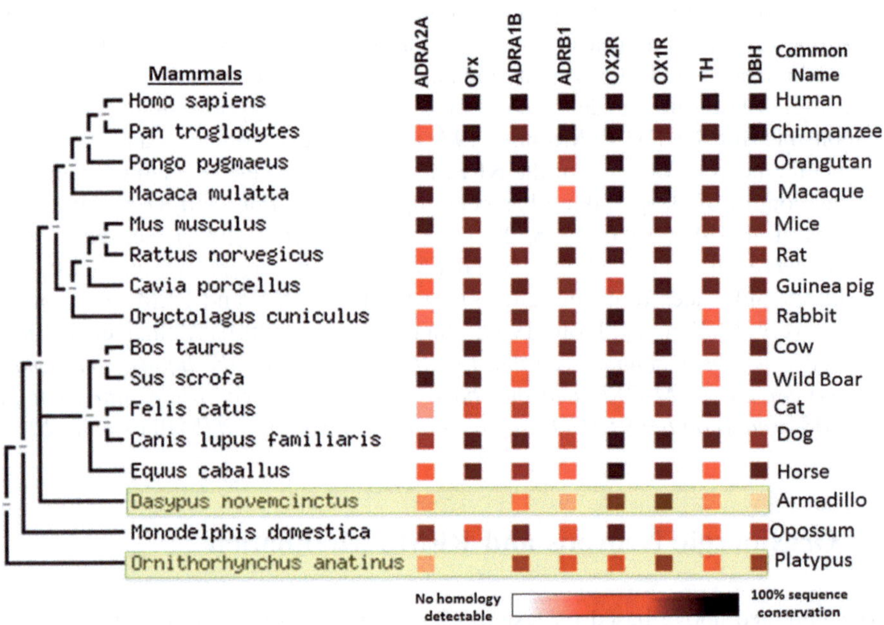

Fig. 1 This figure shows *occurrence view* of STRING analysis for queried genes among different mammals. The following gene viz. Orx, OX1R, OX2R, ADRA1, ADRA2, ADRB1, TH, and DBH, known to be responsible for synthesis of Orx and NA as well as their receptors were submitted for analysis. No homolog of Orx-gene was detected in *Dasypus noveminctus* and *Ornithorhyncus anatinus*. *Abbreviations* α_1 adrenoceptor—ADRA1, α_2 adrenoceptor—ADRA2, β adrenoceptor—ADRB1, tyrosine hydroxylase—TH, and dopamine β hydroxylase—DBH

However, these animals have been reported to express REMS (or like state) along with other mammals studied so far (Allada and Siegel 2008). The REMS characteristics expressed in these animals, although seem to be under brainstem regulation, are obviously without any modulation by Orx. This suggests that Orx may not be directly involved in primary regulation of REMS. Nevertheless, synthesis of sufficient quantity of Orx has been reported in 10–15 post-natal day in the species where Orx is present (Aristakesian 2009; Stoyanova et al. 2010). On the other hand, in related study ontogenic findings suggest that about two post-natal weeks are critical period for the expression of REMS and this has been correlated with development of NA-ergic signalling in specific region in the brain (Mallick et al. 2012). Thus, the Orx-ergic and NA-ergic systems and REMS have comparable temporal ontogenic development schedule. However, for confirmation whether they have cause and effect relationship need further study. These correlations support the school of thought that during development initially there is more REMS like state, which gradually gets fragmented for the expressions of matured form of REMS in animals (Roffwarg et al. 1966; Jouvet-Mounier et al. 1970; Vogel et al. 2000).

2.2 *Support from Clinical Studies (Narcolepsy)*

Narcolepsy is characterized by disorganized sleep-wakefulness cycle and cataplexy (Tsujino and Sakurai 2009; Weinhold et al. 2014) and increased REMS (Chemelli et al. 1999; Willie et al. 2003). Loss of Orx has been considered to be among the primary causes of narcolepsy (Kilduff and Peyron 2000; Sutcliffe and de Lecea 2002). It has been shown that ~90 % of Orx-ergic neurons are lost in human narcolepsy with cataplexy (Burgess and Scammell 2012). Lack of Orx signalling contributes to narcolepsy was further confirmed by the findings from post-mortem brains of human narcolepsy-cataplexy patients that showed a dramatic loss of Orx-mRNA and Orx-immunoreactive neurons (Peyron et al. 2000; Thannickal et al. 2000). Orx-A level is undetectable in cerebro-spinal fluid of most human narcoleptic patients (Nishino et al. 2009). Many of the narcolepsy symptoms, e.g., rapid state-transitions or inability to remain awake for long and sudden episodes of cataplexy during waking behavior are also exhibited by Orx-receptor knockout mice and Orx/ataxin-3 transgenic mice and rats with ablated Orx neurons (Siegel 2004; Ohno and Sakurai 2008). Orx deficiency in humans and animals produces characteristic symptoms of REMS disorder and narcolepsy (Peyron et al. 1998; Chemelli et al. 1999). These findings provide the basis to conclude the Orx is directly or indirectly involved in REMS regulation. Also, the findings suggest that Orx normally facilitates alertness and muscle tone and inhibits REMS and associated atonia.

2.3 Support from Experimental, Anatomical and Physiological Studies

Several lines of evidence support the involvement of Orx neurons in the regulation of sleep-waking-REMS; however, their detail mechanism of action is not clearly understood yet. Administration of Orx (icv) in rats increased arousal (Thakkar et al. 1999; Bourgin et al. 2000). Chemical (Thakkar et al. 1999; Alam and Mallick 2008), electrical (Choudhary et al. 2014) or optogenetic (Carter and de Lecea 2011; Carter et al. 2012, 2013) stimulation of PeF-Orx-ergic neurons induced arousal and reduced REMS. An accepted molecular correlate of neuronal activity, the c-fos expression, increased in PeF-Orx-ergic neurons during extended waking as compared with normal sleep-waking (Estabrooke et al. 2001). REMS duration as well as frequency was significantly increased in Orx-knockout mice (Chemelli et al. 1999; Willie et al. 2003). Orx-A injection into the anterior hypothalamus (Methippara et al. 2000), LC (Bourgin et al. 2000; Smith et al. 2003) and LDT (Bernard et al. 2003, 2006; Nunez et al. 2006) modulated REMS. Also, reduced level of Orx and reduced number of Orx-ergic neurons have been reported in animals showing increased REMS (Chemelli et al. 1999; Nishino et al. 2000; Gerashchenko et al. 2001; Hara et al. 2001; Beuckmann et al. 2004). Orx neurons showed increased activity during waking and are quiescent during NREMS as well as during tonic phase of REMS (Lee et al. 2005; Mileykovskiy et al. 2005). In this context it is expected that Orx level is likely to decrease during REMS and if these neurons are kept active, REMS and NREMS should be prevented and waking should increase. As a corollary, it is also expected that Orx level should increase during increased physical activity and REMS loss. Indeed it has been show that long term activation of the PeF-neurons reduced REMS (Choudhary et al. 2014) and Orx level increased in the LC after REMS deprivation (REMSD) (Mehta et al. 2015).

2.4 Relationship of Orx-ergic Neurons with REMS Regulating Areas

The Orx neurons in the PeF are the primary source of Orx in the brain and project to all parts of the brain, with moderate to dense innervations to LC, the raphe, the tubero-mammillary nucleus (TMN), and the LDT/PPT (Peyron et al. 1998; Greco and Shiromani 2001; Marcus et al. 2001; Gerashchenko et al. 2003); the LC probably receives the maximum inputs (Horvath et al. 1999; Kilduff and Peyron 2000; Sutcliffe and de Lecea 2000). Orx-ergic neurons also receive extensive inputs from the LC, ventral tegmental area (dopaminergic neurons), raphe, TMN; pontine reticular formation; LDT/PPT and the basal forebrain (Yamanaka et al. 2003, 2006; Muraki et al. 2004; Yamanaka 2006).

Normally REMS does not appear during waking and it appears only after a period of NREMS has set in. Thus, REMS may be modulated by direct input on REMS regulatory apparatus, or indirectly by modulating the waking and NREMS regulatory apparatus. Based on several studies we have concluded that cessation of LC-REM-OFF neurons is a necessity for the generation of REMS (Pal and Mallick 2007). Further, simulation studies have shown that REMS does not appear if the waking inputs on the REMS active neurons is not withdrawn (Kumar et al. 2012). Several lines of evidence so far suggest that Orx has profound influence on the muscle activity so that the latter remains active. This helps the animal to keep oneself active (physically as well as mentally/cognitively) so that they can hunt and gather food for survival (Mieda and Yanagisawa 2002; Mieda et al. 2006). This view may be supported by the fact that role of Orx in feeding and hunger (Burdakov et al. 2005; Clark et al. 2009) has also been established. In line with this hypothesis, we proposed and confirmed as explained below that the Orx influences REMS by modulating the LC-neurons, the REMS regulatory apparatus and thereby secondarily influences REMS.

3 Role of NA-LC Neurons in REMS

The neurons in the LC, the major source of NA in the brain, are active during wakefulness, less active during NREMS, and quiescent during REMS (Hobson et al. 1975; Aston-Jones and Bloom 1981) and their cessation is a pre-requisite for REMS regulation (Mallick et al. 2012). Electrical or pharmacological activation of LC neurons reduced REMS (Singh and Mallick 1996; Kaur et al. 2004). NA level is reduced during REMS (Berridge and Abercrombie 1999; Shouse et al. 2000), while it is elevated during REMSD (Porkka-Heiskanen et al. 1995). In a recent study using optogenetic stimulation it has been confirmed that activation of the LC-NA-ergic neurons reduced REMS and increased waking (Carter et al. 2010).

4 Evidences Favouring Interactions Between Orx-Ergic and NA-Ergic Neurons for REMS Regulation

(i) There are to and fro connections between the PeF-Orx-ergic and NA-ergic LC neurons (Baldo et al. 2003; Yamanaka et al. 2003; Espana et al. 2005; Li and van den Pol 2005; Yamanaka 2006); (ii) Orx depolarizes LC neurons (Ivanov and Aston-Jones 2000) and microinjection of Orx into the LC increases wakefulness and reduces NREMS and REMS (Bourgin et al. 2000). Similarly, Orexin-1 receptor (OX1R) knockout in LC increases REMS in mice (Chen et al. 2010); (iii) electrical or chemical stimulation of Orx-ergic neurons modulates waking, NREMS and REMS (Alam and Mallick 2008; Choudhary et al. 2014; Kostin et al. 2014); (iv) NA regulates Orx neuronal activity (Bayer et al. 2005); (v) in vivo studies demonstrated that administration of Orx-A but not Orx-B agonist into the LC of rats suppressed REMS and increased wakefulness at the expense of REMS and NREMS (Bourgin et al. 2000); (vi) excitatory effect of Orx-A on LC neurons in vivo have been demonstrated by increased firing rate in single unit recording from LC after iontophoretic application of Orx-A (Bourgin et al. 2000); (vii) the Orx acting through OX1R modulates NA release (somato-dendritic as well as axon terminal) from LC neurons (Chen et al. 2008) in a dose dependent manner; (viii) the observation in (vii) above may be supported by the findings from in situ hybridization study that the OX1R is exclusively expressed in NA-ergic LC neurons (Mieda et al. 2011); (viii) further, in narcolepsy although the patients can express muscle tone, upon exposure to specific excitement causing factor, there is reduced muscle tone (cataplexy) and there is increased REMS (or like state).

We know that as long as the LC neurons are kept active the REMS does not appear (Mallick et al. 2012) and that the LC neurons are kept active by the wake active brain areas (Thankachan et al. 2001) as well as by Orx (Hagan et al. 1999; Horvath et al. 1999). Therefore, it is likely that Orx-ergic neurons are one of the many components for keeping the LC-neurons active that keeps REMS and its associated muscle atonia at bay during waking. We propose that in normal condition these Orx-ergic neurons possibly are activated by several psychic (excitement) inputs, which keep the LC-neurons active. However, in narcolepsy those excitement related inputs are lost (withdrawn) because the Orx-ergic neurons are known to be lost or reduced in number in narcolepsy (Kilduff and Peyron 2000; Sutcliffe and de Lecea 2002); also we suppose there is likely to be genetic predisposition, which needs to be confirmed. The possible cause of loss of Orx-ergic neurons has been discussed below. So far direct evidence in support of central regulation of Orx-induced muscle tone increase is lacking. However, Orx-ergic activation of LC neurons and muscle tone activation has been reported (Kiyashchenko et al. 2001). NA-ergic systems have been shown to maintain the muscle tone during arousal and this regulation is possibly lost in narcoleptic canines and thus, cataplexy is expressed (Monti et al. 1988; Crochet and Sakai 1999; Wu et al. 1999; Shouse et al. 2000; Kiyashchenko et al. 2001). Also, behavioral instability was observed in NA-deficient models (Monti et al. 1988; Delagrange et al. 1993). Finally, in a recent study in rats we have observed that after REMSD Orx-A level increases in the LC but not in the PPT (Mehta et al. 2015) support our contention.

5 Orexinergic Effects Are Mediated Through the LC-Neurons

So far we discussed the indirect evidence supporting Orx-ergic influence on muscle atonia (and possibly REMS) is mediated through the LC neurons. As direct evidence in a recent in vivo study in rats, we have confirmed the same. In brief, chronic rats were surgically prepared with two pairs of cannulae implanted bilaterally into the PeF and the LC. The rats also had electrodes for recording electrophysiological signals for identification sleep-waking-REMS. In these surgically prepared chronic rats, stimulation of PeF neurons by glutamate microinjection reduced REMS and increased waking. However, the following day in the same rats the effects of PeF stimulation on REMS were prevented if Orx-ergic receptor antagonist (SB-408124) was simultaneously injected into the LC bilaterally (Choudhary et al. 2014) the scheme of the experimentation has been shown in Fig. 2. The finding of this study confirmed that the effect of stimulation of

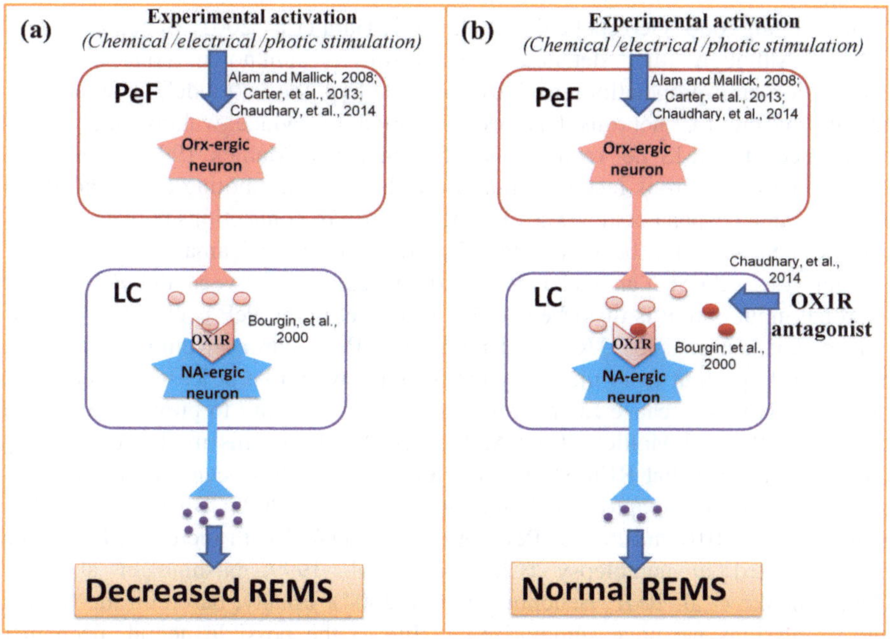

Fig. 2 Neural connection from PeF to LC for modulation of wakefulness and REMS has been illustrated. *Left panel* (**a**) depicts that chemical, electrical or photic stimulation of the PeF neurons have been reported to increase wakefulness and decrease REMS; while the *right panel* (**b**) shows that reduced REMS as in **a** was prevented by simultaneous microinjection of OX1R antagonist into the LC. Abbreviations as in the text

PeF-Orx-ergic neurons on REMS is mediated primarily by activating the LC neurons. Our recent finding that Orx-level in the LC increases after REMSD (Mehta et al. 2015) also support our contention.

6 Justification in Support of Orx-Induced Effects Are Mediated by NA Released from LC Neurons—Validity of Our Contention

So far we have discussed that increased activity of the Orx-ergic neurons reduces REMS by activating the LC neurons. This suggests that increased Orx-induced loss of REMS-associated pathological effects should be mediated by NA released from the LC neurons and reduction of Orx-ergic neurons should increase REMS. We would now attempt to explain some of the processes in support and to validate our contention. It has been reported that increased activity of Orx-ergic neurons is associated with reduced REMS (Alam et al. 2002) and loss of Orx-ergic neurons is associated with narcolepsy, when the REMS is increased (Chemelli et al. 1999; Nishino et al. 2000; Gerashchenko et al. 2001; Hara et al. 2001; Beuckmann et al. 2004). Although the detailed mechanism is unknown, our following explanation-cum-proposition provides a testable hypothesis (model). The increased activity of the LC neurons has been reported to induce waking and prevent occurrence of REMS i.e. reduced REMS (Singh and Mallick 1996; Carter et al. 2010). The LC neurons receive inputs from the wake promoting areas and PeF is one such area (Espana et al. 2005). Further, to and fro connections between the PeF and LC neurons have been reported (Baldo et al. 2003; Yamanaka et al. 2003; Espana et al. 2005; Yamanaka 2006). If the balance between the PeF-LC-PeF neuronal input/output is disturbed so that NA level in the PeF is elevated, it would induce PeF neuronal loss. Reduced number of PeF neurons would withdraw the excitatory PeF input from the LC neurons and it would be at least one reason for increased REMS, because LC neuronal activity is necessary to prevent appearance of REMS (Pal and Mallick 2007; Mallick et al. 2012). This model may be supported by the fact that REMSD induced elevated NA has been reported to cause increased neuronal apoptosis (Majumdar and Mallick 2005; Biswas et al. 2006; Ranjan et al. 2010) and loss of PeF neurons associated with increased REMS has been reported in narcolepsy (Chemelli et al. 1999; Nishino et al. 2000; Gerashchenko et al. 2001; Beuckmann et al. 2004). Finally, based on the findings and discussions we have summarized in Fig. 3 the possible neural mechanism for PeF-induced LC-mediated effect on REMS and its loss-associated patho-physiological changes.

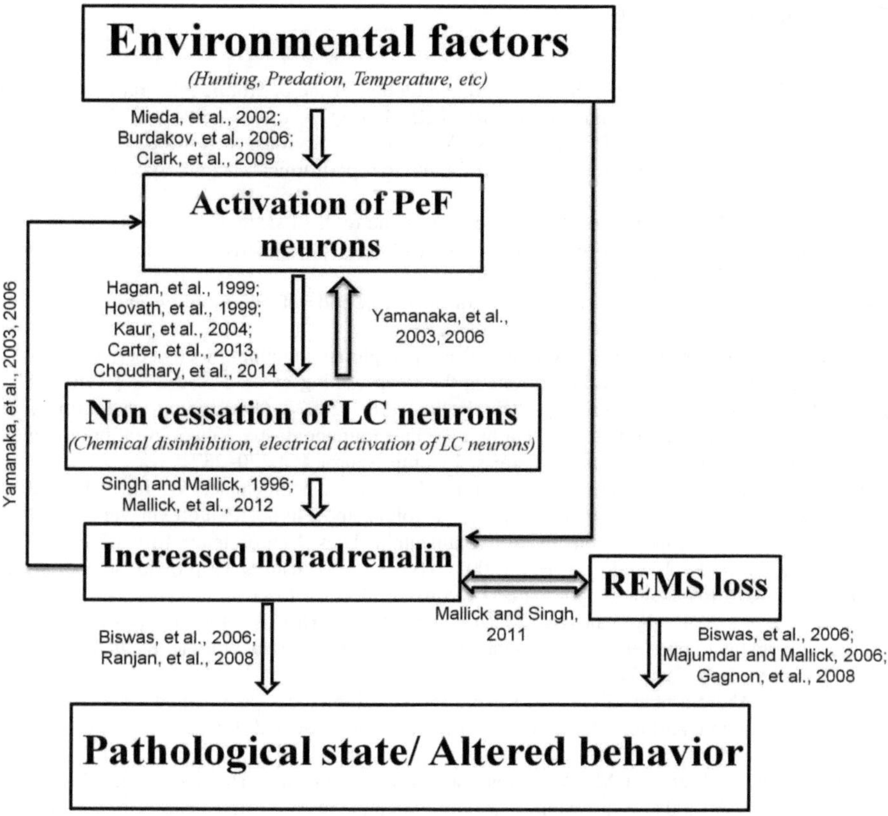

Fig. 3 PeF-induced LC-mediated modulation of REMS and its loss-associated patho-physiological changes have been schematically represented in this figure. Abbreviations as in the text

7 Conclusion

Orx-ergic neurons in the PeF are wake active, there are PeF-LC-PeF connections and Orx stimulates LC neurons. The PeF neurons modulate the LC neurons possibly to keep the animals active in doing normal activities. However, if the input/output of the PeF-LC-PeF is disturbed so as to elevate NA in the PeF, the latter neurons are lost, causing withdrawal of excitatory input from the LC-REM-OFF neurons. The latter is responsible for muscle atonia and increased REMS, the classical symptoms of narcolepsy.

Acknowledgments Research funding from Indian funding agencies viz. CSIR, DBT, BUILDER support, DST, J.C. Bose fellowship and UGC Resource Networking program are acknowledged.

References

Alam MA, Mallick BN (2008) Glutamic acid stimulation of the perifornical-lateral hypothalamic area promotes arousal and inhibits non-REM/REM sleep. Neurosci Lett 439:281–286

Alam MN, Gong H, Alam T, Jaganath R, McGinty D, Szymusiak R (2002) Sleep-waking discharge patterns of neurons recorded in the rat perifornical lateral hypothalamic area. J Physiol 538:619–631

Allada R, Siegel JM (2008) Unearthing the phylogenetic roots of sleep. Curr Biol 18:R670–R679

Aristakesian EA (2009) Evolutionary aspects of sleep and stress interaction: phylo-, ontogenetic approach. Zh Evol Biokhim Fiziol 45:598–611

Aserinsky E, Kleitman N (1953) Regularly occurring periods of eye motility, and concomitant phenomena, during sleep. Science 118:273–274

Aston-Jones G, Bloom FE (1981) Activity of norepinephrine-containing locus coeruleus neurons in behaving rats anticipates fluctuations in the sleep-waking cycle. J Neurosci 1:876–886

Baldo BA, Daniel RA, Berridge CW, Kelley AE (2003) Overlapping distributions of orexin/hypocretin- and dopamine-beta-hydroxylase immunoreactive fibers in rat brain regions mediating arousal, motivation, and stress. J Comp Neurol 464:220–237

Bayer L, Eggermann E, Serafin M, Grivel J, Machard D, Muhlethaler M, Jones BE (2005) Opposite effects of noradrenaline and acetylcholine upon hypocretin/orexin versus melanin concentrating hormone neurons in rat hypothalamic slices. Neuroscience 130:807–811

Bernard R, Lydic R, Baghdoyan HA (2003) Hypocretin-1 causes G protein activation and increases ACh release in rat pons. Eur J Neurosci 18:1775–1785

Bernard R, Lydic R, Baghdoyan HA (2006) Hypocretin (orexin) receptor subtypes differentially enhance acetylcholine release and activate g protein subtypes in rat pontine reticular formation. J Pharmacol Exp Ther 317:163–171

Berridge CW, Abercrombie ED (1999) Relationship between locus coeruleus discharge rates and rates of norepinephrine release within neocortex as assessed by in vivo microdialysis. Neuroscience 93:1263–1270

Beuckmann CT, Sinton CM, Williams SC, Richardson JA, Hammer RE, Sakurai T, Yanagisawa M (2004) Expression of a poly-glutamine-ataxin-3 transgene in orexin neurons induces narcolepsy-cataplexy in the rat. J Neurosci 24:4469–4477

Biswas S, Mishra P, Mallick BN (2006) Increased apoptosis in rat brain after rapid eye movement sleep loss. Neuroscience 142:315–331

Bourgin P, Huitron-Resendiz S, Spier AD, Fabre V, Morte B, Criado JR, Sutcliffe JG, Henriksen SJ, de Lecea L (2000) Hypocretin-1 modulates rapid eye movement sleep through activation of locus coeruleus neurons. J Neurosci 20:7760–7765

Burdakov D, Luckman SM, Verkhratsky A (2005) Glucose-sensing neurons of the hypothalamus. Philos Trans R Soc Lond B Biol Sci 360:2227–2235

Burgess CR, Scammell TE (2012) Narcolepsy: neural mechanisms of sleepiness and cataplexy. J Neurosci 32:12305–12311

Carter ME, de Lecea L (2011) Optogenetic investigation of neural circuits in vivo. Trends Mol Med 17:197–206

Carter ME, Yizhar O, Chikahisa S, Nguyen H, Adamantidis A, Nishino S, Deisseroth K, de Lecea L (2010) Tuning arousal with optogenetic modulation of locus coeruleus neurons. Nat Neurosci 13:1526–1533

Carter ME, Brill J, Bonnavion P, Huguenard JR, Huerta R, de Lecea L (2012) Mechanism for Hypocretin-mediated sleep-to-wake transitions. Proc Natl Acad Sci USA 109:E2635–E2644

Carter ME, de Lecea L, Adamantidis A (2013) Functional wiring of hypocretin and LC-NE neurons: implications for arousal. Front Behav Neurosci 7:43

Chemelli RM, Willie JT, Sinton CM, Elmquist JK, Scammell T, Lee C, Richardson JA, Williams SC, Xiong Y, Kisanuki Y, Fitch TE, Nakazato M, Hammer RE, Saper CB, Yanagisawa M (1999) Narcolepsy in orexin knockout mice: molecular genetics of sleep regulation. Cell 98:437–451

Chen XW, Mu Y, Huang HP, Guo N, Zhang B, Fan SY, Xiong JX, Wang SR, Xiong W, Huang W, Liu T, Zheng LH, Zhang CX, Li LH, Yu ZP, Hu ZA, Zhou Z (2008) Hypocretin-1 potentiates NMDA receptor-mediated somatodendritic secretion from locus ceruleus neurons. J Neurosci 28:3202–3208

Chen L, McKenna JT, Bolortuya Y, Winston S, Thakkar MM, Basheer R, Brown RE, McCarley RW (2010) Knockdown of orexin type 1 receptor in rat locus coeruleus increases REM sleep during the dark period. Eur J Neurosci 32:1528–1536

Choudhary RC, Khanday MA, Mitra A, Mallick BN (2014) Perifornical orexinergic neurons modulate REM sleep by influencing locus coeruleus neurons in rats. Neuroscience 279:33–43

Chu N, Bloom FE (1973) Norepinephrine-containing neurons: changes in spontaneous discharge patterns during sleeping and waking. Science 179:908–910

Clark EL, Baumann CR, Cano G, Scammell TE, Mochizuki T (2009) Feeding-elicited cataplexy in orexin knockout mice. Neuroscience 161:970–977

Crochet S, Sakai K (1999) Alpha-2 adrenoceptor mediated paradoxical (REM) sleep inhibition in the cat. NeuroReport 10:2199–2204

de Lecea L, Kilduff TS, Peyron C, Gao X, Foye PE, Danielson PE, Fukuhara C, Battenberg EL, Gautvik VT, Bartlett FS 2nd, Frankel WN, van den Pol AN, Bloom FE, Gautvik KM, Sutcliffe JG (1998) The hypocretins: hypothalamus-specific peptides with neuroexcitatory activity. Proc Natl Acad Sci USA 95:322–327

Delagrange P, Canu MH, Rougeul A, Buser P, Bouyer JJ (1993) Effects of locus coeruleus lesions on vigilance and attentive behaviour in cat. Behav Brain Res 53:155–165

Dement WC (1958) The occurrence of low voltage fast electroencephalogram patterns during behavioral sleep in the cat. Electroencephalogr Clin Neurophysiol 10:291–296

Espana RA, Reis KM, Valentino RJ, Berridge CW (2005) Organization of hypocretin/orexin efferents to locus coeruleus and basal forebrain arousal-related structures. J Comp Neurol 481:160–178

Estabrooke IV, McCarthy MT, Ko E, Chou TC, Chemelli RM, Yanagisawa M, Saper CB, Scammell TE (2001) Fos expression in orexin neurons varies with behavioral state. J Neurosci 21:1656–1662

Gagnon JF, Petit D, Latreille V, Montplaisir J (2008) Neurobiology of sleep disturbances in neurodegenerative disorders. Curr Pharm Des 14:3430–3445

Gerashchenko D, Kohls MD, Greco M, Waleh NS, Salin-Pascual R, Kilduff TS, Lappi DA, Shiromani PJ (2001) Hypocretin-2-saporin lesions of the lateral hypothalamus produce narcoleptic-like sleep behavior in the rat. J Neurosci 21:7273–7283

Gerashchenko D, Blanco-Centurion C, Greco MA, Shiromani PJ (2003) Effects of lateral hypothalamic lesion with the neurotoxin hypocretin-2-saporin on sleep in Long-Evans rats. Neuroscience 116:223–235

Greco MA, Shiromani PJ (2001) Hypocretin receptor protein and mRNA expression in the dorsolateral pons of rats. Brain Res Mol Brain Res 88:176–182

Hagan JJ, Leslie RA, Patel S, Evans ML, Wattam TA, Holmes S, Benham CD, Taylor SG, Routledge C, Hemmati P, Munton RP, Ashmeade TE, Shah AS, Hatcher JP, Hatcher PD, Jones DN, Smith MI, Piper DC, Hunter AJ, Porter RA, Upton N (1999) Orexin A activates locus coeruleus cell firing and increases arousal in the rat. Proc Natl Acad Sci USA 96:10911–10916

Hara J, Beuckmann CT, Nambu T, Willie JT, Chemelli RM, Sinton CM, Sugiyama F, Yagami K, Goto K, Yanagisawa M, Sakurai T (2001) Genetic ablation of orexin neurons in mice results in narcolepsy, hypophagia, and obesity. Neuron 30:345–354

Hobson JA, McCarley RW, Wyzinski PW (1975) Sleep cycle oscillation: reciprocal discharge by two brainstem neuronal groups. Science 189:55–58

Horvath TL, Peyron C, Diano S, Ivanov A, Aston-Jones G, Kilduff TS, van Den Pol AN (1999) Hypocretin (orexin) activation and synaptic innervation of the locus coeruleus noradrenergic system. J Comp Neurol 415:145–159

Ivanov A, Aston-Jones G (2000) Hypocretin/orexin depolarizes and decreases potassium conductance in locus coeruleus neurons. NeuroReport 11:1755–1758

Jouvet M (1999) Around the discovery of REM sleep in cats. In: Mallick BN, Inoue S (eds) Rapid eye movement sleep. Marcel Dekker, New York

Jouvet-Mounier D, Astic L, Lacote D (1970) Ontogenesis of the states of sleep in rat, cat, and guinea pig during the first postnatal month. Dev Psychobiol 2:216–239

Kaur S, Panchal M, Faisal M, Madan V, Nangia P, Mallick BN (2004) Long term blocking of GABA-A receptor in locus coeruleus by bilateral microinfusion of picrotoxin reduced rapid eye movement sleep and increased brain Na-K ATPase activity in freely moving normally behaving rats. Behav Brain Res 151:185–190

Kilduff TS, Peyron C (2000) The hypocretin/orexin ligand-receptor system: implications for sleep and sleep disorders. Trends Neurosci 23:359–365

Kiyashchenko LI, Mileykovskiy BY, Lai YY, Siegel JM (2001) Increased and decreased muscle tone with orexin (hypocretin) microinjections in the locus coeruleus and pontine inhibitory area. J Neurophysiol 85:2008–2016

Kostin A, Siegel JM, Alam MN (2014) Lack of hypocretin attenuates behavioral changes produced by glutamatergic activation of the perifornical-lateral hypothalamic area. Sleep 37:1011–1020

Kumar R, Bose A, Mallick BN (2012) A mathematical model towards understanding the mechanism of neuronal regulation of wake-NREMS-REMS states. PLoS ONE 7:e42059

Lee MG, Hassani OK, Jones BE (2005) Discharge of identified orexin/hypocretin neurons across the sleep-waking cycle. J Neurosci 25:6716–6720

Li Y, van den Pol AN (2005) Direct and indirect inhibition by catecholamines of hypocretin/orexin neurons. J Neurosci 25:173–183

Majumdar S, Mallick BN (2005) Cytomorphometric changes in rat brain neurons after rapid eye movement sleep deprivation. Neuroscience 135:679–690

Mallick BN, Inoue S (1999) Rapid eye movement sleep. Marcel Dekker, New York

Mallick BN, Singh A (2011) REM sleep loss increases brain excitability: role of noradrenaline and its mechanism of action. Sleep Med Rev 15:165–178

Mallick BN, Mohan Kumar V, Chhina GS, Singh B (1984) Responses of preoptic neurons to stimulation of caudal and rostral brain stem reticular structures. Brain Res Bull 13:353–356

Mallick BN, Kumar VM, Chhina GS, Singh B (1986) Comparison of rostro-caudal brain stem influence on preoptic neurons and cortical EEG. Brain Res Bull 16:121–125

Mallick BN, Thankachan S, Islam F (1998) Differential responses of brain stem neurons during spontaneous and stimulation-induced desynchronization of the cortical EEG in freely moving cats. Sleep Res Online 1:132–146

Mallick BN, Thankachan S, Islam F (2004) Influence of hypnogenic brain areas on wakefulness- and rapid-eye-movement sleep-related neurons in the brainstem of freely moving cats. J Neurosci Res 75:133–142

Mallick BN, Singh A, Khanday MA (2012) Activation of inactivation process initiates rapid eye movement sleep. Prog Neurobiol 97:259–276

Marcus JN, Aschkenasi CJ, Lee CE, Chemelli RM, Saper CB, Yanagisawa M, Elmquist JK (2001) Differential expression of orexin receptors 1 and 2 in the rat brain. J Comp Neurol 435:6–25

McGinty D, Szymusiak R (1971) Neuroanl unit activity patterns in behaving animals: brain stem and limbic system. Annu Rev Psychol 39:135–168

Mehta R, Khanday MA, Mallick BN (2015) REM sleep loss associated changes in orexin-A levels in discrete brain areas in rats. Neurosci Lett 590:62–67

Methippara MM, Alam MN, Szymusiak R, McGinty D (2000) Effects of lateral preoptic area application of orexin-A on sleep-wakefulness. NeuroReport 11:3423–3426

Mieda M, Yanagisawa M (2002) Sleep, feeding, and neuropeptides: roles of orexins and orexin receptors. Curr Opin Neurobiol 12:339–345

Mieda M, Williams SC, Richardson JA, Tanaka K, Yanagisawa M (2006) The dorsomedial hypothalamic nucleus as a putative food-entrainable circadian pacemaker. Proc Natl Acad Sci USA 103:12150–12155

Mieda M, Hasegawa E, Kisanuki YY, Sinton CM, Yanagisawa M, Sakurai T (2011) Differential roles of orexin receptor-1 and -2 in the regulation of non-REM and REM sleep. J Neurosci 31:6518–6526

Mileykovskiy BY, Kiyashchenko LI, Siegel JM (2005) Behavioral correlates of activity in identified hypocretin/orexin neurons. Neuron 46:787–798

Monti JM, D'Angelo L, Jantos H, Barbeito L, Abo V (1988) Effect of DSP-4, a noradrenergic neurotoxin, on sleep and wakefulness and sensitivity to drugs acting on adrenergic receptors in the rat. Sleep 11:370–377

Moruzzi G (1972) The sleep-waking cycle, Ergebnisse der Physiologie, Biologischen Chemie und Experimentellen Pharmakologie 64:1–165

Muraki Y, Yamanaka A, Tsujino N, Kilduff TS, Goto K, Sakurai T (2004) Serotonergic regulation of the orexin/hypocretin neurons through the 5-HT1A receptor. J Neurosci 24:7159–7166

Nambu T, Sakurai T, Mizukami K, Hosoya Y, Yanagisawa M, Goto K (1999) Distribution of orexin neurons in the adult rat brain. Brain Res 827:243–260

Nauta WJH (1946) Hypothalamic regulation of sleep in rats: an experimental study. J Neurophysiol 9:285–316

Nishino S, Sakurai T (eds) (2006) The orexin/hypocretin system: physiology and pathophysiology. Humana Press, New Jersey

Nishino S, Okura M, Mignot E (2000) Narcolepsy: genetic predisposition and neuropharmacological mechanisms. REVIEW ARTICLE. Sleep Med Rev 4:57–99

Nishino S, Sakurai E, Nevsimalova S, Yoshida Y, Watanabe T, Yanai K, Mignot E (2009) Decreased CSF histamine in narcolepsy with and without low CSF hypocretin-1 in comparison to healthy controls. Sleep 32:175–180

Nunez A, Moreno-Balandran ME, Rodrigo-Angulo ML, Garzon M, De Andres I (2006) Relationship between the perifornical hypothalamic area and oral pontine reticular nucleus in the rat. Possible implication of the hypocretinergic projection in the control of rapid eye movement sleep. Eur J Neurosci 24:2834–2842

Ohno K, Sakurai T (2008) Orexin neuronal circuitry: role in the regulation of sleep and wakefulness. Front Neuroendocrinol 29:70–87

Pal D, Mallick BN (2007) Neural mechanism of rapid eye movement sleep generation with reference to REM-OFF neurons in locus coeruleus. Ind J Med Res 125:721–739

Peyron C, Tighe DK, van den Pol AN, de Lecea L, Heller HC, Sutcliffe JG, Kilduff TS (1998) Neurons containing hypocretin (orexin) project to multiple neuronal systems. J Neurosci 18:9996–10015

Peyron C, Faraco J, Rogers W, Ripley B, Overeem S, Charnay Y, Nevsimalova S, Aldrich M, Reynolds D, Albin R, Li R, Hungs M, Pedrazzoli M, Padigaru M, Kucherlapati M, Fan J, Maki R, Lammers GJ, Bouras C, Kucherlapati R, Nishino S, Mignot E (2000) A mutation in a case of early onset narcolepsy and a generalized absence of hypocretin peptides in human narcoleptic brains. Nat Med 6:991–997

Porkka-Heiskanen T, Smith SE, Taira T, Urban JH, Levine JE, Turek FW, Stenberg D (1995) Noradrenergic activity in rat brain during rapid eye movement sleep deprivation and rebound sleep. Am J Physiol 268:R1456–R1463

Prober DA, Rihel J, Onah AA, Sung RJ, Schier AF (2006) Hypocretin/orexin overexpression induces an insomnia-like phenotype in zebrafish. J Neurosci 26:13400–13410

Ranjan A, Biswas S, Mallick BN (2010) Cytomorphometric changes in the dorsal raphe neurons after rapid eye movement sleep deprivation are mediated by noradrenalin in rats. Behav Brain Funct 6:62

Roffwarg HP, Muzio JN, Dement WC (1966) Ontogenetic development of the human sleep-dream cycle. Science 152:604–619

Sakurai T, Amemiya A, Ishii M, Matsuzaki I, Chemelli RM, Tanaka H, Williams SC, Richarson JA, Kozlowski GP, Wilson S, Arch JR, Buckingham RE, Haynes AC, Carr SA, Annan RS, McNulty DE, Liu WS, Terrett JA, Elshourbagy NA, Bergsma DJ, Yanagisawa M (1998) Orexins and orexin receptors: a family of hypothalamic neuropeptides and G protein-coupled receptors that regulate feeding behavior, Cell 92 (1 page following 696)

Shouse MN, Staba RJ, Saquib SF, Farber PR (2000) Monoamines and sleep: microdialysis findings in pons and amygdala. Brain Res 860:181–189

Siegel JM (1975) REM sleep predicts subsequent food intake. Physiol Behav 15:399–403

Siegel JM (2004) The neurotransmitters of sleep. J Clin Psychiatry 65(Suppl 16):4–7

Siegel JM, Manger PR, Nienhuis R, Fahringer HM, Pettigrew JD (1998) Monotremes and the evolution of rapid eye movement sleep. Philos Trans R Soc Lond B Biol Sci 353:1147–1157

Singh S, Mallick BN (1996) Mild electrical stimulation of pontine tegmentum around locus coeruleus reduces rapid eye movement sleep in rats. Neurosci Res 24:227–235

Smith MI, Piper DC, Duxon MS, Upton N (2003) Evidence implicating a role for orexin-1 receptor modulation of paradoxical sleep in the rat. Neurosci Lett 341:256–258

Stickgold R, Walker MP (2005) Sleep and memory: the ongoing debate. Sleep 28:1225–1227

Stoyanova II, Rutten WL, le Feber J (2010) Orexin-A and orexin-B during the postnatal development of the rat brain. Cell Mol Neurobiol 30:81–89

Sutcliffe JG, de Lecea L (1999) Novel neurotransmitters for sleep and energy homeostasis. Results Probl Cell Differ 26:239–255

Sutcliffe JG, de Lecea L (2000) The hypocretins: excitatory neuromodulatory peptides for multiple homeostatic systems, including sleep and feeding. J Neurosci Res 62:161–168

Sutcliffe JG, de Lecea L (2002) The hypocretins: setting the arousal threshold. Nat Rev Neurosci 3:339–349

Thakkar MM, Ramesh V, Cape EG, Winston S, Strecker RE, McCarley RW (1999) REM sleep enhancement and behavioral cataplexy following orexin (hypocretin)-II receptor antisense perfusion in the pontine reticular formation. Sleep Res Online 2:112–120

Thankachan S, Islam F, Mallick BN (2001) Role of wake inducing brain stem area on rapid eye movement sleep regulation in freely moving cats. Brain Res Bull 55:43–49

Thannickal TC, Moore RY, Nienhuis R, Ramanathan L, Gulyani S, Aldrich M, Cornford M, Siegel JM (2000) Reduced number of hypocretin neurons in human narcolepsy. Neuron 27:469–474

Tsujino N, Sakurai T (2009) Orexin/Hypocretin: a neuropeptide at the interface of sleep, energy homeostasis, and reward system. Pharmacol Rev 61:162–176

Vogel GW (1975) A review of REM sleep deprivation. Arch Gen Psychiatry 32:749–761

Vogel GW, Feng P, Kinney GG (2000) Ontogeny of REM sleep in rats: possible implications for endogenous depression. Physiol Behav 68:453–461

Weinhold SL, Seeck-Hirschner M, Nowak A, Hallschmid M, Goder R, Baier PC (2014) The effect of intranasal orexin-A (hypocretin-1) on sleep, wakefulness and attention in narcolepsy with cataplexy. Behav Brain Res 262:8–13

Willie JT, Chemelli RM, Sinton CM, Tokita S, Williams SC, Kisanuki YY, Marcus JN, Lee C, Elmquist JK, Kohlmeier KA, Leonard CS, Richardson JA, Hammer RE, Yanagisawa M (2003) Distinct narcolepsy syndromes in Orexin receptor-2 and Orexin null mice: molecular genetic dissection of Non-REM and REM sleep regulatory processes. Neuron 38:715–730

Wu MF, Gulyani SA, Yau E, Mignot E, Phan B, Siegel JM (1999) Locus coeruleus neurons: cessation of activity during cataplexy. Neuroscience 91:1389–1399

Yamanaka A (2006) Afferent system of orexin neuron. In: Nishino S (ed) The Orexin/Hypocretin system: physiology and pathophysiology. Hamana Press, New Jersey, pp 61–69

Yamanaka A, Muraki Y, Tsujino N, Goto K, Sakurai T (2003) Regulation of orexin neurons by the monoaminergic and cholinergic systems. Biochem Biophys Res Commun 303:120–129

Yamanaka A, Muraki Y, Ichiki K, Tsujino N, Kilduff TS, Goto K, Sakurai T (2006) Orexin neurons are directly and indirectly regulated by catecholamines in a complex manner. J Neurophysiol 96:284–298

Role of Orexin on Sleep: Interactions with Other Neurotransmitter Systems

Pablo Torterolo, Jaime Monti and S.R. Pandi-Perumal

Abstract In 1998, a group of phenotypically distinct neurons were discovered in the postero-lateral hypothalamus, which contained the excitatory neuropeptides orexin-A and orexin-B (also called hypocretin-1 and hypocretin-2). Orexinergic neurons project throughout the central nervous system (CNS) and are involved in the generation and maintenance of wakefulness. The sleep disorder narcolepsy, characterized by hypersomnia and cataplexy, is produced by degeneration of these neurons. We conducted a critical review of the literature on the interactions of the orexinergic system with other neurotransmitter systems involved in the generation and maintenance of wakefulness and sleep. Orexin has an excitatory action over the cholinergic, serotonergic, noradrenergic, histaminergic and dopaminergic neurons that constitute the activating system. Moreover, orexinergic neurons modulate the activity of γ-aminobutyric acid (GABA) and melanin-concentrating hormone (MCH) sleep-promoting neurons. Of note, orexin and MCH have opposite post-synaptic effects. In this respect, the orexinergic neurons are active during active wakefulness and "phasic" rapid-eye movement (REM) sleep, while the MCHergic cells are active during non-rapid-eye movement (NREM) sleep and "tonic" REM sleep. We hypothesize that the interactions of these opposite but complementary hypothalamic systems are critical for the generation of either wakefulness or sleep.

Keywords Hypothalamus · Peptides · REM sleep · Narcolepsy · Cataplexy · Melanin-concentrating hormone · Orexin (hypocretin)

P. Torterolo (✉)
Laboratory of Sleep Neurobiology, Department of Physiology, School of Medicine, University of the Republic, Montevideo, Uruguay
e-mail: ptortero@fmed.edu.uy

J. Monti
Department of Pharmacology and Therapeutics, School of Medicine Clinics Hospital, University of the Republic, Montevideo, Uruguay

S.R. Pandi-Perumal
Division of Health and Behavior, Department of Population Health, NYU Langone Medical Center, New York, NY 10016, USA

© Springer International Publishing Switzerland 2015

T. Sakurai et al. (eds.), *Orexin and Sleep*,
DOI 10.1007/978-3-319-23078-8_11

1 Introduction

Narcolespy is a disabling neurological condition affecting 1 in 2000 individuals. The pathophysiology of narcolepsy spins around two main axes: the difficulty in maintaining wakefulness (hypersomnia, mainly in the form of sleep attacks) and the increase in rapid-eye movement (REM) sleep. This is manifested either by the decrease in REM sleep latency or the intrusion of partial aspects of this state into wakefulness (cataplexy, sleep paralysis, hypnagogic hallucinations) (Mignot 2011). However, in spite of the presence of hypersomnia, night sleep is also disrupted.

Nowadays, based on animal and human research, it is known that the pathogenesis of narcolepsy with cataplexy occurs as a result of the degeneration of the orexinergic (hypocretinergic) neurons within the hypothalamus (Mignot 2011). However, questions remain to be answered in relation to the physiology of the orexinergic system. It was our purpose to review how the orexinergic neurons interrelate with other neurotransmitter systems in order to modulate waking and sleep states.

2 Waking and Sleep Promoting Neuronal Systems

It is presently known that the neuronal networks critical for the generation and maintenance of wakefulness (i.e. activating systems) are located within the dorsal and median raphe nuclei (DRN and MRN), laterodorsal and pedunculopontine tegmental nuclei (LDT-PPT), locus coeruleus (LC), nucleus pontis oralis (NPO), tuberomammillary nucleus (TMN), ventral tegmental area and ventral periaqueductal gray (VTA and vPAG), and basal forebrain (BFB) (Torterolo and Vanini 2010a). These neurons have ample projections towards the thalamus and/or the cortex, structures that support cognitive functions. The activating systems desynchronize the electroencephalogram (EEG), an electrophysiological sign of wakefulness (Torterolo and Vanini 2010a). Some of these neurons also project to premotor or motor nuclei, including the neuronal networks that control breathing, as well as the preautonomic and autonomic nuclei involved in the vegetative aspects of the waking state (Torterolo and Vanini 2010a).

Non-rapid-eye movement (NREM) sleep is induced by neurons located in the preoptic area of the hypothalamus, both in the ventro-lateral preoptic nucleus (VLPO) and the median preoptic nucleus (MnPO); these neurons are active during NREM sleep and inhibit the activating systems (Szymusiak and McGinty 2008; Torterolo et al. 2009; Benedetto et al. 2012). Other regions, such as the caudo-lateral peribrachial area, that may have a role in NREM sleep or in NREM-REM sleep transition, have been recognized also (Torterolo et al. 2011).

Adenosine is a neurotransmitter and a metabolic product involved also in the generation of sleep. Adenosine synaptic levels increase during prolonged wakefulness periods prompting sleep onset (Strecker et al. 2000). This substance promotes sleep by inhibiting activating systems located in the BFB, and disinhibiting the sleep-promoting VLPO (Strecker et al. 2000).

It is well-known that the neuronal network "necessary" and "sufficient" for REM sleep generation is located in the mesopontine tegmentum (Chase 2013; Luppi et al. 2013; Siegel 2005). Some of these mesopontine neurons have a dual role, such that they form part of the activating system and, in addition, have REM sleep generating functions. While monoaminergic (noradrenergic and serotonergic) neurons within this region are Wake-on but REM-off neurons and promote wakefulness, the cholinergic neurons are either Wake-on and REM-on or just REM-on, and are critically involved both in wakefulness and REM sleep occurrence (McCarley 2007). In fact, structural and mathematical models based on the activity of these neurons have tried to explain the dynamic of REM sleep generation (McCarley 2007).

The nucleus pontis oralis (NPO) in the cat (that comprises the peri-locus coeruleus α and a part of the medial pontine reticular formation) or its corresponding nucleus in the rat, which is called sub-laterodorsal nucleus, is constituted by glutamatergic and GABAergic neurons. The nucleus pontis oralis is considered to exert an executive control over the initiation and maintenance of REM sleep, and to be involved in the control of wakefulness as a part of the activating systems (Chase 2013; Siegel 2005; Luppi et al. 2007; Reinoso-Suarez et al. 2001). A single local injection of a cholinergic agonist, such as carbachol, results in the generation of a state with all the behavioral and electrographic signs of REM sleep that occurs with a very short latency (30 s to a few minutes), and can last for up to two hours (Baghdoyan et al. 1987).

Glutamatergic and GABAergic neurons distributed in different areas including the DRN, BFB, LDT-PPT, NPO and vPAG, play also a critical role in the control of the behavioral states (Torterolo and Vanini 2010a, b; Torterolo et al. 2001, 2002; Vanini et al. 2007; Xi et al. 2001). For example, the microinjection of GABA or a $GABA_A$ receptor agonist into the NPO results in a sustained period of arousal. This finding has led to the proposal that GABAergic neurons within this area are involved in the control of wakefulness as part of the activating systems (Xi et al. 2001).

Atypical neurotransmitters such as nitric oxide and the endocannabinoids play also a role in the control of behavioral states (Monti et al. 2013; Murillo-Rodriguez 2008).

The role of the different neurotransmitter systems involved in the control of wakefulness and sleep is summarized in Table 1. Most of these neuronal networks are strongly modulated by orexinergic neurons.

Table 1 Neurotransmitters/neuromodulators that have an active role in the generation of wakefulness and sleep

Neurotransmitter/neuromodulator	Location of the soma	Neuronal pattern of discharge
Wakefulness promoting		
Orexin	Postero-lateral hypothalamus	Wake-on
Acetylcholine	LDT-PPT and BFB	Wake/REM-on
Serotonin	DRN and MRN	Wake-on
Noradrenaline	Locus coeruleus	Wake-on
Dopamine	VTA and SNPc	Bursting discharge during wakefulness and REM sleep[b]
Histamine	TMN	Wake-on
GABA[a]	NPO	Probably Wake-on
Glutamate[a]	Mesopontine reticular formation	Probably Wake-on
NREM sleep promoting		
Melanin-concentrating hormone (MCH)	Postero-lateral hypothalamus	Active in NREM and tonic REM
Adenosine	Probably BFB, POA	[c]
GABA[a]	VLPO, MnPO	NREM-on and NREM-REM-on
REM sleep promoting		
Acetylcholine	LDT-PPT and BFB	Wake/REM-on and REM-on
Melanin-concentrating hormone (MCH)	Postero-lateral hypothalamus	Active in tonic REM sleep
Orexin	Postero-lateral hypothalamus	Active in phasic REM sleep
GABA[a]	vlPAG, DRN	Probably REM-on
Glutamate[a]	NPO	Probably Wake/REM-on

[a]The role of the GABAergic and glutamatergic neurons depends on the location of the neurons; only some examples are listed. [b]The pattern but not the frequency of discharge of the dopaminergic neurons changes along sleep and wakefulness. [c]The release of adenosine increases during prolonged wakefulness; however the origin (metabolic, neurotransmitter) of this substance is still unclear. *BFB* basal forebrain; *DRN* dorsal raphe nucleus; *LDT-PPT* laterodorsal and pedunculopontine tegmental nucleus; *MnPO* median preoptic nucleus; *MRN* median raphe nucleus; *NPO* nucleus pontis oralis; *POA* preoptic area; *SNPc* substantia nigra pars compacta; *TMN* tuberomammillary nucleus of the hypothalamus; *vlPAG* ventrolateral periaqueductal gray; *VLPO* ventrolateral preoptica area; ventral tegmental area

3 Orexinergic Neurons, Orexins and Receptors

Orexin-A and B (also called hypocretin-1 and 2) were discovered in 1998 by two independent groups (de Lecea et al. 1998; Sakurai et al. 1998). These neuropeptides are synthesized by a discrete group of neurons ($\sim$5000 in rodents, $\sim$11,000 in cats and 20–50,000 in humans) in the postero-lateral hypothalamus (Li et al. 2013;

Torterolo et al. 2006). Figure 1 shows the characteristics and distribution of the orexinergic neurons in the postero-lateral hypothalamus of the guinea pig. Orexins exert their biological function through two metabotropic receptors orexin-R1 and orexin-R2 (also known as hypocretin 1 and 2 receptors) that have broad, partially overlapping, but distinct patterns of distribution throughout the brain and body. The

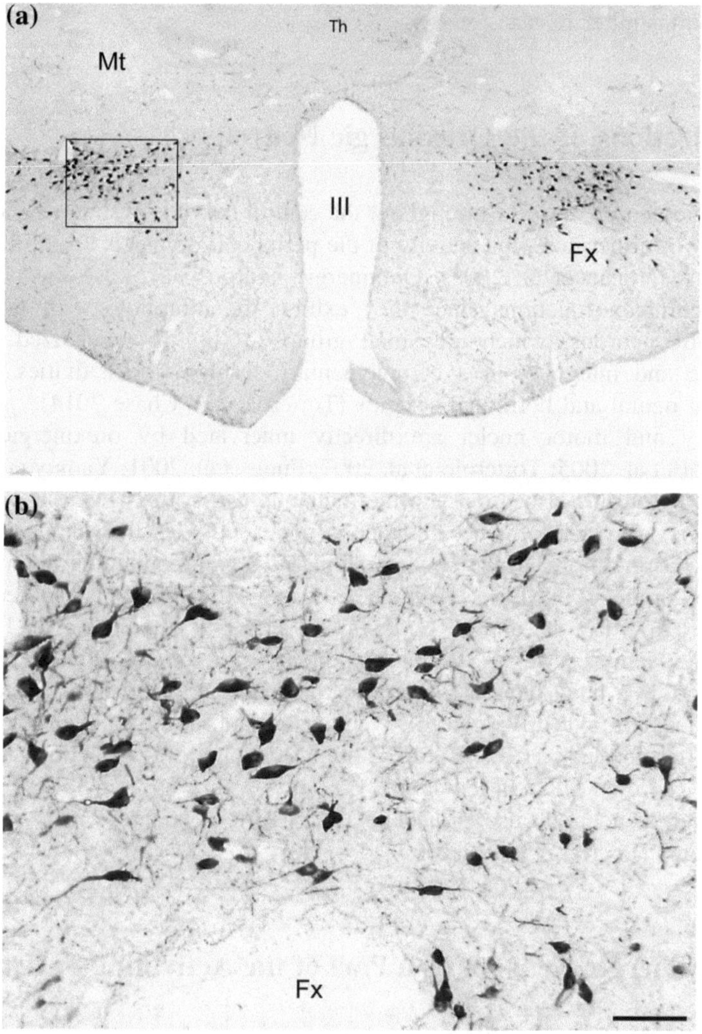

Fig. 1 Orexinergic neurons are located in the postero-lateral hypothalamus. **a** Photomicrograph that shows the distribution of the orexinergic neurons in a coronal plane of the tuberal region of a guinea pig hypothalamus. **b** The inset showed in (**a**), is exhibited in a higher magnification in (**b**). The orexinergic neurons are located close to the fornix. Calibration bar, 50 μm. Sections of 20 μm were processed employing the anti-orexin-B antibodies, ABC method and the DAB-H_2O_2 reaction to detect peroxidase activity

orexin-R1 has 10 to 100 times more affinity for orexin-A than for orexin-B; in contrast, orexin-R2 has the same affinity for both neuropeptides. Through these receptors, orexins produce an excitatory effect at postsynaptic sites (Mignot 2011; Li et al. 2013; van den Pol and Acuna-Goycolea 2006). It has been demonstrated that orexin increases Na^{+}-dependent current, non selective cation currents, and activates the Na^{+}/Ca^{++} exchanger. Depression of K^{+} currents and increase in intracellular Ca^{++} was also demonstrated. At presynaptic sites orexin also increases the neurotransmitter release.

4 Projections of the Orexinergic Neurons

Orexinergic neurons project throughout the central nervous system (CNS) (Peyron et al. 1998); furthermore, the activity of the peripheral organs is also influenced by the orexins (Voisin et al. 2003). Orexinergic neurons have also the potential to mediate complex functions since they exhibit the morphology of prototypical "command" neurons, which are small groups of highly specialized cells that coordinate and integrate, in a complementary fashion, the activities of a vast number of neural and hormonal systems (Torterolo and Chase 2014).

Sensory and motor nuclei are directly innervated by orexinergic neurons (McGregor et al. 2005; Torterolo et al. 2007; Fung et al. 2001; Yamuy et al. 2004). Orexinergic neurons also project to the thalamus and cortex (Peyron et al. 1998), wherein they directly influence thalamo-cortical activities that support cognitive functions. Furthermore, dense concentrations of orexin-containing axon terminals are located in the TMN (Eriksson et al. 2001) as well as in brainstem areas such as the LDT-PPT, VTA, LC and DRN (Peyron et al. 1998; Chemelli et al. 1999; Date et al. 1999; Nambu et al. 1999), that participate in the control of wakefulness and REM sleep. We have demonstrated also that orexinergic neurons project to the NPO, that exerts executive control over the initiation and maintenance of REM sleep (Torterolo et al. 2013). Figure 2 shows orexinergic projections toward the LDT-PPT, LC and DRN in the cat.

Orexinergic neurons also project to the NREM generation areas such as the preoptic area (Peyron et al. 1998).

5 Orexinergic Neurons as a Part of the Activating Systems

The postero-lateral hypothalamic area where the orexinergic neurons are located, is the key brain site that, for decades, has been identified as being responsible for initiating, coordinating and maintaining goal-oriented survival-type behaviors such as fight, flight and food consumption among others. In addition, and in accord with the preceding concept, experimental studies have demonstrated that this area is critically involved in the control of sleep and wakefulness,

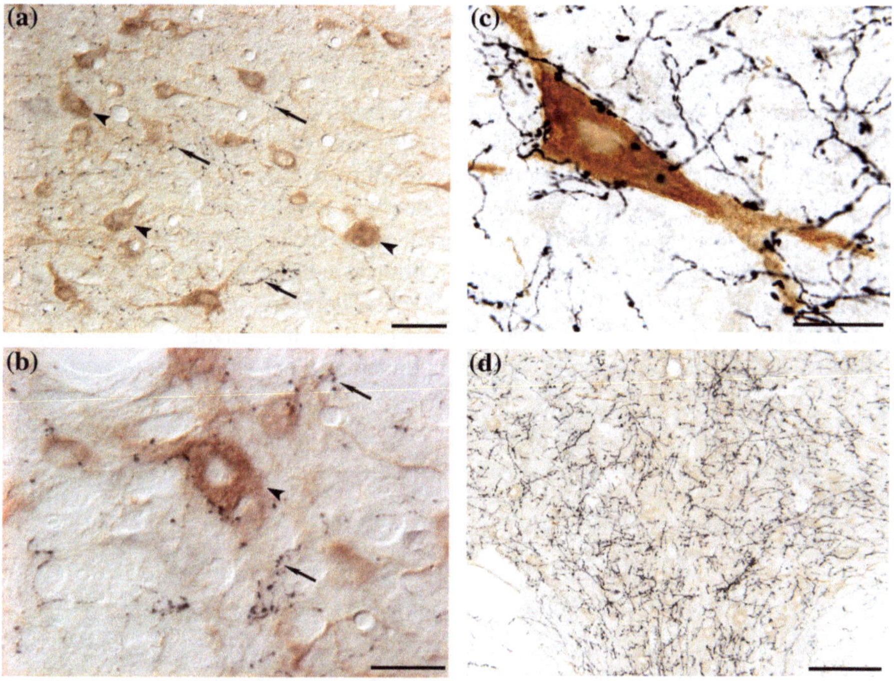

Fig. 2 Examples of the orexinergic projections. **a** and **b** Photomicrographs of the LDT-PPT of the cat. The sections were immunostained for orexin (in *black*, examples of orexienergic fibers/terminals are indicated by *arrows*) and for choline acetyltransferase (in *brown*, examples of cholinergic neurons are signaled by *arrowheads*). Sections were processed utilizing the ABC method and the DAB-H_2O_2 reaction to detect peroxidase activity. This reaction was enhanced with nickel to label orexinergic cells. *Calibration bars* **a** and **b**, 30 and 20 μm respectively. **c** Photomicrographs of the LC of the cat. The sections were immunostained for orexin (in *black*) and tyrosine hydroxylase (in *brown*). Sections were processed utilizing the ABC method and the DAB-H_2O_2 reaction to detect peroxidase activity. This reaction was enhanced with nickel to label orexinergic cells. *Calibration bar* 10 μm. **d** The photomicrograph shows the orexinergic fibers in the dorsal raphe nucleus; this is taken for another section of the experimental series shown in (**c**). *Calibration bar* 100 μm (Color figure online)

somatomotor activity as well as "pleasure" or "reward" (Torterolo and Chase 2014; Chase 2013).

Early studies revealed that the intraventricular injection of orexin induces wakefulness (Hagan et al. 1999; Piper et al. 2000). These data, as well as the fact that the lack of orexinergic neurons in narcoleptic patients induces hypersomnia (Mignot 2011), strongly suggested that this system promotes wakefulness. Genetically modified mice and optogenetic studies also confirmed the role of orexin and orexinergic neurons as a waking promoting substance and system (Chemelli et al. 1999; Hara et al. 2001; Adamantidis et al. 2007). However, by means of Fos technology (the Fos protein is a marker of neuronal activity), we demonstrated in the cat that orexinergic neurons are not active during wakefulness per se (Torterolo et al. 2001) (Fig. 3). A detailed analysis of orexinergic neuronal activity shows that these neurons are strongly activated when animals are exploring an unknown environment (exploratory motor activity) (Torterolo et al. 2011). In the absence of motor activity during alert wakefulness, quiet wakefulness or quiet sleep, the orexinergic neurons are not activated to any significant extent (Torterolo et al. 2003). In fact, the number of Orexin+ Fos+ neurons was approximately 10 times greater during exploratory motor activity than during repetitive motor activity that occurred during forced locomotion, even though in both paradigms there was a comparable amount of motor activity (Torterolo et al. 2011). Therefore, neither wakefulness nor motor activity per se, were critical with respect to the activation of orexinergic neurons. Hence the orexinergic system is engaged when animals are performing

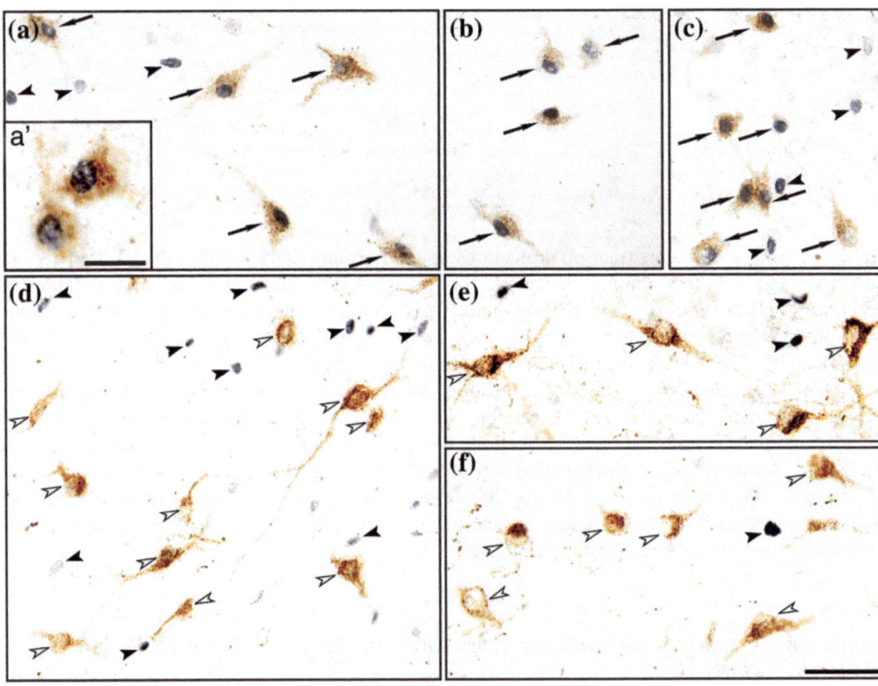

Fig. 3 Photomicrographs containing orexin and Fos immunoreactive neurons in the lateral hypothalamus of the cat during active wakefulness with motor explorative activity (**a**, **a′**, **b** and **c**), alert wakefulness without motor activity (**d**), quiet wakefulness (**e**) and NREM sleep (**f**). Orexinergic neurons are *stained brown*; Fos immunoreactivity, that is restricted to nuclei, is shown in *black*. *Arrows* indicate Orexin+ Fos+ cells; these neurons are exhibited with higher magnification in **a′**. Significant numbers of Orexin+ Fos+ neurons were observed only during active wakefulness with motor explorative activity. *Filled arrowheads* point to non-orexinergic Fos+ neurons, these neurons were observed in all of these behavioral states. *Empty arrowheads* indicate orexinergic neurons that did not express c-fos. *Calibration bars* **a–f** 50 μm; **a′** 20 μm. Modified from Torterolo et al. (2003) (Color figure online)

goal (reward)-directed behaviors. In agreement with our results, it was found that orexin knock-out mice were unable to work for food or water reward during the light phase (McGregor et al. 2011).

Recently, Chase presented the "Unified Survival Theory for the Functioning of the Orexinergic System" (Chase 2013). The basis of this theory is that the main role of the orexinergic system is to initiate, coordinate and maintain survival behaviors and survival-related processes.

In order to promote these waking effects during survival-related behaviors, orexinergic neurons activate different components of the activating systems such as the LC, DRN, LDT-PPT, TMN and BFB.

6 Orexinergic Neurons During Sleep

6.1 NREM Sleep

The orexinergic neurons, as a component of the activating systems (Torterolo and Vanini 2010a), do not actively participate in the occurrence of NREM sleep. In fact, there is a lack of Fos expression in these neurons during this behavioral state (Fig. 3). Subsequently, unit recordings confirmed our early findings (Mileykovskiy et al. 2005; Lee et al. 2005; Takahashi et al. 2008; Kolaj et al. 2008).

6.2 "Tonic" REM Sleep

The orexinergic neurons are considered to be REM-off neurons. The REM-off profile concept of the orexinergic neurons arose based upon electrophysiological recordings of identified orexinergic neurons during "tonic" REM sleep (see below) (Mileykovskiy et al. 2005; Lee et al. 2005; Takahashi et al. 2008).

6.3 "Phasic" REM Sleep

EEG activation, theta activity in the hippocampus and muscle atonia are the classic biomarkers for the identification of REM sleep. Accompanying these "tonic" signs are rapid-eye movements, muscle twitches, PGO waves, breathing irregularities as well as heart rate and blood pressure variations that constitute the phasic events of REM sleep. Other signs such as acceleration of the theta rhythm also correlate with these phasic events (Rowe et al. 1999).

Experimental evidence shows that while orexinergic neurons turn off during "tonic" REM sleep, at least a subset of these neurons discharge in bursts during phasic REM sleep (Mileykovskiy et al. 2005; Lee et al. 2005; Takahashi et al. 2008).

Studies in the cat, an animal that exhibits robust phasic periods of REM sleep (Ursin and Sterman 1981), strongly suggest that there is orexinergic neuronal activity during REM sleep, probably during the phasic events of this state. An increase in the number of Orexin+ Fos+ neurons was observed during REM sleep induced by carbachol microinjections into NPO (Torterolo et al. 2001). In this study, REM sleep was induced by carbachol microinjections in order to generate a state of sufficiently long duration to allow Fos protein to be synthesized in high concentration. During this state, 34 % of the orexinergic neurons were activated according to their Fos-expression (Torterolo et al. 2001). This result was in agreement with the findings by Kiyashchenko et al. (2002), who described an increase in orexin-A release during REM sleep, both in the hypothalamus and BFB in freely moving cats. Hence, this study also indicates that orexinergic neurons are active during REM sleep, probably in conjunction with phasic events.

Microinjection studies suggest also that administration of orexin into critical areas such as the medullary reticular formation of the rat or the NPO of the cat, can induce REM sleep-like atonia or REM sleep, respectively (Mileykovskiy et al. 2002; Xi et al. 2002; Xi and Chase 2010).

If orexinergic neurons are active during "phasic" REM sleep, they are likely to promote the phasic events of REM sleep (Torterolo and Chase 2014). In fact, orexins directly activate motor nuclei, breathing neuronal networks and sympathetic output that controls the cardiovascular system (Yamuy et al. 2004; Shirasaka et al. 2002; Zhang et al. 2005; Williams and Burdakov 2008), as well as the medial septum where the pacemaker for the hypocamppal theta rhythm is located (Gerashchenko et al. 2001). These are typical ergotropic or energy-expending behaviors that are induced by the orexinergic system (Chase 2013).

Then, orexinergic cells in the hypothalamus may be involved in the control of both active wakefulness and active components of REM sleep. As mentioned above, the orexinergic system plays an important role in the regulation of motor activity during motivational states. Is it possible that this system promotes motor activity during wakefulness and atonia during REM sleep? Actually, this pattern of duality of behavioral state control with opposite motor responses is reminiscent of the phenomenon of reticular response-reversal in the NPO (Chase and Babb 1973; Chase and Morales 1990). This phenomenon involves mechanisms that result in the facilitation of wakefulness and somatomotor activation during wakefulness, as well as the generation of REM sleep and its accompanying pattern of motor inhibition during this sleep state. The reticular response reversal determines the auditory stimulation promotion of somatomotor activation during wakefulness, while increasing the hyperpolarization of the motoneurons and atonia during REM sleep (Chase 2013). In fact, microinjections of orexin-A or orexin-B into the NPO of the cat increases the time spent in REM sleep and results in a decrease in the latency to the generation of this state (Xi et al. 2002). Juxtacellular application of orexin-A results in an increase in the excitability of NPO neurons, which is associated with the induction of REM sleep (Xi et al. 2003). In addition, orexin-A increases acetylcholine release in the NPO of the rat (Bernard et al. 2003, 2006). In this respect, it is known that acetylcholine levels within this region increase during REM sleep

(Kodama et al. 1990). On the other hand, an increase in wakefulness accompanied by a decrease in REM sleep has been observed also when orexin-A is microinjected into the NPO of the cat (Moreno-Balandran et al. 2008). Furthermore, the iontophoretic application of orexin-A into the NPO of the rat produces an inhibition of NPO neurons, which can be blocked by previous iontophoretic application of bicuculline, a $GABA_A$ receptor antagonist (Nunez et al. 2006). In fact, it has been shown that orexin increases GABA levels in the NPO of the rat, and orexin and GABA interact within this nucleus to promote wakefulness (Watson et al. 2008; Brevig et al. 2011). The presence of orexin-B receptors on GABAergic neurons within the NPO may be the cellular basis for this effect (Brischoux et al. 2008). The paradoxical or contradictory findings involving the REM sleep and wakefulness promoting actions of orexin within the NPO have been reconciled by Xi and Chase (Xi and Chase 2010). They demonstrated that the microinjections of orexin-A within the NPO generates REM sleep when applied during NREM sleep, but promotes wakefulness when applied during this behavioral state. Thus, the behavioral state of the animal at the time of the application of orexin determines whether REM sleep or wakefulness occur.

7 Afferents to Orexinergic Neurons

Orexinergic neurons receive inputs from several regions such as the allocortex, many hypothalamic nuclei, the periaqueductal gray matter, the DRN, and parabrachial regions, which suggests that these neurons integrate a variety of interoceptive and homeostatic signals (Yoshida et al. 2006). Interestingly, these neurons seem to be weakly influenced by exteroceptive sensory inputs (Mileykovskiy et al. 2005).

8 Effects of Different Neurotransmitters on Orexinergic Neurons

Table 2 summarizes the effects of the main neurotransmitters on orexinergic neurons. In mice, glutamate depolarizes orexinergic neurons acting through AMPA and NMDA receptors, while GABA inhibits these neurons though $GABA_A$ and $GABA_B$ receptors (Yamanaka 2006).

Neurotransmitters used by neurons that form part of the activating systems also modulate the activity of orexinergic neurons. They include the following:

Serotonin. The serotonergic neurons are localized at the raphe nuclei; the DRN is the region with the major concentration of these neurons (Monti 2010a, b). Serotonergic neurons of the DRN and MRN project to the thalamo-cortical systems. These cells are active during wakefulness and a group of them increase the firing rate in relation to automatic movements such as locomotion (Wake-on).

Table 2 Examples of the effects of neurotransmitter/neuromodulators onto orexinergic neurons (in vitro studies)

Neurotransmitter/neuromodulator	Cellular effects on orexineric neuron	References	Animal model
Orexin	despolarization by presynatic facilitation of glutamate release	Li et al. (2002)	Mouse
Acetylcholine	Carbachol depolarizes 27 % and hyperpolarizes 6 % of the neurons	Yamanaka (2006)	Mouse
	Acetylcholine has a predominantly excitatory effect	Bayer et al. (2005)	Rat
Serotonin	Hyperpolarizes through $5HT1_A$ receptors	Yamanaka (2006)	Mouse
Noradrenaline	- Hyperpolarizes through α2 receptors	Yamanaka (2006)	Mouse
	- Depolarization mediated by α1 receptor (in the presence of a α2-receptor antagonist)	Yamanaka (2006)	Mouse
	- Predominant excitatory effect	Bayer et al. (2005)	Rat
Dopamine	- D1 and D2 dopamine receptors have opposing effects on excitatory presynaptic terminals that impinge on orexinergic neurons	Alberto et al. (2006)	Rat
Histamine	- Almost no effect	Yamanaka (2006)	Mouse
GABA	Post-synaptic inhibition though $GABA_A$ and $GABA_B$ receptors	Yamanaka (2006)	Rat
Glutamate	Depolarize acting through AMPA and NMDA receptors	Yamanaka (2006)	Mouse
Melanin-concentrating hormone (MCH)	Attenuates orexin-A induced enhancement of spike frequency and the frequency of miniature excitatory postsynaptic currents	Rao et al. (2008)	Mouse
Adenosine	depresses the amplitude of evoked excitatory postsynaptic potential and the frequency of spontaneous and miniature excitatory postsynaptic currents	Liu and Gao (2007)	Mouse
Endocannabinoids	Agonist of CB1R hyperpolarized and reduced spontaneous firing	Huang et al. (2007)	Mouse

Serotonergic neurons decrease their firing rate during NREM sleep, reaching a nadir during REM sleep (REM-off). An in vitro study in mice has shown that serotonin hyperpolarizes orexinergic neurons through $5HT_{1A}$ receptors (Yamanaka 2006).

Noradrenaline. The noradrenergic neurons are located in the LC. These neurons project to the thalamus and cortex and have a Wake-on, REM-off profile in their firing rate (Torterolo and Vanini 2010). During wakefulness these neurons markedly increase their discharge rate in relation to new stimuli, and have been related with the modulation of attention. An in vitro study in mice has shown that noradrenaline hyperpolarizes orexinergic neurons through α_2 receptors (Yamanaka 2006), while a weak depolarization mediated by α_1 receptor was also observed in the presence of a α_2-receptor antagonist. On the contrary, in rats, noradrenaline has a predominant excitatory effect (Bayer et al. 2005).

Dopamine. These neurons are located in the VTA, the *substancia nigra pars compacta* and the vPAG (Monti and Monti 2007; Leger et al. 2006, 2010). Their projections are restricted to the striatum and the prefrontal cortex. Dopaminergic neurons are known to promote wakefulness and, due to their strong relationship with positive emotions, they may probably be related with the increase in vigilance produced during a motivational state. D1 and D2 dopamine receptors have opposing effects on excitatory presynaptic terminals that impinge on orexinergic neurons (Alberto et al. 2006).

Histamine. Histaminergic neurons are located in the tuberomammillary nucleus of the posterior hypothalamus. These neurons project to the thalamus and cortex and have a Wake-on, REM-off profile in their firing rate (Monti 2011). Interestingly, histamine has almost no effect on orexinergic neurons (Yamanaka 2006).

Acetylcholine. Cholinergic neurons are located in the LDT-PPT and the BFB. Ascending projections of the cholinergic neurons located in the LDT-PPT are directed toward the thalamus, while cholinergic neurons of the BFB project to the cortex and reticular nucleus of the thalamus (Torterolo and Vanini 2010). LDT-PPT cholinergic neurons project also to the NPO, the executive REM sleep generation area, that is critical for REM sleep generation (Chase 2013; Semba 1999). In mice, carbachol depolarizes 27 % and hyperpolarizes 6 % of the population of orexinergic neurons (Yamanaka 2006), and these effects are mediated by different muscarinic receptors. In rats, acetylcholine has a predominantly excitatory effect (Bayer et al. 2005).

The effects of adenosine, nitric oxide and endocannabinoids on orexinergic neurons are listed in Table 2.

The effect of the sleep-promoting factor, melanin-concentrating hormone (MCH) on orexinergic neurons will be reviewed below.

9 Orexin and MCH. Postero-lateral Hypothalamic Duality in the Control of Sleep and Wakefulness

Due to the importance of the MCHergic system in sleep physiology (Monti et al. 2013; Torterolo et al. 2011), it is relevant to examine the interactions between the MCHergic and the orexinergic system. A strong anatomical relationship exists between orexinergic and MCHergic neurons in the hypothalamus. As it is shown in Fig. 4,

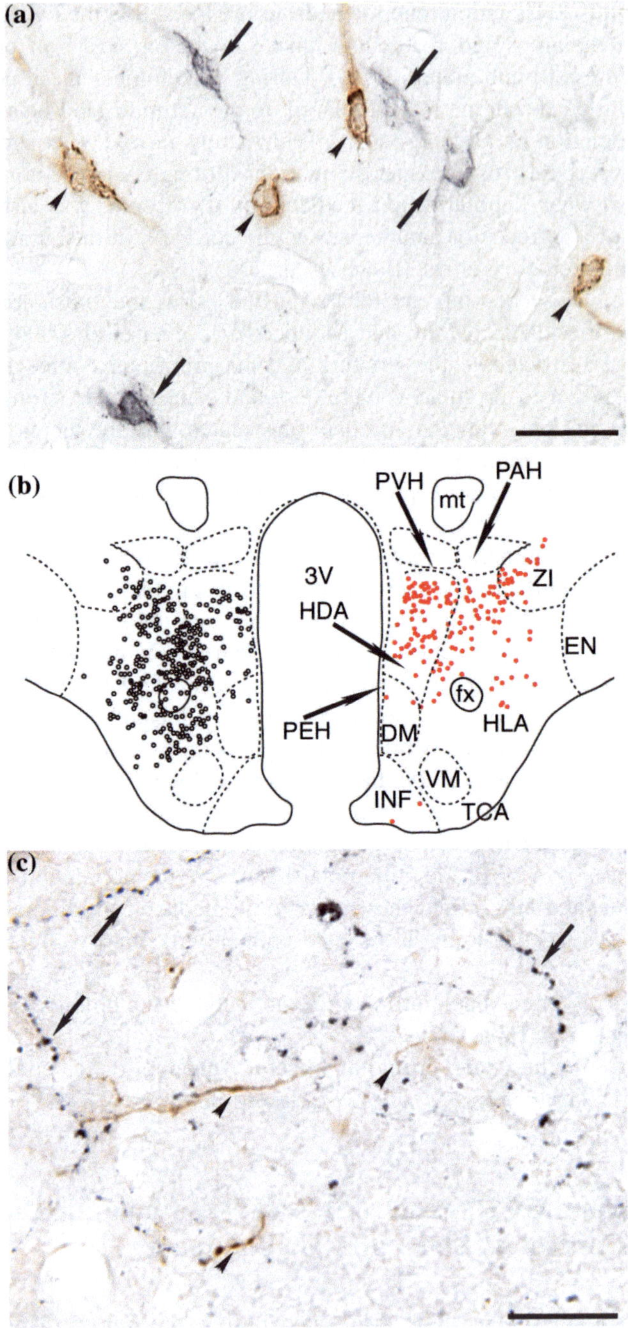
(a)
(b)
PVH
PAH
mt
3V
ZI
HDA
EN
fx
PEH
DM
HLA
VM
INF
TCA
(c)

◀ **Fig. 4** Orexinergic neurons are intermingled with MCHergic neurons in the postero-lateral hypothalamus. **a** Photomicrographs of the postero-lateral hypothalamic area of the cat. The sections were immunostained for orexin (in *black*, *arrows*) and MCH (in *brown*, *arrowheads*). Sections were processed utilizing the ABC method and the DAB-H_2O_2 reaction to detect peroxidase activity. This reaction was enhanced with nickel to label orexinergic cells. *Calibration bars* 50 μm. **b** Location of MCHergic and orexinergic neurons in the postero-lateral hypothalamus of a representative cat. Camera lucida drawings of MCHergic (on the *left*, *black circles*) and orexinergic neuronal bodies (on the *right*, *red circles*) in the postero-lateral hypothalamus. The neurons are from the same hemi-hypothalamus (reflected in the figure). Camera lucida drawings were obtained from adjacent sections that were immunostained for MCH for Orexin-2, respectively; these sections were counterstained with Pyronin-Y. *DM* dorsomedial nucleus; *EN* entopeduncular nucleus; *fx* fornix; *HDA* dorsal hypothalamic area; *HLA* lateral hypothalamic area; *INF* infundibular nucleus; *mt* mammillothalamic tract; *PAH* paraventricular nucleus; *PEH* periventricular complex; *PVH* parvocellular nucleus; *TCA* area of the tuber cinereum; *VM* ventromedial nucleus; *ZI* zona incerta; *3V* third ventricle. Modified from (Torterolo et al. 2006). **c** Photomicrographs of the nucleus pontis oralis of the cat. The sections were immunostained for orexin (in *black*, *arrows*) and MCH (in *brown*, *arrowheads*). Sections were processed utilizing the ABC method and the DAB-H_2O_2 reaction to detect peroxidase activity. This reaction was enhanced with nickel to label orexinergic cells. *Calibration bars* 10 μm (Color figure online)

MCHergic neurons are intermingled with orexin-containing neurons in the postero-lateral hypothalamus, mainly at the tuberal and tuberomammillar levels (Torterolo et al. 2006). MCHergic fibers are in close relationship with orexinergic neurons and viceversa, which suggests the existence of reciprocal synaptic contacts between both types of cells (Torterolo et al. 2006; Guan et al. 2002). This fact, as well as the presence of orexinergic receptors on MCHergic neurons indicates the existence of an important functional interaction between both systems (Backberg et al. 2002). In this respect, orexin increases MCH mRNA expression in hypothalamic neurons, directly excites MCHergic neurons and increases glutamate release onto them (Bayer et al. 2002; van den Pol et al. 2004). Opposite, MCH modulates orexin-mediated effects on behavioral state and synaptic transmission in the lateral hypothalamus (Rao et al. 2008). The efficacy of glutamatergic synapses on orexinergic neurons is enhanced in MCHR1 knockout mice, and orexin A-induced firing is facilitated. On the contrary, in wild-type mice, MCH significantly attenuates orexin-A induced enhancement of spike frequency in orexinergic neurons, but not its basal activity. Furthermore, in these neurons, MCH attenuates orexin-1-induced enhancement of the frequency of miniature excitatory postsynaptic currents. These effects imply that MCH exerts a unique inhibitory influence on orexinergic signaling as a way to fine-tune the output of these neurons.

Interestingly, orexinergic and MCHergic neurons respond in a different way to most homeostatic signals such as glucose (Burdakov et al. 2005) or to waking-related neurotransmitters such as noradrenaline (Bayer et al. 2005). It should be noted that while orexinergic neurons of the rat express α_1 adrenergic receptors, MCHergic neurons express the α_2 adrenergic receptors, which are related to activation or inhibition of their targets, respectively (Modirrousta et al. 2005).

We have shown that both orexinergic and MCHergic neurons project to the NPO (Torterolo et al. 2009, 2013) (Fig. 4). In addition, fibers and terminals of both systems are highly intermingled, which suggests the existence of important interactions between these systems within their mesopontine targets, similar to the anatomical and functional interactions that have been described within the hypothalamus (see above).

Several studies suggest that MCHergic neurons are involved in the generation of sleep, especially REM sleep (Monti et al. 2013; Torterolo et al. 2011). These neurons discharge in a reciprocal manner to orexinergic neurons across the sleep-wake cycle. MCHergic neurons have a high firing rate during tonic REM sleep, but do not increase their firing level during "phasic" REM sleep (Hassani et al. 2009). When MCH is microinjected into the cat NPO (executive REM sleep generation area), it produces an increase in the time the animals spend in REM sleep together with a decrease in the latency to this behavioral state (Torterolo et al. 2009). MCH also exerts its REM sleep promoting functions acting through the DRN (Lagos et al. 2009, 2011), where it shows an inhibitory role on serotonergic neurons (Urbanavicius et al. 2013; Pascovich et al. 2011). A REM sleep promoting effects of MCH have been observed also in the BFB (Lagos et al. 2012). In contrast, bilateral microinjections of MCH into the VLPO induces NREM sleep (Benedetto et al. 2013). Recent studies utilizing optogenetics confirmed the importance of MCH in promoting both NREM and REM sleep (Tsunematsu et al. 2014; Konadhode et al. 2013; Jego et al. 2013).

Interestingly, MCH blunts the central regulation of sympathetic tone, adaptive sympathetic reflexes and decreases metabolism (Saito and Nagasaki 2008; Messina and Overton 2007; Egwuenu et al. 2005). These are trophotropic or energy-conserving effects, which are opposite to the effects produced by the orexinergic system (see above).

The experimental evidence mentioned above is the basis for the hypothesis that MCHergic neurons are active during NREM seep as well as during tonic REM sleep and promote this "quiescent" behavioral state. On the contrary, orexinergic neurons are active during "survival-type" behaviors along wakefulness and during "phasic" REM sleep, and probably induce (at least partially) phasic episodes of this state. Hence, orexinergic neurons play an active role in energy-expending or ergotropic behaviors

10 Conclusions

The orexinergic neurons are a branch of the activating system; they project toward the thalamo-cortical system that supports cognitive functions, to the septum that is the pacemaker of the hippocampal activity, the premotor and motor nuclei including the breathing neuronal network, and the preautonomic and autonomic nuclei. By means of these projections these neurons are capable to integrate different aspects of the waking behavior. As proposed recently by Chase (2013), this system coordinates all facets of "survival" behavior including an increase in the vigilance state.

In addition, these neurons may be involved in the generation of the "active" or "phasic" aspects of REM sleep (Torterolo and Chase 2014).

Orexinergic neurons interrelate with other neuronal networks critical for the generation and maintenance of wakefulness and sleep. These neurons boost the activity of other branches of the activating systems in order to increase vigilance, especially during motivated-behaviors. In addition, it is possible that a group of hypocretinergic neurons promotes the "phasic" aspects of REM sleep.

Yin and yang is a Taoist symbol used to describe how polar or seemingly contrary forces are interconnected and interdependent in the natural world, and how they give rise to each other in turn. In a similar way, the orexinergic and MCHergic systems of the postero-lateral hypothalamus seem to produce either wakefulness or active (phasic) REM sleep (ergotropic or energy-expending function) or NREM sleep and tonic REM sleep (trophotropic or energy-conserving function), respectively.

Acknowledgments This work was partially supported by the "Programa de Desarrollo de Ciencias Básicas, PEDECIBA" and by the ANII-FCE-2-2011-1-7030 grant from Uruguay.

References

Adamantidis AR, Zhang F, Aravanis AM, Deisseroth K, de Lecea L (2007) Neural substrates of awakening probed with optogenetic control of hypocretin neurons. Nature 450:420–424

Alberto CO, Trask RB, Quinlan ME, Hirasawa M (2006) Bidirectional dopaminergic modulation of excitatory synaptic transmission in orexin neurons. J Neurosci 26:10043–10050

Backberg M, Hervieu G, Wilson S, Meister B (2002) Orexin receptor-1 (OX-R1) immunoreactivity in chemically identified neurons of the hypothalamus: focus on orexin targets involved in control of food and water intake. Eur J Neurosci 15:315–328

Baghdoyan HA, Rodrigo-Angulo ML, McCarley RW, Hobson JA (1987) A neuroanatomical gradient in the pontine tegmentum for the cholinoceptive induction of desynchronized sleep signs. Brain Res 414:245–261

Bayer L, Mairet-Coello G, Risold PY, Griffond B (2002) Orexin/hypocretin neurons: chemical phenotype and possible interactions with melanin-concentrating hormone neurons. Regul Pept 104:33–39

Bayer L, Eggermann E, Serafin M, Grivel J, Machard D, Muhlethaler M, Jones BE (2005) Opposite effects of noradrenaline and acetylcholine upon hypocretin/orexin versus melanin concentrating hormone neurons in rat hypothalamic slices. Neuroscience 130:807–811

Benedetto L, Chase M, Torterolo P (2012) GABAergic processes within the median preoptic nucleus promote NREM sleep. Behav Brain Res 232:60–65

Benedetto L, Rodriguez-Servetti Z, Lagos P, D'Almeida V, Monti JM, Torterolo P (2013) Microinjection of melanin concentrating hormone into the lateral preoptic area promotes non-REM sleep in the rat. Peptides 39C:11–15

Bernard R, Lydic R, Baghdoyan HA (2003) Hypocretin-1 causes G protein activation and increases ACh release in rat pons. Eur J Neurosci 18:1775–1785

Bernard R, Lydic R, Baghdoyan HA (2006) Hypocretin (orexin) receptor subtypes differentially enhance acetylcholine release and activate g protein subtypes in rat pontine reticular formation. J Pharmacol Exp Ther 317:163–171

Brevig HN, Watson CJ, Lydic R, Baghdoyan HA (2011) Hypocretin and GABA interact in the pontine reticular formation to increase wakefulness. Sleep 33:1285–1293

Brischoux F, Mainville L, Jones BE (2008) Muscarinic-2 and orexin-2 receptors on GABAergic and other neurons in the rat mesopontine tegmentum and their potential role in sleep-wake state control. J Comp Neurol 510:607–630

Burdakov D, Gerasimenko O, Verkhratsky A (2005) Physiological changes in glucose differentially modulate the excitability of hypothalamic melanin-concentrating hormone and orexin neurons in situ. J Neurosci 25:2429–2433

Chase MH (2013a) Motor control during sleep and wakefulness: clarifying controversies and resolving paradoxes. Sleep Med Rev 17:299–312

Chase MH (2013b) A unified survival theory of the functioning of the hypocretinergic system. J Appl Physiol 115(7):954–971

Chase MH, Babb M (1973) Masseteric reflex response to reticular stimulation reverses during active sleep compared with wakefulness or quiet sleep. Brain Res 59:421–426

Chase MH, Morales FR (1990) The atonia and myoclonia of active (REM) sleep. Annu Rev Psychol 41:557–584

Chemelli RM, Willie JT, Sinton CM, Elmquist JK, Scammell T, Lee C, Richardson JA, Williams SC, Xiong Y, Kisanuki Y et al (1999) Narcolepsy in orexin knockout mice: molecular genetics of sleep regulation. Cell 98:437–451

Date Y, Ueta Y, Yamashita H, Yamaguchi H, Matsukura S, Kangawa K, Sakurai T, Yanagisawa M, Nakazato M (1999) Orexins, orexigenic hypothalamic peptides, interact with autonomic, neuroendocrine and neuroregulatory systems. Proc Natl Acad Sci U S A 96:748–753

de Lecea L, Kilduff TS, Peyron C, Gao X, Foye PE, Danielson PE, Fukuhara C, Battenberg EL, Gautvik VT, Bartlett FS 2nd et al (1998) The hypocretins: hypothalamus-specific peptides with neuroexcitatory activity. Proc Natl Acad Sci USA 95:322–327

Egwuenu EJ, Fong AY, Pilowsky PM (2005) Intrathecal melanin-concentrating hormone reduces sympathetic tone and blocks cardiovascular reflexes. Am J Physiol Regul Integr Comp Physiol 303:R624–R632

Eriksson KS, Sergeeva O, Brown RE, Haas HL (2001) Orexin/hypocretin excites the histaminergic neurons of the tuberomammillary nucleus. J Neurosci 21:9273–9279

Fung SJ, Yamuy J, Sampogna S, Morales FR, Chase MH (2001) Hypocretin (orexin) input to trigeminal and hypoglossal motoneurons in the cat: a double-labeling immunohistochemical study. Brain Res 903:257–262

Gerashchenko D, Salin-Pascual R, Shiromani PJ (2001) Effects of hypocretin-saporin injections into the medial septum on sleep and hippocampal theta. Brain Res 913:106–115

Guan JL, Uehara K, Lu S, Wang QP, Funahashi H, Sakurai T, Yanagisawa M, Shioda S (2002) Reciprocal synaptic relationship between orexin-and melanin-concentrating hormone-containing neurons in the rat lateral hypothalamus: a novel circuit implicated in feeding regulation. Int J Obes Relat Metab Disord 26:1523–1532

Hagan JJ, Leslie RA, Patel S, Evans ML, Wattam TA, Holmes S, Benham CD, Taylor SG, Routledge C, Hemmati P et al (1999) Orexin A activates locus coeruleus cell firing and increases arousal in the rat. Proc Natl Acad Sci USA 96:10911–10916

Hara J, Beuckmann CT, Nambu T, Willie JT, Chemelli RM, Sinton CM, Sugiyama F, Yagami K, Goto K, Yanagisawa M et al (2001) Genetic ablation of orexin neurons in mice results in narcolepsy, hypophagia, and obesity. Neuron 30:345–354

Hassani OK, Lee MG, Jones BE (2009) Melanin-concentrating hormone neurons discharge in a reciprocal manner to orexin neurons across the sleep-wake cycle. Proc Natl Acad Sci USA 106:2418–2422

Huang H, Acuna-Goycolea C, Li Y, Cheng HM, Obrietan K, van den Pol AN (2007) Cannabinoids excite hypothalamic melanin-concentrating hormone but inhibit hypocretin/orexin neurons: implications for cannabinoid actions on food intake and cognitive arousal. J Neurosci 27:4870–4881

Jego S, Glasgow SD, Herrera CG, Ekstrand M, Reed SJ, Boyce R, Friedman J, Burdakov D, Adamantidis AR (2013) Optogenetic identification of a rapid eye movement sleep modulatory circuit in the hypothalamus. Nat Neurosci 16:1637–1643

Kiyashchenko LI, Mileykovskiy BY, Maidment N, Lam HA, Wu MF, John J, Peever J, Siegel JM (2002) Release of hypocretin (orexin) during waking and sleep states. J Neurosci 22:5282–5286

Kodama T, Takahashi Y, Honda Y (1990) Enhancement of acetylcholine release during paradoxical sleep in the dorsal tegmental field of the cat brain stem. Neurosci Lett 114:277–282

Kolaj M, Coderre E, Renaud LP (2008) Orexin peptides enhance median preoptic nucleus neuronal excitability via postsynaptic membrane depolarization and enhancement of glutamatergic afferents. Neuroscience 155:1212–1220

Konadhode RR, Pelluru D, Blanco-Centurion C, Zayachkivsky A, Liu M, Uhde T, Glen WB Jr, van den Pol AN, Mulholland PJ, Shiromani PJ (2013) Optogenetic stimulation of MCH neurons increases sleep. J Neurosci 33:10257–10263

Lagos P, Torterolo P, Jantos H, Chase MH, Monti JM (2009) Effects on sleep of melanin-concentrating hormone microinjections into the dorsal raphe nucleus. Brain Res 1265:103–110

Lagos P, Torterolo P, Jantos H, Chase MH, Monti JM (2011) Immunoneutralization of melanin-concentrating hormone (MCH) in the dorsal raphe nucleus: effects on sleep and wakefulness. Brain Res 1369:112–118

Lagos P, Monti JM, Jantos H, Torterolo P (2012) Microinjection of the melanin-concentrating hormone into the lateral basal forebrain increases REM sleep and reduces wakefulness in the rat. Life Sci 90:895–899

Lee MG, Hassani OK, Jones BE (2005) Discharge of identified orexin/hypocretin neurons across the sleep-waking cycle. J Neurosci 25:6716–6720

Leger L, Sapin E, Goutagny R, Peyron C, Salvert D, Fort P, Luppi PH (2010) Dopaminergic neurons expressing Fos during waking and paradoxical sleep in the rat. J Chem Neuroanat 39:262–271

Li Y, Gao XB, Sakurai T, van den Pol AN (2002) Hypocretin/Orexin excites hypocretin neurons via a local glutamate neuron-A potential mechanism for orchestrating the hypothalamic arousal system. Neuron 36:1169–1181

Li J, Hu Z, de Lecea L (2013) The Hypocretins/orexins: integrators of multiple physiological functions. Br J Pharmacol 171(2):332–350

Liu ZW, Gao XB (2007) Adenosine inhibits activity of hypocretin/orexin neurons by the A1 receptor in the lateral hypothalamus: a possible sleep-promoting effect. J Neurophysiol 97:837–848

Lu J, Jhou TC, Saper CB (2006) Identification of wake-active dopaminergic neurons in the ventral periaqueductal gray matter. J Neurosci 26:193–202

Luppi PH, Gervasoni D, Verret L, Goutagny R, Peyron C, Salvert D, Leger L, Fort P (2007) Paradoxical (REM) sleep genesis: The switch from an aminergic-cholinergic to a GABAergic-glutamatergic hypothesis. J Physiol (Paris) 100:271–283

Luppi PH, Clement O, Fort P (2013) Paradoxical (REM) sleep genesis by the brainstem is under hypothalamic control. Curr Opin Neurobiol 23:786–792

McCarley RW (2007) Neurobiology of REM and NREM sleep. Sleep Med 8:302–330

McGregor R, Damian A, Fabbiani G, Torterolo P, Pose I, Chase M, Morales FR (2005) Direct hypothalamic innervation of the trigeminal motor nucleus: a retrograde tracer study. Neuroscience 136:1073–1081

McGregor R, Wu MF, Barber G, Ramanathan L, Siegel JM (2011) Highly specific role of hypocretin (orexin) neurons: differential activation as a function of diurnal phase, operant reinforcement versus operant avoidance and light level. J Neurosci 31:15455–15467

Messina MM, Overton JM (2007) Cardiovascular effects of melanin-concentrating hormone. Regul Pept 139:23–30

Mignot E (2011) Narcolepsy: pathophysiology and genetic predisposition. In: Krieger MH, Roth T, Dement W (eds) Principles and practices of sleep medicine. Saunders, Philadelphia, pp 938–956

Mileykovskiy BY, Kiyashchenko LI, Siegel JM (2002) Muscle tone facilitation and inhibition after orexin-a (hypocretin-1) microinjections into the medial medulla. J Neurophysiol 87:2480–2489

Mileykovskiy BY, Kiyashchenko LI, Siegel JM (2005) Behavioral correlates of activity in identified hypocretin/orexin neurons. Neuron 46:787–798

Modirrousta M, Mainville L, Jones BE (2005) Orexin and MCH neurons express c-Fos differently after sleep deprivation vs. recovery and bear different adrenergic receptors. Eur J Neurosci 21:2807–2816

Monti JM (2010a) The role of dorsal raphe nucleus serotonergic and non-serotonergic neurons, and of their receptors, in regulating waking and rapid eye movement (REM) sleep. Sleep Med Rev 14:319–327

Monti JM (2010b) The structure of the dorsal raphe nucleus and its relevance to the regulation of sleep and wakefulness. Sleep Med Rev 14:307–317

Monti JM (2011) The role of tuberomammillary nucleus histaminergic neurons, and of their receptors, in the regulation of sleep and waking. In: P-PS Mallick BN, McCarley RW, Morrison AR (ed) REM sleep: regulation and function. Cambridge University Press, pp 223–233

Monti JM, Monti D (2007) The involvement of dopamine in the modulation of sleep and waking. Sleep Med Rev 11:113–133

Monti JM, Torterolo P, Lagos P (2013) Melanin-concentrating hormone control of sleep-wake behavior. Sleep Med Rev 17:293–298

Moreno-Balandran E, Garzon M, Bodalo C, Reinoso-Suarez F, de Andres I (2008) Sleep-wakefulness effects after microinjections of hypocretin 1 (orexin A) in cholinoceptive areas of the cat oral pontine tegmentum. Eur J Neurosci 28:331–341

Murillo-Rodriguez E (2008) The modulatory role of endocannabinoids in sleep. Rev Neurol 46:160–166

Nambu T, Sakurai T, Mizukami K, Hosoya Y, Yanagisawa M, Goto K (1999) Distribution of orexin neurons in the adult rat brain. Brain Res 827:243–260

Nunez A, Moreno-Balandran ME, Rodrigo-Angulo ML, Garzon M, De Andres I (2006) Relationship between the perifornical hypothalamic area and oral pontine reticular nucleus in the rat. Possible implication of the hypocretinergic projection in the control of rapid eye movement sleep. Eur J Neurosci 24:2834–2842

Pascovich C, Devera A, Lagos P, Costa A, Falconi A, Torterolo P (2011) Melanin-concentrating hormone (MCH) decreases presumed serotonergic neuronal activity in the dorsal and median raphe nucleus: Congreso de la Sociedad Argentina de Neurociencias. Huerta Grande, Argentina

Peyron C, Tighe DK, van den Pol AN, de Lecea L, Heller HC, Sutcliffe JG, Kilduff TS (1998) Neurons containing hypocretin (orexin) project to multiple neuronal systems. J Neurosci 18:9996–10015

Piper DC, Upton N, Smith MI, Hunter AJ (2000) The novel brain neuropeptide, orexin-A, modulates the sleep-wake cycle of rats. Eur J Neurosci 12:726–730

Rao Y, Lu M, Ge F, Marsh DJ, Qian S, Wang AH, Picciotto MR, Gao XB (2008) Regulation of synaptic efficacy in hypocretin/orexin-containing neurons by melanin concentrating hormone in the lateral hypothalamus. J Neurosci 28:9101–9110

Reinoso-Suarez F, de Andres I, Rodrigo-Angulo ML, Garzon M (2001) Brain structures and mechanisms involved in the generation of REM sleep. Sleep Med Rev 5:63–77

Rowe K, Moreno R, Lau TR, Wallooppillai U, Nearing BD, Kocsis B, Quattrochi J, Hobson JA, Verrier RL (1999) Heart rate surges during REM sleep are associated with theta rhythm and PGO activity in cats. Am J Physiol 277:R843–R849

Saito Y, Nagasaki H (2008) The melanin-concentrating hormone system and its physiological functions. Results Probl Cell Differ 46:159–179

Sakurai T, Amemiya A, Ishii M, Matsuzaki I, Chemelli RM, Tanaka H, Williams SC, Richardson JA, Kozlowski GP, Wilson S et al (1998) Orexins and orexin receptors: a family of

hypothalamic neuropeptides and G protein-coupled receptors that regulate feeding behavior. Cell 92:573–585
Semba K (1999) The mesopontine cholinergic system: a dual role in REM sleep and wakefulness. In: Lydic R, Baghdoyan H (eds) Handbook of behavioral state control. CRC Press, Washington
Shirasaka T, Kunitake T, Takasaki M, Kannan H (2002) Neuronal effects of orexins: relevant to sympathetic and cardiovascular functions. Regul Pept 104:91–95
Siegel JM (2005) REM Sleep. In: Kryger MH, Roth T, Dement WC (eds) Principles and practices of sleep medicine. Elsevier-Saunders, Philadelphia, pp 120–135
Strecker RE, Morairty S, Thakkar MM, Porkka-Heiskanen T, Basheer R, Dauphin LJ, Rainnie DG, Portas CM, Greene RW, McCarley RW (2000) Adenosinergic modulation of basal forebrain and preoptic/anterior hypothalamic neuronal activity in the control of behavioral state. Behav Brain Res 115:183–204
Szymusiak R, McGinty D (2008) Hypothalamic regulation of sleep and arousal. Ann NY Acad Sci 1129:275–286
Takahashi K, Lin JS, Sakai K (2008) Neuronal activity of orexin and non-orexin waking-active neurons during wake-sleep states in the mouse. Neuroscience 153:860–870
Torterolo P, Chase MH (2014) Role of hypocretin (orexin) in REM sleep generation: a new hypothesis. Sleep Science (in press)
Torterolo P, Vanini G (2010a) New concepts in relation to generating and maintaining arousal. Rev Neurol 50:747–758
Torterolo P, Vanini G (2010b) Involvement of GABAergic mechanisms in the laterodorsal and pedunculopontine tegmental nuclei (LDT-PPT) in the promotion of REM sleep. In: Monti JM, Pandi-Perumal SR, Möhler H (eds) GABA and Sleep: molecular, functional and clinical aspects. Springer, Basel, pp 213–231
Torterolo P, Yamuy J, Sampogna S, Morales FR, Chase MH (2001a) GABAergic neurons of the laterodorsal and pedunculopontine tegmental nuclei of the cat express c-fos during carbachol-induced active sleep. Brain Res 892:309–319
Torterolo P, Yamuy J, Sampogna S, Morales FR, Chase MH (2001) Hypothalamic neurons that contain hypocretin (orexin) express c-fos during active wakefulness and carbachol-induced active sleep. Sleep Res Online 4: 25–32. http://www.sro.org/2001/Torterolo/2025
Torterolo P, Morales FR, Chase MH (2002) GABAergic mechanisms in the pedunculopontine tegmental nucleus of the cat promote active (REM) sleep. Brain Res 944:1–9
Torterolo P, Yamuy J, Sampogna S, Morales FR, Chase MH (2003) Hypocretinergic neurons are primarily involved in activation of the somatomotor system. Sleep 1:25–28
Torterolo P, Sampogna S, Morales FR, Chase MH (2006) MCH-containing neurons in the hypothalamus of the cat: searching for a role in the control of sleep and wakefulness. Brain Res 1119:101–114
Torterolo P, Vanini G, Zhang J, Sampogna S, Chase M (2007) Hypocretinergic fibers and receptors in the inferior colliculus. Sleep 30:A15
Torterolo P, Benedetto L, Lagos P, Sampogna S, Chase MH (2009a) State-dependent pattern of Fos protein expression in regionally-specific sites within the preoptic area of the cat. Brain Res 1267:44–56
Torterolo P, Sampogna S, Chase MH (2009b) MCHergic projections to the nucleus pontis oralis participate in the control of active (REM) sleep. Brain Res 1268:76–87
Torterolo P, Sampogna S, Chase MH (2011a) A restricted parabrachial pontine region is active during non-rapid eye movement sleep. Neuroscience 190:184–193
Torterolo P, Ramos OV, Sampogna S, Chase MH (2011b) Hypocretinergic neurons are activated in conjunction with goal-oriented survival-related motor behaviors. Physiol Behav 104:823–830
Torterolo P, Lagos P, Monti JM (2011c) Melanin-concentrating hormone (MCH): a new sleep factor? Front Neurol 2:1–12
Torterolo P, Sampogna S, Chase MH (2013) Hypocretinergic and non-hypocretinergic projections from the hypothalamus to the REM sleep executive area of the pons. Brain Res 1491:68–77

Tsunematsu T, Ueno T, Tabuchi S, Inutsuka A, Tanaka KF, Hasuwa H, Kilduff TS, Terao A, Yamanaka A (2014) Optogenetic manipulation of activity and temporally controlled cell-specific ablation reveal a role for MCH neurons in sleep/wake regulation. J Neurosci 34:6896–6909

Urbanavicius J, Lagos P, Torterolo P, Scorza C (2013) Role of melanin concentrating hormone (MCH) on the dorsal raphe nucleus (DRN): its relevance for depression. J Neurochem 125:265

Ursin R, Sterman M (1981) Manual for standardized scoring of sleep and waking states in adult cats. University of California, Los Angeles, BIS/BRI

van den Pol A, Acuna-Goycolea C (2006) Neuronal responses to hypocretin/orexin. In: Nishino S, Sakurai T (eds) The orexin/hypocretin system. Humana Press, Totowa, New Jersey, Physiology and pathophysiology, pp 45–60

van den Pol AN, Acuna-Goycolea C, Clark KR, Ghosh PK (2004) Physiological properties of hypothalamic MCH neurons identified with selective expression of reporter gene after recombinant virus infection. Neuron 42:635–652

Vanini G, Torterolo P, McGregor R, Chase MH, Morales FR (2007) GABAergic processes in the mesencephalic tegmentum modulate the occurrence of active (rapid eye movement) sleep in guinea pigs. Neuroscience 145:1157–1167

Voisin T, Rouet-Benzineb P, Reuter N, Laburthe M (2003) Orexins and their receptors: structural aspects and role in peripheral tissues. Cell Mol Life Sci 60:72–87

Watson CJ, Soto-Calderon H, Lydic R, Baghdoyan HA (2008) Pontine reticular formation (PnO) administration of hypocretin-1 increases PnO GABA levels and wakefulness. Sleep 31:453–464

Williams RH, Burdakov D (2008) Hypothalamic orexins/hypocretins as regulators of breathing. Expert Rev Mol Med 10:e28

Xi M, Chase MH (2010) The injection of hypocretin-1 into the nucleus pontis oralis induces either active sleep or wakefulness depending on the behavioral state when it is administered. Sleep 33:1236–1243

Xi MC, Morales FR, Chase MH (2001) Induction of wakefulness and inhibition of active (REM) sleep by GABAergic processes in the nucleus pontis oralis. Arch Ital Biol 139:125–145

Xi MC, Fung SJ, Yamuy J, Morales FR, Chase MH (2002) Induction of active (REM) sleep and motor inhibition by hypocretin in the nucleus pontis oralis of the cat. J Neurophysiol 87:2880–2888

Xi MC, Fung SJ, Yamuy J, Morales FR, Chase MH (2003) Hypocretinergic facilitation of synaptic activity of neurons in the nucleus pontis oralis of the cat. Brain Res 976:253–258

Yamanaka A (2006) Afferent system of orexin neurons. In: Nishino S, Sakurai T (eds) The orexin/hypocretin system. Humana Press, Totowa, New Jersey, Physiology and pathophysiology, pp 61–70

Yamuy J, Fung SJ, Xi M, Chase MH (2004) Hypocretinergic control of spinal cord motoneurons. J Neurosci 24:5336–5345

Yoshida K, McCormack S, Espana RA, Crocker A, Scammell TE (2006) Afferents to the orexin neurons of the rat brain. J Comp Neurol 494:845–861

Zhang W, Fukuda Y, Kuwaki T (2005) Respiratory and cardiovascular actions of orexin-A in mice. Neurosci Lett 385:131–136

Animal Models of Narcolepsy

Takeshi Sakurai

Abstract The importance of orexin in the regulation of sleep/wakefulness states, especially in the maintenance of wakefulness, was initially revealed by animal studies which showed the relationship between orexin and narcolepsy. Familial narcolepsy of dogs was shown to be caused by functionally-null mutations in the orexin 2 receptor (*OX2R* or *hcrtr2*) gene (Lin et al. 1999). At the almost same time, orexin gene disruption in mice was shown to cause narcolepsy (Chemelli et al. 1999). Orexin neuron-ablated (*orexin/ataxin-3*-transgenic) mice, and *OX1R/OX2R* double-deficient mice were subsequently showed to have the same phenotype with $orexin^{-/-}$ mice (Hara et al. 2001; Willie et al. 2003), characterized by behavioral arrests that are similar to cataplexy, occasional direct transitions to REM sleep from wakefulness, and highly fragmented sleep-wake cycles, all of which are important elements of narcolepsy. These phenotypes have strong parallels to the human narcolepsy. In this chapter, I will discuss animal models of narcolepsy.

1 Dog Narcolepsy

Narcolepsy has been well described in several mammalian species including dogs, mice and humans. Similar to human narcolepsy, main symptoms of dog narcolepsy are excessive daytime sleepiness (EDS) and cataplexy. Similar to that of humans, dog cataplexy is characterized by sudden muscle atonia without loss of consciousness. The animals usually remain alert and can follow movement with their eyes throughout thc episode. The episodes typically last from several seconds to 30 min, often occurring when the dogs are running, eating, playing, excited, doing sexual activity. Food and play with other dogs are the well-documented paradigms used to trigger cataplexy. During cataplexy, dogs collapse onto its side or belly, and

T. Sakurai (✉)
Department of Molecular Neuroscience and Integrative Physiology, Faculty of Medicine, Kanazawa University, 13-1 Takaramachi, Kanazawa, Ishikawa 920-8640, Japan
e-mail: tsakurai@med.kanazawa-u.ac.jp

© Springer International Publishing Switzerland 2015

T. Sakurai et al. (eds.), *Orexin and Sleep*,
DOI 10.1007/978-3-319-23078-8_12

all physical movement ceases. Dogs usually come out of an episode in response to external stimuli, such as loud sounds. Occasionally, dogs fall asleep after attacks. Narcoleptic dogs sometimes suffer from sudden loss of consciousness or sleep attack, phenotypes that suggest excessive daytime sleepiness.

Like human cases, most dog narcolepsy is sporadic, and CSF orexin A levels are low or undetectable, suggesting that postnatal degeneration of orexin neurons due to similar pathophysiology as human cases. However, rare hereditary cases were found in Labrador retrievers, poodles, dachshunds, and Doberman pinschers.

Autosomal recessive transmission with full penetrance for canine narcolepsy was first established in Labrador retrievers and Doberman pinschers (Foutz et al. 1979). Although one of the predisposing genetic factors of human narcolepsy is a HLA-DQ allele, HLA-DQB1*0602, genetic linkage analysis had excluded the linkage between the dog narcolepsy and the canine major histocompatibility complex (Mignot et al. 1991).

In 1999, by screening BAC clones containing a genetic marker a team of Stanford University led by Mignot showed that loss of function mutations in *OX2R (hcrtr-2)* gene are responsible for familial narcolepsy-cataplexy in dogs, supporting the idea that deficiency in orexin (hypocretin) signaling might have an important role in pathophysiology of narcolepsy.

The *OX2R* gene of narcoleptic Dobermans contained a 226-bp insertion corresponding to a common canine short interspersed elements (SINE) repeat element located 35 bp upstream of the exon 4. The insertion of the SINE displaces a putative lariat branchpoint sequence ocated at position −40 through −46 upstream of the 3'-splice site. The mRNA potentially encodes a protein with 38 amino acids deletion within the fifth transmembrane domain followed by a frameshift, making the transcript functionally null. In narcoleptic Labrador retrievers, a separate line, G to A transition in the 5'-splice junction consensus sequence was found. Therefore, the exon 5 is spliced directly to exon 7, omitting exon 6. The C terminus of the OX2R of narcoleptic Labrador was truncated and did not include a seventh transmembrane domain. In both cases, these structural abnormalities of OX2R disrupt proper function of OX2R, suggesting the importance of OX2R-mediated signaling in the stabilization of wakefulness.

2 Rodent Narcolepsy

To date, several genetically-modified narcoleptic rodent models have been established, including *orexin*$^{-/-}$ mice, *orexin-ataxin-3* mice/rats, *OX2R*-deficient mice, *OX1R/OX2R* -double knockout mice, and *orexin-tTA;TetO DTA* mice. Sleep/wakefulness sates of these mice are similar, although each shows some distinguishing characteristics. In this section, I will firstly mention about their overall phenotypes, and then discuss about phenotypes of each model (Figs. 1 and 2).

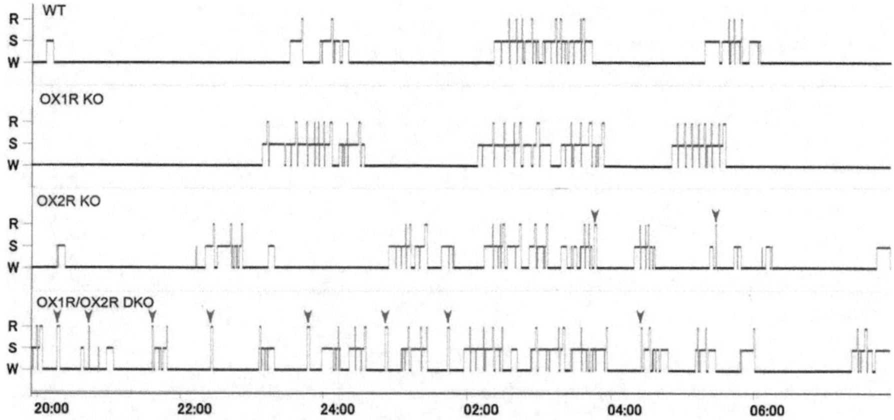

Fig. 1 Representative hypnograms of wild-type, *OX1R*$^{-/-}$, *OX2R*$^{-/-}$ and double receptor deficient mice over 12 h of the dark phase obtained by concatenating 16 s epoch EEG/EMG stage scores. The height of the *horizontal line* above *baseline* indicates the vigilance state of the mouse at the time. *Baseline*, *W* represents periods of wakefulness; *S* non-REM sleep; *R* REM sleep. *Arrowheads* highlight direct transitions from wakefulness to REM sleep

2.1 Phenotypes of Narcoleptic Mice

Orexin$^{-/-}$ mice, *orexin/ataxin-3* mice, and OX1R/OX2R double orexin receptor deficient (*OX1R*$^{-/-}$; *OX2R*$^{-/-}$) mice showed the almost identical sleep/wakefulness characteristics, including abrupt behavioral arrest (cataplexy) and fragmentation of sleep/wakefulness states. This section describes these phenotypes (Table 1).

These phenotypes caused by genetic modifications are strikingly similar to symptomes seen in human narcolepsy. However, it should be noted that familial transmission of human narcolepsy is very rare in humans, and even in these rare cases penetrance is far less than 100 %. No mutation has been found so far either in the *prepro-orexin* or *orexin receptor* genes of human narcolepsy patients, except an unusually severe, early onset case, which is associated with mutation in the *prepro-orexin* gene that impairs peptide trafficking and processing (Peyron et al. 2000).

Behavioral arrests (Cataplexy): These narcoleptic (*orexin*$^{-/-}$, *orexin/ataxin-3*, and *OX1R*$^{-/-}$;*OX2R*$^{-/-}$) mice exhibited substantial numbers of behavioral arrests during the dark periods. They were specifically recognized by the abrupt cessation of purposeful motor activity associated with sudden, sustained change in posture that was maintained throughout the episode, ending abruptly with complete resumption of purposeful motor activity. Characteristics of abrupt arrests were very different from those of quiet behavioral states or normal transitions into sleep. Each episode usually lasted for a short period, mostly less than a minute. Occasionally, gait disturbance lasting several seconds was observed immediately before episodes. Side-to-side rocking, without change in overall posture, frequently occurred several seconds after the start of the arrest.

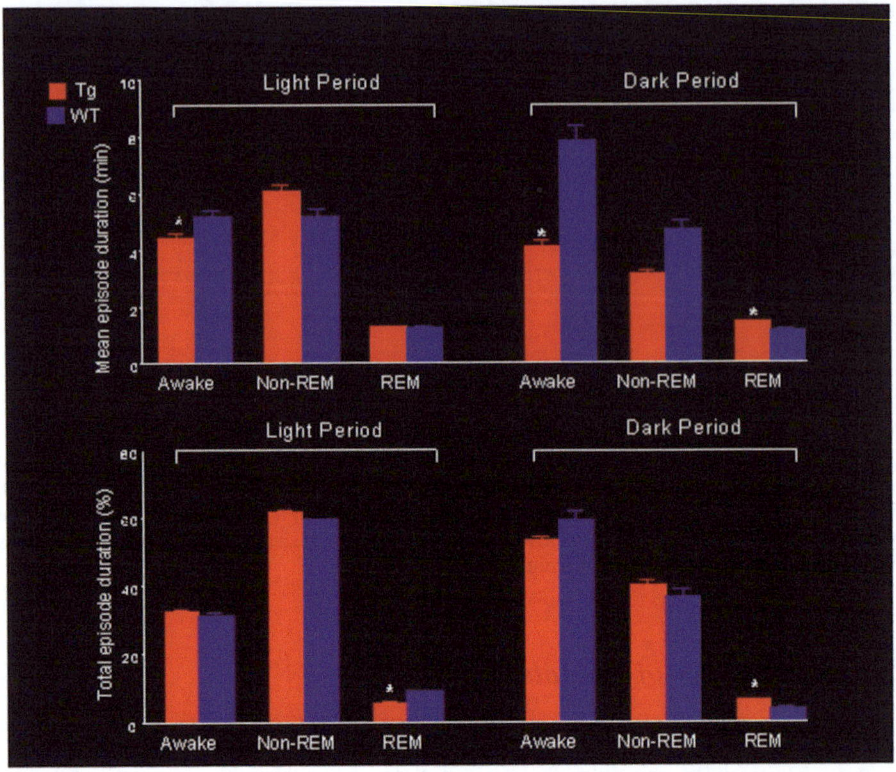

Fig. 2 Durations (*upper panels*) and total amounts (*lower panels*) of awake, non-REM sleep, and REM sleep states (means and SEM) for wild type and *orexin/ataxin-3* (Tg) mice in light and dark periods. Narcoleptic mice consistently show significantly increased REM sleep times during the dark phase compared to normal mice. *, $p < 0.05$

Detailed observations of behaviors during EEG/EMG recordings found that abrupt arrests in narcoleptic mice occurred at timing of the direct transitions from wakefulness to REM sleep or during pre-REM phase immediately after a waking period. (The pre-REM phase shows EEG with high-amplitude spindle oscillations superimposed on non-REM sleep background, and these spindles are observed only during the transition phase immediately prior to REM sleep in wild-type mice.) The direct or very rapid transitions from wakefulness to REM sleep are the most prominent characteristic of EEG/EMG in narcoleptic mice, which were observed almost exclusively in the dark phase, and were never observed in wild-type mice.

Abrupt arrests in *orexin*$^{-/-}$ mice were ameliorated by systemic administration of clomipramine, an agent used for treatment of human cataplexy. While administration of caffeine, a psychostimulant used to treat excessive sleepiness in human narcolepsy, produce a mild exacerbation of abrupt arrest frequency.

These behavioral, electrophysiological, and pharmacological characteristics of abrupt arrests in narcoleptic mice suggest that these attacks are the counterpart of

Table 1 Phenotypes of rodent narcolepsy models produced by genetic engineering

	Sleep/wake state abnormality	Other phenotypes	Ref.
Prepro-orexin knockout	Cataplexy (+), sleep attacks (+) Sleep/wake fragmentation (severe)	Slight decrease in food intake, mild tendency for obesity (dependent on genetic background)	Chemelli et al. (1999); Hara et al. (2005)
OX1R knockout	Cataplexy (−), sleep attacks (+) Sleep/wake fragmentation	Impairment in emotional memory formation	Willie et al. (2001); Soya et al. (2013)
OX2R knockout	Cataplexy (+), sleep attacks (+) Sleep/wake fragmentation (severe)	ND	Willie et al. (2003)
Orexin/ataxin-3 mouse	Cataplexy (+), sleep attacks (+) Sleep/wake fragmentation (severe)	Decrease in food intake, mild tendency for obesity (dependent on genetic background), lack of food entrainable activity, lack of fasting-induced increase in wake time	Hara et al. (2001)
Orexin/ataxin-3 rat	Cataplexy (+), sleep attacks (+) Sleep/wake fragmentation (severe)	ND	Beuckmann et al. (2004)
orexin-tTA; TetO DTA mice	Cataplexy (+) (depending on numbers of orexin neurons ablated), sleep attacks (+) Sleep/wake fragmentation (severe)	Late onset obesity	Tabuchi et al. (2014)

cataplexy observed in human narcolepsy patients. Cataplexy has been well known to be triggered by strong emotions such as laughing, anger, fear, surprise, and excitement in human narcolepsy patients. In narcoleptic mice, abrupt arrests are very often preceded by purposeful motor activity such as excited ambulation, grooming, burrowing, and climbing. Palatable foods such as chocolates can also trigger cataplexy. The dramatic increase in the number of abrupt arrests noted in the group-housed mice as compared with the individually-housed littermates suggests that social interaction may enhance this attack. Chasing, tail biting, and social grooming were often observed to immediately precede the attacks in the group-housed mice.

Only approximately one-third of human patients experience full loss of muscle tone causing collapse to the floor with the majority having partial cataplexy evidenced by jaw sagging, head bobbing, arm dropping, ptosis, or dysarthria (Parkes et al. 1974). Unambiguous full postural collapse was frequently observed in young *orexin*$^{-/-}$ mice, while adults tended to collapse onto their ventral surface at odd angles, suggesting some residual muscle tone. Cataplexy is not always instantaneous in human patients but can progress over several seconds with some patients experiencing gait disturbance known as "zig-zag walking". Similarly, gait ataxia immediately preceding almost one third of abrupt arrests in narcoleptic mice.

Sleepiness and sleep attack: One of the prominent features of sleep/wakefulness patterns in narcoleptic mice was shortened durations of both wakefulness and NREM sleep in the dark phase, causing increased fragmentation of sleep/wake cycle. The shortened durations of wakefulness suggests sleepiness, which is the cardinal symptom of narcolepsy. *OX2R*$^{-/-}$ mice also showed sleep/wake fragmentation, while occurrence of REM sleep-related abnormalities was very rare as compared to *orexin*$^{-/-}$, *orexin/ataxin-3* or double orexin receptor knockout mice. This fragmentation was accompanied by statistically insignificant tendency toward reduced amounts of wakefulness and increased amounts of NREM sleep during the dark phase.

Presumptive excessive sleepiness in *orexin*$^{-/-}$ mice was also shown by detailed analyses of "gradual arrests" in *orexin*$^{-/-}$ and *OX2R*$^{-/-}$ mice, which can be interpreted analogous to "sleep attacks" in human narcolepsy. Gradual arrests typically began during quiet wakefulness and could be easily distinguished from the normal onset of resting behavior by lack of stereotypic preparation for sleep (e.g., nesting and/or assumption of a curled or hunched posture, with limbs drawn under the body) and a characteristic ratchet-like "nodding" of the head over a period of several seconds, with a transition to a collapsed posture. Gradual arrests in both *orexin*$^{-/-}$ and *OX2R*$^{-/-}$ mice resemble the sleep attacks in human narcolepsy. Unlike cataplexy, the gradual arrests are not associated with strong emotions or muscle atonia, and are reduced by psychostimulants such as amphetamines, modafinil, and caffeine.

Systemic administration of caffeine dose-dependently suppressed gradual arrests, while administration of an anticataplectic agent clomipramine did not affect the frequency of gradual arrests in both *orexin*$^{-/-}$ and *OX2R*$^{-/-}$ mice.

EEG/EMG recordings with simultaneous video-capture further differentiated gradual arrests from abrupt ones. As described above, abrupt arrests were accompanied by direct transition from wakefulness to REM sleep in narcoleptic mice. In contrast, EEG/EMG correlates of gradual arrests in narcoleptic mice invariably revealed the onset of attenuated muscle tone, but not atonia, and an EEG transition from wakefulness to NREM sleep. Gradual arrests were occasionally accompanied by apparent automatic behavior, which is continuation of semipurposeful motor activity after the onset of light sleep such as stereotypic chewing of food.

2.2 *Characteristic of Each Model*

2.2.1 *Pepeo-Orexin* Gene Knock Out (*Orexin*$^{-/-}$) Mice

Conventional behavioral tests revealed that *orexin*$^{-/-}$ mice did not show any overt abnormalities during the light period, except mild decrease in food intake (Chemelli et al. 1999). However, infrared video-recordings at the dark period to examine abnormalities on feeding behavior of *orexin*$^{-/-}$ mice at night, when mice are normally most active, found frequent periods of behavioral arrest. No serum electrolyte imbalance or hypoglycemia was observed in *orexin*$^{-/-}$ mice. Electroencephalograph/electromyographic (EEG/EMG) recordings from *orexin*$^{-/-}$ mice showed no evidence of epileptic seizure during the attack. Instead, these EEG/EMG recordings revealed abnormal intrusions of REM sleep-like episodes into wakefulness and fragmentation of sleep/wakefulness cycle (Chemelli et al. 1999). Markedly reduced latency to REM sleep and increase in REM sleep during the dark phase were also observed. These characteristics of *orexin*$^{-/-}$ mice were strikingly similar to characteristics human narcolepsy.

2.2.2 *Orexin-Ataxin-3* Mice

Orexin levels in the CSF and postmortem brains of narcolepsy patients was dramatically decreased (Nishino et al. 2000; Peyron et al. 2000; Thannickal et al. 2000). However, in most cases of human narcolepsy, onset of symptoms occurs during adolescence. Based on the late onset, as well as a strong association of human narcolepsy with certain HLA alleles (Kadotani et al. 1998; Singh et al. 2013), it has been thought that narcolepsy result from acquired, selective autoimmune degeneration of orexin neurons.

In order to mimic the postnatal degeneration of orexin neurons in human narcolepsy, we generated transgenic mice (*orexin/ataxin-3* transgenic mice) in which orexin neurons are ablated by orexinergic-specific expression of a truncated Machado-Joseph disease gene (ataxin-3) fragment with expanded polyglutamine stretch as a toxic trans gene (Hara et al. 2001). Number of orexin neurons decreased after their birth, and at 12 weeks of age, over 99 % of orexin neurons were lost. *Orexin/ataxin-3* transgenic mice exhibited behavioral and electrophysiological defects that are essentially the same as those displayed by *orexin*$^{-/-}$ mice, including cataplexy-like behavioral arrests, direct transitions from wakefulness to REM sleep, shortened latency to REM sleep, fragmentation of sleep/wakefulness states, and increases in REM sleep time and duration in the dark period. However, the behavioral arrests seen in *orexin/ataxin-3* transgenic mice typically began about 6 weeks of age, while arrests were observed in some *orexin*$^{-/-}$ mice earlier than 3 weeks of age. Thus, *orexin/ataxin-3* transgenic mice have etiology and the course of disease similar to human narcolepsy. This model has been widely used as an useful narcoleptic model.

The fact that *orexin/ataxin-3* transgenic mice exhibited similar phenotype with that of *orexin*$^{-/-}$ mice also revealed that orexin is the primarily important factor for the maintenance of the sleep/wakefulness state by orexin neurons, although orexin neurons produce other neurotransmitters such as glutamate, dynorphins, and neurotensin (Chou et al. 2001; Rosin et al. 2003; Furutani et al. 2013).

We utilized this mouse model of narcolepsy to rescue its narcoleptic phenotype in *orexin/ataxin-3* transgenic mice by genetic (overexpression of orexin peptides throughout the brain) and pharmacological (i.c.v. administration of orexin A) means (Mieda et al. 2004), demonstrating that mice retain the ability to respond to orexin neuropeptides even if they lack endogenous orexin neurons and that a temporally regulated and spatially targeted secretion of orexins is not necessary to prevent narcoleptic symptoms. This finding suggests that orexin receptor agonists would be of potential value for treating human narcolepsy.

2.2.3 *Orexin/Ataxin-3* Rats

Transgenic rats made with the same transgene as *orexin/ataxin-3* transgenic mice was also generated (*orexin/ataxin-3* transgenic rats) (Beuckmann et al. 2004). Similar to *orexin/ataxin-3* mice, number of orexin neurons was gradually reduced after their birth, and after 17 weeks of age orexin in the hypothalamus became undetectable. *Orexin/ataxin-3* transgenic rats showed a typical narcoleptic phenotype, with a decreased latency to REM sleep, increased REM sleep time in dark period, direct transitions from wakefulness to REM sleep, and a marked fragmentation of vigilance states. Brief episodes of muscle atonia and postural collapse resembling cataplexy were also observed while rats maintained the EEG characteristics of wakefulness, suggesting they were conscious during these episodes as human patients are during cataplexy.

2.2.4 OX2R-Deficient Mice

In infrared videotaping studies in the dark phase revealed that *OX2R*$^{-/-}$ mice also exhibited behavioral arrests, or cataplexy (Willie et al. 2003). However, its frequency was much less than in *orexin*$^{-/-}$ mice (31-fold lower frequency in *OX2R*$^{-/-}$ mice as compared to *orexin*$^{-/-}$ mice). *OX2R*$^{-/-}$ mice appeared to exhibit different type behavioral arrests with onsets that were more gradual in nature (gradual arrests). O*rexin*$^{-/-}$ mice and *orexin/ataxin*-3 mice also found to exhibit the gradual arrests with the frequency similar to *OX2R*$^{-/-}$ mice in addition to numbers of cataplexy-like, abrupt arrests.

OX2R$^{-/-}$ mice and *orexin*$^{-/-}$ mice are similarly affected by gradual arrests ("sleep attacks") (Willie et al. 2003), but *OX2R*$^{-/-}$ mice show a lower degree of disrupted wakefulness compared with double receptor-deficient (*OX1R*$^{-/-}$; *OX2R*$^{-/-}$) mice (Willie et al. 2003; Sakurai 2007; Hondo et al. 2010). In particular, *OX2R*$^{-/-}$ mice are only mildly affected by cataplexy and direct transitions to REM sleep from awake states, as compared with *orexin*$^{-/-}$ mice and double receptor KO mice(Chemelli et al. 1999; Willie et al. 2003).

2.2.5 OX1R/OX2R Gene Double Knockout Mice

Orexin$^{-/-}$ mice and double receptor-knockout mice have the virtually same phenotype (Willie et al. 2003). Considering the fact that *OX2R*$^{-/-}$ mice are only mildly affected by cataplexy and direct transitions to REM sleep from awake states, these observations suggest that *OX2R* plays a pivotal role in maintain wakefulness, but *OX1R* has additional effects on sleep-wake regulation. The gating of REM sleep is likely to critically involve both receptors. The contributions of both receptors in the sleep/wakefulness states regulation are described in detail in our separate chapter (Mieda).

2.2.6 Orexin-TTA;TetO DTA Mice

In this model, diphtheria toxin A (DTA) was expressed in orexin neurons under control of the Tet-off system. Doxycycline removal from the diet of adult orexin-tTA;TetO DTA mice induce orexin neurodegeneration with 80 % cell loss after one week of tet-off condition, and resulted in apparent sleep/wakefulness fragmentation. However, cataplexy was only observed 14 d after the removal of doxycycline when less than 5 % of the orexin neurons remained. Cataplexy frequency increased for at least 11 weeks after doxycycline removal. These suggest that very low level of orexin could inhibit cataplexy. Body weight was reported to be increased without a change in food consumption, as in human narcolepsy and *orexin/ataxin*-3 mice.

References

Beuckmann CT, Sinton CM, Williams SC, Richardson JA, Hammer RE, Sakurai T, Yanagisawa M (2004) Expression of a poly-glutamine-ataxin-3 transgene in orexin neurons induces narcolepsy-cataplexy in the rat. J Neurosci 24:4469–4477

Chemelli RM, Willie JT, Sinton CM, Elmquist JK, Scammell T, Lee C, Richardson JA, Williams SC, Xiong Y, Kisanuki Y, Fitch TE, Nakazato M, Hammer RE, Saper CB, Yanagisawa M (1999) Narcolepsy in orexin knockout mice: molecular genetics of sleep regulation. Cell 98:437–451

Chou TC, Lee CE, Lu J, Elmquist JK, Hara J, Willie JT, Beuckmann CT, Chemelli RM, Sakurai T, Yanagisawa M, Saper CB, Scammell TE (2001) Orexin (hypocretin) neurons contain dynorphin. J Neurosci 21, RC168

Foutz AS, Mitler MM, Cavalli-Sforza LL, Dement WC (1979) Genetic factors in canine narcolepsy. Sleep 1:413–421

Furutani N, Hondo M, Kageyama H, Tsujino N, Mieda M, Yanagisawa M, Shioda S, Sakurai T (2013) Neurotensin co-expressed in orexin-producing neurons in the lateral hypothalamus plays an important role in regulation of sleep/wakefulness states. PLoS ONE 8:e62391

Hara J, Beuckmann CT, Nambu T, Willie JT, Chemelli RM, Sinton CM, Sugiyama F, Yagami K, Goto K, Yanagisawa M, Sakurai T (2001) Genetic ablation of orexin neurons in mice results in narcolepsy, hypophagia, and obesity. Neuron 30:345–354

Hara J, Yanagisawa M, Sakurai T (2005) Difference in obesity phenotype between orexin-knockout mice and orexin neuron-deficient mice with same genetic background and environmental conditions. Neurosci Lett 380:239–242

Hondo M, Nagai K, Ohno K, Kisanuki Y, Willie JT, Watanabe T, Yanagisawa M, Sakurai T (2010) Histamine-1 receptor is not required as a downstream effector of orexin-2 receptor in maintenance of basal sleep/wake states. Acta Physiol (Oxf) 198:287–294

Kadotani H, Faraco J, Mignot E (1998) Genetic studies in the sleep disorder narcolepsy. Genome Res 8:427–434

Lammers GJ, Middelkoop HA, Smilde-Van Den Doel DA, Mourtazaev M, Van Dijk JG (1997) Sleep scoring at a lower resolution. Sleep 20:641–644

Lin L, Faraco J, Li R, Kadotani H, Rogers W, Lin X, Qiu X, De Jong PJ, Nishino S, Mignot E (1999) The sleep disorder canine narcolepsy is caused by a mutation in the hypocretin (orexin) receptor 2 gene. Cell 98:365–376

Mieda M, Willie JT, Hara J, Sinton CM, Sakurai T, Yanagisawa M (2004) Orexin peptides prevent cataplexy and improve wakefulness in an orexin neuron-ablated model of narcolepsy in mice. Proc Natl Acad Sci USA 101:4649–4654

Mignot E, Wang C, Rattazzi C, Gaiser C, Lovett M, Guilleminault C, Dement WC, Grumet FC (1991) Genetic linkage of autosomal recessive canine narcolepsy with a mu immunoglobulin heavy-chain switch-like segment. Proc Natl Acad Sci USA 88:3475–3478

Nishino S, Ripley B, Overeem S, Lammers GJ, Mignot E (2000) Hypocretin (orexin) deficiency in human narcolepsy. Lancet 355:39–40

Parkes JD, Fenton G, Struthers G, Curzon G, Kantamaneni BD, Buxton BH, Record C (1974) Narcolepsy and cataplexy. Clinical features, treatment and cerebrospinal fluid findings. Q J Med 43:525–536

Peyron C, Faraco J, Rogers W, Ripley B, Overeem S, Charnay Y, Nevsimalova S, Aldrich M, Reynolds D, Albin R, Li R, Hungs M, Pedrazzoli M, Padigaru M, Kucherlapati M, Fan J, Maki R, Lammers GJ, Bouras C, Kucherlapati R, Nishino S, Mignot E (2000) A mutation in a case of early onset narcolepsy and a generalized absence of hypocretin peptides in human narcoleptic brains. Nat Med 9:991–997

Rosin DL, Weston MC, Sevigny CP, Stornetta RL, Guyenet PG (2003) Hypothalamic orexin (hypocretin) neurons express vesicular glutamate transporters VGLUT1 or VGLUT2. J Comp Neurol 465:593–603

Sakurai T (2007) The neural circuit of orexin (hypocretin): maintaining sleep and wakefulness. Nat Rev Neurosci 8:171–181

Singh AK, Mahlios J, Mignot E (2013) Genetic association, seasonal infections and autoimmune basis of narcolepsy. J Autoimmun 43:26–31

Soya S, Shoji H, Hasegawa E, Hondo M, Miyakawa T, Yanagisawa M, Mieda M, Sakurai T (2013) Orexin receptor-1 in the locus coeruleus plays an important role in cue-dependent fear memory consolidation. J Neurosci 33:14549–14557

Tabuchi S, Tsunematsu T, Black SW, Tominaga M, Maruyama M, Takagi K, Minokoshi Y, Sakurai T, Kilduff TS, Yamanaka A (2014) Conditional ablation of orexin/hypocretin neurons: a new mouse model for the study of narcolepsy and orexin system function. J Neurosci 34:6495–6509

Thannickal TC, Moore RY, Nienhuis R, Ramanathan L, Gulyani S, Aldrich M, Cornford M, Siegel JM (2000) Reduced number of hypocretin neurons in human narcolepsy. Neuron 27:469–474

Willie JT, Chemelli RM, Sinton CM, Tokita S, Williams SC, Kisanuki YY, Marcus JN, Lee C, Elmquist JK, Kohlmeier KA, Leonard CS, Richardson JA, Hammer RE, Yanagisawa M (2003) Distinct narcolepsy syndromes in Orexin receptor-2 and Orexin null mice: molecular genetic dissection of Non-REM and REM sleep regulatory processes. Neuron 38:715–730

Willie JT, Chemelli RM, Sinton CM, Yanagisawa M (2001) To eat or to sleep? Orexin in the regulation of feeding and wakefulness. Annu Rev Neurosci 24:429–458

Symptomatic Narcolepsy or Hypersomnia, with and Without Orexin (Hypocretin) Deficiency

T. Kanbayashi, A. Imanishi, Y. Ohmori, Y. Sagawa, Y. Takahashi, M. Omokawa, M. Sato, Y. Hishikawa, T. Shimizu and S. Nishino

Abstract The symptoms of narcolepsy can occur during the course of other neurological conditions (i.e. symptomatic narcolepsy). We define symptomatic narcolepsy as cases that meet the International Sleep Disorders Narcolepsy Criteria, which are also associated with a significant underlying neurological disorder that accounts for excessive daytime sleepiness (EDS) and temporal associations. By 2005, we have counted 116 symptomatic cases of narcolepsy reported in the literature. As several authors previously reported, inherited disorders (n = 38), tumors (n = 33), and head trauma (n = 19) are the three most frequent causes for symptomatic narcolepsy. A review of these cases (especially those with brain tumors), illustrates a clear picture that the hypothalamus is most often involved. Reduced CSF orexin-A levels were seen in most symptomatic narcolepsy cases of EDS, with various etiologies including brain tumors, head trauma, and immune-mediated neurological conditions, such as neuromyelitis optica (NMO), and EDS in these cases is sometimes reversible with an improvement of the causative neurological disorder and an improvement of orexin status. It is also noted that some symptomatic EDS cases (with Parkinson diseases and the thalamic infarction) were observed,, but were not linked with orexin ligand deficiency. In this chapter, we first provide an overview of cases of symptomatic narcolepsy and EDS and then extend our discussions to the roles of the orexin system in EDS disorders associated with various neurological conditions.

T. Kanbayashi (✉) · A. Imanishi · Y. Ohmori · Y. Sagawa · Y. Takahashi · M. Omokawa · M. Sato · Y. Hishikawa · T. Shimizu
Department of Neuropsychiatry, Akita University School of Medicine, Akita, Japan
e-mail: yuya@med.akita-u.ac.jp

T. Kanbayashi · T. Shimizu
International Institute for Integrative Sleep Medicine (WPI-IIIS), University of Tsukuba, Tsukuba, Japan

S. Nishino
Sleep and Circadian Neurobiology Laboratory, Center for Narcolepsy, Stanford University School of Medicine, Palo, Alto CA 94304, USA

© Springer International Publishing Switzerland 2015

T. Sakurai et al. (eds.), *Orexin and Sleep*,
DOI 10.1007/978-3-319-23078-8_13

Keywords Narcolepsy · Cataplexy · Sleepiness · Orexin · Hypocretin · Cerebrospinal fluid · Tumor · Multiple sclerosis · Neuromyelitis optic · Aquaporin-4

1 Introduction

Narcolepsy is a chronic sleep disorder characterized by excessive daytime sleepiness (EDS), cataplexy, hypnagogic hallucinations (HH) and sleep paralysis (SP) (i.e. narcolepsy tetrad) (Nishino et al. 2000; Mignot et al. 2002). A major breakthrough in narcolepsy research was recently made through the identification of orexin deficiency in narcolepsy-cataplexy (Mignot et al. 2002; Nishino et al. 2000, 2001; Dalal et al. 2001; Kanbayashi et al. 2002; Krahn et al. 2002; Bassetti et al. 2003; Ebrahim et al. 2003). Orexins are hypothalamic neuropeptides involved in various fundamental hypothalamic functions including sleep-wake control, energy homeostasis, autonomic and neuroendocrine functions (de Lecea et al. 1998; Sakurai et al. 1998; Willie et al. 2001). Orexin containing neurons are located exclusively in the lateral hypothalamic area (LHA). Since orexin deficiency in narcolepsy is also tightly associated with human leukocyte antigen (HLA) DR2/DQ6 (DQB1*0602) positivity, an acquired cell loss of orexin containing neurons with autoimmune process are suggested in "idiopathic" cases of narcolepsy (Mignot et al. 2002; Kanbayashi et al. 2002). "Idiopathic narcolepsy" is defined as narcolepsy cases unassociated with apparent radiographical or clinical evidence of brain pathology apart from sleep-related abnormalities. Orexin deficiency in the brain can be determined clinically via cerebrospinal fluid (CSF) orexin-A measures with CSF orexin-A levels in healthy subjects above 200 pg/ml regardless of gender, age (from neonatal to 70s), and time of the CSF collections (Nishino et al. 2000, 2001; Kanbayashi et al. 2002). Due to the specificity and sensitivity of low CSF orexin-A levels (less than 110 pg/ml or 30 % of the mean normal levels) narcolepsy-cataplexy is high among various sleep disorders (Mignot et al. 2002; Ripley et al. 2001; Kanbayashi et al. 2003), CSF orexin measures was a diagnostic criteria for narcolepsy-cataplexy in the 2nd edition of international classification of sleep disorders (ICSD-2) (Medicine AAoS 2005). In the 3rd edition of ISCD (ICSD III), narcolepsy was reclassified depending on orexin deficiency status (i.e., Type I and Type II narcolepsy) (Medicine AAoS 2014).

Impaired orexin systems may also be observed in some neurological disorders affecting the LHA (where orexin cell bodies locate) and/or orexin projection pathways. Indeed, Ripley et al. (2001) had measured CSF orexin levels in 235 neurological patients and shown that a subset of subjects with acute or sub-acute neurological disorders (i.e. intracranial tumors, cerebrovascular events, craniocerebral trauma, central nervous system (CNS) infections and Guillain-Barré Syndrome [GBS]) had decreased CSF orexin-A levels, although CSF orexin-A levels in the majority of patients with chronic neurological conditions, such as Alzheimer's disease and Parkinson disease, were not significantly reduced. Arii et al. (2004) also studied CSF orexin-A levels in 132 pediatric neurological conditions. The results were consistent with Ripley's study (Ripley et al. 2001), and only a limited number of neurological conditions besides narcolepsy showed reduced CSF orexin-A levels. These included intracranial tumors, craniocerebral trauma, autoimmune and post-infectious diseases [GBS and acute disseminated encephalomyelitis (ADEM)], and some inherited disorders, such as Niemann-Pick disease, type C (NPC) and Prader-Willi syndrome (PWS) (Arii et al. 2004).

These findings are particularly interesting since these neurological conditions are often associated with acutely disturbed consciousness, lethargy, sleepiness, and/or residual sleep disturbances.

In rare cases, symptoms of narcolepsy can be seen during the course of a neurological disease process (i.e. symptomatic narcolepsy). By 2005, we have counted 116 symptomatic cases of narcolepsy reported in the literature and inherited disorders (n = 38), tumors (n = 33), and head trauma (n = 19) are the three most frequent causes for symptomatic narcolepsy (Nishino and Kanbayashi 2005). Involvements of the hypothalamic structures in symptomatic narcoleptic cases have been emphasized repeatedly for many decades (von Economo 1930; Adie 1926), and an impaired orexin system may also be involved in some symptomatic narcolepsy cases. Association with EDS/cataplexy in some inherited neurological diseases (such as NPC, PWS, or myotonic dystrophy) is also known (Kandt et al. 1982; Parkes 1999; Martinez-Rodriguez et al. 2003). An impaired orexin system may thus also be involved in these sleep-related symptoms of these neurological conditions.

In this chapter, we first overview cases of symptomatic narcolepsy reported in literature. Since EDS without other narcolepsy symptoms can also occur with a variety of neurological disorders and are not usually an indication of narcolepsy, we will also extend our discussion on the roles of orexin system in EDS associated with various neurological conditions.

Since data of CSF hypoocretin-1 measures are available for some recent symptomatic narcolepsy and/or EDS cases, we will focus on these cases and discuss the roles of orexin status in these disorders. For this purpose, we categorized the cases as follows: (I) symptomatic narcolepsy-cataplexy associated with focal/generalized CNS invasion, such as cerebral tumors, vascular diseases (Sect. 3.1), and neurodegenerative disorders (Sect. 3.2), (II) hypersomnia associated with (IIa) focal/generalized CNS invasion, such as cerebral tumors, brain infections, vascular diseases, neurodegenerative disorders (AD and PD) and head trauma (Sect. 4.1), and (IIb) with CNS diseases mediated with neuroimmune mechanisms, such as inflammatory and demyelinating diseases (Sect. 4.2). Non-narcoleptic hypersomnia categories include less defined EDS cases, and likely consists of heterogeneous conditions. This is partially due to the fact that applying standardized polygraphic assessments [all night polygraphic recordings followed by multiple sleep latency test (MSLT)] was often difficult in these neurological conditions. However, since prevalence of these hypersomnia cases appeared to be much higher than that of symptomatic narcolepsy, we believe that the discussion on the roles of the orexin system in less well-defined EDS cases also have valuable clinical implications.

2 Definition of Symptomatic Narcolepsy and Its Overview

Symptoms of narcolepsy can sometimes be seen during the course of a neurological disease process. In such instances, the term "symptomatic narcolepsy" is used, implying that the narcolepsy is a symptom of the underlying process rather than being idiopathic. For these cases, the signs and symptoms of narcolepsy must be temporally associated with the underlying neurological process. "Symptomatic narcolepsy" and "secondary narcolepsy" are used more or less indiscriminately, even though they have different meanings. We recommend the use of symptomatic narcolepsy/EDS, since "secondary EDS" has also been used to describe EDS associated with sleep apnea and restless leg syndrome.

In the ICSD-3 (Medicine AAoS 2014), "Narcolepsy due to Medical Condition" is reclassified under "Narcolepsy Type 1 or Type 2 Due to a Medical Condition" depending on the hypocretin deficiency status, and the criteria for "Narcolepsy Type 1 Due to a Medical Condition" is "The condition must fulfill criteria for narcolepsy type 1 (i.e., hypocretin deficient narcolepsy_and be attributable to another medical disorder)." The criteria for "Hypersomnia Due to Medical Condition" has changed to "Hypersomnia Due to a Medical Disorder". The following are the criteria:

Narcolepsy type 1 (Criteria A and B must be met)

A. The patient has daily periods of irrepressible need to sleep or daytime lapses into sleep occurring for at least three months.
B. The presence of one or both of the following:

    1. Cataplexy (as defined under Essential Features) *and* a mean sleep latency of ≤8 min and two or more sleep onset REM periods (SOREMPs) on an MSLT performed according to standard techniques. A SOREMP (within 15 min of sleep onset) on the preceding nocturnal polysomnogram may replace one of the SOREMPs on the MSLT.
    2. CSF hypocretin-1 concentration, measured by immunoreactivity, is either ≤110 pg/mL or <1/3 of mean values obtained in normal subjects with the same standardized assay.

"Narcolepsy Type 2 Due to a Medical Condition" is "[a] condition [that] fulfills criteria for narcolepsy type 2 and is attributable to another medical disorder." The following are the criteria:

Narcolepsy type 2 (Criteria A–E must be met)

A. The patient has daily periods of irrepressible need to sleep or daytime lapses into sleep occurring for at least three months.
B. A mean sleep latency of ≤8 min and two or more sleep onset REM periods (SOREMPs) are found on a MSLT performed according to standard techniques. A SOREMP (within 15 min of sleep onset) on the preceding nocturnal polysomnogram may replace one of the SOREMPs on the MSLT.
C. Cataplexy is absent.
D. *Either* CSF hypocretin-1 concentration has not been measured *or* CSF hypocretin-1 concentration measured by immunoreactivity is either >110 pg/mL *or* >1/3 of mean values obtained in normal subjects with the same standardized assay.
E. The hypersomnolence and/or MSLT findings are not better explained by other causes such as insufficient sleep, obstructive sleep apnea, delayed sleep phase disorder, or the effect of medication or substances or their withdrawal.

"Hypersomnia Due to a Medical Disorder" fulfills the following criteria.
Hypersomnia Due to a Medical Disorder (Criteria A–D must be met)

A. The patient has daily periods of irrepressible need to sleep or daytime lapses into sleep occurring for at least three months.
B. The daytime sleepiness occurs as a consequence of a significant underlying medical or neurological condition.
C. If an MSLT is performed, the mean sleep latency is ≤8 min, and fewer than two sleep onset REM periods (SOREMPs) are observed.
D. The symptoms are not better explained by another untreated sleep disorder, a mental disorder, or the effects of medications or drugs.

2.1 *Anatomical Substrate for the Symptoms of Narcolepsy*

It is important to understand what mechanisms and which brain sites are involved in the occurrence of symptomatic narcolepsy, especially in relation to the orexin system. Although it is not simple to discuss mechanisms uniformly for symptomatic narcolepsy associated with various genetic disorders, analysis of symptomatic narcolepsy with tumor cases showed clearly that the lesions were most often (about 70 % of cases) involved in the hypothalamus and adjacent structures (the pituitary, supraseller or optic chiasm). Impairments in the hypothalamus are noted in most symptomatic cases of narcolepsy which also suggests a possible involvement of impaired orexin neurotransmission.

Lumber CSF orexin-A measurements were carried out in neurological condtions possibly associated with symptomatic cases of narcolepsy/EDS using lumber CSF. The ventricular CSF data of PD in Drouot et al. (2003) were not included. The

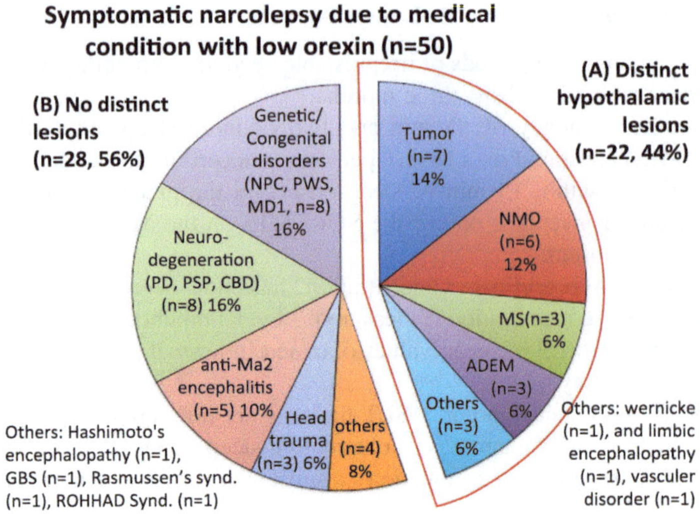

Fig. 1 Category of medical conditions associated with low orexin symptomatic narcolepsy. 50 cases of narcolepsy due to medical conditions with low orexin are included. The percentage of each medical condition was displayed. Tumors (n = 7, 14 %), demyelinating disorders (NMO (n = 6, 12 %), MS (n = 3, 6 %), ADEM, (n = 3, 6 %), genetic/congenital disorders (n = 8, 16 %) and neuro-degeneration (n = 8, 16 %) are the four most frequent causes. Several categories showed distinct hypothalamic lesions (*A*, n = 22, 44 %), including tumors and demyelinating disorders, while genetic/congenital disorders, neuro-degeneration, paraneoplastic autoimmune syndromes (anti-Ma associated encephalitis) and head trauma did not show distinct lesions (*B*, n = 28, 56 %)

following is a breakdown of the 346 cases measured by 2014, with incidence of low orexin-A level out of total cases per neurological condition: narcolepsy/EDS associated with tumors (9 out of 14 had low levels of orexin), head trauma (3 out of 7), vascular disorders (1 out of 8), encephalopathies (4 out of 6), neuro degeneration (7 out of 209), immune-mediated demyelinating disorder (11 out of 19), immune-mediated polyneuropathy (1 out of 23), paraneoplastic autoimmune syndrome (2 out of 8), genetic/congenital disorders (7 out of 66) and others (2 out of 4) (Appendix Table). Among the 50 low orexin-A cases, 22 (44 %) cases had hypothalamic lesions and 28 (56 %) cases had no distinct lesions (Fig. 1).

Recently, we have reported a new possible pathophysiology of symptomatic narcolepsy/EDS in patients with MS and its related disorders (Kanbayashi et al. 2009). These cases often show unique bilateral symmetric hypothalamic lesions associated with significant orexin ligand deficiency. Interestingly, these patients often share clinical characteristics with neuromyelitis optica (NMO) patients, including optic neuritis or spinal cord lesions and the presence of NMO-IgG (anti aquaporin-4 (AQP4) antibodies) (Fig. 2) (Kanbayashi et al. 2009). AQP4 is highly expressed in the hypothalamic periventricular regions (Amiry-Moghaddam et al. 2003; Pittock et al. 2006), thus an immune attack to AQP4 may possibly be responsible for the bilateral and hypothalamic lesions and orexin deficiency in

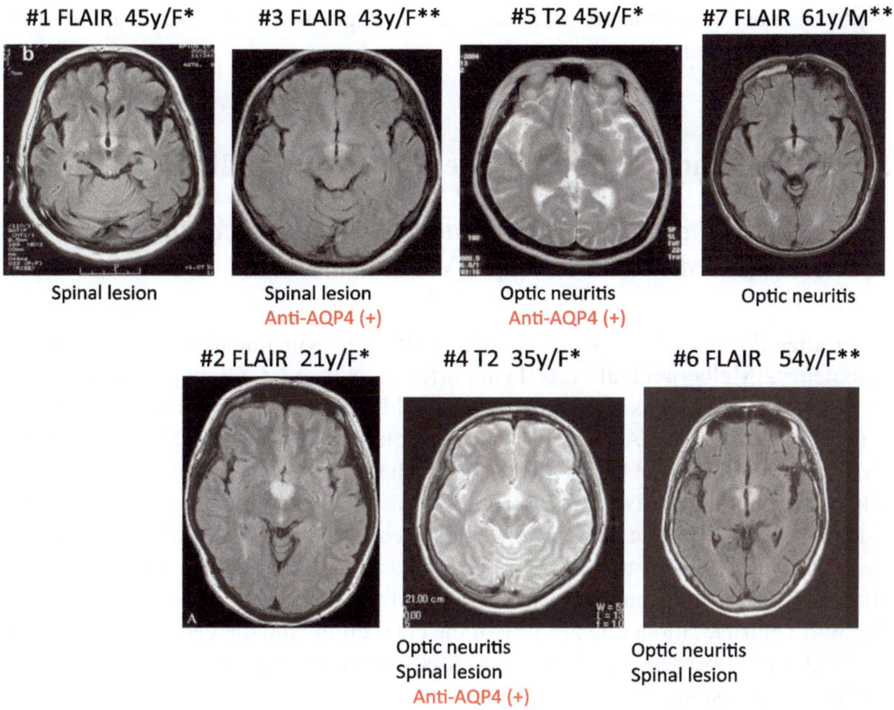

Fig. 2 MRI findings (FLAIR or T2) in multiple sclerosis (*MS*)/neuromyelitis optica (*NMO*) patients with orexin deficiency and excessive daytime sleepiness. A typical horizontal slice including the hypothalamic, periventricular area from each case is presented. All cases were female. *, met with ICSD-2 criteria for narcolepsy caused by a medical condition; **, met with ICSD-2 criteria for hypersomnia caused by a medical, condition. All cases were initially diagnosed as MS. Cases 3–7 had optic neuritis and/or spinal cord lesions, and cases 4, 5, and 7 are seropositive for anti-AQP4 antibody and thus were diagnosed as NMO. (Data from Kanbayashi et al. (Kanbayashi et al. 2009)

narcolepsy/EDS associated with these diseases. As AQP4 is found in nonneuronal structures such as astrocytes and ependymocytes, impairments of the orexin neurons are likely to be secondary to changes in their surrounding regions (Kanbayashi et al. 2009). None of these cases exhibited cataplexy, but some exhibited REM sleep abnormalities, and thus some of these cases meet the ICSD narcolepsy criteria (Fig. 2). It should also be noted that many earlier narcolepsy-cataplexy cases were associated with MS (5 out of 6 cases reported before 1970). Considering most recent cases were treated with steroids (or other immunosuppressants) at the early stage of the disease and EDS and orexin deficiency was often recovered, chronic impairments of the orexin system may be required for the occurrences of cataplexy (see Nishino and Kanbayashi 2005).

Although detailed mechanisms of orexin impairments in these NMO subjects need to be further explored, these new finding also confirm the importance of the hypothalamus, where the orexin neurons are located, as the brain structure involved in symptomatic narcolepsy.

3 Orexin Status in Various Neurological Conditions

3.1 Orexin Status in Symptomatic Narcolepsy-Cataplexy Associated with Distinct CNS Lesions

Soon after the discovery of the involvement of orexin impairments in idiopathic narcolepsy, Melberg et al. (2001) reported a reduced CSF hypoocretin-1 level (96 pg/ml) in a previously reported 51-year-old male case with autosomal dominant cerebellar ataxia (ADCA), deafness and narcolepsy (DN). In this Swedish pedigree (ADCA-DN; OMIM, Online Mendelian Inheritance in Man, accession number 604121), four out of five ADCA subjects are affected with narcolepsy-cataplexy (Melberg et al. 1999), and CSF previously collected from one of theses subjects (patients III-2) was available for orexin measures. The patient was negative for HLA-DR2. Since this case is a heredodegenerative disease with an enlargement of the third ventricle, moderate atrophy of the cerebellum and the cerebral hemispheres by MRI were observed, and we listed this case under narcolepsy associated with distinct CNS lesions.

Scammell et al. (2001) subsequently reported a 23-year-old male who developed narcolepsy-cataplexy due to a large hypothalamic stroke after a craniopharyngioma resection. This lesion included 2/3 of the caudal hypothalamus except for the most lateral component on the right and extended into the mediodorsal thalamus bilaterally, the left amygdala, and parts of the basal forebrain and the rostral midbrain. Postoperative course was complicated by panhypopituitarism, staphylococcal meningitis and hydrocephalus. He experienced HH. He became obese with a body mass index (BMI) of 31.7. Sleep latency was 0.5 min according to MSLT, and REM latency was 3.5 min. An overnight polysomnography (PSG) showed 1 min and 1.5 min of SL and REM latency respectively, without significant sleep apnea. HLA was negative for DQB1*0602, and CSF orexin level was 167 pg/ml.

Nokura et al. (2004) reported one case with narcolepsy and cataplexy-like phenomena in a 66-year old female with hypersomnia due to a hypothalamic tumor. She showed EDS and cataplexy-like symptoms, such as abrupt falling without loss of consciousness. An MRI revealed lesions with high signal intensities in the hypothalamus, thalamus and midbrain bilaterally. This case was accompanied with mild anterior hypopituiarism and a SOREMP in a daytime polysomnography. Orexin-A level was 61 pg/ml. Symptoms improved with tumor reduction after radiotherapy and the intravenous administrations of nimustine hydrochloride and interferon beta.

The lesions in these cases had different etiologies: degeneration, infarction and tumor. Although the number of cases is still limited, the hypothalamic lesions were noted in all cases. Moderate reduction of CSF orexin levels (low in two cases and intermediate in one) also confirmed the functional impairment of the hypothalamus. A massive impairment of orexin projections and projection sites are likely involved in the mentioned case with hypothalamic stroke after craniopharyngioma resection, implying that a more severe orexin neurotransmission impairment than the intermediate CSF orexin-A level implies, may exist. Although these results are consistent with the hypothesis of hypothalamic orexinergic involvement in symptomatic cases of narcolepsy, it is not certain if all cases with low orexin levels associated with hypothalamic damage develop narcoleptic symptoms.

3.2 *Orexin Status in Symptomatic Narcolepsy-Cataplexy and/or EDS Associated with Inherited Disorders*

There are clusters of cases of genetic or congenital disorders associated with primary central hypersomnolence and/or cataplexy, and CSF orexin-A has also been assessed in several patients with Prader-Willi syndrome (PWS), Niemann-Pick type C disease (NPC) and myotonic dystrophy.

3.2.1 Prader-Willi Syndrome (PWS)

EDS is a common symptom in PWS (Vela-Bueno et al. 1984; Helbing-Zwanenburg et al. 1993; Vgontzas et al. 1996). Sleep disordered breathing (SDB) and narcoleptic traits such as SOREMPs and cataplexy have also been reported in these subjects (Manni et al. 2001; Tobias et al. 2002). If SDB exists, primary hypersomnia should only be diagnosed if excessive daytime sleepiness does not improve after adequate treatment of sleep-disordered breathing. Mignot et al. (2002) reported a 16 year old male with the following: EDS, HLA-DQB1*0602 positive, 109 pg/ml orexin-A, obese (BMI = 48.1), documented 15q11-13 deletion, limited number of sleep disordered breathing events (apnea hypoxia index [AHI] = 5.6), no cataplexy, SL = 3.0 min, no SOREMPs by MSLT., Nevsimalova et al. (2004) also measured CSF orexin-A in another three PWS cases. One subject exhibited EDS (AHI = 3.1., age = 10) had low orexin-A levels (130 pg/ml) and DQB1 *0602, but the other two who did not exhibit EDS had intermediate (191 pg/ml) or normal (226 pg/ml) orexin-A with (AHI = 46.8, age = 26) and (AHI = 0, age = 6), respectively. All three subjects were obese and did not exhibit cataplexy. Interestingly, AHI in these PWS subjects were correlated with age and BMI, but not with CSF orexin-A levels and EDS.

Additional reports suggested the possibility that EDS in PWS may also be attributed to the orexin system, not necessarily to sleep disordered breathing caused by obesity. First, Arii et al. (2004) reported a two week-old PWS male with severe hypotonia, poor feeding, documented 15q11-12 deletion and intermediate orexin level (192 pg/ml). Then, Terashima et al. (2012) reported a 11 year-old PWS female with the following: EDS, mildly obesity (BMI = 19.9, %BMI = 108), no sleep disordered breathing events, no cataplexy; SL = 3.0 min, SOREMPs confirmed by MSLT, orexin 60 pg/ml. Dr. Nevsimalova also proposed that PWS cases may be a model for congenital dysfunction/developmental failure of the orexin system (Nevsimalova et al. 2004).

However, no decrease in the number of orexin-containing neurons was observed in post-mortem human adult and infant brains (Fronczek et al. 2005), which suggests a lack of involvement of orexin in the pathogenesis of the disorder. In a larger context, this result underlies the need for larger studies to determine whether decreased CSF orexin-A remains anecdotal in inherited neurological conditions.

3.2.2 Niemann-Pick Type C Disease (NPC)

NPC is an autosomal recessive and congenital neurological disorder characterized by the accumulation of cholesterol and glycosphingolipids in the peripheral tissues and of the glycosphingolipids in the brain. Classic NPC symptoms include hepatosplenomegaly, vertical supranuclear gaze palsy, ataxia, dystonia, and dementia. Subjects with NPC have been reported to frequently display narcolepsy-like symptoms, including cataplexy (Kandt et al. 1982; Autret et al. 1994; Vankova et al. 2003; Vanier 1983; Kanbayashi et al. 2003). This condition is remarkable as cataplexy is often triggered by typical emotions (laughing) and responsive to anticataplectic treatments.

Kanbayashi et al. (2003) measured CSF orexin levels in two NPC cases with and without cataplexy. In the first case (male, age 5), cataplexy and an intermediate orexin level (142 pg/ml) was detected. Cataplexy was triggered by laughter since the age 2. EDS was not claimed by the patient, and SL (16.5 min) was normal without SOREMPs (Philip et al. 1997). No abnormalities in the hypothalamus were detected by MRI scans. He was negative for HLA DR2. In the second case (female, age 3), a normal orexin level (299 pg/ml) was detected. Neurological symptoms such as tremor, ataxia and akathisia were present, cataplexy nor EDS was present.

Vankova et al. (2003) reported five patients with juvenile NPC. Deterioration of intellectual function, the presence of pyramidal, dystonic and cerebellar signs, and splenomegaly were observed in all cases as well as disrupted sleep in nocturnal polysomnography. Total sleep time, sleep efficiency, REM sleep, and delta sleep amounts were decreased when compared to age-matched controls. Cataplexy was reported in one patient. Shortened mean sleep latencies were observed in three patients during the MSLT, but SOREMPs were observed only in the case with cataplexy, and this case met with the criteria of symptomatic cases of narcolepsy. This patient was HLA DQB1*0602 positive, while the other subjects were HLA DQB1*0602 negative. CSF orexin-A levels were reduced in patients (190 pg/ml and 157 pg/ml in the subject with cataplexy) while in the two other patients, the CSF orexin-A were at the lower end of normal (226 pg/ml, 245 pg/ml). The authors speculated that lysozomal storage abnormalities in NPC patients may also have the impact on the hypothalamus including, area orexin-containing cells are located.

Oyama et al. (2006) reported a Japanese patient with NPC caused by a homozygous c.2974 G > T mutation of the NPC1 gene, a well-known NPC1 gene mutation that causes a unique phenotype of NPC, which has been limited to a single Acadian ancestor in Nova Scotia, Canada. The patient characteristically started presenting with cataplexy at the age of 9 years, and the level of orexin-A was moderately low, 174 pg/ml.

Eto et al. (2015) reported NPC case complicated by cataplexy (age 4, male) with orexin level was 106 pg/ml. He had compound heterozygous *NPC1* mutations: a novel missense mutation (G9D) in exon 1 and a known missense mutation (R1186H) in exon 23.

Soda et al. (2011) reported a NPC and narcolepsy-cataplexy case (age, 24) with orexin level of 88 pg/ml. Short sleep latency (1 min) and 3 SOREMPs in 4 naps were observed by MSLT. HLA typing was not typical for narcolepsy.

In these five reports, all of the NPC patients with cataplexy have an association with reduced orexin-A levels, while CSF orexin-A levels in the NPC cases without cataplexy are in the lower limit of normal, suggesting a degree of impairments of the orexin system may contribute the occurrence of cataplexy in this inherited diffuse CNS impairment condition.

3.2.3 Myotonic Dystrophy (MYD)

Myotonic dystrophy type 1 (MD1) is a multisystem disorder with myotonia, muscle weakness, cataracts, endocrine dysfunction, and intellectual impairment (Coccagna et al. 1975) (Park and Radtke 1995) (Gibbs et al. 2002). This disorder is caused by a CTG triplet expansion in the 3′ untranslated region of the DMPK gene on 19q13. The expansion resides within ubiquitously expressed genes and when transcribed, accumulates in the nuclei as RNA expansions. This induces the sequestration of muscleblind proteins (Mbnl 1,2,3—RNA binding proteins selective for UG-rich domains) and upregulation of CUG-binding protein/Elav-like family (ex. CELF) resulting in altered splicing of Mbnl-regulated transcripts and causing major aspects of DM (Charizanis et al. 2012; Kanadia et al. 2003, 2006; Hao et al. 2008; Wang et al. 2012; Charizanis et al. 2012). MD1 is frequently associated with EDS and the presence of SOREMPs, which are sleep abnormalities similar to narcolepsy, during the MSLT (Cirignotta et al. 1987; Finnimore et al. 1994; Begin et al. 1997; van der Meche et al. 1994; Guilleminault et al. 1998; Dauvilliers and Laberge 2012) (van Hilten et al. 1993; Gibbs et al. 2002; Hansotia and Frens 1981; Park and Radtke 1995; Yu et al. 2011). The disease is also often associated with SDB, and thus this may also account for appearances of SOREMPs. However, adequate treatment of sleep-disordered breathing does not always eliminate the EDS (van der Meche et al. 1994) (Guilleminault et al. 1998). As many DM1 patients with no sign of sleep apnea or chronic alveolar hypoventilation also exhibit EDS, some authors believe that a central dysfunction is primarily involved in the EDS in DM1 (Dauvilliers and Laberge 2012; van Hilten et al. 1993; Gibbs et al. 2002; Hansotia and Frens 1981).

Martinez-Rodriguez et al. (2003) reported six patients with MYD1 complaining of EDS. The mean sleep latency on MSLTs was abnormal in all patients (<5 min in two, <8 min in four) and two SOREMPs were observed in two subjects, meeting the criteria for symptomatic narcolepsy. It should be noted that these two cases also had SDB. All patients were HLA-DQB1*0602 negative. Orexin-A levels (181 pg/ml) were significantly lower in patients versus controls (340 pg/ml); the one case with two SOREMPs had orexin-A levels in the low range (<110 pg/ml) generally observed in narcolepsy. Three cases had intermediate levels (110–200 pg/ml). The authors suggested that a dysfunction of the hypothalamic orexin system may mediate sleepiness and abnormal MSLT results in patients with MD1.

In one case of late-onset congenital hypoventilation syndrome, a disorder with reported hypothalamic abnormalities (Katz et al. 2000), Martinez-Rodriguez found very low CSF orexin-A levels in an individual with otherwise unexplained sleepiness and cataplexy-like episodes (Martinez-Rodriguez et al. 2003). Excellent response to anti-cataplectic medication was observed in this case.

Iwata et al. (2009) and Yasui et al. (2010) each reported a case of MYD1 with narcolepsy due to medical condition. Both patient had no cataplexy and were HLA-DQB1*0602 negative. In the former case, orexin was markedly decreased to <40 pg/ml. The size of the CTG repeat was markedly increased in the 3′ untranslated region of the DMPK gene at 1800–2400 repeats and PSG revealed severe sleep apnea (AHI = 59/h, BMI = 27.7) and chronic alveolar hypoventilation indicating severe disesase. Since her nocturnal sleeping time was extended to 18 h per day, MSLT revealed normal sleep latencies and no SOREMPs. In the latter case, patient was using Bipap due to nocturnal hypoxia (BMI = 27.3).

However, a larger study failed to confirm these results (Ciafaloni et al. 2008). Orexin-A concentrations did not correlate clinically with disease severity or duration, nor with subjective or objective reports of sleepiness. Because CSF orexin concentrations are often only slightly decreased in some patients, a functional abnormality that causes sleepiness and SOREMPs in myotonic dystrophy type 1 is unlikely to be a common occurrence.

It should also be pointed out that EDS in DM1 is distinctive (from such as that of narcolepsy), and a recent comprehensive sleep evaluations in forty DM1 patients (Yu et al. 2011) demonstrated that unlike in narcolepsy, most patients did not show shortened sleep latency in MSLT (DM1: 14. 2 min (2.8–20 min) versus Control: 14.2 min (8.2–20 min)), although most of them claimed moderate to severe subjective daytime sleepiness (79.5 % vs. 17.1 %, $p < 0.002$) or fatigue (62.2 % vs. 17.1 %, $p < 0.002$). The current international criteria for sleep disorders sets the cut off for the MSLT mean sleep latency as less than 8 min, and thus most of these sleepy DM1 patients do not even fit in the diagnostics category of hypersomnias (Medicine AAoS 2005). Occurrence of cataplexy was also never reported in DM1 (Cirignotta et al. 1987; Finnimore et al. 1994; Begin et al. 1997; van der Meche et al. 1994; Guilleminault et al. 1998; Dauvilliers and Laberge 2012; van Hilten et al. 1993; Gibbs et al. 2002; Hansotia and Frens 1981; Park and Radtke 1995; Yu et al. 2011).

Thus, the pathophysiology of EDS in DM1 is truly mysterious.

Recent animal studies using the mouse model of DM demonstrated a selective and robust increase in REM sleep propensity (Charizanis et al. 2012). Mbnl1 KO and Mbnl2 KO mice were recently generated and shown to develop muscle and other DM symptoms, and thus these KO mice are informative animal models of DM (Kanadia et al. 2003; Hao et al. 2008). As Mbnl2 plays a more important role as a splicing regulator during brain development compared to Mbnl1 (Charizanis et al. 2012; Suenaga et al. 2012), the sleep phenotype of Mbnl2 KO mice have been evaluated (Charizanis et al. 2012); Mbnl2 KO mice showed an increase of REM sleep amounts associated with increased EEG theta power. This change was most notable during the dark period when mice are normally awake. Interestingly, a larger portion of these dark period REM sleep episodes in Mbnl2 KOs exhibited a short latency from the proceeding wake episodes, but they did not exhibit cataplexy. A more profound REM sleep rebound after 6 h sleep deprivation was also observed in KOs, compared to wild-type (WT) mice. These sleep changes were REM sleep specific, as no changes in wake and non-REM sleep was seen in these KO mice at the baseline and during sleep rebound, suggesting that Mbnl2 KO mice exhibit selective increases in REM sleep propensity. Based on these results and the fact that selective REM sleep deprivation in human induces a significant increase in REM sleep propensity and sleepiness during daytime (Endo et al. 1998), we hypothesize that abnormal REM sleep propensity may primarily cause EDS in DM1, and altered splicing of Mbnl-regulated transcripts can induce REM sleep abnormalities in DM1.

4 Orexin Status in Hypersomnia in Various Neurological Conditions

4.1 Focal/Generalized CNS Invasion

Symptomatic narcolepsy is relatively rare, but sleepiness without other narcoleptic symptoms can often occur with a variety of neurological disorders; they are more likely to be due to multifocal or global disturbances of the brainstem, diencephalon and cerebral cortex. Recently, several clinical studies also suggested that the disruption of the hypothalamic orexin system in EDS associated with various neurological conditions.

4.1.1 Cerebral Tumors

Cases with EDS seen along with various cerebral tumors have been reported. Six of these cases we reviewed presented low orexin-A levels (Arii et al. 2001; Marcus et al. 2002; Marcus and Mignot 2003; Tachibana et al. 2005; Dauvilliers et al. 2007; Sakuta et al. 2012; Uchida et al. 2014).

(Case 1) Hypersomnia seen after removal of a hypothalamic suprasellar Grade II pilocystic astrocytoma: MRI showed the bilateral, medial and lateral hypothalamic areas and right posterior hypothalamus were damaged. Orexin-A levels was 104 pg/ml and HLA-DR2 negative. Symptoms included diabetes insipidus (DI), hypothyroidism, weight gain, no cataplexy. MSLT: sleep latency: 1.7 min, no SOREMPs (Arii et al. 2001).

(Case 2) EDS in a patient in a vegetative state following astrocytoma resection and CNS hemorrhage: MRI revealed a large suprasellar mass that extended into the sella inferiorly and was displaced posteriorly. Orexin-A was undetectably low and HLA-DR2 and DQB1*0602 was negative. Nocturnal EEG study showed fragmented sleep with 16 short REM cycles. The daytime EEG showed frequent REM periods. EDS improved with 200 mg modafinil and 5 mg methylphenidate (Marcus et al. 2002; Marcus and Mignot 2003).

(Case 3) Hypersomnolence in patient with extensive hypothalamic damage after removal of a craniopharyngioma: CSF orexin-A level (93 pg/ml) was low with negative HLA DQB1*0602 typing.

Short sleep latency and SOREMPs during a MSLT suggested a diagnosis of symptomatic narcolepsy, indicated destruction of orexin-producing neurons in the hypothalamus (Tachibana et al. 2005).

(Case 4) Severe symptomatic narcolepsy in a patient with primary CNS B-cell lymphoma whose symptoms reversed after chemotherapy: MRI revealed an infiltrative hyperintensity in the left basal ganglia, thalamus, cerebral pedunculus, splenium of the corpus callosum, and the right internal temporal lobe. CSF orexin-A level was undetectable and DQB1*0602 typing was negative. Patient had entered a permanent hypersomnia status, related to a coma-like state. After IV and intrathecal chemotherapy, hypersomnia resolved completely and without any abnormal REM sleep manifestation. Eight months later, a 24-h polysomnography was normal without daytime sleep episodes. Brain MRI and FDG PET scans as well as orexin-A level (244 pg/mL), were normal after treatment (Dauvilliers et al. 2007).

(Case 5) A 19-year-old woman suffered from severe EDS accompanied with long sleep episodes both in the daytime and nighttime and frequent episodes of cataplexy shortly after the removal of craniopharyngioma in the intrasellar space (Fig. 3). MSLT showed a typical finding of narcolepsy, and CSF orexin concentration (71 pg/ml) was below the narcolepsy cut-off value. MRI-tractography showed a clear lack of neuronal fiber connections from the hypothalamus to the frontal lobe (Sakuta et al. 2012).

(Case 6) A 13-year-old girl suffered from severe hypersomonolence in the daytime and nighttime and several episodes of cataplexy after the removal of craniopharyngioma. She had also DI and hypothalamic-pituitary dysfunction. Her orexin level was <40 pg/ml (Uchida et al. 2014).

We also reviewed 3 case reports in which orexin-A levels were normal to high.

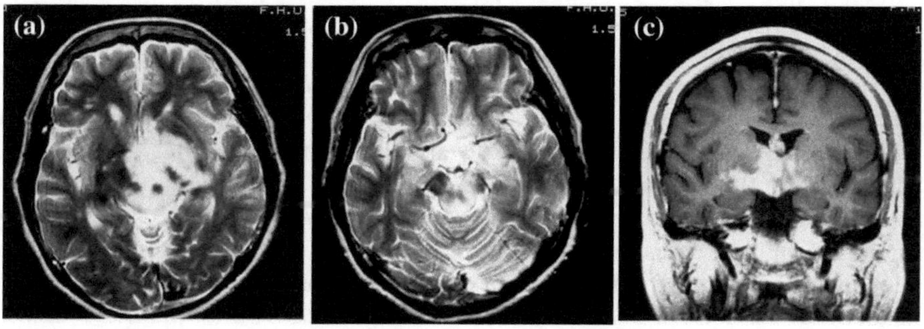

Fig. 3 A narcolepsy-cataplexy case with hypothalamic tumor and low orexin level (61 pg/ml). A 66-year old female case with hypothalamic tumor. **a**, **b** Axial T2-weighted image of MRI at admission exhibits high signal intensities in the midbrain, hypothalamus, and thalamus. **c** Coronal T1-weighted image with gadolinium exhibits enhancement in the same lesion. This case also accompanied with mild anterior hypopituitarism. Her symptoms and MRI findings were improved with reduction of the tumor after 46 Gy radiation and nimustine hydrochloride and interferon beta were administered intravenously (Nokura et al. 2004)

(Case 7–11) EDS in a total of 5 patients who underwent relatively extensive surgeries involving the hypophysis and hypothalamus, for craniopharyngioma (n = 3), germ cell tumor (1), and thalamic arachnoid cyst (n = 1): The craniopharyngiomas and germ cell tumor were located in the hypothalamus-hypophysis region, and the arachnoid cyst was in the thalamic region. All patients received hormone replacement therapies. Mean orexinn-1 (133 pg/ml) was as same as their control range. The mean sleep latency by MSLT in the five patients was 10.3 min. Two patients were morbidly obese and had obstructive sleep apnea, and although treatment with continuous positive airway pressure resulted in complete resolution of their sleep-disordered breathing, daytime somnolence was unchanged (Snow et al. 2002).

(Case 12) A patient who developed a narcoleptic-like sleep disorder immediately following pinealectomy for a choroid plexus carcinoma of the pineal gland: The patient also underwent chemotherapy and radiation treatment. Immediately after surgery, the patient developed EDS that she attributed to severe insomnia and an irregular sleep/wake rhythm. SP and HH were present but not cataplexy. An increased percentage of REM sleep was seen in nocturnal polysomnography, and three out of four SOREMPs were seen during the MSLT. CSF orexin level (518 pg/ml) was normal, and patient was negative for HLA-DQB1*0602. The author proposed that her symptoms be caused by an unknown mechanism unrelated to orexin depletion (Krahn et al. 2002).

(Case 13) Narcolepsy-cataplexy that developed in acromegaly patient, two weeks after completing radiotherapy for a pituitary adenoma: Orexin-A was normal (275 pg/ml) and HLA was not typical for narcolepsy. Both HH and SP were present. Sleep latency by MSLT was 6.4 min and REM latency was 9 min (3 SOREMPs/5 naps). He was obese (BMI: 35) and his AHI was 17/h. The authors have speculated that the radiotherapy of the tumor was associated with damage to a locus rich in orexin receptors (Dempsey et al. 2003).

Overall, we reviewed 7 symptomatic cases with EDS with low orexin-A levels (one case in Sect. 3.1) *and 7 cases in 3 case reports with normal to high orexin-A.* All the cases with low CSF orexin-A levels were either HLA-DR2 or HLA-DR2 and DQB1*0602 negative, thus EDS in these cases are likely secondary to the orexin deficiency caused by the tumors/tumor removal. Other mechanism likely cause EDS in the 7 cases with normal or high orexin-A levels, although impairment in orexin projections, terminals or postsynaptic receptors may also be caused by the tumors.

4.1.2 Infarctions

EDS has been reported in cerebra infraction cases. Bassetti et al. reported two cases with EDS and cerebral infarction. In a thalamic infarction case, mean sleep latency was 9 min and orexin level was 265 pg/ml. In a ponto-medullary infarction case, sleep latency was 1 min and orexin level was 316 pg/ml (Bassetti et al. 2003).

Two hypersomnia cases with bilateral paramedian thalamic infarctions were also independanlty reported (Nokura et al. 2004; Tohyama et al. 2004). The paramedian thalamus believed to play an important role in the regulation of sleep, and disturbances of sleep regulation are known to occur in paramedian thalamic stroke (Guilleminault et al. 1993; Bassetti et al. 1996). The first case suffered from bilateral paramedian thalamic infarctions and had EDS with SOREMPs (two times in four naps). His orexin-A level was 312 pg/ml (Nokura et al. 2004), the symptoms met with the criteria for Narcolepsy Type 2 Due to a Medical Condition. The second case suffered from bilateral paramedian thalamic infarctions and hypersomnia. His orexin level was 274 pg/ml (Tohyama et al. 2004) (Fig. 4). Since the lesions of infarctions did not include the orexin cell bodies, their orexin levels seemed to be normal. However, orexin projection could still be impaired. Guilluminault et al. has pointed out that patients with bilateral paramedian thalamic lesions do not present a typical hypersomnia but a de-arousal or subwakefullness with an inability to develop sleep outside the normal circadian boundary (pseudo-hypersomnia) (Guilleminault et al. 1993). Indeed these patients showed reduced latency to stage 1 during MSLT, but did not develop other normal non REM sleep and REM sleep status during the daytime. It may also be possible that orexin deficiency is not involved in so-called pseudo-hypersomnia associated with bilateral paramedian thalamic lesions, and other pathophysiology needs to be considered for these unique sleep symptoms.

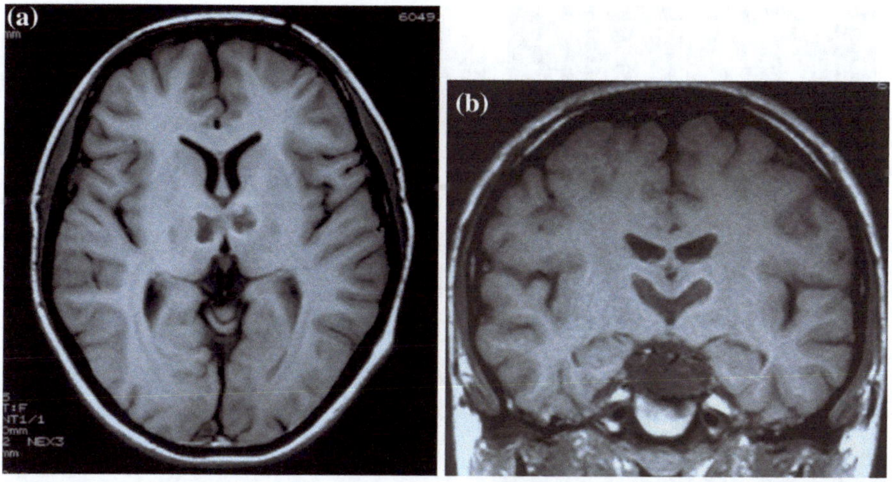

Fig. 4 **a**, **b** A 15 year old case with paramedian thalamic infarctions and normal orexin level (274 pg/ml) (Tohyama et al. 2004). Dauvilliers et al. (2007) reported a case with paramedian thalamic infractions and normal orexin level (274 pg/ml). A 15-year old male with EDS due to bilateral paramedian thalamic infarctions. Patients with bilateral paramedian thalamic lesions are known to often exhibit atypical hypersomnia (i.e. de-arousal or subwakefullness) (Sakuta et al. 2012). The lateral hypothalamus (where orexin cell bodies locate) was not affected, and CSF orexin level was in the normal range. It is not known whether the other orexin systems (projections or receptive sites) are still involved in EDS with paramedian thalamic infractions

4.1.3 Encephalopathies

Wernicke's Encephalopathy

A five-year-old female with Wernicke's Encephalopathy (Fig. 5) was reported to have gradually developed sleepiness and an abnormal sleep/wake schedule (Kashiwagi et al. 2004). She slept 15–20 h per day and fell asleep frequently even while eating. She developed ocular and neurological symptoms (such as involuntary movements, hemiparesis, depression of speech and global confusional state). An MRI revealed lesions in the bilateral hypothalamus in addition to the dorso-medial nucleus of thalamus and mammillary bodies and periaqueductal gray and floor of 4th ventricle. Vitamin B1 levels were low (38.7 ng/ml, normal range: 52–176 ng/ml) and the level of orexin of CSF was decreased (<40 pg/ml). Her sleepiness and MRI findings gradually improved with thiamine therapy. Six months after the onset of sleepiness, both MRI lesion and CSF orexin level (158 pg/ml) recovered to some degree. It is not clear whether Wernicke's encephalopathy affect the orexin system directly or indirectly. It is also not fully studied whether the change in the orexin neurotransmission is solely responsible for the occurrence of the EDS. The dysfunction of orexin neuron due to hypothalamic lesion would be caused by damages for AQP4 water channel (Chan et al. 2004).

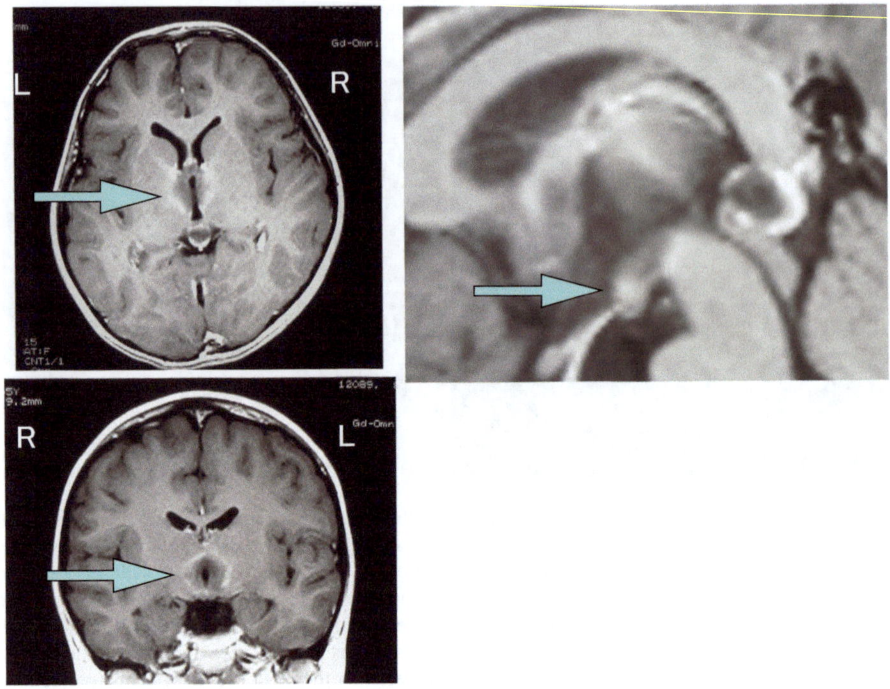

Fig. 5 Hypersomnia due to Wernicke encephalopathy. A 5 year old case with Wernicke's encephalopathy. Her sleep time was 15–20 h per day and she fell asleep frequently even while eating. Gd-enhanced MRI revealed lesions in the bilateral hypothalamus in addition to dorso-medial nucleus of thalamus and mammillary bodies and periaqueductal gray and floor of IVth ventricle. The level of vitamin B1 was low. The level of orexin was decreased (<40 pg/ml). Her sleepiness and MRI findings gradually improved with replacement of vitamin B1 (Kashiwagi et al. 2004)

Limbic Encephalopathy

Chronic progressive hypersomnia was seen in a patient with non-paraneoplastic immune-mediated limbic encephalitis. Orexin-A concentration was low (87 pg/ml). An MRI of the brain showed bilateral signal abnormalities in the medial temporal lobes and the hypothalamus, but systemic examinations for malignant tumors were negative. Acyclovir treatment failed to amend his condition. Subsequent steroid treatment improved his hypersomnia and reduced the extent of abnormal signals on MRI. The CSF orexin concentration increased to 148 pg/ml in 23 days after (Yamato et al. 2004).

Rasmussen's Syndrome

Lagrange et al. reported a case of narcolepsy and Rasmussen's syndrome in a previously healthy 40 year-old man. Severe EDS, cataplexy, HH, and SP developed over the course of a few months. Brain MRI was normal and polysomnography with MSLT confirmed a diagnosis of narcolepsy (SL: 1.6 min, three SOREMPs in four naps). His HLA haplotype is DQB1*0602, and CSF analysis showed no detectable orexin. Approximately 18 months later, he developed complex partial seizures. Further MRI showed a progressively enlarging lesion involving the left frontotemporal and insular areas. Pathology from a partial resection samples was consistent with Rasmussen's syndrome. Evaluation for tumor, infectious, and paraneoplastic etiologies was negative. There was no further progression of the residual lesion on serial MRI (Lagrange et al. 2003).

Although the pathophysiological bases of narcolepsy and Rasmussen's syndrome are unknown, the author speculated the possibility of a common underlying disease processes related to autoimmune mechanism. However, whether or not this case highlighted a temporal relationship between orexin deficiency and the onset of the disease is not known. It nay also be possible for Rasmussen's syndrome to be the comorbidity with idiopathic narcolepsy, since the subject is HLA positive, and late onset cases of idiopathic narcolepsy are also reported.

Brain Stem Encephalitis

Mathis et al. described a case of a previously healthy young man who concurrently developed a narcoleptic syndrome and a fullblown REM sleep related behavior disorder (RBD) after an acute brain stem encephalitis with an isolated inflammatory lesion in the dorsomedial pontine tegmentum. Presenting with hypersomnia, sleep paralysis, hypnagogic hallucinations and SOREMPs, the patient fulfilled the criteria of narcolepsy, although cataplexy was mild and rare. CSF orexin was normal (266 pg/ml) and HLA haplotypes were not typically associated with narcolepsy and RBD (DQB1*0602, DQB1*05) (Mathis et al. 2007).

Hashimoto's Encephalopathies

Castillo et al. (2006) reported a 65 years old male patient with steroid-responsive encephalopathy associated with autoimmune thyroiditis (SREAT) and hypersomnolence and then coma. He had undetectable levels of orexin-A in the CSF during symptomatic period.

4.1.4 Neurodegenerative Disorders

Parkinson's Disease (PD)

Thirty percent of patients with Parkinson disease (PD) have been reported to have EDS. Sleep problems are often related to the disease itself (e.g., difficulties in maintaining sleep because of motor disabilities), but they can also occur secondary to pharmacological treatments, especially with dopamine D2/3 agonists. Ripley et al. initially reported that CSF orexin-A in 7 PD subjects were in the normal range, but sleep abnormalities of these subjects were not assessed (Ripley et al. 2001). In a separate study, CSF orexin levels were also normal in all three PD patients with EDS (Overeem et al. 2002).

However, subsequent studies reported that patients with late-stage PD had low ventricular CSF orexin-A levels (n = 16: < 50-97 pg/ml, N = 3: 138-169 pg/ml) (Drouot et al. 2003). Orexin-A levels decreased with increasing disease severity. The author described that CSF orexin-A levels may reflect the size of the orexin neuron pool, and a decrease in orexin-A level may indicate degeneration of orexin neurons in PD. The sleepiness of the patients was assessed by Epworth sleepiness scale (ESS). The mean ESS of these PD patients (11 ± 1) was significantly higher than that of controls (4 ± 1), but orexin-A level was not correlated with ESS among PD subjects.

Two recent studies reported significant (50 %) orexin cell loss in post-mortem hypothalami of patients with Parkinson's disease, and the presence of Lewy bodies in some orexin-producing cells (Fronczek et al. 2007; Thannickal et al. 2007). In Parkinson's disease, orexin cell loss is 23–62 % and correlates with disease severity (Thannickal et al. 2007), as measured on the Hoehn and Yahr scale (Hoehn and Yahr 1967). However, orexin cell loss was not specific, and nearby neurons containing melanin-concentrating hormone were similarly lost (12–74 %) in proportion to disease severity (Thannickal et al. 2007). Furthermore, CSF orexin-A levels were normal in Parkinson's disease (Ripley et al. 2001; Dauvilliers et al. 2003; Overeem et al. 2002; Yasui et al. 2006; Asai et al. 2009; Compta et al. 2009), even if associated with severe sleepiness (Overeem et al. 2002; Baumann et al. 2005), in most studies, excluding a few (Drouot et al. 2003, 2011; Maeda et al. 2006; Mizuno et al. 2011; Wakai and Kanbayashi 2011;) (Teraoka et al. 2013). Interestingly however, significant reductions in the number of orexin cells in the hypothalamus (Fronczek et al. 2007; Thannickal et al. 2007) and decrease in orexin-A concentrations in the ventricles (Drouot et al. 2003; Fronczek et al. 2007) are evident, and thus moderate orexin deficiency that are not detectable by lumbar CSF orexin-A measures likely exist in a subset of PD subjects. Specificity of these finding and functional correlations, especially with EDS are still unknown.

Dementia with Lewy Bodies (DLB)

Dementia with Lewy bodies (DLB) is the second major type of senile, degenerative dementia, after Alzheimer's disease (AD). DLB shares many features with PD. EDS, hallucinations and REM sleep behavior disorder are symptoms reported in both DLB and narcolepsy. However, Baumann et al. (2004) reported that patients with DLB had normal orexin-A levels. No histological studies focusing on orexin neurons in DLB was available.

CSF orexin-A concentrations have also been assessed in multiple system atrophy, DLB and corticobasal degeneration (Ripley et al. 2001; Dauvilliers et al. 2003; Friedman et al. 2007; Baumann et al. 2004; Yasui et al. 2006; Martinez-Rodriguez et al. 2007; Abdo et al. 2008). In almost all cases, CSF orexin-A concentrations were normal.

Progressive Supranuclear Palsy (PSP)

EDS was reported in probable progressive supranuclear palsy (PSP) in a 74 year-old female. The EDS mimicked narcolepsy without cataplexy (MSLT showed short latencies of less than 2 min without SOREMPs), HLA was positive for DR2/DQB1 and CSF orexin-A concentration was undetectable (Hattori et al. 2003). It is not clear if the co-occurrence of these disorders is due to a common process or comorbidity. The author speculated that the existence of neuropathological changes, such as neurofibrillary tangles in hypothalamus of the patient with PSP might cause decreased orexin neurotransmission.

Narcolepsy with cataplexy has also been reported in probable PSP, in a 74 year-old male (Sugiura et al. 2007). Patient medical history revealed cataplexy and EDS symptoms fluctuating since age 20, but by age 69, cataplexy and EDS had returned. PSP symptoms such as dysarthria, difficulty in writing, gait disturbance, appeared from age 70. At the time of detailed examinations, ESS was 14 and patient showed the sleep paralysis even during eating and the cataplexy induced by laughter. Other symptoms included the following: vertical gaze limitation, Meyerson sign, small voice, a masked face, wide gait, easily falling backwards, mild muscular rigidity in neck and wrists, bradykinesia in the extremities, no resting tremor, these symptoms were agreeable to diagnose as PSP. In the MRI, the third ventricular enlargement, midbrain tegmentum atrophy and mild frontal lobe atrophy were detected. CSF orexin-A was less than 40 pg/ml and HLA DR2 and DQB1 were positive. In PSG, total sleep time was short with 274 min, 43.1 % wake after sleep onset were present. The AHI was 14/h. Mean sleep latency was shortened to less than 2.9 min and SOREMPs were present in all four naps. L-Dopa and Amantadine were slightly effective for gait, but never effective for other motor symptoms. Methylphenidate (20 mg/day) was effective for daytime sleepiness and Clomipramine was effective for cataplexy.

Yasui et al. also reported that orexin levels were significantly lower in the PSP group compared to PD ($p < 0.001$) and that orexin levels were inversely correlated with duration of morbidity in PSP but not in the other conditions studied (Yasui et al. 2006). They speculated that loss of orexin neurons or impaired orexin neurotransmission might exist as a part of the neurodegeneration associated with advanced PSP with long duration of morbidity. Considering the aforementioned two case reports by Hattori et al. (2003) and Sugiura et al. (2007), PSP may be a susceptibility factor for EDS and/or symptomatic narcolepsy associated with orexin deficiency. However, more cases are needed to address this question.

Alzheimer's Disease (AD)

CSF orexin-A levels in 24 patients with Alzheimer disease (AD) were reported normal (Ripley et al. 2001). AD was known with established sleep abnormalities (Bliwise et al. 2002). In AD subjects, dysfunction of other neurochemical systems, for example cholinergic systems, may be more directly involved in sleep abnormalities.

Subsequently, several studies have explored orexin abnormalities in Alzheimer's disease. Studies in older rats have suggested a very slight orexin cell loss and significantly decreased CSF orexin-A concentrations (Desarnaud et al. 2004). By contrast, lumber CSF orexin-A concentrations have been shown to be normal in all studied patients with Alzheimer's disease (Dauvilliers et al. 2003; Ebrahim et al. 2003; Friedman et al. 2007), although wake fragmentation was correlated with lower CSF orexin-A concentrations in one study (Friedman et al. 2007). No histological studies focusing on orexin neurons in Alzheimer's disease were available.

Hungtington's Disease

In Huntington's disease, disrupted orexin transmission was first suggested through the study of R6/2 mice, a murine model of Huntington's disease with accelerated disease progression. Low CSF orexin-A concentrations and decreased orexin cell counts were reported in these mice (Petersen et al. 2005). Huntington's disease is an autosomal dominant disorder with impaired motor coordination, caused by a CAG triplet repeat extension in the Huntington's disease gene (HTT). Huntington's disease is not associated with hypersomnia, cataplexy, or SOREMPs. Widespread cell loss occurs in Huntington's disease, including in the hypothalamus (Petersen and Bjorkqvist 2006). A slight (27 %) loss of orexin neurons was also reported in post-mortem human brains (Petersen et al. 2005). More recent studies have shown that the cell loss is not associated with low CSF orexin-A concentrations (Baumann et al. 2006; Bjorkqvist et al. 2006; Gaus et al. 2005; Meier et al. 2005). Functional roles of orexin cell loss in Huntington's disease is not known, but may not strong. Indeed, studies in rats have shown that decreased CSF orexin occurs only when more than 50 % of cells are lost or affected (Gerashchenko et al. 2003; Zhang et al. 2007).

4.1.5 Head Trauma

The association of narcolepsy/EDS with head injury is controversial. Most people with hypersomnolence after closed head injury do not have narcolepsy (Guilleminault et al. 1983), but some patients with narcolepsy report that their symptoms began after a head injury (Lankford et al. 1994; Good et al. 1989; Maccario et al. 1987; Francisco and Ivanhoe 1996; Maeda et al. 1995; Bruck and Broughton 2004). Lankford et al. (1994) reported 9 detailed cases with narcolepsy (5 HLA positive, 2 HLA negative and 2 undetermined), but orexin-A levels were not measured. Later, low to undetectable CSF orexin-A concentrations have been found in many patients with acute brain trauma or post-CNS haemorrhage (Ripley et al. 2001; Baumann et al. 2005; Dohi et al. 2008). Because adding blood to CSF in vitro does not alter CSF orexin-A concentrations, the possibility of a functional connection has been raised.

Dauvilliers et al. (2003) reported that a patient severely affected with post-traumatic hypersomnia with brain lesions (determined by MRI) had an intermediate CSF orexin-A level (176 pg/ml, HLA negative), while another severely affected patient had a normal level (503 pg/ml, HLA positive). These two patients had no cataplexy but had shortened sleep latencies (4.5 min, 3.0 min, respectively) without SOREMPs by MSLT.

Arii et al. (2004) reported a 15-year-old male affected with post-traumatic hypersomnia with an intermediate orexin-A level. His Glasgow scale at 48 h after injury was 12 (E2V4M6). An MRI showed severe cerebral contusion of the bilateral basalis of the fronto-temporal lobe and medial part of right occipital lobe with CSF leakage. One year after injury, he needed more than nine hours nocturnal sleep and one or two 1–3 h naps daily. The orexin-A level was 151 pg/ml. MRI showed atrophies in the basalis of temporal lobe and medial part of right occipital lobe. The hypothalamus showed moderate atrophy with dilatation of third ventricle but no localized lesion.

Baumann et al. (2005) reported abnormally low CSF orexin-A concentrations immediately after traumatic brain injury in approximately 95 % of patients with severe-to-moderate brain injury. However, orexin-A concentrations improved to normal in most patients 6 months after traumatic brain injury, suggesting a functional alteration rather than neuronal loss (Baumann et al. 2007). Further studies are assessing the prevalence of residual hypersomnia and narcolepsy in correlation with CSF orexin-A concentration and areas of focal damage. A temporary decrease in CSF orexin-A could indicate a decrease in orexin tone (e.g., if CSF flow dynamics or dilution occurs) and/or contribute to changes in consciousness in patients with traumatic brain injury.

Baumann reported two male patients in whom MSLT revealed >2SOREMP's and abnormally short mean sleep latencies (6.3 and 2.9 min, respectively) (Baumann et al. 2007). ESS scores were 13 and 9, respectively. In both patients, Ullanlinna and Swiss Narcolepsy Scales were normal. Neither patient had cataplexy-like episodes, hypnagogic hallucinations, or sleep paralysis. CSF orexin-A levels in the acute phase were 63 and 83 pg/ml. Six months after TBI, levels were normal (468 pg/ml) and low (289 pg/ml), respectively. HLA typing was negative for both patients. In the younger patient, TBI was mild, but severe in the 26-year-old patient. Brain CT scans did not reveal hypothalamic lesions. These patients were asymptomatic before TBI. Based on the MSLT findings and according to the international classification of sleep disorders, these two patients can be diagnosed as narcolepsy without cataplexy [ICSD2] (Medicine AAoS 2005).

One male patient (22 years old) reported hypnagogic hallucinations and cataplexy-like episodes (subjective weakness in both knees with laughter), which did not fulfill the criteria of cataplexy (Anic-Labat et al. 1999). ESS was 11, Ullanlinna Narcolepsy Scale 15, Swiss Narcolepsy Scale normal, mean sleep latency 5.6 min and there were no SOREMP's. This patient with a narcolepsy-cataplexy-like phenotype reported that he had not observed these symptoms prior to TBI. CSF orexin-A was low 6 months after TBI (225 pg/ml). There were two other patients with a low CSF orexin-A level 6 months after TBI (besides one patient with narcolepsy, and one patient with a narcolepsy-like phenotype, see earlier). In a 58-year-old patient (211 pg/ml), PSG revealed a moderate sleep apnea syndrome (apnea hypopnea index: 25/h), and a short mean sleep latency on MSLT (2.5 min). In a 19-year-old patient (234 pg/ml), all findings were normal.

EDS appearing during the first year following a head injury may be considered as post-traumatic (Billiard 2003). This typically presents itself as extended night sleep and episodes of daytime sleep. Sleepiness is usually associated with other characteristics such as headaches, difficulties in concentration or memory disorder. Radioimaging studies may reveal several possibilities: lesions affecting the hypothalamic region or brainstem, midbrain or pontine tegmentum, or more often than not, the absence of any significant lesions. Sleepiness should be objectively evaluated by a MSLT but is often is not in clinical situation. Cases with hypersomnia after head or brain trauma associated with sleep apnea syndrome were also reported (Guilleminault et al. 1983).

Although two out of three patients with post-traumatic EDS had decreased CSF orexin-A levels moderately, it is not known whether all post-traumatic subjects with declined CSF orexin-A levels exhibit EDS. Similarly, it has not been studied whether more pronounced degree of orexin-A impairments is evident for the post-traumatic symptomatic narcolepsy.

4.2 CNS Diseases Mediated with Neuroimmune Mechanisms

In this section, we will specifically discuss neuro-immunological disorders that meet the ICSD3 criteria of "Narcolepsy Due to Medical Conditions". There are three reasons for discussing this topic; (1) The etiology of idiopathic (orexin deficient) narcolepsy is not yet known, but an involvement of neuroimmune-interaction is suggested, (2) functional significance of CSF orexin levels in symptomatic narcolepsy and symptomatic EDS has not been evaluated systematically, and (3) our recent study suggests an existence of a new clinical syndrome, symptomatic EDS associated with neuromyelitis optica (NMO) and with anti-aquaporin 4 (AQP4) antibody, with low CSF orexin-A levels. Symptomatic narcolepsy cases with NMO and/or MS cases with anti-aquaporin 4 (AQP4) antibody cases are extremely interesting both in the clinical practice and research. Some of these cases were previously categorized as a Multiple Sclerosis (MS) subtype, and our findings may explains why some of MS cases show EDS and selective lesions in the paramedian hypothalamus and periventricular area.

We will include clinical data of 'Narcolepsy Due to medical Condition' from the following three subcategories, (1) Acute disseminated encephalomyelitis (ADEM) (2) Multiple Sclerosis (MS), (3) Neuromyelitis optica (NMO) and Anti-aquaporin 4 (AQP4) antibody.

4.2.1 Acute Disseminated Encephalomyelitis (ADEM)

Symptomatic narcolepsy was recently reported in four ADEM cases (Kubota et al. 2002; Gledhill et al. 2004; Yoshikawa et al. 2004; Mizuno et al. 2011; Yano et al. 2004). All these cases associated with EDS had hypothalamic lesions and low CSF orexin-A levels, suggesting an involvement of the hypothalamic orexin system in these conditions.

Improvement of sleepiness and increase in orexin-A levels after treatment has also been reported in ADEM patients. A 38 year-old female had hypersomnia, but no REM related symptoms such as cataplexy, hypnagogic hallucinations, or sleep paralysis. An MRI revealed lesions in the hypothalamus, walls of 3rd ventricle, corona radiata, floor of the aqueduct, and raphe nuclei. She was positive for DR2/DQB1*0602 and orexin-A levels were 87 pg/ml. After treatement with high-dose steroids, MRI showed smaller and fewer lesions. Six months later, her subjective sleepiness was partially improved and orexin-A level was 148 pg/ml. One year after her initial examination, her sleepiness persisted and the results of MSLT were almost unchanged (Gledhill et al. 2004).

In a 7-year-old girl with ADEM, visual symptoms, and hypersomnia, MRI revealed bilateral lesions in the white matter, basal ganglia, and hypothalamus. CSF orexin-A level was intermediate (146 pg/ml) at admission, and with steroid plus treatment, the orexin level gradually recovered to the normal range (263 pg/ml) within 47 days, and excessive sleepiness was reduced. Decreased hypothalamic orexin neurotransmission may be involved in this symptomatic case of hypersomnia associated with a clinical course of ADEM, and interestingly, double vision was also noted in this case during the course of the disease (Yoshikawa et al. 2004).

4.2.2 Demyelinating Diseases: Multiple Sclerosis (MS) and Neuromyelitis Optica (NMO)

Symptomatic narcolepsy in patients with MS have been reported for several decades. Because both MS and narcolepsy are associated with the HLA-DR2 positivity, an autoimmune target on the same brain structures has been proposed to be a common cause for both diseases, (Poirier et al. 1987). However, the discovery of the selective loss of hypothalamic orexin neurons in narcolepsy indicates that narcolepsy coincidently occurs in patients with MS when MS plaques appear in the hypothalamic area and secondarily damage the orexin neurons. Supporting this interpretation, the orexin system is not impaired in patients with MS who do not exhibit narcolepsy (Ripley et al. 2001). Nevertheless, a subset of patients with MS predominantly shows EDS and REM sleep abnormalities, and it is likely that specific immune-mediated mechanisms may be involved in these cases.

Kanbayashi et al. recently reported 7 cases of EDS occurring in patients initially diagnosed with MS with symmetric hypothalamic inflammatory lesions and orexin ligand deficiency that contrasts with the typical MRI image of MS (Fig. 2). CSF orexin measures revealed that marked (<110 pg/mL, n = 3) or moderate (110-200 pg/mL, n = 4) orexin deficiency was observed in all 7 cases (Kanbayashi et al. 2009). Four of these cases met with ICSD-2 criteria (Medicine AAoS 2005) for narcolepsy caused by a medical condition, and 3 cases met criteria for hypersomnia caused by a medical condition. HLA was negative for DQB1*0602 in the 2 cases evaluated for it. Orexin evaluation was repeated in 6 cases, and CSF orexin-A levels became normal or significantly increased, along with marked improvements of EDS and hypothalamic lesions in all cases (Kanbayashi et al. 2009). Because four cases had clinical characteristics of neuromyelitis optica (NMO) (either optic neuritis and spinal cord lesions, or both, were present) anti-AQP4 antibody was evaluated, and 3 cases came back positive; these were diagnosed as NMO-related disorder.

AQP4, a member of the AQP superfamily, is an integral membrane protein that forms pores in the membrane of biologic cells (Amiry-Moghaddam et al. 2003). Aquaporins selectively conduct water molecules in and out of the cell while preventing the passage of ions and other solutes, and are known as water channels. AQP4 is expressed throughout the central nervous system, especially in periaqueductal and periventricular regions (Amiry-Moghaddam et al. 2003; Pittock et al. 2006), and is found in nonneuronal structures such as astrocytes and ependymocytes, but is absent from neurons. NMO-IgG, which can be detected in the serum of patients with NMO, has been shown to selectively bind to AQP4 (Lennon et al. 2005). Because AQP4 is enriched in the periventricular regions of the hypothalamus, where orexin-containing neurons are primarily located, symmetric hypothalamic lesions associated with reduced CSF orexin-A levels in our 3 NMO cases with anti-AQP4 antibody might be caused by the immune-attack to the AQP4 that secondarily affect the orexin neurons. However, as described earlier, Kanbayashi et al. also had 4 MS cases with EDS and orexin deficiency that tested negative for anti-AQP4 antibody,, which leaves a possibility that other antibody-mediated mechanisms are additionally responsible for the bilateral symmetric hypothalamic damage causing EDS in the MS/NMO subjects. There is also a possibility that these 4 MS cases could be NMO, because anti-AQP4 antibody was tested only once for each subject during the course of the disease and the assay was not standardized among the institutes (Kanbayashi et al. 2009).

It is thus essential to further determine the immunologic mechanisms that cause the bilateral hypothalamic lesions with orexin deficiency and EDS, and their association with NMO and AQP4. This effort may lead to establishment of a new clinical entity, and the knowledge is essential to prevent and treat EDS associated with MS and its related disorders. None of these cases had cataplexy, contrary to the 9 out of 10 symptomatic narcoleptic MS cases reported in the past (Nishino and Kanbayashi 2005). Early therapeutic intervention with steroids and other immunosuppressants may thus prevent irreversible damage of orexin neurons and chronic sleep-related symptoms.

4.2.3 Guillain-Barre's Syndrome (GBS)

Guillain-Barré syndrome (GBS) is an acute autoimmune polyradiculoneuritis with sensory and motor impairment. Since GBS may also cause autonomic dysfunction, aspiration pneumonia, and respiratory failure, some patients undergo intensive care including invasive ventilation. Although GBS is generally restricted to the peripheral nervous system, clinically and pathologically, central dysfunctions have also been documented (Cochen et al. 2005). These include hyponatremia caused by abnormal antidiuretic hormone secretion (Hochman et al. 1982), rapid eye movement sleep (REM sleep) motor behavior disorders (Schenck et al. 1986), EDS (Guilleminault and Mondini 1986), and abnormally low CSF orexin-A levels (Ripley et al. 2001; Kanbayashi et al. 2002; Nishino et al. 2003). A subset of Miller-Fisher syndrome subjects, but not chronic inflammatory demyelinating polyneuropathy (CIDP) subjects, also has significantly low CSF orexin-A (Nishino et al. 2003).

Undetectably low CSF orexin-A levels were found in seven cases of GBS in the Japanese population (Ripley et al. 2001; Kanbayashi et al. 2002; Nishino et al. 2003). Reduced CSF orexin-A levels in GBS are not likely due to secondary effects of the treatment or associated health conditions, since two GBS patients showed undetectable levels at the time of admission to the hospital (before treatment), but only exhibited general fatigue and/or lower limb weakness, with no increase in CSF protein 1evels (Nishino et al. 2003).

This finding was rather unexpected, since GBS is a presumed autoimmune disorder of peripheral polyradiculoneuropathy. However, additional CNS involvements (i.e. hypothalamus), such as occurrence of syndrome of inappropriate antidiuretic hormone (ADH) secretion and diabetes insipidus have also been suggested in severe cases. Interestingly, all these GBS subjects with low orexin-A were severe cases and developed tetraplegia, bulbar symptoms, and/or respiratory failure shortly after the disease onset. Since the clinical picture of these subjects is quite different from that of narcolepsy, any diagnostic confusion by measurement of CSF orexin-A levels between the two is unlikely (Nishino et al. 2003).

However, these findings were not confirmed in two studies of Caucasian patients (Dauvilliers et al. 2003; Cochen et al. 2005; Baumann and Bassetti 2004). In one study, CSF orexin-A concentrations were lowered but within the normal range in GBS patients with hypnagogic-like hallucinations and severely disturbed sleep (Cochen et al. 2005). One possible explanation of this discrepancy is the difference in ethnic origin of the recruited patients (Japanese vs. Caucasian patients) and thus the possibility of different pathophysiological pathways (Baumann and Bassetti 2004). Griffin et al. suggested that GBS in northern China, which is acute motor axonal neuropathy (AMAN) associated with Campylobacter infection,, is a different disease than GBS seen in western countries (Griffin et al. 1995; Ho et al. 1995). The seven Japanese GBS cases with undetectable CSF orexin-A exhibited severe and rapid onsets with frequent respiratory involvement; this may suggest a link between AMAN and orexin deficiency, but low orexin levels were not associated with neither anti-ganglioside antibodies nor antecedent infection. Orexin deficiency in the brain, as observed in idiopathic orexin-deficient narcolepsy, has not yet been confirmed in GBS subjects with low CSF orexin-A levels. For these reasons, future studies regarding the mechanism of low CSF orexin levels in a subset of GBS subjects is important.

4.2.4 Paraneoplastic Syndrome

A recent report described four Anti-Ma2 associated encephalitis patients with EDS and undetectable CSF orexin-A levels. Interestingly, orexin-A levels of two other patients who did not exhibit EDS, were in the normal range (Overeem et al. 2004). MRI showed abnormalities involving medial temporal lobes, hypothalamus, basal ganglia or upper brainstem in the four patients with EDS. The author concluded that anti-Ma antibodies is an example of an immune-mediated cause that may result in EDS and low orexin-A levels.

In contrast to MS and ADEM, distinct CNS lesions were not observed in GBS and neoplastic syndromes. Nevertheless, orexin deficiency was observed in both conditions. This suggests that the orexin deficiency in these conditions may occur at the neuron or ligand levels. Considering that the autoimmune hypothesis is the most popular theory for orexin cell death in narcolepsy (Lecendreux et al. 2003; Mignot 2004), but no clear inflammation was observed in the hypothalamus (Peyron et al. 2000), a subset of GBS and Ma2 antibody positive paraneoplastic syndromes that is associated with orexin deficiency, may be important models for studying possible autoimmune cell damage/ligand deficiency in narcolepsy.

5 Others

Behçet's disease

Recently, a 31-year-old man with Behçet's disease was reported with acute diplopia, hypersomnia (>12 h sleep/day) and sleepiness (ESS: 14) (Baumann et al. 2010). Cranial MRI revealed diencephalic lesions with left-sided subthalamic gadolinium enhancement. CSF Orexin-A was decreased (215 pg/ml). Following treatment with prednisone and azathioprine for 1 month,, diplopia, sleepiness and hypersomnia disappeared within 2 months. Hyperintense lesions vanished on cranial MRI, and orexin-A increased to a normal level (400 pg/ml).

A 19 y old male with Behçet's disease, with EDS and intermediate orexin level (131 pg/ml) was also reported (Itoh et al. 2007).

Rapid-onset obesity with hypothalamic dysfunction, hypoventilation, and autonomic dysregulation (ROHHAD) syndrome.

EDS has also been newly been reported in ROHHAD syndrome. Firstly, a 7-year-old girl with ROHHAD syndrome displayed the classic features of narcolepsy with cataplexy (Dhondt et al. 2013). ROHHAD is a rare and complex pediatric syndrome, essentially caused by dysfunction of 3 vital systems regulating endocrine, respiratory, and autonomic nervous system functioning. Her nocturnal polysomnography revealed sleep fragmentation and a sleep-onset REM period characteristic for narcolepsy. The diagnosis was confirmed by undetectable CSF orexin-A.

In addition, a 11-year-old girl with ROHHAD syndrome was reported. She had EDS and an intermediate orexin level (151 pg/ml) (Sato et al. 2014).

6 Conclusion

Symptoms of narcolepsy can occur during the course of neurological conditions. Although it is difficult to rule out the comorbility of idiopathic narcolepsy in some cases, literature review reveals numerous unquestionable cases of symptomatic narcolepsy. These include cases with HLA negative and/or late onset, and cases in which narcoleptic symptoms occur in parallel with the rise and fall of the causative disease. Symptomatic narcolepsy cases are most often associated with brain tumors and inherited disease followed by head trauma. Cases associated with vascular diseases, degeneration and autoimmune/immune-mediated diseases are also reported. Review of these cases, especially with brain tumors, illustrates a clear picture that the hypothalamus is most often involved. Several cases of symptomatic cataplexy (without EDS) are also reported. In contrast, symptomatic cataplexy appeared to be often associated with non-hypothalamic structures.

Recently, it was revealed that the pathophysiology of idiopathic narcolepsy was linked to orexin ligand deficiency. CSF orexin-A measures were also carried out in some symptomatic cases of narcolepsy/EDS. Reduced CSF orexin-A levels were seen in most symptomatic cases of narcolepsy/EDS with various etiologies, and EDS in these cases were sometimes reversible with an improvement of the causative neurological disorder and also the orexin status. It is also notable that some symptomatic EDS cases with Parkinson diseases or thalamic infarction were not linked with orexin ligand deficiency, though non-specific reduction of orexin neurons may occur in a subset of PD patients.

Since CSF orexin measures are still experimental, cases with sleep abnormalities/cataplexy are habitually selected for CSF orexin measures. Therefore, it is still not known whether all or a large majority of cases with low CSF orexin-A levels with CNS intervention exhibit EDS/cataplexy.

Occurrences of cataplexy in idiopathic narcolepsy cases are tightly associated with orexin ligand deficiency. However, this link is less clear in symptomatic cases. Since none of the acute and subacute symptomatic cases (such as NMO, GBS, and ADEM) with undetectable CSF orexin-A levels developed cataplexy, chronic orexin deficiency may be required to express cataplexy. Even when a very strict criterion for cataplexy is applied, approximately 10 % of narcolepsy-cataplexy patients have normal CSF orexin-A (Mignot et al. 2002; Nishino et al. 2001; Krahn et al. 2002). Whether or not orexin neurotransmission is abnormal in these rare cases is unknown. Considering the fact that orexin production and orexin neurons appeared to be normal in orexin receptor 2-mutated narcoleptic Dobermans (Ripley et al. 2001), it is possible that deficiencies in orexin receptors and a downstream pathway may exist in some of these patients. However, this cannot be tested currently. Similarly, it is not known whether narcoleptic subjects without cataplexy simply have milder neuropathology. Narcoleptic subjects without cataplexy may have sufficient orexin production to maintain normal CSF levels and stave off cataplexy, but the partial loss may still be great enough to produce sleepiness (see Thannickal et al. 2009).

A large number of HLA DR2/DQ6 (DQB1*0602) negative symptomatic narcolepsy/EDS (53 % (31/59) in narcolepsy and 87 % (13/15) in EDS) were found (see Sect. 2.1 and Ref. Nishino and Kanbayashi 2005). The brain system critical for these sleep abnormalities (i.e. the orexin system) could be damaged by certain neurological conditions such as tumors, vascular diseases. These cases are often associated with detectably low or intermediate CSF orexin levels, in contrast to the undetectable idiopathic narcoleptic cases.

Nevertheless, increased HLA DR2/DQ6 (DQB1*0602) positively (47 % (28/59)) was still observed in symptomatic narcoleptic cases. Although some HLA positive orexin-deficient symptomatic cases may be due to simple comorbilities of idiopathic narcolepsy, HLA may also play a role(s) in other cases: brain insult may trigger/facilitate the HLA-mediated orexin cell damage in which the mechanism may also be shared with that in the orexin deficient idiopathic cases of narcolepsy. Regarding orexin deficiency among immune-mediated neurological conditions, orexin deficiency with the hypothalamic lesions was noted in some NMO and ADEM cases. In contrast, no clear local lesions were noted in orexin deficiency in GBS and Ma2 positive paraneoplastic syndromes. Thus, it appears that orexin ligand deficiency in GBS and Ma2 may possibly be more selective at the cellular or ligand level, and the mechanism involved in these conditions should be further studied.

Finally, further studies of the involvement of the orexin system in symptomatic narcolepsy and EDS are helpful to understand the pathophysiological mechanisms for occurrence of EDS and cataplexy. Measuring CSF orexin-A may be also useful to choose treatment options such as wake-promoting compounds, anticataplectic medications and ultimately, for starting treatment with orexin agonists when they become available.

Acknowledgements We would like to thank Ms. Mari Matsumura for writing and editing English for this paper.

Appendix

Symptomatic narcolepsy, hypersomnia or EDS with orexin measurements (n = 346)

Classification	Diagnosis	Location	Age	Gender	Narcoleptic symptoms		SOREMP	CA	HLA	orexin		Note	References
					EDS	Sleep latency							
	Tumors (n = 14)	Lesion	Low orexin cases (7/14)		Narcolepsy due to medical condition (9/14)				DR2/DQB1*0602 (DQw1)				
Type 1, NM	Astrocytoma resection	Hypothalamus	16	f	+	1.7 m/MSLT	–	–	–	Low	104 pg/ml		Arii et al. (2001)
Type 1, NM	Astrocytoma resection	Suprasellar	11	m	+	?	+	–	–	Low	<40 pg/ml		Marcus et al. (2002, Marcus and Mignot 2003)
HM/SE (n = 5)	Craniopharyngioma (n = 3), germinoma (n = 1), arachnoid cyst (n = 1)	Hypothalamus, thalamus	Mean 15	m = 2, f = 3	+	Mean: 10.3 m/MSLT	?	–	?	Control level	Mean = 133 pg/ml		Snow et al. (2002)
Type 2, NM	Choroid plexus carcinoma resection	Pineal gland, thalamus	28	f	+	7.5 m/MSLT	+	–	–	Normal	518 pg/ml		Krahn et al. (2002)
Type 1, NM	Adenoma	Pituitary, hypothalamus	60	m	+	6.4 m/MSLT	+	+	–	Normal	275 pg/ml	Acromegaly+	Dempsey et al. (2003)
Type 1, NM	Tumor	Hypothalamus	65	f	+	5 m/1EEG	+	+	?	Low	61 pg/ml		Nokura et al. (2004)
Type 1, NM	Craniopharyngioma	Hypothalamus	11	f	+	1.4 m/MSLT	+	–	–	Low	93 pg/ml		Tachibana et al. (2005)
Type 1, NM	CNS lymphoma	Left basal ganglia, thalamus, cerebral pedunculus, splenium of the corpus callosum, right internal temporal lobe	46	m	+	?	?	–	–	Low	<40 pg/ml–>244 pg/ml		Dauvilliers et al. (2007)
Type 1, NM	Craniopharyngioma	Expansion of the 3rd ventricle and a cavity forming in the whole hypothalamus	19	f	+	1 m/MSLT	+	+	–	Low	71 pg/ml		Sakuta et al. (2012)
Type 1, NM	Craniopharyngioma	Bilateral thlamus, hypothalamus	12	f	+	0.1 m/MSLT	+	+	–	Low	<40 pg/ml	DI, hypothalamic-pituitary dysfunction	Uchida et al. (2014)
	Head trauma (n = 7)	Head injury location	Low orexin cases (3/7)	Narcolepsy due to medical condition (4/7)									
HM (n = 2)	Head trauma (n = 2)	Non-specific	23, 21	m	+	4.5 m, 3 m/MSLT	–	–	(−) = 1, (+) = 1	Intermediate, normal	176, 503 pg/ml		Dauvilliers et al. (2003)
HM (n = 1)	Head trauma	Base of skull	15	m	+	2 m/1EEG	–	–	?	Intermediate	151 pg/ml	GCS12	Arii et al. (2004)
Type 1, NM (n = 3), Type 2, NM (n = 1)	Head trauma (n = 4)	Non-specific	18, 26, 22, 58	m = 3, f = 1	+	2.5–6.3 m/MSLT	+	(−) = 3, (±) = 1	(−) = 2, ? = 2	Low: n = 3, Normal: n = 1, (33 % of controls)	211, 225, 289, 468 pg/ml	AHI: 25/h(n = 1)	Baumann et al. (2005)

(continued)

(continued)

Classification	Diagnosis	Location	Age	Gender	Narcoleptic symptoms		SOREMP	CA	HLA	orexin		Note	References
					EDS	Sleep latency							
	Vascular disorders (n = 8)		Low orexin cases (1/8)		Narcolepsy due to medical condition (3/8)								
Type 1, NM	Infarction	Hypothalamus	23	m	+	0.5 m/MSLT	+	+	–	Intermediate	167 pg/ml		Scammell et al. (2001)
HM (n = 1), SE (n = 1)	Infarction (n = 2)	Thalamus, ponto-medullary	34, 40	m	+	9 m, 1 m/MSLT	–	–	?	Normal	265, 316 pg/ml		Bassetti et al. (2003)
Type 2, NM	Infarction	Thalamus	45	m	+	5 m/MSLT	+	–	?	Normal	312 pg/ml		Nokura et al. (2004)
HM (n = 1)	Infarction	Thalamus	15	m	+	?	–	–	?	Normal	274 pg/ml		Tohyama et al. (2004)
HM (n = 2)	Infarction (n = 2)	Thalamus	61, 82	f	+	?	?	?	?	Normal	293, 184 pg/ml		Miyamoto et al. (2004)
Type 1, NM	Infarction	Thalamus	83	f	+	?	?	?	?	Low	109 to >323 pg/ml		Adachi et al. (2013)
	Encephalopathies (n = 6)		Low orexin cases (4/6)		Narcolepsy due to medical condition (5/6)								
Type 1, NM	Rasmussen's syndrome	Left frontotemporal and insular	40	m	+	1.6 m/MSLT	+	+	+	low	<40 pg/ml		Lagrange et al. (2003)
Type 1, NM	Wernicke encephalitis	Hypothalamus	5	f	+	?	–	–	?	low	<40 pg/ml		Kashiwagi et al. (2004)
Type 1, NM	Limbic encephalitis	Limbic, hypothalamus	65	m	+	?	–	–	?	Low	87 pg/ml		Yamato et al. (2004)
Type 1, NM	Brain stem encephalitis	Pontine tegmentum adjacent and rostral to both fourth nerve nuclei	30	m	+	<3 m/MSLT	+	+	–	Normal	266 pg/ml	RBD	Mathis et al. (2007)
HM	Meningo-encephalitis	Hypothalamus, 4th ventricle	36	m	+	?	?	–	–	Intermediate	122 to >191 pg/ml	Mycoplasma pneumoniae	Sugeno et al. (2012)
Type 1, NM	Hashimoto's encephalopathy	Non-specific	65	m	+	?	–	–	?	Low	<40 pg/ml	Full recover by steroid	Castillo et al. (2006)
	Neuro Degeneration (lumber CSF; n = 190, ventricular CSF; n = 19)				Low orexin cases (8/190)		Narcolepsy due to medical condition (8/190)						
HM (n = 3)	Parkinson's disease (n = 3)	Non-specific	52, 64, 69	m	+	4.4, 4.9, 6.1 m/MSLT	(−) = 2, (+)=1	–	–	Normal	253, 307, 319 pg/ml		Overeem et al. (2002)
Type 1, NM (n = 16)	PD (n = 16), ventricular CSF	Non-specific	?	?	±	?	?	–	?	Low	<50–97 pg/ml	Ventricular CSF	Drouot et al. (2003)
HM (n = 3)	PD (n = 3), ventricular CSF	Non-specific	?	?	+	?	?	–	?	Intermediate	138–169 pg/ml		Drouot et al. (2003)
Type 1, NM	PD	Non-specific	58	m	+	2 m/MSLT	+	+	+	Low	86 pg/ml		Maeda et al. (2006)
HM/REF (n = 62)	PD (n = 62)	Non-specific	Mean = 70	m:23,f:39	±	?	?	?	?	Intermediate: n = 1, normal: n = 61	Mean = 302 pg/ml		Yasui et al. (2006)

(continued)

(continued)

Classification	Diagnosis	Location	Age	Gender	Narcoleptic symptoms		SOREMP	CA	HLA	orexin		Note	References
					EDS	Sleep latency							
HM/REF (n = 25)	PD (n = 25)	Non-specific	Mean = 66	m:14,f:11	±	?	?	?	?	Normal	Mean = 285 pg/ml		Asai et al. (2009)
HM (n = 21)	PD (n = 21)	Non-specific	Mean = 69	m:12,f:9	+	?	?	?	?	Normal	Mean = 301 pg/ml		Compta et al. (2009)
HM (n = 20)	PD with dementia (n = 20)	Cortical atrophy	Mean = 73	m:9,f:11	+	?	?	?	?	Normal	Mean = 310 pg/ml		Compta et al. (2009)
Type 1, NM (n = 2), HM (n = 6)	PD (n = 8)	Non-specific	48–69	m:4,f:4	+	?	?	?	?	Low: n = 2, intermediate: n = 2, normal: n = 4	39–50 pg/ml, 138–156 pg/m, 200–450 pg/mll		Drouot et al. (2011)
Type 1, NM	PD	Non-specific	57	m	+	5.4 m/MSLT	?	?	+	Low	<40 pg/ml		Wakai and Kanbayashi (2011)
Type 1, NM	PD	Non-specific	83	f	+	?	–	–	–	Low	100 pg/ml		Teraoka et al. (2013)
HM (n = 10)	Dementia with Lewy bodies (n = 10)	Cortical atrophy	69–82	m = 7, f = 3	+	?	–	–	?	Normal	382–667 pg/ml		Baumann et al. (2004)
HM/REF (n = 13)	DLBD (n = 13)	Cortical atrophy	Mean = 76	m = 7, f = 6	±	?	?	?	?	Normal	Mean = 297 pg/ml		Yasui et al. (2006)
Type 1, NM	Progressive supranuclear palsy	Enlargement of the 3rd V	74	f	+	2 m/MSLT	–	–	+	low	<40 pg/ml		Hattori et al. (2003)
Type 1, NM	PSP	Enlargement of the 3rd V	74	m	+	2.9 m/MSLT	+	+	+	low	<40 pg/ml		Sugiura et al. (2007)
HM/REF (n = 16)	PSP (n = 16)	Enlargement of the 3rd V	Mean = 72	m = 11, f = 5	±	?	?	?	?	Intermediate: n = 2, normal: n = 14	Mean = 258 pg/ml		Yasui et al. (2006)
Type 1, NM (n = 1), HM/REF (n = 6),	Corticobasal degeneration (n = 7)	Unilateral atrophy of the cerebral hemisphere	Mean = 71	m = 3, f = 4	±	?	?	?	?	Normal: n = 6, low: n = 1	Mean = 246 pg/ml		Yasui et al. (2006)
	Immnune-mediated Demyelinating disorders (n20)			Low orexin cases (12/20)		Narcolepsy due to medical condition (14/20)							
Type 1, NM	Multiple sclerosis (MS)	Hypothalamus	22	f	+	2.8 m/MSLT	+	–	–	Low	<40 pg/ml	AQP4-	Iseki et al. (2002, Oka et al. 2004)
Type 1, NM	MS	Hypothalamus	45	f	+	?	–	–	?	Low	<40 pg/ml	AQP4-	Kato et al. (2003)
Type 1, NM	MS	No Hypothalamic lesion	21	m	+	1.5 m/MSLT	+	+	DQB1*0602	Low	<40 pg/ml	AQP4–	Vrethem et al. (2012)
HM	NMO spectrum disorders	Hypothalamus	43	f	+	?	–	–	?	Intermediate	191 pg/ml	AQP4+	Nozaki et al. (2004)
Type 1, NM	NMO spectrum disorders	Hypothalamus	48	f	+	?	–	–	?	Low	106 pg/ml	AQP4+	Nakamura et al. (2005)
Type 1, NM	Neuromyelitis optica	Hypothalamus	49	f	+	?	–	–	?	Low (33 % of control)	158 pg/ml	AQP4+	Carlander et al. (2008)
Type 1, NM	Neuromyelitis optica	Hypothalamus	35	f	+	6 m/MSLT	+	–	–	low	91 to >290 pg/ml	AQP4+	Baba et al. (2009)

(continued)

(continued)

Classification	Diagnosis	Location	Age	Gender	Narcoleptic symptoms		SOREMP	CA	HLA	orexin		Note	References
					EDS	Sleep latency							
Type 2, NM	Neuromyelitis optica	Hypothalamus	41	f	+	4.8 m/MSLT	+	–	?	Intermediate	177 pg/ml	AQP4+, Anhidrosis	Sekiguchi et al. (2011)
HM	Neuromyelitis optica	Hypothalamus	31	f	+	?	?	–	?	intermediate	187 to >230 pg/ml	AQP4+, SIADH	Nakano et al. (2011)
HM	Neuromyelitis optica	Hypothalamus	36	f	+	?	–	–	?	Intermediate	118 to >195 pg/ml	AQP4+	Deguchi et al. (2012)
Type 1, NM	Neuromyelitis optica	Hypothalamus	21	f	+	1 m/MSLT	+	–	–	Low	92 pg/ml	AQP4+, SIADH	Suzuki et al. (2012)
Type 2, NM	Neuromyelitis optica	Hypothalamus, basal ganglia, white matter	26	f	+	2.5 m/MSLT	+	–	–	Intermediate	166 to >185 pg/ml	AQP4+	Miyagawa et al. (2013)
HM	Neuromyelitis optica	Bilateral cerebral cortex, and right posterior limb of internal capsule and the central part of pons	36	f	+	?	?	–	?	Intermediate	164 pg/ml	AQP4+, SIADH	Sakai et al. (2014)
Type 1, NM	Neuromyelitis optica	Bilateral hypothalamus and the left caudate nucleus	46	f	+	?	?	–	–	Low	86 to >131 to >262 pg/ml	AQP4+, Hypothalamic-pituitary dysfunction	Kume et al. (2014)
Type 1, NM	Neuromyelitis optica?	Bilateral lesions in the thalamus and basal ganglia	39	f	+	?	?	–	?	Low	<40 pg/ml	AQP4−	Saito et al. (2014)
Type 1, NM	Acute disseminated encephalomyelitis	Hypothalamus	12	f	+	4.5 m/MSLT	–	–	–	low	102 pg/ml		Kubota et al. (2002)
Type 1, NM	ADEM	Hypothalamus, coronaradiata, aqueduct, raphe	38	f	+	4.4 m/MSLT	+	–	+	Low	87 pg/ml		Gledhill et al. (2004)
HM	ADEM	Hypothalamus	7	f	+	?	–	–	–	Intermediate	146 pg/ml		Yoshikawa et al. (2004)
Type 1, NM	ADEM	Hypothalamus	0.9	f	+	?	–	–	?	Low	<40 pg/ml		Yano et al. (2004)
HM	ADEM	Hypothalamus	6	m	+	?	?	–	?	Intermediate	124 pg/ml		Mizuno et al. (2011)
	Immune-mediated polyneuropathy (n = 23)		Low orexin cases (1/23)		Narcolepsy due to medical condition (1/23)								
Type 1, NM	Guillain-Barre syndrom	Non-specific	28	m	+	0.7 m/TNST	–	–	?	Low	<40 pg/ml		Nishino et al. (2003)
HM	Guillain-Barre syndrom	Non-specific	19	m	+	0.8 m/TNST	–	–	?	Intermediate	151 pg/ml		Nishino et al. (2003)
SE, REF (n = 20)	Guillain-Barre syndrom (n = 20)	Non-specific	Mean = 48	m = 57 %	case by case, ±	?	?	?	?	Normal	555 versus 664 pg/ml		Cochen et al. (2005)
HM	Bickerstaff's brainstem encephalitis	Brainstem	33	f	+	?	–	–	?	Intermediate	128 pg/ml		Saji et al. (2007)

(continued)

(continued)

Classification	Diagnosis	Location	Age	Gender	Narcoleptic symptoms		SOREMP	CA	HLA	orexin		Note	References
					EDS	Sleep latency							
	Paraneoplastic autoimmune syndromes (n = 6)		Low orexin cases (5/7)		Narcolepsy due to medical condition (5/7)								
Type 1, NM (n = 4)	Anti-Ma associated encephalitis (n = 4)	Multiple area in the brain including hypothalamus	45, 22, 67, 38	m = 3, f = 1	+	?	?	–	?	Low	<100 pg/ml	Tumor of the testis (n = 3), adenocartinoma of the lung (n = 1)	Overeem et al. (2004)
REF (n = 2)	Anti-Ma associated encephalitis (n = 2)	Brainstem, periventricular region, basal ganglia, non-specific	82, 53	f	–	?	?	–	?	Normal	237, 218 pg/ml	Lung cancer, Adenocarcinoma of ovary	Overeem et al. (2004)
Type 1, NM	Anti-Ma associated encephalitis	Lesions surrounding the third ventricle in the thalamus, hypothalamus, and mammillary bodies	63	m	+	7.2	+	+	?	Low	<40 pg/ml	Ma-2 positive, no tumor detected	Dauvilliers et al. (2013)
	Genetic/congenital disorders (n = 66)		Low orexin cases (7/66)		Narcolepsy due to medical condition (13/66)								
Type 1, NM	Prader-Willi syndrome	Non-specific	16	m	+	3 m/MSLT	–	–	–	Low	109 pg/ml	BMI 48.1, AHI 5.6	Mignot et al. (2002)
HM	PWS	Non-specific	10	m	+	6 m/MSLT	–	–	?	Intermediate	130 pg/ml	BMI29.8, AHI 3.1	Nevsimalova et al. (2004)
REF	PWS	Non-specific	23	m	–	?	–	–	?	Intermediate	191 pg/ml	BMI49, AHI 46.8	Nevsimalova et al. (2004)
REF	PWS	Non-specific	6	m	–	?	–	–	?	Normal	226 pg/ml	BMI 25.8, AHI0	Nevsimalova et al. (2004)
REF	PWS	Non-specific	0.5 m	m	?	?	–	–	?	Intermediate	192 pg/ml		Arii et al. (2004)
Type 1, NM	PWS	Non-specific	11	f	+	3 m/MSLT	+	–	?	Low	60 pg/ml	BMI 19.9, (% BMI = 108)	Terashima et al. (2012)
Symptomatic Cataplexy	Niemann-Pick type C	Non-specific	5	m	–	16.5 m/TNST	–	+	–	Intermediate	142 pg/ml		Kanbayashi et al. (2003)
Type 1, NM	NPC	Non-specific	14	f	+	5.1 m/MSLT	+	+	+	Intermediate	157 pg/ml		Vankova et al. (2003)
HM (n = 2), SE (n = 1)	NPC (n = 3)	Non-specific	24, 25, 31	m = 2, f = 1	+	3.2, 3.5, 10.7 m/MSLT	–	–	–	Intermediate: n = 1, normal: n = 2	190, 226, 245 pg/ml		Vankova et al. (2003)
Symptomatic Cataplexy	NPC	Non-specific	10	f	–	?	?	+	?	Intermediate	174 pg/ml		Oyama et al. (2006)
Type 1, NM	NPC	Atophy in whole cerebrum.	24	m	+	1 m/MSLT	+	+	–	Low	88 pg/ml		Soda et al. (2011)
Type 1, NM	NPC	Atrophy in cerebrum, cerebellum and corpus callosum.	4	m	?	?	?	+	?	Low	106 pg/ml		Eto et al. (2015)
Type 1, NM	Myotonic dystrophy	Non-specific	46	?	+	4.7 m/MSLT	+	–	–	Low	91 pg/ml		Martinez-Rodriguez et al. (2003)
Type 2, NM	MYD	Non-specific	50	?	+	1.8 m/MSLT	+	–	–	Normal	206 pg/ml		Martinez-Rodriguez et al. (2003)

(continued)

(continued)

Classification	Diagnosis	Location	Age	Gender	Narcoleptic symptoms		SOREMP	CA	HLA	orexin		Note	References
					EDS	Sleep latency							
HM (n = 4)	MYD (n = 4)	Non-specific	19, 25, 47, 68	?	+	5.7 m, 5.7 m, 7 m, 8 m/MSLT	–	–	–	Intermediate: n = 2, normal: n = 2	167, 187, 200, 235 pg/ml		Martinez-Rodriguez et al. (2003)
Type 1, NM	MYD	Non-specific	23	m	+	6.4 m/MSLT	+	–	–	Normal	401 pg/ml		Dauvilliers and Laberge (2012)
Type 2, NM (n = 3), SE (n = 14), REF (n = 21)	MYD (n = 38)	Nonspecific	Mean 43, 24-72	m = 23, f = 15	+	9.1 m/MSLT	5 cases	–	3 cases: DQB10602(+)	Normal	Mean = 277 pg/ml		Ciafaloni et al. (2008)
Type 1, NM	MYD	Nonspecific	60	f	+	20 m/MSLT	–	–	–	Low	<40 pg/ml	BMI:27.7,AHI:59.2	Iwata et al. (2009)
Type 1, NM	MYD	Nonspecific	49	m	+	?	?	?	–	Low	<40 pg/ml	BMI:27.3	Yasui et al. (2010)
	Heredogenerative disorders (n = 1)		Low orexin cases (1/1)		Narcolepsy due to medical condition (1/1)								
Type 1, NM	ADCA-DN	Enlargement of the 3rd V, brainstem atrophy	51	m	+	?	?	+	–	Low	96 pg/ml		Melberg et al. (2001)
	Others (n = 5)		Low orexin cases (1/5)		Narcolepsy due to medical condition (1/5)								
REF	Whipple's disease	?	53	m	?	?	–	–	?	Intermediate	113 pg/ml		Voderholzer et al. (2002)
Type 1, NM	ROHHAD syndrome	Nonspecific? Hypothalamic?	7.5	f	+	8 m/PSG	+	+	?	Low	<40 pg/ml	Intubated	Dhondt et al. (2013)
SE	ROHHAD syndrome	Nonspecific? Hypothalamic?	11	f	+	11 m/PSG	–	–	?	Intermediate	152 pg/ml	AHI 4.5/h	Sato et al. (2014)
HM	Behcet's disease	Thalamus	31	m	+	?	?	?	?	Intermediate (low: 33 % of control)	215 pg/ml		Baumann et al. (2010)
HM	Behcet's disease	Thalamus, Hypothalamus	19	m	+	?	?	?	B51+, DQB1?	Intermediate	131 pg/ml		Itoh et al. (2007)

Abbreviations used: *Type 1* Narcolepsy, type 1; *Type 2* Narcolepsy, type 2; *NM* Narcolepsy due to a medical condition; *EDS* Excessive daytime sleepiness; + Present; – Absent; *V* Ventricle; *?* Not assessed; *MSLT* Multiple sleep latency test; *ADCA-DN* Autosomal dominant cerebellar ataxia, deafness and narcolepsy; *O* Other, *TNST* Two-nap sleep test; *REF* Reference cases (no EDS); *EEG* Electroencephalogram

Vascular: Miyamoto et al. (2004), Adachi et al. (2013); Encephalitis: Sugeno et al. (2012); NMO: Iseki et al. (2002), Oka et al. (2004), Kato et al. (2003), Vrethem et al. (2012), Nozaki et al. (2004), Nakamura et al. (2005), Carlander et al. (2008), Baba et al. (2009), Sekiguchi et al. (2011), Nakano et al. (2011), Deguchi et al. (2012), Suzuki et al. (2012), Miyagawa et al. (2013), Sakai et al. (2014), Kume et al. (2014), Saito et al. (2014); GBS: Saji et al. (2007); Ma2: Dauvilliers et al. (2013); Whipple's: Voderholzer et al. (2002)

References

Abdo WF, Bloem BR, Kremer HP, Lammers GJ, Verbeek MM, Overeem S (2008) CSF hypocretin-1 levels are normal in multiple-system atrophy. Parkinsonism Relat Disord 14 (4):342–344

Adachi T, Maeda H, Yoshimizu A, Iwata N, Yoshimura J, Sakai S et al (2013) A case with acute onset hypersomnolence. Mem Yamaguchi Med Soc 47:172

Adie J (1926) Idiopathic narcolepsy: a disease sui generis, with remarks on the mechanism of sleep. Brain J Neurol 49:257–306

Amiry-Moghaddam M, Otsuka T, Hurn PD, Traystman RJ, Haug FM, Froehner SC et al (2003) An alpha-syntrophin-dependent pool of AQP4 in astroglial end-feet confers bidirectional water flow between blood and brain. Proc Natl Acad Sci USA 100(4):2106–2111

Anic-Labat S, Guilleminault C, Kraemer HC, Meehan J, Arrigoni J, Mignot E (1999) Validation of a cataplexy questionnaire in 983 sleep-disorders patients. Sleep 22(1):77–87

Arii J, Kanbayashi T, Tanabe Y, Ono J, Nishino S, Kohno Y (2001) A hypersomnolent girl with decreased CSF hypocretin level after removal of a hypothalamic tumor. Neurology 56 (12):1775–1776

Arii J, Kanbayashi T, Tanabe Y, Sawaishi Y, Kimura S, Watanabe A et al (2004a) CSF hypocretin-1 (orexin-A) levels in childhood narcolepsy and neurologic disorders. Neurology 63(12):2440–2442

Arii J, Kanbayashi T, Miyamoto H, Suzuki H, Kondo H, Ishii M (2004b) A case with post-trumatic hypersomnia and low hypocretin/orexin level. JSSR Abstract 257

Asai H, Hirano M, Furiya Y, Udaka F, Morikawa M, Kanbayashi T et al (2009) Cerebrospinal fluid-orexin levels and sleep attacks in four patients with Parkinson's disease. Clin Neurol Neurosurg 111(4):341–344

Autret A, Lucas B, Henry-Lebras F, de Toffol B (1994) Symptomatic narcolepsies. Sleep 17(8 Suppl):S21–S24

Baba T, Nakashima I, Kanbayashi T, Konno M, Takahashi T, Fujihara K et al (2009) Narcolepsy as an initial manifestation of neuromyelitis optica with anti-aquaporin-4 antibody. J Neurol 256 (2):287–288

Bassetti C, Mathis J, Gugger M, Lovblad K, Hess C (1996) Hypersomnia following paramedian thalamic stroke: a report of 12 patients. Ann Neurol 39(4):471–480

Bassetti C, Gugger M, Bischof M, Mathis J, Sturzenegger C, Werth E et al (2003) The narcoleptic borderland: a multimodal diagnostic approach including cerebrospinal fluid levels of hypocretin-1 (orexin A). Sleep Med 4(1):7–12

Baumann CR, Bassetti CL (2004) CSF hypocretin levels in Guillain-Barre syndrome and other inflammatory neuropathies. Neurology 62(12):2337

Baumann CR, Dauvilliers Y, Mignot E, Bassetti CL (2004) Normal CSF hypocretin-1 (orexin A) levels in dementia with Lewy bodies associated with excessive daytime sleepiness. Eur Neurol 52(2):73–76

Baumann C, Ferini-Strambi L, Waldvogel D, Werth E, Bassetti CL (2005a) Parkinsonism with excessive daytime sleepiness–a narcolepsy-like disorder? J Neurol 252(2):139–145

Baumann CR, Stocker R, Imhof HG, Trentz O, Hersberger M, Mignot E et al (2005b) Hypocretin-1 (orexin A) deficiency in acute traumatic brain injury. Neurology 65(1):147–149

Baumann CR, Hersberger M, Bassetti CL (2006) Hypocretin-1 (orexin A) levels are normal in Huntington's disease. J Neurol 253(9):1232–1233

Baumann CR, Werth E, Stocker R, Ludwig S, Bassetti CL (2007) Sleep-wake disturbances 6 months after traumatic brain injury: a prospective study. Brain J Neurol 130(Pt 7):1873–1883

Baumann CR, Bassetti CL, Hersberger M, Jung HH (2010) Excessive daytime sleepiness in Behcet's disease with diencephalic lesions and hypocretin dysfunction. Eur Neurol 63(3):190

Begin P, Mathieu J, Almirall J, Grassino A (1997) Relationship between chronic hypercapnia and inspiratory-muscle weakness in myotonic dystrophy. Am J Respir Crit Care Med 156 (1):133–139

Billiard M (2003) Other hypersomnias. In: Billiard M, Santo JB (eds) Sleep: physiology, investigations, and medicine. Kluwer Academic, New York

Bjorkqvist M, Petersen A, Nielsen J, Ecker D, Mulder H, Hayden MR et al (2006) Cerebrospinal fluid levels of orexin-A are not a clinically useful biomarker for Huntington disease. Clin Genet 70(1):78–79

Bliwise D, Rye D, Dihenia B, Gurecki P (2002) Greater daytime sleepiness in subcortical stroke relative to Parkinson's disease and Alzheimer's disease. J Geriatr Psychiatry Neurol 15(2):61–67

Bruck D, Broughton R (2004) Diagnostic ambiguities in a case of post-traumatic narcolepsy with cataplexy. Brain Inj 18(3):321–326

Carlander B, Vincent T, Le Floch A, Pageot N, Camu W, Dauvilliers Y (2008) Hypocretinergic dysfunction in neuromyelitis optica with coma-like episodes. J Neurol Neurosurg Psychiatry 79(3):333–334

Castillo P, Woodruff B, Caselli R, Vernino S, Lucchinetti C, Swanson J et al (2006) Steroid-responsive encephalopathy associated with autoimmune thyroiditis. Arch Neurol 63 (2):197–202

Chan H, Butterworth RF, Hazell AS (2004) Primary cultures of rat astrocytes respond to thiamine deficiency-induced swelling by downregulating aquaporin-4 levels. Neurosci Lett 366(3):231–234

Charizanis K, Lee KY, Batra R, Goodwin M, Zhang C, Yuan Y et al (2012) Muscleblind-like 2-mediated alternative splicing in the developing brain and dysregulation in myotonic dystrophy. Neuron 75(3):437–450

Ciafaloni E, Mignot E, Sansone V, Hilbert JE, Lin L, Lin X et al (2008) The hypocretin neurotransmission system in myotonic dystrophy type 1. Neurology 70(3):226–230

Cirignotta F, Mondini S, Zucconi M, Barrot-Cortes E, Sturani C, Schiavina M et al (1987) Sleep-related breathing impairment in myotonic dystrophy. J Neurol 235(2):80–85

Coccagna G, Mantovani M, Parchi C, Mironi F, Lugaresi E (1975) Alveolar hypoventilation and hyperosmnia in myotonic dystrophy. J Neurol Neurosurg Psychiatry 38(10):977–984

Cochen V, Arnulf I, Demeret S, Neulat ML, Gourlet V, Drouot X et al (2005) Vivid dreams, hallucinations, psychosis and REM sleep in Guillain-Barre syndrome. Brain J Neurol 128(Pt 11): 2535–2545

Compta Y, Santamaria J, Ratti L, Tolosa E, Iranzo A, Munoz E et al (2009) Cerebrospinal hypocretin, daytime sleepiness and sleep architecture in Parkinson's disease dementia. Brain J Neurol 132(Pt 12):3308–3317

Dalal M, Schuld A, Haack M, Uhr M, Geisler P, Eisensehr I et al (2001) Normal plasma levels of orexin A (hypocretin-1) in narcoleptic patients. Neurology 56(12):1749–1751

Dauvilliers YA, Laberge L (2012) Myotonic dystrophy type 1, daytime sleepiness and REM sleep dysregulation. Sleep Med Rev 16(6):539–545

Dauvilliers Y, Baumann C, Carlander B, Bischof M, Blatter T, Lecendreux M et al (2003) CSF hypocretin-1 levels in narcolepsy, Kleine-Levin syndrome, and otherhypersomnias and neurological conditions. J Neurol Neurosurg Psychiatry 74(12):1667–1673

Dauvilliers Y, Abril B, Charif M, Quittet P, Bauchet L, Carlander B et al (2007) Reversal of symptomatic tumoral narcolepsy, with normalization of CSF hypocretin level. Neurology 69 (12):1300–1301

Dauvilliers Y, Bauer J, Rigau V, Lalloyer N, Labauge P, Carlander B et al (2013) Hypothalamic immunopathology in anti-Ma-associated diencephalitis with narcolepsy-cataplexy. JAMA Neurol 70(10):1305–1310

de Lecea L, Kilduff T, Peyron C, Gao X, Foye P, Danielson P et al (1998) The hypocretins: hypothalamus-specific peptides with neuroexcitatory activity. Proc Natl Acad Sci USA 95 (1):322–327

Deguchi K, Kono S, Deguchi S, Morimoto N, Ikeda M, Kurata T et al (2012) A patient with anti-aquaporin 4 antibody presenting hypersomnolence as the initial symptom and symmetrical hypothalamic lesions. J Neurol Sci 312(1–2):18–20

Dempsey O, McGeoch P, de Silva R, Douglas N (2003) Acquired narcolepsy in an acromegalic patient who underwent pituitary irradiation. Neurology 61(4):537–540

Desarnaud F, Murillo-Rodriguez E, Lin L, Xu M, Gerashchenko D, Shiromani SN et al (2004) The diurnal rhythm of hypocretin in young and old F344 rats. Sleep 27(5):851–856
Dhondt K, Verloo P, Verhelst H, Van Coster R, Overeem S (2013) Hypocretin-1 deficiency in a girl with ROHHAD syndrome. Pediatrics 132(3):e788–e792
Dohi K, Ripley B, Fujiki N, Ohtaki H, Yamamoto T, Goto Y et al (2008) CSF orexin-A/hypocretin-1 concentrations in patients with intracerebral hemorrhage (ICH). Regul Pept 145(1–3):60–64
Drouot X, Moutereau S, Nguyen J, Lefaucheur J, Creange A, Remy P et al (2003) Low levels of ventricular CSF orexin/hypocretin in advanced PD. Neurology 61(4):540–543
Drouot X, Moutereau S, Lefaucheur JP, Palfi S, Covali-Noroc A, Margarit L et al (2011) Low level of ventricular CSF orexin-A is not associated with objective sleepiness in PD. Sleep Med 12(9):936–937
Ebrahim I, Sharief M, de Lacy S, Semra Y, Howard R, Kopelman M et al (2003a) Hypocretin (orexin) deficiency in narcolepsy and primary hypersomnia. J Neurol Neurosurg Psychiatry 74 (1):127–130
Ebrahim IO, Semra YK, De Lacy S, Howard RS, Kopelman MD, Williams A et al (2003b) CSF hypocretin (Orexin) in neurological and psychiatric conditions. J Sleep Res 12(1):83–84
Endo T, Roth C, Landolt HP, Werth E, Aeschbach D, Achermann P et al (1998) Selective REM sleep deprivation in humans: effects on sleep and sleep EEG. Am J Physiol 274(4 Pt 2):R1186–R1194
Eto K, Ito Y, Nishikawa A, Saito K, Kanbayashi T, Inoue T et al (2015) Use of clomipramine for cataplexy associated with Niemann-Pick disease type C. (Submitted)
Finnimore AJ, Jackson RV, Morton A, Lynch E (1994) Sleep hypoxia in myotonic dystrophy and its correlation with awake respiratory function. Thorax 49(1):66–70
Francisco G, Ivanhoe C (1996) Successful treatment of post-traumatic narcolepsy with methylphenidate: a case report. Am J Phys Med Rehabil 75(1):63–65
Friedman LF, Zeitzer JM, Lin L, Hoff D, Mignot E, Peskind ER et al (2007) In Alzheimer disease, increased wake fragmentation found in those with lower hypocretin-1. Neurology 68(10):793–794
Fronczek R, Lammers GJ, Balesar R, Unmehopa UA, Swaab DF (2005) The number of hypothalamic hypocretin (orexin) neurons is not affected in Prader-Willi syndrome. J Clin Endocrinol Metab 90(9):5466–5470
Fronczek R, Overeem S, Lee SY, Hegeman IM, van Pelt J, van Duinen SG et al (2007) Hypocretin (orexin) loss in Parkinson's disease. Brain J Neurol 130(Pt 6):1577–1585
Gaus SE, Lin L, Mignot E (2005) CSF hypocretin levels are normal in Huntington's disease patients. Sleep 28(12):1607–1608
Gerashchenko D, Murillo-Rodriguez E, Lin L, Xu M, Hallett L, Nishino S et al (2003) Relationship between CSF hypocretin levels and hypocretin neuronal loss. Exp Neurol 184 (2):1010–1016
Gibbs JW 3rd, Ciafaloni E, Radtke RA (2002) Excessive daytime somnolence and increased rapid eye movement pressure in myotonic dystrophy. Sleep 25(6):662–665
Gledhill R, Bartel P, Yoshida Y, Nishino S, Scammell T (2004) Narcolepsy caused by acute disseminated encephalomyelitis. Arch Neurol 61(5):758–760
Good J, Barry E, Fishman P (1989) Posttraumatic narcolepsy: the complete syndrome with tissue typing. Case report. J Neurosurg 71(1):765–767
Griffin JW, Li CY, Ho TW, Xue P, Macko C, Gao CY et al (1995) Guillain-Barre syndrome in northern China. The spectrum of neuropathological changes in clinically defined cases. Brain J Neurol 118(Pt 3):577–595
Guilleminault C, Mondini S (1986) Mononucleosis and chronic daytime sleepiness. A long-term follow-up study. Arch Intern Med 146(7):1333–1335
Guilleminault C, Faull K, Miles L (1983) Post-traumatic daytime sleepiness: a review of 20 patients. Neurology 33:1584–1589
Guilleminault C, Quera-Salva M, Goldberg M (1993) Pseudo-hypersomnia and pre-sleep behaviour with bilateral paramedian thalamic lesions. Brain J Neurol 116(6):1549–1563

Guilleminault C, Philip P, Robinson A (1998) Sleep and neuromuscular disease: bilevel positive airway pressure by nasal mask as a treatment for sleep disordered breathing in patients with neuromuscular disease. J Neurol Neurosurg Psychiatry 65(2):225–232

Hansotia P, Frens D (1981) Hypersomnia associated with alveolar hypoventilation in myotonic dystrophy. Neurology 31(10):1336–1337

Hao M, Akrami K, Wei K, De Diego C, Che N, Ku JH et al (2008) Muscleblind-like 2 (Mbnl2)-deficient mice as a model for myotonic dystrophy. Dev Dyn (an official publication of the American Association of Anatomists) 237(2):403–410

Hattori Y, Hattori T, Mukai E, Mori H, Mizuno Y, Kayukawa Y et al (2003) Excessive daytime sleepiness and low CSF orexin-A/hypocretin-I levels in a patient with probable progressive supranuclear palsy. No to Shinkei 55(12):1053–1056

Helbing-Zwanenburg B, Kamphuisen H, Mourtazaev M (1993) The origin of excessive daytime sleepiness in the Prader-Willi syndrome. J Intellect Disabil Res 37(6):533–541

Ho TW, Mishu B, Li CY, Gao CY, Cornblath DR, Griffin JW et al (1995) Guillain-Barre syndrome in northern China. Relationship to Campylobacter jejuni infection and anti-glycolipid antibodies. Brain J Neurol 118(Pt 3):597–605

Hochman MS, Kobetz SA, Handwerker JV (1982) Inappropriate secretion of antidiuretic hormone associated with Guillain-Barre syndrome. Ann Neurol 11(3):322–323

Hoehn MM, Yahr MD (1967) Parkinsonism: onset, progression and mortality. Neurology 17 (5):427–442

Iseki K, Mezaki T, Oka Y, Terada K, Tomimoto H, Miki Y et al (2002) Hypersomnia in MS. Neurology 59(12):2006–2007

Itoh T, Ishimaru Y, Yatsuda K (2007) A new onset and spontaneous remission case with Behcet's disease during Infliximab treatment. Abstract of Japan College of Rheumatology 51:412

Iwata T, Suzuki N, Mizuno H, Nakashima I, Kanbayashi T, Itoyama Y (2009) A marked decrease of orexin in the cerebrospinal fluid in a patient with myotonic dystrophy type 1 showing an excessive daytime sleepiness. Rinsho Shinkeigaku (Japanese) 49(7):437–439

Kanadia RN, Johnstone KA, Mankodi A, Lungu C, Thornton CA, Esson D et al (2003) A muscleblind knockout model for myotonic dystrophy. Science 302(5652):1978–1980

Kanadia RN, Shin J, Yuan Y, Beattie SG, Wheeler TM, Thornton CA et al (2006) Reversal of RNA missplicing and myotonia after muscleblind overexpression in a mouse poly(CUG) model for myotonic dystrophy. Proc Natl Acad Sci USA 103(31):11748–11753

Kanbayashi T, Inoue Y, Chiba S, Aizawa R, Saito Y, Tsukamoto H et al (2002a) CSF hypocretin-1 (orexin-A) concentrations in narcolepsy with and without cataplexy and idiopathic hypersomnia. J Sleep Res 11(1):91–93

Kanbayashi T, Ishiguro H, Aizawa R, Saito Y, Ogawa Y, Abe M et al (2002b) Hypocretin-1 (orexin-A) concentrations in cerebrospinal fluid are low in patients with Guillain-Barre syndrome. Psychiatry Clin Neurosci 56(3):273–274

Kanbayashi T, Inoue Y, Kawanishi K, Takasaki H, Aizawa R, Takahashi K et al (2003a) CSF hypocretin measures in patients with obstructive sleep apnea. J Sleep Res 12(4):339–341

Kanbayashi T, Abe M, Fujimoto S, Miyachi T, Takahashi T, Yano T et al (2003b) Hypocretin deficiency in niemann-pick type C with cataplexy. Neuropediatrics 34(1):52–53

Kanbayashi T, Shimohata T, Nakashima I, Yaguchi H, Yabe I, Nishizawa M et al (2009) Symptomatic narcolepsy in patients with neuromyelitis optica and multiple sclerosis: new neurochemical and immunological implications. Arch Neurol 66(12):1563–1566

Kandt R, Emerson R, Singer H, Valle D, Moser H (1982) Cataplexy in variant forms of Niemann-Pick disease. Ann Neurol 12(3):284–288

Kashiwagi M, Tanabe T, Hara K, Suzuki S, Wakamiya E, Shimizu T, et al (2004) Sleepiness due to Wernicke's encephalopathy wirh bilateral hypothalamic lesion in a 5-year-old girl. Sleep, suppl:A315

Kato T, Kanbayashi T, Yamamoto K, Nakano T, Shimizu T, Hashimoto T et al (2003) Hypersomnia and low CSF hypocretin-1 (orexin-A) concentration in a patient with multiple sclerosis showing bilateral hypothalamic lesions. Int Med 42(8):743–745

Katz E, McGrath S, Marcus C (2000) Late-onset central hypoventilation with hypothalamic dysfunction: a distinct clinical syndrome. Pediatr Pulmonol 29(1):62–68

Krahn L, Pankratz V, Oliver L, Boeve B, Silber M (2002a) Hypocretin (orexin) levels in cerebrospinal fluid of patients with narcolepsy: relationship to cataplexy and HLA DQB1*0602 status. Sleep 25(7):733–736

Krahn L, Boeve B, Oliver L, Silber M (2002b) Hypocretin (orexin) and melatonin values in a narcoleptic-like sleep disorder after pinealectomy. Sleep Med 3(6):521–523

Kubota H, Kanbayashi T, Tanabe Y, Takanashi J, Kohno Y (2002) A case of acute disseminated encephalomyelitis presenting hypersomnia with decreased hypocretin level in cerebrospinal fluid. J Child Neurol 17(7):537–539

Kume K, Deguchi K, Ikeda K, Takata T, Kokudo Y, Kamada M et al (2014) Neuromyelitis optica spectrum disorder presenting with repeated hypersomnia due to involvement of the hypothalamus and hypothalamus-amygdala linkage. Rinsho Shinkeigaku (Japanese) 54(4):364

Lagrange A, Blaivas M, Gomez-Hassan D, Malow B (2003) Rasmussen's syndrome and new-onset narcolepsy, cataplexy, and epilepsy in an adult. Epilepsy Behav 4(6):788–792

Lankford D, Wellman J, O'Hara C (1994) Posttraumatic narcolepsy in mild to moderate closed head injury. Sleep 17(8 Suppl):S25–S28

Lecendreux M, Maret S, Bassetti C, Mouren M, Tafti M (2003) Clinical efficacy of high-dose intravenous immunoglobulins near the onset of narcolepsy in a 10-year-old boy. J Sleep Res 12 (4):347–348

Lennon VA, Kryzer TJ, Pittock SJ, Verkman AS, Hinson SR (2005) IgG marker of optic-spinal multiple sclerosis binds to the aquaporin-4 water channel. J Exp Med 202(4):473–477

Maccario M, Ruggles K, Meriwether M (1987) Post-traumatic narcolepsy. Mil Med 152(7):370–371

Maeda M, Tamaoka A, Hayashi A, Mizusawa H, Shoji S (1995) A case of HLA-DR2, DQw1 negative post-traumatic narcolepsy. Rinsho shinkeigaku Clin Neurol 35(7):811–813

Maeda T, Nagata K, Kondo H, Kanbayashi T (2006) Parkinson's disease comorbid with narcolepsy presenting low CSF hypocretin/orexin level. Sleep Med 7(8):662

Manni R, Politini L, Nobili L, Ferrillo F, Livieri C, Veneselli E et al (2001) Hypersomnia in the Prader Willi syndrome: clinical-electrophysiological features and underlying factors. Clin Neurophysiol 112(5):800–805

Marcus C, Mignot E (2003) Letter to the editor regarding our previous publication: "Secondary narcolepsy in children with brain tumors". Sleep 26(2):228

Marcus C, Trescher W, Halbower A, Lutz J (2002) Secondary narcolepsy in children with brain tumors. Sleep 25(4):435–439

Martinez-Rodriguez J, Lin L, Iranzo A, Genis D, Marti M, Santamaria J et al (2003) Decreased hypocretin-1 (Orexin-A) levels in the cerebrospinal fluid of patients with myotonic dystrophy and excessive daytime sleepiness. Sleep 26(3):287–290

Martinez-Rodriguez JE, Seppi K, Cardozo A, Iranzo A, Stampfer-Kountchev M, Wenning G et al (2007) Cerebrospinal fluid hypocretin-1 levels in multiple system atrophy. Mov Disord 22 (12):1822–1824

Mathis J, Hess CW, Bassetti C (2007) Isolated mediotegmental lesion causing narcolepsy and rapid eye movement sleep behaviour disorder: a case evidencing a common pathway in narcolepsy and rapid eye movement sleep behaviour disorder. J Neurol Neurosurg Psychiatry 78(4):427–429

Medicine AAoS (2005) International classification of sleep disorders. Diagnostic and coding manual, 2nd edn. American Academy of Sleep Medicine, Westchester

Medicine AAoS (2014) International classification of sleep disorders—(ICSD-3), 3rd edn. AASM Resource Library (online)

Meier A, Mollenhauer B, Cohrs S, Rodenbeck A, Jordan W, Meller J et al (2005) Normal hypocretin-1 (orexin-A) levels in the cerebrospinal fluid of patients with Huntington's disease. Brain Res 1063(2):201–203

Melberg A, Dahl N, Hetta J, Valind S, Nennesmo I, Lundberg P et al (1999) Neuroimaging study in autosomal dominant cerebellar ataxia, deafness, and narcolepsy. Neurology 53(9):2190–2192

Melberg A, Ripley B, Lin L, Hetta J, Mignot E, Nishino S (2001) Hypocretin deficiency in familial symptomatic narcolepsy. Ann Neurol 49(1):136–137

Mignot E (2004) Sleep, sleep disorders and hypocretin (orexin). Sleep Med 5(Suppl 1):S2–S8

Mignot E, Lammers G, Ripley B, Okun M, Nevsimalova S, Overeem S et al (2002) The role of cerebrospinal fluid hypocretin measurement in the diagnosis of narcolepsy and other hypersomnias. Arch Neurol 59(10):1553–1562

Miyagawa S, Kitamura E, Uchino A, Kanbayashi T, Ogino M (2013) A case of 27 year old woman who developed hypersomnia due to a large lesion of hypothalamus with anti-AQP4-antibody. Rinsho Shinkeigaku (Japanese) 53(2):169

Miyamoto T, Miyamoto M, Hirata K (2004) Comparison of the three cases of nocturnal polysomnography in bilateral paramedian thalamic infarction. Jpn J Clin Neurophysiol 32 (5):550

Mizuno K, Kiba Y, Yokoyama T, Tosho Y, Inoue M, Maeba H et al (2011) A case of ADEM with hypersomnolence associated with low CSF hypocretin-1 level. J Jpn Pediatr Soc 115(2):330

Nakamura M, Nishii M, Maki S, Nakamuara M, Suenaga T (2005) A MS access with EDS and bilateral hypothlamic lesions. Clin Neurol (in Japanese) 45(2):187

Nakano T, Fujimoto T, Fukuda Y, Takahashi T, Kanbayashi T (2011) Neuromyelitis optica with syndrome of inappropriate secretion of antidiuretic hormone and hypersomnia associated with bilateral hypothalamic lesions: a case report. Rinsho shinkeigaku Clin Neurol 51(8):599–602

Nevsimalova S, Vankova J, Stepanova I, Seemanova E, Mignot E, Nishino S (2004) Hypocretin deficiency in Prader-Willi syndrome. J Sleep Res

Nishino S, Kanbayashi T (2005) Symptomatic narcolepsy, cataplexy and hypersomnia, and their implications in the hypothalamic hypocretin/orexin system. Sleep Med Rev 9(4):269–310

Nishino S, Okura M, Mignot E (2000a) Narcolepsy: genetic predisposition and neuropharmacological mechanisms. Sleep Med Rev 4(1):57–99

Nishino S, Ripley B, Overeem S, Lammers G, Mignot E (2000b) Hypocretin (orexin) deficiency in human narcolepsy. Lancet 355(9197):39–40

Nishino S, Ripley B, Overeem S, Nevsimalova S, Lammers G, Vankova J et al (2001) Low cerebrospinal fluid hypocretin (Orexin) and altered energy homeostasis in human narcolepsy. Ann Neurol 50(3):381–388

Nishino S, Kanbayashi T, Fujiki N, Uchino M, Ripley B, Watanabe M et al (2003) CSF hypocretin levels in Guillain-Barre syndrome and other inflammatory neuropathies. Neurology 61(6):823–825

Nokura K, Kanbayashi T, Ozeki T, Koga H, Senda M, Zettsu T et al (2004) Hypersomnia secondary to hypothalamic damage and csf orexin-A level in four cases. Sleep 27(suppl):A249

Nozaki H, Katada S, Sato M, Tanaka K, Nishizawa M (2004) A case with hypersomnia and paresthesia due to diffuse MS lesions from hypothalamus to spine. Rinsho shinkeigaku Clin Neurol 44(1):59

Oka Y, Kanbayashi T, Mezaki T, Iseki K, Matsubayashi J, Murakami G et al (2004) Low CSF hypocretin-1/orexin-A associated with hypersomnia secondary to hypothalamic lesion in a case of multiple sclerosis. J Neurol 251(7):885–886

Overeem S, van Hilten J, Ripley B, Mignot E, Nishino S, Lammers G (2002) Normal hypocretin-1 levels in Parkinson's disease patients with excessive daytime sleepiness. Neurology 58(3):498–499

Overeem S, Dalmau J, Bataller L, Nishino S, Mignot E, Verschuuren J et al (2004) Hypocretin-1 CSF levels in anti-Ma2 associated encephalitis. Neurology 62(1):138–140

Oyama K, Takahashi T, Shoji Y, Oyamada M, Noguchi A, Tamura H et al (2006) Niemann-Pick disease type C: cataplexy and hypocretin in cerebrospinal fluid. Tohoku J Exp Med 209 (3):263–267

Park J, Radtke R (1995) Hypersomnolence in myotonic dystrophy: demonstration of sleep onset REM sleep. J Neurol Neurosurg Psychiatry 58(4):512–513

Parkes J (1999) Genetic factors in human sleep disorders with special reference to Norrie disease, Prader-Willi syndrome and Moebius syndrome. J Sleep Res 8(suppl 1):14–22
Petersen A, Bjorkqvist M (2006) Hypothalamic-endocrine aspects in Huntington's disease. Eur J Neurosci 24(4):961–967
Petersen A, Stewenius Y, Bjorkqvist M, Gisselsson D (2005) Euploidy in somatic cells from R6/2 transgenic Huntington's disease mice. BMC Cell Biol 6:34
Peyron C, Faraco J, Rogers W, Ripley B, Overeem S, Charnay Y et al (2000) A mutation in a case of early onset narcolepsy and a generalized absence of hypocretin peptides in human narcoleptic brains. Nat Med 6(9):991–997
Philip P, Ghorayeb I, Leger D, Menny J, Bioulac B, Dabadie P et al (1997) Objective measurement of sleepiness in summer vacation long-distance drivers. Electroencephalogr Clin Neurophysiol 102(5):383–389
Pittock SJ, Lennon VA, Krecke K, Wingerchuk DM, Lucchinetti CF, Weinshenker BG (2006a) Brain abnormalities in neuromyelitis optica. Arch Neurol 63(3):390–396
Pittock SJ, Weinshenker BG, Lucchinetti CF, Wingerchuk DM, Corboy JR, Lennon VA (2006b) Neuromyelitis optica brain lesions localized at sites of high aquaporin 4 expression. Arch Neurol 63(7):964–968
Poirier G, Montplaisir J, Dumont M, Duquette P, Decary F, Pleines J et al (1987) Clinical and sleep laboratory study of narcoleptic symptoms in multiple sclerosis. Neurology 37(4):693–695
Ripley B, Overeem S, Fujiki N, Nevsimalova S, Uchino M, Yesavage J et al (2001a) CSF hypocretin/orexin levels in narcolepsy and other neurological conditions. Neurology 57:2253–2258
Ripley B, Fujiki N, Okura M, Mignot E, Nishino S (2001b) Hypocretin levels in sporadic and familial cases of canine narcolepsy. Neurobiol Dis 8(3):525–534
Saito N, Inoue M, Hasuo K, Kanbayashi T, Murayama S, Takeuchi S (2014) A 39 years old woman responding to modafinil with bilateral hypothalamic lesion associated with hyperthermia and hypersomnia: a case report. Rinsho shinkeigaku Clin Neurol 54(7):550–555
Saji N, Kawarai T, Tadano M, Shimizu H, Kita Y, Susuki K et al (2007) Does CSF hypocretin-1 decrease in Bickerstaff's brainstem encephalitis? Clin Neurol Neurosurg 109(6):547–548
Sakai W, Matsui N, Fujita K, Izumi Y, Nishida Y, Takahashi T et al (2014) A case of neuromyelitis optica spectrum disorder associated with central pontine and extrapontine myelinolysis preceded by syndrome of inappropriate antidiuretic hormone secretion. Rinsho Shinkeigaku (Japanese) 54(7):556–560
Sakurai T, Amemiya A, Ishii M, Matsuzaki I, Chemelli R, Tanaka H et al (1998) Orexins and orexin receptors: a family of hypothalamic neuropeptides and G protein-coupled receptors that regulate feeding behavior. Cell 92(4):573–585
Sakuta K, Nakamura M, Komada Y, Yamada S, Kawana F, Kanbayashi T et al (2012) Possible mechanism of secondary narcolepsy with a long sleep time following surgery for craniopharyngioma. Intern Med 51(4):413–417
Sato T, Muroya K, Kanbayashi T (2014) Components of excessive daytime sleepiness in patients with rapid-onset obesity with hypothalamic dysfunction, hypoventilation, and autonomic dysregulation. Pediatrics (online): http://pediatrics.aappublications.org/content/132/3/e788.long/reply-pediatrics_el_56631
Scammell T, Nishino S, Mignot E, Saper C (2001) Narcolepsy and low CSF orexin (hypocretin) concentration after a diencephalic stroke. Neurology 56(12):1751–1753
Schenck CH, Bundlie SR, Ettinger MG, Mahowald MW (1986) Chronic behavioral disorders of human REM sleep: a new category of parasomnia. Sleep 9(2):293–308
Sekiguchi T, Ishibashi T, Kubodera T, Yokota T, Mizusawa H (2011) A, 41 year old female with anti-AQP4 related encephalitis had narcolepsy and autonomic disturbance at onset. Rinsho shinkeigaku Clin Neurol 51(7):516
Snow A, Gozal E, Malhotra A, Tiosano D, Perlman R, Vega C et al (2002) Severe hypersomnolence after pituitary/hypothalamic surgery in adolescents: clinical characteristics and potential mechanisms. Pediatrics 110(6):e74

Soda M, Kanno K, Oguchi H, Matsumoto J, Itagaki S, Kobayashi N et al (2011) A case of Niemann-Pick disease type C with epilepsy, autistic symptoms, and narcolepsy associated with low CSF orexin level. Fukushima J Med Sci 61(3):184

Suenaga K, Lee KY, Nakamori M, Tatsumi Y, Takahashi MP, Fujimura H et al (2012) Muscleblind-like 1 knockout mice reveal novel splicing defects in the myotonic dystrophy brain. PLoS ONE 7(3):e33218

Sugeno N, Kawaguchi N, Hasegawa T, Kuroda T, Nakashima I, Kanbayashi T et al (2012) A case with anti-galactocerebroside antibody-positive Mycoplasma pneumoniae meningoencephalitis presenting secondary hypersomnia. Neurol Sci 33(6):1473–1476

Sugiura K, Kanbayashi T, Hattori N, Inoue Y (2007) A case with narcolepsy which symptoms were exacerbated by comorbidity of progressive supranuclear palsy. Suimin Iryou 2(1):101–104

Suzuki K, Nakamura T, Hashimoto K, Miyamoto M, Komagamine T, Nagashima T et al (2012) Hypothermia, hypotension, hypersomnia, and obesity associated with hypothalamic lesions in a patient positive for the anti-aquaporin 4 antibody: a case report and literature review. Arch Neurol 69(10):1355–1359

Tachibana N, Taniike M, Okinaga T, Ripley B, Mignot E, Nishino S (2005) Hypersomnolence and increased REM sleep with low cerebrospinal fluid hypocretin level in a patient after removal of craniopharyngioma. Sleep Med 6(6):567–569

Teraoka H, Yasui K, Nomura T, IIto S, Watanabe Y, Kowa H et al (2013) A case of Parkinson's disease with hypersomnia and low CSF orexin level. Rinsho Shinkeigaku (Japanese) 53(5):389

Terashima H, Hoshino H, Ota S, Kashiwai H, Kubota M, Horikawa R et al (2012) A case of Prader-Willi Syndrome with temporary hypersomnic symptoms. No to Hattatsu 44(2):155

Thannickal TC, Lai YY, Siegel JM (2007) Hypocretin (orexin) cell loss in Parkinson's disease. Brain J Neurol 130(Pt 6):1586–1595

Thannickal TC, Nienhuis R, Siegel JM (2009) Localized loss of hypocretin (orexin) cells in narcolepsy without cataplexy. Sleep 32(8):993–998

Tobias E, Tolmie J, Stephenson J (2002) Cataplexy in the Prader-Willi syndrome. Arch Dis Child 87(2):170

Tohyama J, Kanazawa O, Akasaka N, Kamimura T (2004) A case of bilateral paramedian thalamic infarction in childhood with the sensory disturbance and the sensory loss of taste. No to Hattatsu 36(1):65–69

Uchida T, Kanbayashi T, Tanabe R, Nagasawa K, Ishii M (2014) A case of a 13 year old girl who was diagnosed as narcolepsy with hypersomnia after surgery on craniopharyngioma. No to Hattatsu 46:S423

van der Meche FG, Bogaard JM, van der Sluys JC, Schimsheimer RJ, Ververs CC, Busch HF (1994) Daytime sleep in myotonic dystrophy is not caused by sleep apnoea. J Neurol Neurosurg Psychiatry 57(5):626–628

van Hilten JJ, Kerkhof GA, van Dijk JG, Dunnewold R, Wintzen AR (1993) Disruption of sleep-wake rhythmicity and daytime sleepiness in myotonic dystrophy. J Neurol Sci 114 (1):68–75

Vanier M (1983) Maladie de Niemann-Pick: etude biochimique de 107 cas (97 familles). Manifestations clinique et approche physiopathologique du type C, Lyon

Vankova J, Stepanova I, Jech R, Elleder M, Ling L, Mignot E et al (2003) Sleep disturbances and hypocretin deficiency in Niemann-Pick disease type C. Sleep 26(4):427–430

Vela-Bueno A, Kales A, Soldatos C, Dobladez-Blanco B, Campos-Castello J, Espino-Hurtado P et al (1984) Sleep in the Prader-Willi syndrome. Clinical and polygraphic findings. Arch Neurol 41(3):294–296

Vgontzas A, Bixler E, Kales A, Centurione A, Rogan P, Mascari M et al (1996) Daytime sleepiness and REM abnormalities in Prader-Willi syndrome: evidence of generalized hypoarousal. Int J Neurosci 87(3–4):127–139

Voderholzer U, Riemann D, Gann H, Hornyak M, Juengling F, Schumacher M et al (2002) Transient total sleep loss in cerebral Whipple's disease: a longitudinal study. J Sleep Res 11 (4):321–329

von Economo C (1930) Sleep as a problem of localization. J Nerv Ment Dis 71(3):249–259

Vrethem M, Malmgren K, Lindh J (2012) A patient with both narcolepsy and multiple sclerosis in association with Pandemrix vaccination. J Neurol Sci 321(1–2):89–91

Wakai M, Kanbayashi T (2011) Narcolepsy due to Parkinson's disease with a decrease in cerebrospinal-fluid orexin: a case report. Sleep Biol Rhythms 9(2):2

Wang ET, Cody NA, Jog S, Biancolella M, Wang TT, Treacy DJ et al (2012) Transcriptome-wide regulation of pre-mRNA splicing and mRNA localization by muscleblind proteins. Cell 150 (4):710–724

Willie J, Chemelli R, Sinton C, Yanagisawa M (2001) To eat or to sleep? Orexin in the regulation of feeding and wakefulness. Ann Rev Neurosci 24:429–458

Yamato H, Tanozaki M, Seino Y, Wakasaya Y, Suda T, Baba M (2004) A hypersomnia case due to limbic encephalitis with decreased orexin level. Rinsho shinkeigaku 44(7):482

Yano T, Kanbayashi T, Sawaishi Y, Shimizu T, Takada G (2004) An infant case of hypersomnia with acute disseminated encephalomyelitis due to hypothalamic dysfunction. Sleep 27(suppl): A238

Yasui K, Inoue Y, Kanbayashi T, Nomura T, Kusumi M, Nakashima K (2006) CSF orexin levels of Parkinson's disease, dementia with Lewy bodies, progressive supranuclear palsy and corticobasal degeneration. J Neurol Sci 250(1–2):120–123

Yasui K, Saito T, Tastumi M, Saito T, Inoyama A, Toyooka K et al (2010) A case of myotonic dystrophy type 1 with recurrent sleep attack and low CSF orexin level. Rinsho Shinkeigaku (Japanese) 50(7):501

Yoshikawa S, Suzuki S, Yamamoto K, Shichiri M, Miyamoto R, Kuroda K et al (2004) A case of acute disseminated encephalomyelitis associated with hypersomnia and low CSF hypocretin levels. Pediatr Neurol

Yu H, Laberge L, Jaussent I, Bayard S, Scholtz S, Raoul M et al (2011) Daytime sleepiness and REM sleep characteristics in myotonic dystrophy: a case-control study. Sleep 34(2):165–170

Zhang S, Lin L, Kaur S, Thankachan S, Blanco-Centurion C, Yanagisawa M et al (2007) The development of hypocretin (orexin) deficiency in hypocretin/ataxin-3 transgenic rats. Neuroscience 148(1):34–43

Narcolepsy and Idiopathic Hypersomnia

Seiji Nishino

Keywords Narcolepsy · Hypocretin · Orexin · EDS · Idiopathic hypersomnia · Symptomatic narcolepsy · Symptomatic EDS · Histamine · GABA · CSF

1 Introduction

Recent progress for understanding the pathophysiology of excessive daytime sleepiness (EDS) particularly owes itself to the discovery of narcolepsy genes (i.e., hypocretin receptor and peptide genes) in animals in 1999 and the subsequent discovery in 2000, of hypocretin ligand deficiency (i.e., loss of hypocretin neurons in the brain) in idiopathic cases of human narcolepsy-cataplexy. The hypocretin deficiency can be clinically detected by cerebrospinal fluid (CSF) hypocretin-1 measures; low CSF hypocretin-1 levels are seen in over 90 % of narcolepsy-cataplexy patients. Since the specificity of the CSF finding is also high (no hypocretin deficiency was seen in patients with idiopathic hypersomnia), low CSF hypocretin-1 levels have been included in the 2nd revision of the International Classifications of Sleep Disorder (ICSD-2) as a positive diagnosis for narcolepsy-cataplexy (ICSD-2 2005), only 5 years after the discovery of hypocretin deficiency in human narcolepsy. In the most recent revision of the ICSD, [i.e., ICSD-3, published in 2014], narcolepsy-cataplexy was renamed as narcolepsy Type 1 or hypocretin deficient syndrome, while narcolepsy without cataplexy was renamed as narcolepsy Type 2 (hypocretin non-deficient) by attempting to emphasize the pathophysiological basis of the diseases.

Narcolepsy-cataplexy is tightly associated with human leukocyte antigen (HLA) DQB1*0602. Hypocretin deficiency in narcolepsy-cataplexy is also tightly

S. Nishino (✉)
Stanford University Sleep and Circadian Neurobiology Laboratory,
Department of Psychiatry and Behavioral Sciences, Stanford University School of Medicine,
3155 Porter Drive, Rm2106, Palo Alto, CA 94304, USA
e-mail: nishino@stanford.edu

© Springer International Publishing Switzerland 2015
T. Sakurai et al. (eds.), *Orexin and Sleep*,
DOI 10.1007/978-3-319-23078-8_14

associated with HLA positivity, suggesting an involvement of immune-mediated mechanisms in the loss of hypocretin neurons. However the specificity of HLA positivity for narcolepsy-cataplexy is much lower than that of low CSF hypocretin-1 levels, as up to 30 % of the general population shares this HLA haplotype.

The prevalence of primary hypersomnia, such as narcolepsy and idiopathic hypersomnia is not high at about 0.05 and 0.005 %, respectively, but the prevalence of symptomatic hypersomnia (Hypersomnia Due to a Medical Condition) may be much higher. For example, approximately several million subjects in the USA suffer from chronic brain injury, and 75 % of those people have sleep problems and about half of them claim sleepiness (Verma et al. 2007). Symptomatic narcolepsy has also been reported, but the prevalence of symptomatic narcolepsy is much smaller, and only about 130 cases have been reported in the literature in the past 30 years (Nishino and Kanbayashi 2005) (see the chapter for symptomatic narcolepsy). The meta-analysis of these symptomatic cases indicates that hypocretin deficiency may also partially explain the neurobiological mechanisms of EDS associated with symptomatic cases of narcolepsy and hypersomnia (Nishino and Kanbayashi 2005).

Anatomical and functional studies demonstrate that the hypocretin systems integrate and coordinate the multiple wake-promoting systems, such as monoamine and acetylcholine systems to keep subjects alert and prevent abnormal REM sleep manifestations (Marcus et al. 2001), suggesting that understanding of the roles of hypocretin peptidergic systems in sleep regulation in normal and pathological conditions is important, as alterations of these systems may also be responsible not only for narcolepsy but also for other less well-defined hypersomnia.

Since a large majority of patients with EDS are currently treated with pharmacological agents, new knowledge about the neurobiology of EDS will likely lead to the development of new diagnostic tests as well as new treatments and managements of patients with hypersomnia with various etiologies.

This review focuses on pathophysiological mechanisms and nosological aspects of narcolepsy (and idiopathic hypersomnia). For the treatments of these conditions, the readers are referred to more specific publications available (Hirai and Nishino 2011; Nishino and Mignot 2011; Nishino and Kotorii 2009; Nishino 2009).

2 Symptoms of Narcolepsy

2.1 Excessive Daytime Sleepiness (EDS)

EDS and cataplexy are considered to be the two primary symptoms of narcolepsy, with EDS often being the most disabling symptom (Nishino and Mignot 1997). The EDS most typically mimics the feeling that people experience when they are severely sleep deprived but may also manifest itself as a chronic tiredness or fatigue. Narcoleptic subjects generally experience a permanent background of baseline sleepiness that easily leads to actual sleep episodes in monotonous sedentary situations. This feeling is most often relieved by short naps (15–30 min), but in most cases the refreshed sensation only lasts a short time after awaking. The refreshing value of short naps is of considerable diagnostic value. Sleepiness also occurs in irresistible waves in these patients, a phenomenon best described as "sleep attacks". Sleep attacks may occur in very unusual circumstances, such as in the middle of a meal, a conversation or riding a bicycle. These attacks are often accompanied by microsleep episodes (Guilleminault 1987) where the patient "blanks out". The patient may then continue his or her activity in a semi-conscious manner (writing incoherent phrases in a letter, speaking incoherently on the phone, etc.), a phenomenon called automatic behavior (Guilleminault 1987; Broughton 1976; Dement 1976). Learning problems and impaired concentration are frequently associated with these symptomes (Guilleminault 1987; Broughton and Ghanem 1976; Dement 1976; Cohen and Smith 1989; Rogers and Rosenberg 1990), but psychophysiological testing is generally normal.

Sleepiness is usually the first symptom to appear, followed by cataplexy, sleep paralysis and hypnagogic hallucinations (Billiard et al. 1983; Honda 1988; Parkes et al. 1975; Roth 1980; Yoss and Daly 1957). Cataplexy onset occurs within five years after the occurrence of daytime somnolence in approximately two thirds of the cases (Honda 1988; Roth 1980). Less frequently, cataplexy appears many years after the onset of sleepiness. The mean age of onset of sleep paralysis and hypnagogic hallucinations is also two to seven years later than that of sleepiness (Billiard et al. 1983; Kales et al. 1982).

In most cases, EDS and irresistible sleep episodes persist throughout the lifetime although they often improve after retirement (possibly due to better management of activities), daytime napping and adjustment of nighttime sleep.

2.2 *Cataplexy*

Cataplexy is distinct from EDS associated with narcolepsy and pathognomonic of the disease (Guilleminault et al. 1974). The importance of cataplexy for the diagnosis of narcolepsy has been recognized since its description (Henneberg 1916; Löwenfeld 1902) and in subsequent reviews on narcolepsy (Daniels 1934; Wilson 1927). Most authors now recognize patients with recurring sleepiness and cataplectic attacks as a homogeneous clinical entity, and this is now shown to be tightly associated with hypocretin-deficiency (see the section on the pathophysiology of the disease). Cataplexy is defined as a sudden episode of muscle weakness triggered by emotional factors (see Fig. 1 for cataplexy in narcoleptic Dobermans), most often in the context of positive emotions (such as laughter, having good cards at card games, the pull of the fishing rod with a baiting fish, the perfect hit at baseball), and less frequently by negative emotions (most typically anger or frustration). All antigravity muscles can be affected leading to a progressive collapse of the subject, but respiratory and eye muscles are not affected. The patient is typically awake at

Fig. 1 Cataplectic attacks in Doberman pinschers. Emotional excitations, appetizing food or playing, easily elicit multiple cataplectic attacks in these animals. Most cataplexy attacks are bilateral (97.9 %). Atonia initiated partially in the hind legs (79.8 %), front legs (7.8 %) neck/face (6.2 %), or whole body/complete attacks (6.2 %) Progression of attacks was also seen (49 % of all attacks) (Fujiki et al. 2002)

the onset of the attack but may experience blurred vision or ptosis. The attack is almost always bilateral and usually lasts a few seconds. Neurological examination performed at the time of an attack shows a suppression of the patellar reflex and sometimes that of a Babinski's sign.

Cataplexy is an extremely variable clinical symptom (Gelb et al. 1994). Most often, it is mild and occurs as a simple buckling of the knees, head dropping, facial muscle flickering, sagging of the jaw or weakness in the arms. Slurred speech or mutism is also frequently associated. It is often imperceptible to the observer and may even be only a subjective feeling difficult to describe, such as a feeling of warmth or that somehow time is suspended (Wilson 1927; Gelb et al. 1994). In other cases, it escalates to actual episodes of muscle paralysis that may last up to a few minutes. Falls and injury are rare and most often the patient will have time to find support or will sit down while the attack is occurring. Long episodes occasionally blend into sleep and may be associated with hypnagogic hallucinations. Children with narcolepsy often (about 1/3 of patients) present with a previously unrecognized description of cataplexy that we coined "cataplectic facies," consisting of a state of semipermanent eyelid and jaw weakness (Serra et al. 2008).

Patients may also experience "status cataplecticus". This rare manifestation of narcolepsy is characterized by subintrant cataplexy that lasts several hours per day and confines the subject to bed. It can occur spontaneously or more often upon withdrawal from anticataplectic drugs (Parkes et al. 1975; Passouant 1970; Hishikawa 1995).

Cataplexy often improves with advancing age. In rare cases it disappears completely but in most patients it is better controlled (probably after the patient has learned to control their emotions) (Billiard et al. 1983; Rosenthal et al. 1990).

2.3 Sleep Paralysis

Sleep paralysis is present in 20–50 % of all narcoleptic subjects (Roth 1980; Hishikawa 1976; Parkes et al. 1974; Yoss and Daly 1960). It is often associated with hypnagogic hallucinations. Sleep paralysis is best described as a brief inability to perform voluntary movements at the onset of sleep, upon awakening during the night, or in the morning. Contrary to simple fatigue or locomotion inhibition, the patient is unable to perform even a small movement, such as lifting a finger. Sleep paralysis may last a few minutes and is often finally interrupted by noise or other external stimuli. The symptom is occasionally bothersome in narcoleptic subjects, especially when associated with frightening hallucinations (Rosenthal 1939).

Whereas excessive daytime sleepiness and cataplexy are the cardinal symptoms of narcolepsy, sleep paralysis occurs frequently as an isolated phenomenon, affecting 5–40 % of the general population (Dahlitz and Parkes 1993; Fukuda et al. 1987; Goode 1962). Occasional episodes of sleep paralysis are often seen in adolescence and after sleep deprivation, thus prevalence is high for single episodes.

2.4 Hypnagogic and Hypnopompic Hallucinations

Abnormal visual (most often) or auditory perceptions that occur while falling asleep (hypnagogic) or upon waking up (hypnopompic) are frequently observed in narcoleptic subjects (Ribstein 1976). These hallucinations are often unpleasant and are typically associated with a feeling of fear or threat (Hishikawa 1976; Rosenthal 1939).

Polygraphic studies indicate that these hallucinations occur most often during REM sleep (Hishikawa 1976; Chetrit et al. 1994). These episodes are often difficult to distinguish from nightmares or unpleasant dreams, which also occur frequently in narcolepsy.

Hypnagogic hallucinations are most often associated with sleep attacks and their content is described by the patient. The hallucinations are most often complex, vivid, dream-like experiences ("half sleep" hallucinations) and may follow episodes of cataplexy or sleep paralysis, a feature that is not uncommon in severely affected patients. These hallucinations are usually easy to distinguish from hallucinations observed in schizophrenia or related psychotic conditions.

Compared to cataplexy, hypnagogic and hypnopompic hallucinations themselves do not have specificity for the diagnosis of narcolepsy, as up to 20–30 % of normal people exhibit these symptoms in various situations (Fukuda et al. 1987; Takeuchi et al. 1992). The folklore over the world describes similar phenomenon, including "Kanashibari" in Japan, literally meaning "bound or fastened in metal", "Pinyin" in China, literally meaning" ghost pressing on body".

2.5 Other Important Symptoms

One of the most frequently associated symptoms is insomnia, best characterized as a difficulty to maintain nighttime sleep. Typically, narcoleptic patients fall asleep easily, only to wake up after a short nap unable to fall asleep again before an hour or so. Narcoleptic patients do not usually sleep more than normal individuals over the 24 h cycle but frequently have a very disrupted nighttime sleep (Hishikawa et al. 1976; Broughton et al. 1988; Montplaisir et al. 1978). This symptom often develops later in life and can be very disabling.

Frequently associated problems are periodic leg movements (Godbout 1985; Mosko et al. 1984), REM behavior disorder, other parasomnias (Mayer et al. 1993; Schenck and Mahowald 1992), and obstructive sleep apnea (Mosko et al. 1984; Chokroverty 1986.

Narcolepsy was reported to be associated with changes in energy homeostasis several decades ago. Narcolepsy patients are frequently (1) obese (Honda et al. 1986; Schuld et al. 2000), (2) more often have insulin-resistant diabetes mellitus (Honda et al. 1986), (3) exhibit reduced food intake (Lammers et al. 1996) and (4) have lower blood pressure and temperature (Mayer et al. 1997; Sachs and Kaisjer 1980). These findings however, had not received much attention since they were believed to be secondary to sleepiness or inactivity during the daytime. More recently however, it was shown that these metabolic changes may be found more specifically in hypocretin deficient patients (Nishino et al. 2001; Kok et al. 2003; Hara et al. 2001), suggesting a direct pathophysiological link.

Narcolepsy is a very incapacitating disease. It interferes with every aspect of life. The negative social impact of narcolepsy has been extensively studied. Patients experience impairments in driving and there is a high prevalence of either car or machine-related accidents. Narcolepsy also interferes with professional performance, leading to unemployment, frequent changes of employment, working disability or early retirement (Broughton et al. 1981; Aldrich 1989; Alaila 1992). Several subjects also develop symptoms of depression, although these symptoms are often masked by anticataplectic medications (Broughton 1976; Broughton et al. 1981; Roth and Nevsimalova 1975).

3 The Prevalence of Narcolepsy

The prevalence of narcolepsy has been investigated in several ethnic groups and countries. One of the most sophisticated prevalence studies was a Finnish cohort study consisting of 11,354 twin individuals (Hublin et al. 1994). All subjects who responded to a questionnaire with answers suggestive of narcolepsy were contacted by telephone. Clinical interviews were performed and polysomnographic recordings were then conducted in five subjects considered to be narcoleptic. Sleep monitoring finally identified three narcoleptic subjects with cataplexy, thus leading to a prevalence of 0.026 % (Hublin et al. 1994). All 3 subjects were dizygotic twins and the co-twins were not affected (Hublin et al. 1994). Other prevalence studies have led to similar prevalence values (0.02–0.067 %) in Great Britain (Ohayon et al. 1996), France (Ondzé et al. 1999), the Czech Republic (Roth 1980), 5 European countries (Ohayon et al. 2002) and in the United States (Dement et al. 1973; Silber et al. 2002).

A study performed in 1945 in African American navy recruits also led to 0.02 % in this ethnic group for narcolepsy-cataplexy (Solomon 1945). Narcolepsy-cataplexy may be more frequent in Japan and less frequent in Israel. Two population-based prevalence studies led to a 0.16 and 0.18 % prevalence figure in Japan (Honda 1979; Tashiro 1994). However, these studies used only questionnaires and interviews but not polysomnography to confirm the diagnosis. In Israel, only a few narcoleptic patients have been identified when compared to the large population of subjects recruited into sleep clinics (Lavie and Peled 1987). This has led to the suggestion that the prevalence of narcolepsy could be as low as 0.002 % in this ethnic group.

The age of onset varies, from early childhood to the fifties, with two peaks, a larger one that occurs at around 15 years of age and a smaller peak at approximately 36 years of age (Dauvilliers et al. 2001). Similar results were found in two different populations but the reasons for this bimodal distribution remains obscure. Incidence of the disease was reported to be 1.37/100.000 per year (1.72 for men and 1.05 for women) in Olmsted County in Minnesota, and the incidence rate was highest in the second decade, followed in descending order by the 3rd, 4th and 1st decade (Silber et al. 2002).

4 Genetic and Environmental Factors in Nacrolepsy

A familial tendency for narcolepsy has been recognized since its description in the late 19th century (Westphal 1877). Starting in the 1940s, several studies were published that investigated the familial history of small cohorts of narcoleptic probands (Yoss and Daly 1960; Krabbe and Magnussen 1942; Nevsimalova-Bruhova and Roth 1972; Kessler et al. 1979; Baraitser and Parkes 1978; Honda et al. 1983). Using standard diagnostic criteria, more recent studies have showed rates of familial cases as 4.3 % in Japan (Honda 1988), 6 % in the USA (Guilleminault et al. 1989), 7.6 % in France, and 9.9 % in Canada (Dauvilliers et al. 2001). In addition to subjects who fulfill all diagnostic criteria for narcolepsy–cataplexy, other relatives may report isolated recurrent sleep episodes and unexplained EDS; these subjects may suffer from an incomplete and milder form of the disorder (Billiard et al 1994). Studies also revealed that the risk of a first-degree relative developing narcolepsy–cataplexy is 1–2.0 %, 10–40 times higher than the risk in the general population (Billiard et al. 1994; Mignot 1998; Hublin et al. 1994).

A limited number of twin cases studies have been published. Of 16 monozygotic (MZ) twin pairs with at least one affected twin, only four (or five depending on the criteria used for concordance) were concordant (Mignot 1998). Although genetic predisposition is likely to be involved in the development of narcolepsy, the relatively low rate of concordance in narcoleptic MZ twins indicates that environmental factors also play a role in the development of the disorder. The nature of the possible environmental factors involved is not known. Frequently cited factors are head trauma (Lankford et al. 1994), sudden change in sleep/wake habits (Orellana et al. 1994), and various infections (Roth 1980). Although these factors may be involved, there are no documented studies demonstrating increased frequency when compared to multiple control groups.

Recently, incidences of narcolepsy-cataplexy after pH1N1 influenza vaccination were reported in Northern Europe (Zarocostas 2011; Dauvilliers et al. 2010) and some discussions are made in the sect. 8 below.

5 Neurobiology of Wakefulness

In order to understand the neurobiology of hypersomnia, we will first discuss current understandings of the neurobiology of wakefulness. Sleep/wake is a complex physiology regulated by brain activity, and multiple neurotransmitter systems such as monoamines, acetylcholine, excitatory and inhibitory amino acids, peptides, purines, and neuronal and non-neuronal humoral modulators (i.e., cytokines and prostaglandins) (Nishino and Kotorii 2009; Jones 2005) are likely to be involved. Monoamines are perhaps the first neurotransmitters recognized to be involved in wakefulness (Nishino 2009; Jouvet 1972) and the monoaminergic systems have been the most common pharmacological targets for wake-promoting compounds in the past years. On the other hand, prototypical hypnotics target the gammaaminobutyric acid (GABA)-ergic system, a main inhibitory neurotransmitter system in the brain (Nishino and Mignot 1997; Nishino 2004).

Cholinergic neurons also play critical roles in cortical activation during wakefulness (and during REM sleep) (Nishino 2009; Jones 2005). Brainstem cholinergic neurons originating from the laterodorsal and pedunculopontine tegmental nuclei activate thalamocortical signaling, and cerebral cortical activation is further reinforced by direct cholinergic projections from the basal forebrain. However, currently no cholinergic compounds are used in sleep medicine, perhaps due to the complex nature of the systems and prominent peripheral side effects.

Monoaminergic neurons, such as norepinephrine (NE) containing locus coeruleus neurons, serotonin (5-HT) containing raphe neurons, and histamine containing tuberomammillary neurons (TMN) are wake-active and act directly on cortical and subcortical regions to promote wakefulness (Nishino and Kotorii 2009; Jones 2005). In contrast to the focus on these wake-active monoaminergic systems, researchers have often underestimated the importance of dopamine (DA) in promoting wakefulness. Most likely, this is because the firing rates of midbrain DA-producing neurons (ventral tegmental area [VTA] and substantia nigra) do not have an obvious variation according to behavioral states (Steinfels et al. 1983). In addition, DA is produced by many different cell groups (Björklund and Lindvall 1984), and which of these promote wakefulness remains undetermined. Nevertheless, DA release is greatest during wakefulness (Trulson 1985), and DA neurons increase discharge and tend to fire bursts of action potentials in association with significant sensory stimulation, purposive movement, or behavioral arousal (Ljungberg et al. 1992). Lesions that include the dopaminergic neurons of the VTA reduce behavioral arousal (Jones et al. 1973). Furthermore, recent work has also identified a small wake-active population of dopamine-producing neurons in the ventral periaqueductal grey that project to other arousal regions (Lu et al. 2006). People with DA deficiency from Parkinson's disease are often sleepy (Moller et al. 2000), and dopamine antagonists are frequently sedating. These physiologic and clinical evidences clearly demonstrate that DA also plays a role in wakefulness.

Wakefulness (and various physiological systems associated with wakefulness) is essential for the survival of creatures and thus is likely to be regulated by multiple systems, each may have a distinct but a complementary role. Some arousal systems may have essential roles for cortical activation, attention, cognition, or neuroplasticity during wakefulness while others may only be active during specific times to promote particular physiology during wakefulness. Some of the examples may be motivated-behavioral wakefulness or wakefulness in emergency states. Wakefulness may thus likely be maintained by many systems with differential roles to be fully alert. Similarly, the wake-promoting mechanism of some drugs may not be able to be explained by a single neurotransmitter system.

6 Sleep Physiology and Narcolepsy Symptomes

Since narcolepsy is a prototypical EDS disorder and since the major pathophysiology of narcolepsy-cataplexy (i.e. deficient in hypocretin neurotransmission) has recently been revealed, the discussion of neurophysiological aspects of narcolepsy will help understand neurobiology of EDS and cataplexy, a pathognomonic symptom of hypocretin deficient narcolepsy.

Narcolepsy patients manifest symptoms specifically related to dysregulation of REM sleep (Nishino and Mignot 1997). In the structured, cyclic process of normal sleep, two distinct states—REM and 3 stages (S1, S2, S3) of non-REM (NREM) sleep—alternate sequentially every 90–110 min in a cycle repeating 4–5 times per night (Nishino et al. 2004). As EEG signals in humans indicate, NREM sleep, characterized by slow oscillation in thalamocortical neurons (detected as cortical slow waves) and muscle tonus reduction, precedes REM sleep, when complete muscle atonia occurs. Slow wave NREM predominates during the early phase of normal sleep, followed by a predominance of REM during the later phase (Nishino et al. 2004).

Notably, sleep and wake are highly fragmented in narcolepsy, and affected subjects could not maintain long bouts of wake and sleep. Normal sleep physiology is currently understood as dependent upon coordination of the interactions of facilitating sleep centers and inhibiting arousal centers in the brain, such that stable sleep and wake states are maintained for specific durations (Nishino et al. 2004). An ascending arousal pathway, running from the rostal pons and through the midbrain reticular formation, promotes wakefulness (Nishino et al. 2004a, b; Saper et al. 2005). As discussed earlier, this arousal pathway may be composed of neurotransmitters (acetylcholine, NE, DA, excitatory amino acids), produced by brainstem and hypothalamic neurons (hypocretin/orexin, and histamine) and also be linked to muscle tonus control during sleep (Nishino et al. 2004; Saper et al. 2005). Whereas full alertness and cortical activation require coordination of these arousal networks, effective sleep requires suppression of arousal by the hypothalamus (Saper et al. 2005). Narcolepsy patients may experience major neurological malfunction of this control system originated in the hypothalamus.

Narcoleptics (both type 1 and type 2) exhibit a phenomenon, termed short REM sleep latency or sleep onset REM period (SOREMP), in which they enter REM sleep more immediately upon falling asleep than normal (Nishino and Mignot 1997). In some cases, NREM sleep is completely bypassed and the transition to REM sleep occurs instantly (Nishino and Mignot 1997). SOREMS are not observed in idiopathic hypersomnia, suggesting a distinct etiology from narcolepsy.

Moreover, intrusion of REM sleep into wakefulness may explain cataplexy, sleep paralysis, and hypnogogic hallucinations, which are symptoms of narcolepsy. Although sleep paralysis and hallucinations manifest in other sleep disorders (sleep apnea syndromes and disturbed sleep patterns in the normal population) (Aldrich et al. 1997) (see also "sleep paralysis and hallucinations section) (Aldrich et al. 1997), cataplexy is pathognomic for narcolepsy (Nishino and Mignot 1997).

As such, identifying cataplexy's unique pathophysiological mechanism emerged to be potentially crucial in understanding overall pathophysiology of narcolepsy.

7 Discovery of Hypocretin Ligand Deficiency in Human Narcolepsy

The significant roles, first, of hypocretin deficiency and, subsequently, of postnatal cell death of hypocretin neurons as the major pathophysiological process underlying narcolepsy with cataplexy, were established from a decade of investigation in, employing both animal and human models. In 1998, the simultaneous discovery of a novel hypothalamic peptide neurotransmitter by two independent research groups proved pivotal (Sakurai et al. 1998; De Lecea et al. 1998). One group called the peptides "hypocretin" because of their primary hypothalamic localization and similarities with the hormone "secretin" (De Lecea et al. 1998). The other group called the molecule "orexin" (after the meaning of appetite in Greek) after observing that central administration of these peptides increased appetite in rats (Sakurai et al. 1998). These neurotransmitters are produced exclusively by thousands of neurons, which are localized in the lateral hypothalamus, and project broadly to specific cerebral regions and more densely to others (Peyron et al. 1998).

Within a year, Stanford researchers, using positional cloning of a naturally-occuring familial canine narcolepsy model, identified an autosomal recessive mutation of hypocretin receptor 2 (Hcrtr 2) responsible for canine narcolepsy, characterized by cataplexy, reduced sleep latency, and SOREMPs (Lin et al. 1999). This finding coincided with the simultaneous observation of the narcolepsy phenotype, characterized by cataplectic behavior and sleep fragmentation, in hypocretin ligand deficient mice (prepro-orexin gene knockout mice) (Chemelli et al. 1999). Together, these findings confirmed hypocretins as principal sleep/wake-modulating neurotransmitters and prompted investigation of the hypocretin system's involvement in human narcolepsy.

Although screening of high risk patients with cataplexy (i.e., familial, early onset and/or HLA negative cases) did not reveal hypocretin-related gene mutation as a major cause of human narcolepsy, narcoleptic patients did exhibit low CSF hypocretin-1 levels (Nishino et al. 2000) (Fig. 2). Post-mortem brain tissue of narcoleptic patients, assessed with immunochemistry, radioimmunological peptide assays, and in situ hybridization, revealed hypocretin peptide loss and undetectable levels of hypocretin peptides or prepro-hypocretin RNA (Fig. 2). Further, melanin-concentrating hormone (MCH) neurons, which are normally located in the same brain region (Peyron et al. 2000), were observed intact, thus indicating that damage to hypocretin neurons and its production is selective in narcolepsy, rather than due to generalized neuronal degeneration.

As a result of these findings, a diagnostic test for narcolepsy, based on clinical measurement of CSF hypocretin-1 levels for detecting hypocretin ligand deficiency,

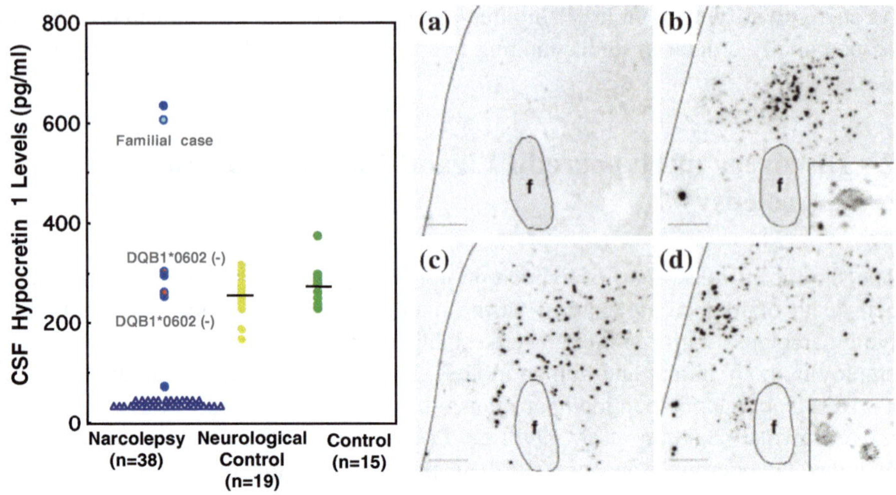

Fig. 2 Hypocretin deficiency in narcoleptic subjects. **a** CSF hypocretin-1 levels are undetectably low in most narcoleptic subjects (84.2 %). Note that two HLA DqB1*0602-negative and one familial case have normal or high CSF hypocretin-1 levels. **b** Preprohypocretin transcripts are detected in the hypothalamus of control (**b**) but not in narcoleptic subjects (**a**). Melanin-concentrating hormone (MCH) transcripts are detected in the same region in both control (**d**) and narcoleptic (**c**) sections. f and fx, fornix. *Scale bar* represents 10 mm (**a–d**) [from Peyron et al. (2000)]

became available (ICSD-2 2005). Whereas CSF hypocretin-1 concentrations above 200 pg/ml almost always occur in controls and patients with other sleep and neurological disorders, concentrations below 110 pg/ml are 94 % predictive of narcolepsy with cataplexy (Mignot et al. 2002) (Fig. 3). As this represents a more specific assessment than the multiple sleep latency test (MSLT), CSF hypocretin-1 levels below 110 pg/ml are indicated in the ICSD-2 (2005) as diagnostic of narcolepsy with cataplexy (ICSD-2 2005).

Moreover, separate coding of "narcolepsy with cataplexy" and "narcolepsy without cataplexy" in the ICSD-2 underscores how discovery of specific diagnostic criteria now informs our understanding of narcolepsy's nosology; narcolepsy with cataplexy, as indicated by low CSF hypocretin-1, appears etiologically homogeneous and distinct from most narcolepsy without cataplexy cases, exhibiting normal hypocretin-1 levels (Mignot et al. 2002). In the 3rd revision, narcolepsy was reclassified as hypocretin deficient/Type 1 and hypocretin non-deficient/Type 1 narcolepsy, and this classification is solely based on the pathophysiological findings. The potential of hypocretin receptor agonists (or cell transplantation) in narcolepsy treatment is currently being explored, and identifying hypocretin deficient status may be useful in identifying appropriate patients as candidates for a novel therapeutic option, namely hypocretin replacement therapy.

Soon after the discovery of human hypocretin deficiency, researchers identified specific substances and genes, such as dynorphin and neuronal activity-regulated

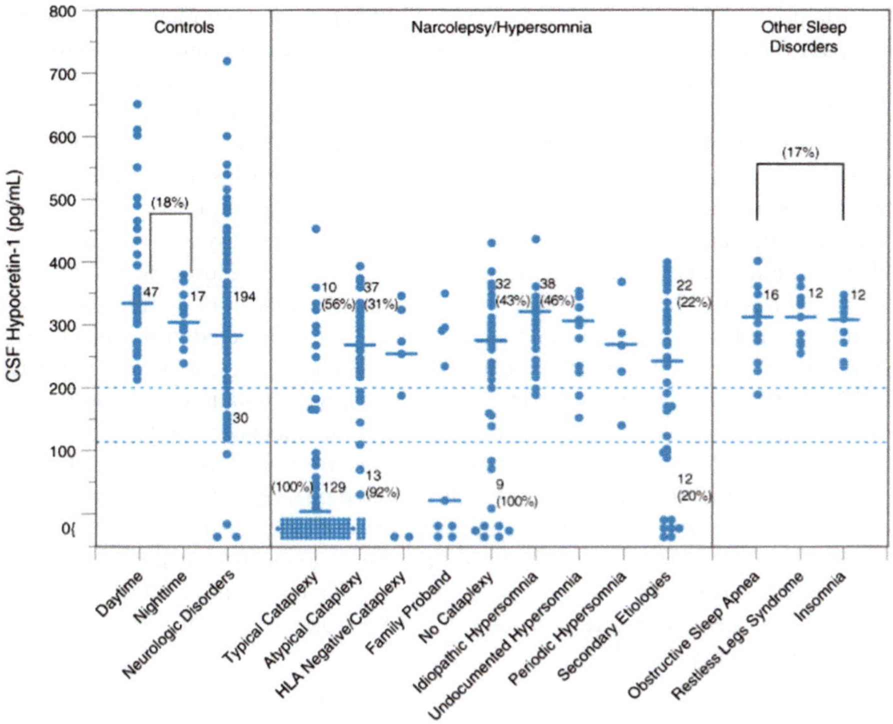

Fig. 3 CSF hypocretin-1 levels in individuals across various control and sleep disorders. Each point represents the crude concentration of Hypocretin-1 in a single person. The cutoffs for normal (>200 pg/mL) and low (<110 pg/mL) Hypocretin-1 concentrations are shown. Also noted is the total number of subjects in each range, and the percentage human leukocyte antigen (HLA)-DQB1*0602 positivity for a given group in a given range is parenthetically noted for certain disorders. Note that control carrier frequencies for DQB1*0602 are 17 to 22 % in healthy control subjects and secondary narcolepsy, consistent with control values reported in whites. In other patient groups, values are higher, with almost all Hypocretin-deficient narcolepsy being HLA DQB1*0602 positive. The median value in each group is shown as a *horizontal bar*. [Updated from previously published data in (82)] (Mignot et al. 2002)

pentraxin (NARP) (Crocker et al. 2005) and, most recently, insulin-like growth factor binding protein 3 (IGF BP3) (Honda et al. 2009), which colocalize in neurons containing hypocretin. These findings underscored selective hypocretin cell death as the cause of hypocretin deficiency (as opposed to transcription/biosynthesis or hypocretin peptide processing problems) because these substances are also deficient in postmortem brain in lateral hypothalamic area (LHA) of hypocretin deficient narcoleptic patients (Crocker et al. 2005; Honda et al. 2009). Furthermore, these findings, in view of the generally late onsets of sporadic narcolepsy compared with those of familial cases, suggest that postnatal cell death of hypocretin neurons constitutes the major pathophysiological process in human narcolepsy with cataplexy.

A large kindred of familial narcolepsy (12 affected members) has been reported in Spain (Hor et al. 2011). Affected members do not exhibit any symptoms suggesting symptomatic cases of narcolepsy, and were diagnosed as familial diopathic narcolepsy-cataplexy. The family includes a pair of dizygotic twins concordant for narcolepsy-cataplexy in the third generation; the distribution of the disorder indicates an autosomal-dominant transmission of the disease-causing gene. Hor et al. recently performed linkage analysis and sequenced coding regions of the genome (exome sequencing) of three affected members with narcolepsy and cataplexy, and identified a missense mutation in the second exon of oligodendrocyte glycoprotein MOG (Hor et al. 2011). A c.398C > G mutation was present in all affected family members but absent in unaffected members and 775 unrelated control subjects (Hor et al. 2011). Affected members were hypocretin deficient, but association with HLA DQB1*0602 was not observed (Hor et al. 2011). The mutation may induce secondary hypocretin deficiency with or without immune mediated mechanisms. MOG has been linked to various neuropsychiatric disorders and is considered a key autoantigen in multiple sclerosis and in its animal model, experimental autoimmune encephalitis (Clements et al. 2003), and thus autoimmune mechanisms may also be involved in these cases. However, even if autoimmune mechanisms are involved in these cases, it is possible that the primary target for the immune attack is not the hypocretin system. These results suggest the heterogeneity of the etiology of idiopathic narcolepsy-cataplexy.

8 What Causes Loss of Hypocretin Neurons?

Because of its strong association with certain human leukocyte antigen (HLA) alleles (i.e., HLA DQB1*0602 and DQA1*0102) (Mignot et al. 1997), it has long been speculated that narcolepsy results from an autoimmune-mediated mechanism (see Table 1). Increase in Antistreptolysin O [ASO] titer in narcolepsy, especially in the period close to the disease onset, was reported (Aran et al. 2009). However, attempts to identify specific autoantibodies have been unsuccessful for decades. Recently, Tribbles homolog 2 (Trib2) was reported as a candidate antigen involved in the destruction of hypocretin neurons in narcolepsy (Cvetkovic-Lopes et al. 2010). Trib2 was shown to be abundantly expressed in hypocretin neurons, and levels of Trib2-specific antibodies were much higher in patients with narcolepsy, especially shortly after the disease onset. Thus, Trib2 is the first antibody specifically associated with hypocretin deficient narcolepsy (Cvetkovic-Lopes et al. 2010). However, it is still unknown if Trib2-specific antibodies are directly involved in cell death, or if the antibody production is a consequence of cell damage by other unknown mechanisms (Lim and Scammell 2010).

Recent large scale genome wide association studies (GWAS) showed that susceptibility to narcolepsy is associated with single nucleotide polymorphisms (SNPs) in the T-cell receptor alpha gene locus (Hallmayer et al. 2009). Other SNPs identified are located between *carnitine palmitoyl-transferase 1B* (CPT 1B) and *choline*

Table 1 Clinical characteristics Type 1 and Type 2 narcolepsy, and idiopathic hypersomnia

	Daytime sleepiness duration awaken-refreshed	Other symptoms	MSLT		HLADQB1*0602 positivity	Low CSF hypocretin levels (<110 pg/ml)
			Sleep latency	SOREMPS		
Narcolepsy type 1						
Narcolepsy-cataplexy	Short (<30 min) (+)	Cataplexy REM sleep-related symptoms	<8 min[a]	≥2	>90 %	85–90 > 90 % in HLA positive
Narcolepsy type2						
Narcolepsy w/o cataplexy	Short (<30 min) (+)	Cataplexy (−) REM sleep-related symptoms	<8 min[a]	≥2	40–50 %	10–20 % (almost all HLA positive)
Secondary narcolepsy (due to the medical condition)	Varied (±)	with and without Cataplexy REM sleep-related symptoms	<8 min[a]	≥2	Not systematically studied	Most cases studied[b]
Idiopathic hypersomnia with long sleep time	Long (>30 min) (−)	Cataplexy (−) prolonged nighttime sleep (≥ 10 h) autonomic nervous dysfunction	<8 min[a]	≤1	No consistent results	Normal
Idiopathic hypersomnia without long sleep time	Varied (−)	Cataplexy (−) no prolonged nighttime sleep (<10 h) autonomic nervous dysfunction	<8 min[a]	≤1	No consistent results	Normal

[a]Less than 8 min is considered for both narcolepsy and idiopathic hypersomnia in the 2nd and 3rd revision of ICSD

[b]CSF hypocretin-1 measures wee done only in a limited number of secondary narcolepsy. Although many cases associated EDS with and without cataplexy exhibited moderately reduced CF hypocretin-1 levels, cases with sleep abnormalities/cataplexy are habitually selected for CSF hypocretin-1 measures, and it is not known if all or a large majority of cases with low CSF hypocretin-1 levels with CNS interventions, exhibit EDS/cataplexy. *CSF* lumbar sac cerebrospinal fluid; *hcrt* hypocretin; *HLA* human leukocyte antigen; *MSLT* multiple sleep latency test; *SOREMP* sleep-onset rapid eye movement period

kinase beta (CHKB) (Miyagawa et al. 2008) and those on *purinergic receptor P2Y11* (Kornum et al. 2011), *Cathepsin H* (*CTSH*) and *Tumor necrosis factor (ligand) superfamily member 4 (TNFSF4*, also called *OX40L*) (Kornum et al. 2011). These genes may be involved either in degeneration of hypocretin neurons or enhancing narcolepsy symptoms. Note that the association with the T-cell receptor alpha locus may be of importance, as the interactions between HLA molecules on antigen presenting cells and T cell receptors on T cells play critical roles in self/non-self discrimination by the immune system. Many of these genes, including *P2Y11*, *CTSH* and OX40L are involved in T cell activation and/or antibody processing (Faraco et al. 2013).

Recently, incidences of narcolepsy-cataplexy after pH1N1 influenza vaccination were reported in Northern Europe (Zarocostas 2011; Dauvilliers et al. 2010). These narcoleptic cases are typically HLA DQB1*0602 positive and hypocretin ligand deficient (Zarocostas 2011; Dauvilliers et al. 2010). In these countries, AS03-adjuvanted vaccination was used. Since the incidence associated with pH1N1 influenza vaccination was less or unchanged in other countries using other adjuvant (or without adjuvant in Japan), it is likely that enhancement of immunization may induce narcolepsy. However, it is also reported in China that a seasonal variation of the narcolepsy onset (most frequent in April and least frequent in November) and an increased incidence of narcolepsy (a 3-fold) following the 2009 H1N1 winter influenza pandemic were observed (Han et al. 2011). As large majority of these subjects did not receive influenza vaccinations suggesting influenza infection itself may enhance the incidence of narcolepsy.

These epidemiological data further suggest the involvement of immunological mechanisms responsible for the loss of orexin-producing neurons in narcolepsy.

9 How Does Hypocretin Ligand Deficiency Cause the Narcolepsy Symptomes?

Since hypocretin deficiency is a major pathophysiological mechanism for narcolepsy-cataplexy, it is important to know how the hypocretin ligand deficiency can cause the narcolepsy phenotype.

9.1 Hypocretin/Orexin System and Sleep Regulation

Hypocretins/orexins (hypocretin-1 and hypocretin-2/Orexin A and Orexin B) are cleaved from a precursor preprohypocretin (prepro-orexin) peptide (Sakurai et al. 1998; De Lecea et al. 1998; Sakurai 2002)). Hypocretin-1 with 33 residues contains four cysteine residues forming two disulfide bonds. Hypocretin-2 consists of 28 amino acids and shares similar sequence homology especially at the C-terminal side but has no disulfide bonds (a linear peptide) (Sakurai et al. 1998). There are two G-protein-coupled hypocretin receptors, Hcrtr 1 and Hcrtr 2, also called orexin receptor 1 and 2 (OX_1R and OX_2R), and distinct distribution of these receptors in the brain is known. Hcrtr 1 is abundant in the locus coeruleus (LC) while Hcrtr 2 is found in the TMN and basal forebrain. Both receptor types are found in the midbrain raphe nuclei and mesopontine reticular formation (Marcus et al. 2001).

Hypocretins-1 and -2 are produced exclusively by a well-defined group of neurons localized in the lateral hypothalamus. The neurons project to the olfactory bulb, cerebral cortex, thalamus, hypothalamus, and brainstem, particularly the LC, raphe nucleus, and to the cholinergic nuclei (the laterodorsal tegmental and pedunculopontine tegmental nuclei), and cholinoceptive sites (such as pontine reticular formation) (Peyron et al. 1998; Sakurai 2002). All of these projection sites are thought to be important for sleep and wake regulation.

A series of recent studies have now shown that the hypocretin system is a major excitatory system that affects the activity of monoaminergic (DA, NE, 5-HT and histamine) and cholinergic systems with major effects on vigilance states (Sakurai 2002; Willie et al. 2001). It is thus likely that a deficiency in hypocretin neurotransmission induces an imbalance among these classical neurotransmitter systems, with primary effects on sleep-state organization and vigilance.

Many measurable activities (brain and body) and compounds manifest rhythmic fluctuations over a 24-h period. Whether or not hypocretin tone changes with zeitgeber time was assessed by measuring extracellular hypocretin-1 levels in the rat brain CSF across 24-h periods, using in vivo dialysis (Fujiki et al. 2001). The results demonstrate the involvement of a diurnal pattern of hypocretin neurotransmission regulation (as in the homeostatic and/or circadian regulation of sleep). Hypocretin levels increase during the active periods and are highest at the end of the active period, and the levels decline with the onset of sleep. Furthermore, sleep deprivation increases hypocretin levels (Fujiki et al. 2001).

Recent electrophysiological studies have shown that hypocretin neurons are active during wakefulness and reduce the activity during slow wave sleep (Lee et al. 2005). The neuronal activity during REM sleep is the lowest, but intermittent increases in the activity associated with body movements or phasic REM activity are observed (Lee et al. 2005). In addition to this short-term change, the results of microdialysis experiments also suggest that basal levels of hypocretin neurotransmission fluctuate across the 24-h period and slowly build up toward the end of the active period. Adrenergic LC neurons are typical wake-active neurons involved in vigilance control and firing rates rapidly change during short sleep/wake cycles, but it has been demonstrated that basic firing activity of wake-active LC neurons also significantly fluctuates across various circadian times (Aston-Jones et al. 2001).

Several acute manipulations such as exercise, low glucose utilization in the brain, and forced wakefulness increase hypocretin levels (Willie et al. 2001; Fujiki et al. 2001). It is therefore hypothesized that a build-up/acute increase of hypocretin levels may counteract homeostatic sleep propensity that typically increases during the daytime and during forced wakefulness (Yoshida et al. 2001).

9.2 *Hypocretin/Orexin Deficiency and Narcoleptic Symptoms*

Human studies have demonstrated that the occurrence of cataplexy is closely associated with hypocretin deficiency (Mignot et al. 2002). Furthermore, the hypocretin deficiency was already observed at very early stages of the disease (just after the onset of EDS), even before the occurrences of clear cataplexy. Occurrences of cataplexy are rare in acute symptomatic cases of EDS associated with a significant hypocretin deficiency [see Nishino and Kanbayashi (2005)]; therefore, it appears that a chronic and selective deficit of hypocretin neurotransmission may be required for the occurrence of cataplexy. The possibility of involvement of a secondary neurochemical change for the occurrence of cataplexy still cannot be ruled out. If some of these changes are irreversible, hypocretin supplement therapy may only have limited effects on cataplexy.

Sleepiness in narcolepsy is most likely due to the difficulty in maintaining wakefulness as normal subjects do. The sleep pattern of narcoleptic subjects is also fragmented; they exhibit insomnia (frequent wakening) at night. This fragmentation occurs across 24 h, thus the loss of hypocretin signaling is likely to play a role in this vigilance stage stability (see Saper et al. 2001), but other mechanism may also be involved in EDS in narcoleptic subjects. One of the most important characteristics of EDS in narcolepsy is that sleepiness is reduced and patients feel refreshed after a short nap, but this does not last long as they become sleepy within a short period of time. Hypocretin-1 levels in the extracellular space and in the CSF of rats significantly fluctuate across 24 h and build up toward the end of the active periods (Yoshida et al. 2001). Several manipulations (such as sleep deprivation, exercise, and long-term food deprivation) are also known to increase hypocretin tonus (Fujiki et al. 2001; Yoshida et al. 2001). Thus, the lack of this hypocretin increase caused by circadian time and by various alerting stimulations may also play a role for EDS associated with hypocretin-deficient narcolepsy.

Mechanisms for cataplexy and REM sleep abnormalities associated with impaired hypocretin neurotransmission have been studied. Hypocretin strongly inhibit REM sleep and activate brainstem REM-off LC and raphe neurons and REM-on cholinergic neurons as well as stimulation of local GABAnergic neurons. Therefore disfacilitation of REM-off monoaminergic neurons together with stimulation of REM-on cholinergic neurons, possibly mediated through disfacilitation of inhibitory GABAnergic local interneurons caused by impaired hypocretin neurotransmission are proposed for abnormal manifestations of REM sleep.

10 The Pathophysiology of Narcolepsy Type 2 (Hypocretin Non-deficient Narcolepsy)

There are debates about the pathophysiology of narcolepsy with normal hypocretin levels. Over 90 % patients with narcolepsy without cataplexy show normal CSF hypocretin levels, yet they show apparent REM sleep abnormalities (i.e., SOREMS). Furthermore, even if the strict criteria for narcolepsy-cataplexy are applied, up to 10 % of patients with narcolepsy-cataplexy show normal CSF hypocretin levels. Considering the fact that occurrence of cataplexy is tightly associated with hypocretin deficiency, impaired hypocretin neurotransmission is still likely involved in narcolepsy-cataplexy with normal CSF hypocretin levels. Conceptually, there are two possibilities to explain these mechanisms: (1) specific impairment of hypocretin receptor and their downstream pathway and (2) partial/localized loss of hypocretin ligand (yet exhibit normal CSF levels). A good example for the first hypothesis is Hcrtr 2 mutated narcoleptic dogs; they exhibit normal CSF hypocretin-1 levels (Ripley et al. 2001), while having full blown narcolepsy. Thannickal et al. recently reported one narcolepsy patient without cataplexy (HLA typing was unknown) who had an overall loss of 33 % of hypocretin cells compared to normal, with maximal cell loss in the posterior hypothalamus (Thannickal et al. 2009). This result favors the second hypothesis, but studies with more cases are needed.

11 Idiopathic Hypersomnia, Another Hypocretin Non-deficient Central Hypersomnia

With the clear definition of narcolepsy (cataplexy and dissociated manifestations of REM sleep), it became apparent that some patients with hypersomnia suffer from a different disorder. Bedrich Roth was the first in the late 1950s and early 1960s to describe a syndrome characterized by EDS, prolonged sleep, and sleep drunkenness, and by the absence of "sleep attacks," cataplexy, sleep paralysis, and hallucinations. The terms "independent sleep drunkenness" and "hypersomnia with sleep drunkenness" were initially suggested (Roth 1962), but now this syndrome is categorized as idiopathic hypersomnia with and without long sleep time (ICSD-2 2005). Idiopathic hypersomnia should therefore not be considered synonymous with hypersomnia of unknown origin.

In the absence of systematic studies, the prevalence of idiopathic hypersomnia is unknown. Nosologic uncertainty causes difficulty in determining the epidemiology of the disorder. Recent reports from large sleep centers reported the ratio of idiopathic hypersomnia to narcolepsy to be 1:10 (Bassetti and Aldrich 1997). The age of onset of symptoms varies, but it is frequently between 10 and 30 years. The condition usually develops progressively over several weeks or months. Once established, symptoms are generally stable and long lasting, but spontaneous improvement in EDS may be observed in up to one quarter of patients (Bassetti and Aldrich 1997).

The pathogenesis of idiopathic hypersomnia is unknown. Hypersomnia usually starts insidiously. Occasionally, EDS is first experienced after transient insomnia, abrupt changes in sleep-wake habits, overexertion, general anesthesia, viral illness, or mild head trauma (Bassetti and Aldrich 1997). Despite reports of an increase in HLA DQ1,11 DR5 and Cw2, and DQ3, and a decrease in Cw3, no consistent findings have emerged (Bassetti and Aldrich 1997).

The most recent attempts to understand the pathophysiology of idiopathic hypersomnia relate to the investigation of potential role of the hypocretins. However, most studies suggest normal CSF levels of hypocretin-1 in idiopathic hypersomnia (Mignot et al. 2002; Bassetti et al. 2003). Recent studies suggest that changes in histamine productions and those of GABAA receptor modulator are involved. (see sect. 13, Histamine and GABAA Receptor Modulator in Narcolepsy and Idiopathic Hypersomnia).

12 Nosological and Diagnostic Considerations of Central Hypersomnias

Narcolepsy-cataplexy (Type 1), narcolepsy without cataplexy (Type 2), and idiopathic hypersomnia are diagnosed mostly by sleep phenotypes, especially by occurrences of cataplexy and SOREMPS (Fig. 4) (ICSD-2). Discovery of hypocretin deficiency in narcolepsy-cataplexy was a breakthrough, but also brought a new nosological and diagnostic uncertainty of the primary hypersomnias. Up to 10 % of patients with narcolepsy-cataplexy show normal CSF hypocretin-1 levels (Fig. 3). As discussed above, altered hypocretin neurotransmissions may still be involved in some of these cases. However, up to 10 % of patients with narcolepsy without cataplexy instead show low CSF hypocretin-1 levels, suggesting a substantial pathophysiological overlap between narcolepsy-cataplexy and narcolepsy without cataplexy, and the hypocretin deficient status (measured in CSF) does not completely separate these two disease conditions (Figs. 3 and 4). The situation is still the same under the ICSD-3 classification.

Similarly, concerns about the nosology of narcolepsy without cataplexy and idiopathic hypersomnia should also be addressed. Since patients with typical cases of idiopathic hypersomnia exhibit unique symptomatology, such as long hours of sleep, no refreshment from naps, and generally resistance to stimulant medications, the pathophysiology of idiopathic hypersomnia may be distinct from that of narcolepsy without cataplexy. However, current diagnostic criteria are not specific enough to diagnose these disorders; the test- retest reliability of numbers of SOREMS during MSLT is relatively low (less than 70 %) in narcolepsy and idiopathic hypersomnia (Trotti et al. 2013).

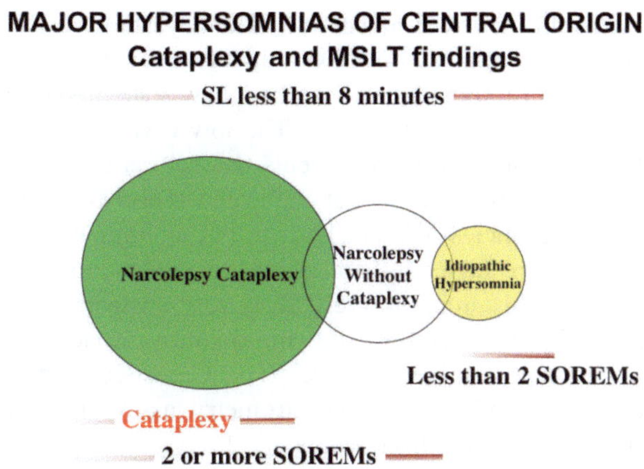

Fig. 4 Nosological and diagnostic considerations of major primary hypersomnias Narcolepsy-cataplexy, narcolepsy without cataplexy and idiopathic hypersomnia are diagnosed by occurrences of cataplexy and SOREMPS. A pathophysiogy based marker, low CSF hypocretin levels are included in the ICSD2 for the positive diagnosis for narcolepsy-cataplexy. However, up to 10 % of patients with narcolepsy-cataplexy show normal CSF hypocretin levels. In contrast, up to 10 % of patients with narcolepsy without cataplexy show low CSF hypocretin-1 levels. These results suggest a substantial pathophysiological overlap between narcolepsy-cataplexy and narcolepsy without cataplexy. Similarly, a substantial overlap likely exists between narcolepsy without cataplexy and idiopathic hypersomnia, as these disorders are diagnosed by the occurrences of SOREMs (2 or more). However, test-retest reliability of detecting number of SOREMs in these conditions has not been systematically evaluated. In the most recent revision of the ICSD, (i.e., ICSD-3), narcolepsy-cataplexy was renamed as narcolepsy Type 1 or hypocretin deficient syndrome, while narcolepsy without cataplexy was renamed as narcolepsy Type 2 (hypocretin non-deficient) by attempting to emphasize the pathophysiological basis of the diseases. However, the ambiguity of the diagnostic classification is still the same under the ICSD-3

13 Histamine and GABAA Receptor Modulator in Narcolepsy and Idiopathic Hypersomnia

Although pathophysiology of hypocretin non-deficient hypersomnia is largely unknown, neurochemical changes in these conditions, namely, reduced CSF histamine contents and increased activity of GABAA receptor modulator in the CSF, have recently been reported by two groups (Nishino et al. 2009; Kanbayashi et al. 2009; Ryer et al. 2012).

Histamine is one of these wake-active monoamines (Brown et al. 2001), and low CSF histamine levels are also found in narcolepsy with hypocretin deficiency (Nishino et al. 2009; Kanbayashi et al. 2009). Since hypocretin neurons project and excite histamine neurons in the posterior hypothalamus, it is conceivable that impaired histamine neurotransmission may mediate sleep abnormalities in hypocretin deficient narcolepsy. However, low CSF histamine levels were also observed in narcolepsy with normal hypocretin levels and in idiopathic hypersomnia, and decreased histamine neurotransmission may be involved in a broader category of

EDS than in hypocretin deficient narcolepsy (Kanbayashi et al. 2009). Since CSF histamine levels are normalized in EDS patients treated with wake-promoting compounds, low CSF histamine levels may be a new state marker for hypersomnia of central origin (Kanbayashi et al. 2009). The low CSF histamine levels in EDS was confirmed in a smaller sample of patients (Bassetti et al. 2010), but in a much larger sample size, the results were not replicated (Dauvilliers et al. 2012). The reason for these discrepancies was not known, but an animal study indicated that CSF histamine levels are under influences of the vigilance state/diurnal changes (Soya et al. 2008), and histamine in the CSF is very unstable, and thus the highly controlled study design is crucial for drawing definitive conclusion. The methodological differences in histamine measures (HPLC/post-column derivatization vs liquid chromatography-tandem mass spectrometry assay) may also be another factor to be considered for the discrepancy.

It should be also noted that 2 independent research group reported that increased number of histaminergic neurons in the postmortem human hypocretin deficient narcolepsy (Type 1) (John et al. 2013: Valko et al. 2013), the results difficult to reconcile with the reports of low CSF histamine levels in narcolepsy (Nishino et al. 2009; Kanbayashi et al. 2009). These human postmortem studies include only small numbers of patients with the long-term history of stimulant mediations, and thus the influence of the past medications on the findings may not be excluded. It is however, one of these two studies (Valko et al. 2013) also reported an increase in number of histamine neurons in the mouse model of hypocretin deficient (drug naïve) narcolepsy (i.e., orexin/hypocretin knock out mice), and thus the increases in number of histamine neurons in narcolepsy in human and mice are not likely to the influences of the previous medications. However, it is still unknown if the neurogenesis of histamine neurons specifically occur in the adult human brain or the histidine decarboxylase (HDC) induction are enhanced and this lead to the increased in HDC positive (i.e., histaminergic) neurons. It is also important to know if the histamine release from these neurons are intact or not, and this may be the key to explain the controversial results.

Rye et al. recently reported that activities of substance in CSF that augments inhibitory GABA signaling are enhanced in hypersomnia (Rye 2012). The authors demonstrated that in the presence of GABA (10 μM), CSF can stimulate GABAA receptor function in vitro (measures of GABAAR-mediated chloride currents in recombinant pentameric human GABAAR-expressed cultured cells). Interestingly, stimulations of GABAA receptor function by CSF from hypersomnolent patients (idiopathic hypersomnia with and without long sleep, long sleepers and narcolepsy without cataplexy) are significantly enhanced compared to those by CSF from control subjects (84.0 vs. 35.8 %) (Rye 2012). This bioactive CSF component had a mass of 500 to 3000 daltons and was neutralized by trypsin. Flumazenil, a benzodiazepine receptor antagonist, reversed enhancement of GABAA signaling by hypersomnolent CSF in vitro, and flumazenil normalized vigilance in all seven hypersomnolent patients who underwent the drug challenge (Rye 2012). The authors conclude that a naturally occurring substance in CSF augments inhibitory GABA signaling, revealing a new pathophysiology associated with EDS. These results are especially interesting, as GABAAR has never been targeted for the treatment of hypersomina. It is still unknown if these changes are primary or secondary to the changes in other neurotransmitter systems. In this regard, it is critical to test if the same change is observed in hypocretin-deficient narcolepsy-cataplexy.

These new findings are interesting as they are some of the first biomarkers for idiopathic hypersomnia, and these finding may lead to development of new treatments for somewhat treatment-resistant hypersomnia. However these markers do not discriminate the types of hypersomnia, and similar changes were observed in various types of hypersomnia.

14 Conclusion

Idiopathic narcolepsy-cataplexy (narcolepsy Type 1) is likely to be a clinical entity and be caused by postnatal selective cell death of hypocretin neurons. This is likely to be mediated by autoimmune process (i.e., autoantibody presentation by specific human leukocyte antigen subtypes and T cell receptors), but exact mechanism of the disease is not yet revealed.

Symptomatic narcolepsy has also been reported, but the prevalence of symptomatic narcolepsy is much smaller than the idiopathic type. The meta-analysis of these symptomatic cases indicates that hypocretin deficiency may also partially explain the neurobiological mechanisms of EDS associated with symptomatic cases of narcolepsy.

The pathophysiology of hypocretin non-deficient narcolepsy (narcolepsy Type 2) is debated, and the pathophysiology of idiopathic hypersomnia is largely unknown, but hypocretin deficiency is not likely to be involved in this condition. Altered histaminergic neurotransmission is observed in narcolepsy but the resuls are still cotrovercial. It is also still not known these changes are primary or secondary and of importance of the pethophysologuiical mechanis of narcolepsy (if these mediate sleepiness or passively reflect sleepiness). Another study reported that activities of substance in CSF that augments inhibitory GABA signaling are enhanced in hypersomnia with various etiologies. Functional significances of these new findings need further evaluation, since if new findings play significant roles in the pathopsyiology of hypersomnia, the findings likely to lead the development of new treatment.

Although much progress has been made regarding the pathophysiology of EDS, this new knowledge has yet to be incorporated into the development of new treatments, and further research is critical.

References

Alaila SL (1992) Life effects of narcolepsy: measures of negative impact, social support and psychological well-being. In: M Goswanmi et al. (eds) Loss, grief and care: psychosocial aspects of narcolepsy. Haworth Press, New York. p 1–22

Aldrich MS (1989) Automobile accidents in patients with sleep disorders. Sleep 12:487–494

Aldrich MS, Chervin RD, Malow BA (1997) Value of the multiple sleep latency test (MSLT) for the diagnosis of narcolepsy. Sleep 20(8):620–629

Aran A et al (2009) Elevated anti-streptococcal antibodies in patients with recent narcolepsy onset. Sleep 32(8):979–983

Aston-Jones G et al (2001) A neural circuit for circadian regulation of arousal. Nature Neurosci. 4(7):732–738

Baraitser M, Parkes JD (1978) Genetic study of narcoleptic syndrome. J Med Genet 15:254–259

Bassetti C, Aldrich MS (1997) Idiopathic hypersomnia. A series of 42 patients. Brain 120(Pt 8):1423–1435

Bassetti C et al (2003) The narcoleptic borderland: a multimodal diagnostic approach including cerebrospinal fluid levels of hypocretin-1 (orexin A). Sleep Med 4(1):7–12

Bassetti CL et al (2010) Cerebrospinal fluid histamine levels are decreased in patients with narcolepsy and excessive daytime sleepiness of other origin. J Sleep Res 19(4):620–623

Billiard M, Besset A, Cadilhac J (1983) The clinical and polygraphic development of narcolepsy. In: C Guilleminault, E Lugaresi (eds) Sleep/wake disorders: natural history, epidemiology and longterm evolution. Raven Press, New York, p 171–185

Billiard M et al (1994) Family studies in narcolepsy. Sleep 17: S-54–S-59

Björklund A, Lindvall O (1984) Dopamine-containing systems in the CNS. In: A Björklund, T Hökfelt (eds) Handbook of chemical neuroanatomy, vol 2, classical transmitter in the CNS, Part I. Elsevier, Amsterdam. pp 55–121

Broughton R, Q Ghanem (1976) The impact of compound narcolepsy on the life of the patient. In: C Guilleminault, WC Dement, P Passouant (eds) Narcolepsy. Spectrum, New York, p 201–220

Broughton R et al (1981) Life effects of narcolepsy in 180 patients from North America, Asia and Europe compared to matched controls. Le J Can Sci Neurol 8(4):299–303

Broughton R et al (1988) Ambulatory 24 hour sleep-wake monitoring in narcolepsy-cataplexy compared to matched control. Electroenceph Clin Neurophysiol 70:473–481

Brown RE, Stevens DR, Haas HL (2001) The physiology of brain histamine. Prog Neurobiol 63(6):637–672

Chetrit M et al (1994) Hypnogogic hallucinations associated with sleep onset REM period in narcolepsy-cataplexy. J Sleep Res 3(Suppl 1):43

Chemelli RM et al (1999) Narcolepsy in orexin knockout mice: molecular genetics of sleep regulation. Cell 98:437–451

Chokroverty S (1986) Sleep apnea in narcolepsy. Sleep 9(1):250–253

Clements CS et al (2003) The crystal structure of myelin oligodendrocyte glycoprotein, a key autoantigen in multiple sclerosis. Proc Natl Acad Sci USA 100(19):11059–11064

Cohen FL, Smith KM (1989) Learning and memory in narcoleptic patients and controls. Sleep Res 18:117

Crocker A et al (2005) Concomitant loss of dynorphin, NARP, and orexin in narcolepsy. Neurology 65(8):1184–1188

Cvetkovic-Lopes V et al (2010) Elevated Tribbles homolog 2-specific antibody levels in narcolepsy patients. J Clin Invest 120(3):713–719

Dahlitz M, Parkes JD (1993) Sleep paralysis. Lancet 341:406–407

Daniels LE (1934) Narcolepsy. Medicine 13(1):1–122

Dauvilliers Y et al (2001) Age at onset of narcolepsy in two large populations of patients in France and Quebec. Neurology 57(11):2029–2033

Dauvilliers Y et al (2010) Post-H1N1 narcolepsy-cataplexy. Sleep 33(11):1428–1430

Dauvilliers Y et al (2012) Normal cerebrospinal fluid histamine and tele-methylhistamine levels in hypersomnia conditions. Sleep 35(10):1359–1366

Dement WC (1976) Daytime sleepiness and sleep “attacks”. In: C Guilleminault WC Dement, P Passouant (eds) Narcolepsy, Spectrum Publications Inc, New York. p 17–42

Dement WC, Carskadon M, Ley R (1973) The prevalence of narcolepsy II. Sleep Res 2:147

De Lecea L et al (1998) The hypocretins: hypothalamus-specific peptides with neuroexcitatory activity. Proc Natl Acad Sci USA 95:322–327

Faraco J et al (2013) ImmunoChip study implicates antigen presentation to T cells in narcolepsy. PLoS Genet 9(2):e1003270

Fujiki N et al (2001) Changes in CSF hypocretin-1 (orexin A) levels in rats across 24 hours and in response to food deprivation. NeuroReport 12(5):993–997

Fujiki N et al (2002) Analysis of onset location, laterality and propagation of cataplexy in canine narcolepsy. Psychiatry Clin Neurosci 56(3):275–276

Fukuda K et al (1987) High prevalence of isolated sleep paralysis: kanashibari phenomenon in Japan. Sleep 10(3):279–286

Gelb M et al (1994) Stability of cataplexy over several months-information for the design of therapeutic trials. Sleep 17:265–273

Godbout R, Montplaisir J (1985) Comparison of sleep parameters in narcoleptics with and without periodic movements of sleep, in Sleep ‘84. In: WP Koella, E Ruther, H Schulz (eds) Fischer, Gustav, p 380–382

Goode B (1962) Sleep paralysis. Arch Neurol 6(3):228–234

Guilleminault C, Wilson RA, Dement WC (1974) A study on cataplexy. Arch Neurol 31:255–261

Guilleminault C (1987) Narcolepsy and its differential diagnosis, in Sleep and it disorders in children. In: C Guilleminault (ed) Raven Press, New York, p 181–194

Guilleminault C, Mignot E, Grumet FC (1989) Familial patterns of narcolepsy. Lancet 2(8676):1376–1379

Hallmayer J et al (2009) Narcolepsy is strongly associated with the T-cell receptor alpha locus. Nat Genet 41(6):708–711

Han F et al (2011) Narcolepsy onset is seasonal and increased following the 2009 H1N1 pandemic in China. Ann Neurol 70(3):410–417

Hara J et al (2001) Genetic ablation of orexin neurons in mice results in narcolepsy, hypophagia, and obesity. Neuron 30(2):345–354

Henneberg R (1916) Uber genuine narkolepsie. Neurol. Zbl. 30:282–290

Hirai N, Nishino S (2011) Recent advances in the treatment of narcolepsy. Curr Treat Options Neurol 13(5):437–457

Hishikawa Y (1976) Sleep paralysis. In: C Guilleminault, WC Dement, Passpouant P (eds). Spectrum, New York, pp 97–124

Hishikawa Y, Shimizu T (1995) Physiology of REM sleep, cataplexy, and sleep paralysis, in Negative Motor Phenomena. In: S Fahn et al. (ed). Lippincot-Raven Publishers, Philadelphia, p 245–271

Hishikawa Y et al (1976) Sleep satiation in narcoleptic patients. Electroencephalogr Clin Neurophysiol 41:1–18

Honda Y (1979) Census of narcolepsy, cataplexy and sleep life among teen-agers in Fujisawa city. Sleep Res 8:191

Honda Y (1988) Clinical features of narcolepsy. In: Y Honda, T Juji (eds) HLA in narcolepsy. Springer, Berlin. p 24–57

Honda Y et al. (1983) A genetic study of narcolepsy and excessive daytime sleepiness in 308 families with a narcolepsy or hypersomnia proband. In: C Guilleminault, E Lugaresi (eds) Sleep/wake disorders: natural history, epidemiology and long term evolution. Raven Press, New York

Honda Y et al (1986) Increased frequency of non-insulin-dependent diabetes mellitus among narcoleptic patients. Sleep 9(1):254–259

Honda M et al (2009) IGFBP3 colocalizes with and regulates hypocretin (orexin). PLoS ONE 4 (1):e4254

Hor H et al (2011) A missense mutation in myelin oligodendrocyte glycoprotein as a cause of familial narcolepsy with cataplexy. Am J Hum Genet 89(3):474–479
Hublin C et al (1994a) The prevalence of narcolepsy: an epidemiological study of the Finnish twin cohort. Ann Neurol 35:709–716
Hublin C et al (1994b) Epidemiology of narcolepsy. Sleep 17:S7–S12
ICSD-2 (ed) (2005) ICSD-2-International classification of sleep disorders, 2nd ed: diagnostic and coding manual. A.A.o.S. Medicine. American Academy of Sleep Medicine Westchester, Illinois
ICSD-3 (ed) (2014) international classification of sleep disorders, 3rd edn. A.A.o.S. Medicine. American Sleep Disorders Association: Rochester, MN
John J et al (2013) Greatly increased numbers of histamine cells in human narcolepsy with cataplexy. Ann Neurol 74(6):786–793
Jones BE et al (1973) The effect of lesions of catecholamine-containing neurons upon monoamine content of the brain and EEG and behavioral waking in the cat. Brain Res 58:157–177
Jones BE (2005) Basic mechanism of sleep-wake states. In: MH Kryger, T Roth, WC Dement (eds) Principles and practice of sleep medicine, 4th edn. Elsvier Saunders, Philadelphia, pp 136–153
Jouvet M (1972) The role of monoamines and acethylcholine-containing neurons in the regulation of the sleep-waking cycle. Ergebn Physiol 64:166–307
Kales A, Soldates CR, Bixler EO (1982) Narcolepsy-cataplexy II Psychosocial consequences and associated psychopathology. Arch Neurol 39:169–171
Kanbayashi T et al (2009) CSF histamine contents in narcolepsy, idiopathic hypersomnia and obstructive sleep apnea syndrome. Sleep 32(2):181–187
Kessler S, Guilleminault C, Dement WC (1979) A family study of 50 REM narcoleptics. Acta Neurol Scandinav 50:503–512
Kok SW et al (2003) Hypocretin deficiency in narcoleptic humans is associated with abdominal obesity. Obes Res 11(9):1147–1154
Kornum BR et al (2011) Common variants in P2RY11 are associated with narcolepsy. Nat Genet 43(1):66–71
Krabbe E, Magnussen G (1942) Familial narcolepsy. Acta Psychiatr Scand 17:149–173
Lammers GJ et al (1996) Spontaneous food choice in narcolepsy. Sleep 19(1):75–76
Lankford DA, Wellman JJ, O'Hara C (1994) Posttraumatic narcolepsy in mild to moderate closed head injury. Sleep 17:S25–S28
Lavie P, Peled R (1987) Narcolepsy is a rare disease in Isreal. Sleep 10(6):608–609
Lee MG, Hassani OK, Jones BE (2005) Discharge of identified orexin/hypocretin neurons across the sleep-waking cycle. J Neurosci 25(28):6716–6720
Lim AS, Scammell TE (2010) The trouble with Tribbles: do antibodies against TRIB2 cause narcolepsy? Sleep 33(7):857–858
Lin L et al (1999) The sleep disorder canine narcolepsy is caused by a mutation in the hypocretin (orexin) receptor 2 gene. Cell 98(3):365–376
Ljungberg T, Apicella P, Schultz W (1992) Responses of monkey dopamine neurons during learning of behavioral reactions. J Neurophysiol 67(1):145–163
Löwenfeld L (1902) Uber Narkolepsie. Munch Med Wochenschr 49:1041–1045
Lu J, Jhou TC, Saper CB (2006) Identification of wake-active dopaminergic neurons in the ventral periaqueductal gray matter. J Neurosci 26(1):193–202
Marcus JN et al (2001) Differential expression of orexin receptors 1 and 2 in the rat brain. J Comp Neurol 435(1):6–25
Mayer G et al (1993) Zur Einschätzung des Behinderungsgrades bei Narkolepsie. Gesundh-Wes 55:337–342
Mayer G et al (1997) Circadian temperature and activity rhythms in unmedicated narcoleptic patients. Pharmacol Biochem Behav 58(2):395–402
Mignot E (1998) Genetic and familial aspects of narcolepsy. Neurology 50(suppl 1):S16–S22
Mignot E et al (1997) HLA DQB1*0602 is associated with cataplexy in 509 narcoleptic patients. Sleep 20(11):1012–1020

Mignot E et al (2002) The role of cerebrospinal fluid hypocretin measurement in the diagnosis of narcolepsy and other hypersomnias. Arch Neurol 59(10):1553–1562

Miyagawa T et al (2008) Variant between CPT1B and CHKB associated with susceptibility to narcolepsy. Nat Genet 40(11):1324–1328

Montplaisir J et al (1978) Twenty-four-hour recording in REM-narcoleptics with special reference to nocturnal sleep disruption. Biol Psych 13(1):78–89

Mosko SS, Shampain DS, Sassin JF (1984) Nocturnal REM latency and sleep disturbance in narcolepsy. Sleep 7:115–125

Moller JC et al (2000) "Sleep attacks" in Parkinson patients. A side effect of nonergoline dopamine agonists or a class effect of dopamine agonists? Nervenarzt 71(8):670–676

Nevsimalova-Bruhova S, Roth B (1972) Heredofamilial aspects of narcolepsy and hypersomnia. Schweiz Arch Neurol Neurochir Psychiat 110:45–54

Nishino S. Modes of action of drugs related to narcolepsy: pharmacology of wake-promoting compounds and anticataplectics. In: M Goswami, SR Pandi-Perumal, MJ Thorpy (eds) (2009) Narcolepsy Humana, Totowa. p (in press)

Nishino S, Mignot E (1997) Pharmacological aspects of human and canine narcolepsy. Prog Neurobiol 52(1):27–78

Nishino S, Kanbayashi T (2005) Symptomatic narcolepsy, cataplexy and hypersomnia, and their implications in the hypothalamic hypocretin/orexin system. Sleep Med Rev 9(4):269–310

Nishino S, Kotorii N (2009) Overview of management of narcolepsy. In: M Goswami, SR Pandi-Perumal, MJ Thorpy (eds) Narcolepsy. Humana: Totowa. p (in press)

Nishino S, Mignot E (2011) Narcolepsy and cataplexy. Handb Clin Neurol 99:783–814

Nishino S et al (2001) Low CSF hypocretin (orexin) and altered energy homeostasis in human narcolepsy. Ann Neurol 50:381–388

Nishino S, E Mignot, WC Dement (2004a) Sedative-hypnotics, in textbook of psychopharmacology. In: AF Schatzberg, CB Nemeroff (eds). American Psychiatric Press, Washington DC, p 651–684

Nishino S et al. (2004b) The neurobiology of sleep in relation to mental illness. In: NE Charney DS (ed) Neurobiology of mental illness. Oxford University Press, New York. pp 1160–1179

Nishino S et al (2000) Hypocretin (orexin) deficiency in human narcolepsy. Lancet 355(9197):39–40

Nishino S et al (2009) Decreased CSF histamine in narcolepsy with and without low CSF hypocretin-1 in comparison to healthy controls. Sleep 32(2):175–180

Ohayon MM et al (1996) Hypnagogic and hypnopompic hallucinations: pathological phenomena? Br J Psychiatry 169(4):459–467

Ohayon MM et al (2002) Prevalence of narcolepsy symptomatology and diagnosis in the European general population. Neurology 58(12):1826–1833

Ondzé B et al (1999) Frequency of narcolepsy in the population of a French departement. Sleep 22 (Supplement):S121

Orellana C et al (1994) Life events in the year preceding the onset of narcolepsy. Sleep 17:S50–S53

Parkes JD et al (1974) Narcolepsy and Cataplexy. Clinical features, treatment and cerebrospinal fluid findings. Q J Med 172:525–536

Parkes JD et al (1975) Natural history, symptoms and treatment of the narcoleptic syndrome. Acta Neurol Scand 52:337–353

Passouant P, Baldy-Moulinier M, Assilloux C (1970) Eata de mal cataplectique au cours d'ume maladie de Gelineau, influence de la clomipramine. Rev Neurol 123:56–50

Peyron C et al (1998) Neurons containing hypocretin (orexin) project to multiple neuronal systems. J Neurosci 18(23):9996–10015

Peyron C et al (2000) A mutation in a case of early onset narcolepsy and a generalized absence of hypocretin peptides in human narcoleptic brains. Nat Med 6(9):991–997

Ribstein M (1976) Hypnagogic hallucinations. In: C Guilleminault, WC Dement, P Passouant (eds) Narcolepsy. Spectrum, New York. pp 145–160

Ripley B et al (2001) Hypocretin levels in sporadic and familial cases of canine narcolepsy. Neurobiol of Dis 8(3):525–534
Rogers AE, Rosenberg RS (1990) Test of memory in narcoleptics. Sleep 13:42–52
Rosenthal C (1939) Uber das aufreten von halluzinatorisch-kataplektischem angstsyndrom, wachanfallen und ahnlichen storungen bei schizophrenen. Mschr Psychiat 102:11
Rosenthal L et al (1990) Subjective and polysomnographic characteristics of patients diagnosed with narcolepsy. Gen Hosp Psychiatry 12:191–197
Roth B (1962) Narkolepsie und hypersomnie. Berlin, VEB Verlag Volk und Gesundheit. p 428
Roth B (1980) Narcolepsy and Hypersomnia. In: B Roth, W Broughton. Basel, Karger. p 310
Roth B, Nevsimalova S (1975) Depression in narcolepsy and hypersomnia. Schweitz Arch Neurol Neurochir Psychiat 116:291–300
Rye DB et al (2012) Modulation of vigilance in the primary hypersomnias by endogenous enhancement of GABAA receptors. Sci Transl Med 4(161):161ra151
Ryer EJ et al (2012) Revascularization for acute mesenteric ischemia. J Vasc Surg 55(6):1682–1689
Sachs C, Kaisjer L (1980) Autonomic control of cardiovascular reflexes in Narcolepsy. J Neurol Neurosurg Psychiatry 43:535–539
Schenck CH, Mahowald MW (1992) Motor Dyscontrol in narcolepsy: Rapid-Eye-Movement (REM) sleep without atonia and REM Sleep Behavior Disorder. Annals of Neurology 32(1):3–10
Schuld A et al (2000) Increased body-mass index in patients with narcolepsy. Lancet 355 (9211):1274–1275
Sakurai T (2002) Roles of orexins in regulation of feeding and wakefulness. NeuroReport 13 (8):987–995
Sakurai T et al (1998) Orexins and orexin receptors: a family of hypothalamic neuropeptides and G protein-coupled receptors that regulate feeding behavior. Cell 92:573–585
Saper CB, Chou TC, Scammell TE (2001) The sleep switch: hypothalamic control of sleep and wakefulness. Trends Neurosci 24(12):726–731
Saper CB, Scammell TE, Lu J (2005) Hypothalamic regulation of sleep and circadian rhythms. Nature 437(7063):1257–1263
Serra L et al (2008) Cataplexy features in childhood narcolepsy. Mov Disord 23(6):858–865
Silber MH et al (2002) The epidemiology of narcolepsy in Olmsted County, Minnesota: a population-based study. Sleep 25(2):197–202
Solomon P (1945) Narcolepsy in Negroes. Dis Nerv Sys 6:179–183
Soya A et al (2008) CSF histamine levels in rats reflect the central histamine neurotransmission. Neurosci Lett 430(3):224–229
Steinfels GF et al (1983) Behavioral correlates of dopaminergic activity in freely moving cats. Brain Res 258:217–228
Takeuchi T et al (1992) Isolated sleep paralysis elicited by sleep interruption. Sleep 15(3):217–225
Tashiro T, Kanbayashi T, Hishikawa Y (1994) An epidemiological study of narcolepsy in Japanese. In: The 4th international symposium on narcolepsy. Tokyo, Japan
Thannickal TC, Nienhuis R, Siegel JM (2009) Localized loss of hypocretin (orexin) cells in narcolepsy without cataplexy. Sleep 32(8):993–998
Trotti LM, Staab BA, Rye DB (2013) Test-retest reliability of the multiple sleep latency test in narcolepsy without cataplexy and idiopathic hypersomnia. J Clin Sleep Med 9(8):789–795
Trulson ME (1985) Simultaneous recording of substantia nigra neurons and voltametric release of dopamine in the caudate of behaving cats. Brain Res Bull 15:221–223
Valko PO et al (2013) Increase of histaminergic tuberomammillary neurons in narcolepsy. Ann Neurol 74(6):794–804
Verma A, Anand V, Verma NP (2007) Sleep disorders in chronic traumatic brain injury. J Clin Sleep Med 3(4):357–362
Westphal C (1877) Eigenthümliche mit Einschläfen verbundene Anfälle. Arch Psychiat 7:631–635
Willie JT et al (2001) To eat or to sleep? Orexin in the regulation of feeding and wakefulness. Annu Rev Neurosci 24:429–458
Wilson SAK (1927) The narcolepsies. Ann Congress Assoc Phys June 3 p 63–109

Yoss RE, Daly DD (1957) Criteria for the diagnosis of the narcoleptic syndrome. Proc Staff Meet Mayo 32:320–328

Yoshida Y et al (2001) Fluctuation of extracellular hypocretin-1 (orexin A) levels in the rat in relation to the light-dark cycle and sleep-wake activities. Eur J Neurosci 14(7):1075–1081

Yoss RE, Daly DD (1960) Narcolepsy. Med Clin North Am 44(4):953–967

Zarocostas J (2011) WHO backs further probes into possible link between H1N1 vaccine and narcolepsy in children. BMJ 342:d909

Hypocretin/Orexin Pathology in Human Narcolepsy with and Without Cataplexy

Thomas C. Thannickal and Jerome M. Siegel

Abstract Hypocretin (Hcrt, also called orexin) neurons have been implicated in the pathology underlying narcolepsy. The number of Hcrt cells in normal humans ranges from 51,000–83,000. Human narcolepsy is correlated with a greatly reduced number of Hcrt containing neurons and axons, and an elevated level of hypothalamic gliosis. Narcolepsy with cataplexy is characterized by a loss of approximately 90 % of Hcrt neurons. However, more than a quarter of narcoleptics do not have cataplexy and have normal levels of Hcrt in their cerebrospinal fluid. Narcolepsy without cataplexy has an overall a loss of 33 % of Hcrt cells compared to normal, with maximal cell loss in the posterior hypothalamus. A better understanding of the pattern of damage to Hcrt containing somas and axons and of the gliosis occurring in narcolepsy should clarify the nature of the pathological process responsible for this disorder.

Keywords Hypocretin • Orexin • Narcolepsy • Cataplexy • Neurodegeneration

1 Introduction

As often happens in medical sciences, it is research on the possible treatment for another disease, obesity, which led to the discovery of two peptides expressed in the hypothalamus and named "hypocretins" or "orexins". The discovery of these peptides and of their receptors opened the door to the most current understanding of narcolepsy (de Lecea et al. 1998; Sakurai et al. 1988). In 1999, the positional

T.C. Thannickal (✉) · J.M. Siegel
Department of Psychiatry and Biobehavioral Sciences, University of California
Los Angeles, Los Angeles, CA 90095, USA
e-mail: thomastc@ucla.edu

T.C. Thannickal · J.M. Siegel
Veterans Administration Greater Los Angeles Healthcare System,
Neurobiology Research (151A3), North Hills, CA 91343, USA

© Springer International Publishing Switzerland 2015

T. Sakurai et al. (eds.), *Orexin and Sleep*,
DOI 10.1007/978-3-319-23078-8_15

cloning of the canine narcolepsy gene and its identification as the Hcrt receptor 2 gene was another milestone in the field (Lin et al. 1999). A mouse knockout model for the Hcrt gene was also found to display narcolepsy–like symptoms (Chemelli et al. 1999). In 2000, these discoveries were followed by the report that most cases of human narcolepsy-cataplexy are associated with hypocretin deficiency (Nishino et al. 2000; Peyron et al. 2000; Thannickal et al. 2000).

The hypocretins (Hcrt) are two peptides, Hcrt-1 (orexin-A) and Hcrt-2 (orexin-B), generated from a single preprohypocretin molecule and synthesized by a small number of neurons restricted to the lateral, dorsomedial and perifornical hypothalamus (de Lecea et al. 1998; Sakurai et al. 1988). In contrast, Hcrt axons are found throughout the CNS, with innervations of the hypothalamus, locus coeruleus, raphe, midline thalamus, all levels of spinal cord, sympathetic and parasympathetic centers, and many other brain regions (Peyron et al. 1998). Two G protein–coupled receptors that respond to the hypocretins have been identified (Sakurai et al. 1988). In parallel to the wide distribution of axons, the two Hcrt receptors show a widespread and heterogeneous pattern of expression throughout the CNS (Trivedi et al. 1998).

2 Narcolepsy and Hypocretin System

Human narcolepsy is a lifelong sleep disorder characterized by excessive daytime sleepiness, disrupted nocturnal sleep, rapid eye movement (REM) sleep occurring at the onset of sleep, and cataplexy. The presence of cataplexy is distinctively characteristic for narcolepsy. Narcolepsy affects approximately 1 in 2000 individuals in the United States. Importantly, however, the prevalence of the milder form of narcolepsy, narcolepsy without cataplexy, could be substantially higher.

The potential importance of Hcrt neurons in preventing narcolepsy was suggested by the finding that narcoleptic dogs have mutations in Hcrt receptor 2 (Lin et al. 1999). Although different mutations were found in each of two narcoleptic breeds of dogs, Dobermans and Labradors, in each breed the mutation was localized to the Hcrt receptor 2, rendering it non-functional. Parallel studies in Hcrt knockout mice revealed a similar narcoleptic phenotype (Sakurai et al. 1988) indicating that loss of either the peptide or one of the two peptide receptors results in narcolepsy. These findings became even more exciting and relevant when hypocretin was found in the cerebrospinal fluid of eight normal humans but could not be detected in seven of nine narcoleptics (Nishino et al. 2000).

3 Hypocretin Cell Loss in Narcolepsy with Cataplexy

Hcrt immunoreactive cells were found in anterior (AH), dorsal (DH), dorsomedial (DMH), posterior (PH) and lateral (LH) hypothalamic nuclei of both normal and narcoleptic human brains. (Figs. 1, 2 and 3). Cell number was significantly reduced in narcoleptics compared to normals. The percentage cell loss was significantly more severe in certain nuclei (Fig. 3). In the normal brain, Hcrt cell density was highest in DMH and lowest in AH. In narcoleptics, the maximum percentage loss of Hcrt cells occurred in the posterior hypothalamus (97 %). The minimum percent loss of Hcrt cells was seen in AH (74 %). Overall, narcoleptics had a mean 90 % reduction in Hcrt cell number compared to the average number seen in normals (Thannickal et al. 2003).

The axon density reduction in Hcrt innervated nuclei in narcoleptics was positively correlated with the density of Hcrt axons in normal humans. The total number of Hcrt axons in all the structures analyzed was reduced by 67 % compared to an 89 % reduction of Hcrt soma count in the same brains, suggesting that Hcrt cells

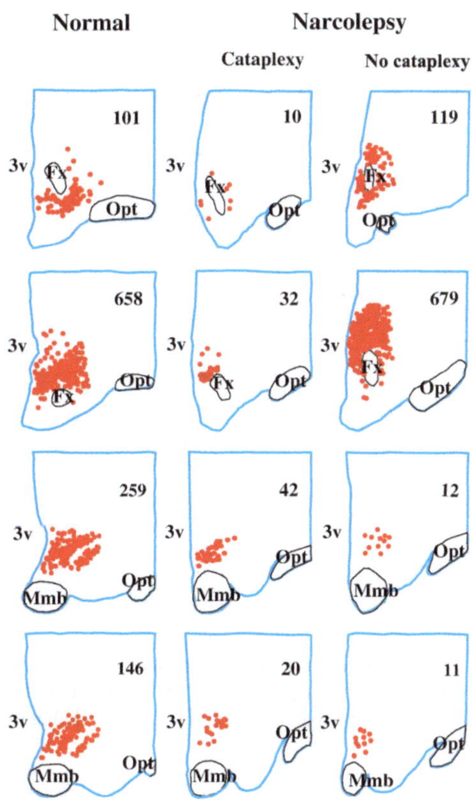

Fig. 1 Neurolucida mapping of Hcrt cells in normal and narcolepsy. The cell counts are listed in each section. *3v* third ventricle, *Fx* fornix, *Mmb* mammillary body, *Opt* optic tract

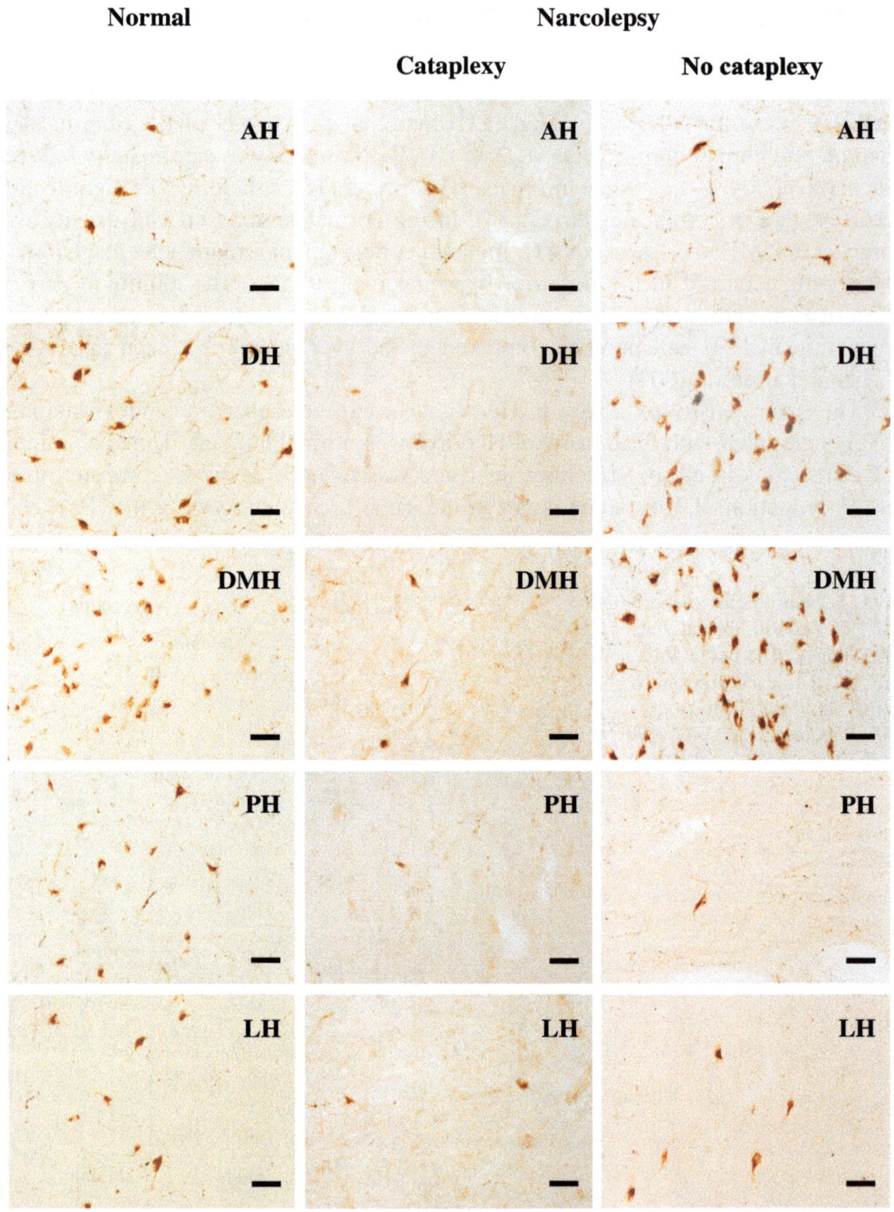

Fig. 2 Hcrt cells in the hypothalamic nuclei of normal and narcolepsy with and without cataplexy. *AH* anterior hypothalamus, *DH* dorsal hypothalamus, *DMH* dorsomedial hypothalamus, *PH* posterior hypothalamus, *LH* lateral hypothalamus, Scale bar–50 μm

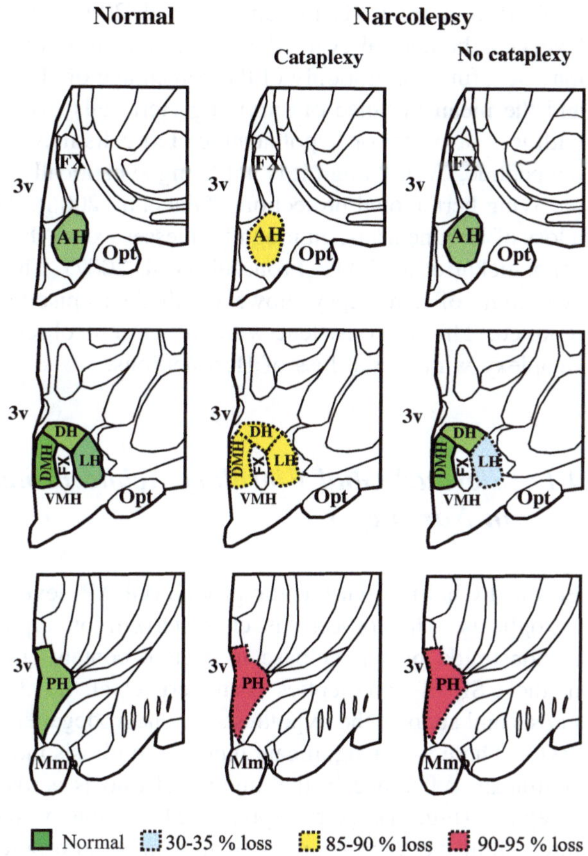

Fig. 3 Diagrammatic representation of the location of hcrt cells in the hypothalmus of normal brain and differential loss of Hcrt cells in narcolepsy with and without cataplexy. In normal, Hcrt cell somas are localized in AH, DH, DMH, PH and LH nuclei. In narcolepsy with cataplexy, cell loss was found in AH, DH, DMH, PH and LH nuclei, whereas, in narcolepsy without cataplexy cell loss was limited to LH and PH nuclei. *AH* anterior hypothalamus, *DH* dorsal hypothalamus, *DMH* dorsomedial hypothalamus, *PH* posterior hypothalamus, *LH* lateral hypothalamus, *3v* third ventricle, *Fx* fornix, *Mmb* mammillary body, *Opt* optic tract, *VMH* ventromedial hypothalamus

with smaller axonal fields are lost to a greater extent in narcolepsy or that axonal sprouting occures in surviving Hcrt cells. (Thannickal et al. 2003). 99 % of Hcrt neurons are Narp positive (Reti et al. 2002).The number of Narp-positive neurons was reduced by 89 % in these areas of the narcoleptic hypothalamus (Blouin et al. 2005; Crocker et al. 2005).

4 Hypocretin Cell Loss in Narcolepsy Without Cataplexy

The narcolepsy without cataplexy patient whose complete brain was available for study had an overall loss of 33 % of Hcrt cells compared to normals, with maximal cell loss in the posterior hypothalamus. The maximum percentage reduction occurred in the posterior hypothalamic nucleus (95 % loss). There was no Hcrt cell loss in anterior, dorsal and dorsomedial nuclei of the narcolepsy without cataplexy patient (Figs. 1, 2 and 3). There was no reduction in the number of MCH neurons in either type of narcolepsy (Thannickal et al. 2009).

Work in animals (Gerashchenko et al. 2003) has shown that cerebrospinal Hcrt levels can be normal even when there is a substantial loss of Hcrt cells. CSF levels may be a function not only of the percentage of Hcrt cells lost, but also the activity and the mean distance of surviving Hcrt cells from the ventricles. Thus, patients with loss of posterior hypothalamic Hcrt cells may still have normal CSF levels of the peptide, even though it is not being synaptically delivered to the cells normally receiving hcrt axonal projections (Oka et al. 2006). Parkinson disease patients have a loss of Hcrt neurons, although to a lesser extent than in narcolepsy with cataplexy (Thannickal et al. 2007; Fronczek et al. 2007). These patients have many of the symptoms of narcolepsy; however, distinct episodes of cataplexy have not been reported. This is consistent with the current observations of narcolepsy without cataplexy with partial loss of Hcrt neurons.

4.1 Elevated Glial Fibrillary Acidic Protein Levels in Narcolepsy

Elevated Glial fibrillary acidic protein (GFAP) levels is an established indicator of astrogliosis. This process is characterized by rapid synthesis of GFAP and is demonstrated by an increase in protein content (Eng et al. 2000). GFAP levels in the CSF of narcoleptics (Feneberg et al. 2013) may represent hypothalamic gliosis and support the hypothesis of a neurodegenerative process (Thannickal et al. 2000). There was a significant increase in gliosis indicated by GFAP staining and a significant difference in the amount of gliosis across hypothalamic nuclei in narcoleptics (Fig. 4). Narcoleptics with cataplexy had increased GFAP staining throughout the hypothalamus, with a maximum percentage GFAP increase in the posterior hypothalamic nucleus. The number of Hcrt axons in the anterior hypothalamus was normal in narcolepsy without cataplexy (Thannickal et al. 2009). We speculate that high GFAP levels may be a differential biomarker in some sleep diseases such as secondary hypersomnia in Prader–Willi syndrome or neurologic

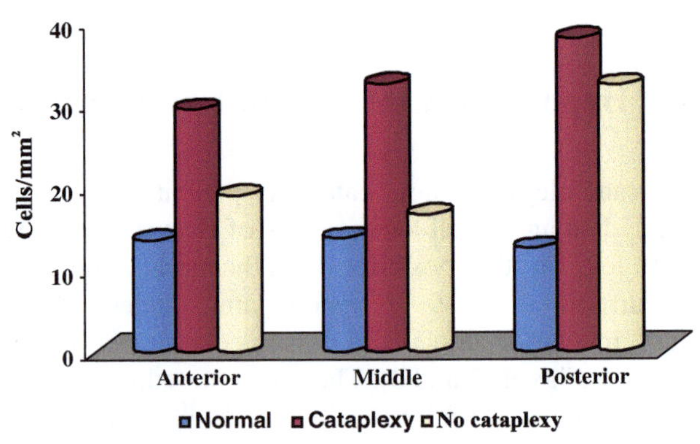

Fig. 4 Gliosis in narcolepsy. Glial fibrillary acidic protein labeled astrocytes (GFAP) density (cells/mm^2) in normal and narcoleptic brains of the hypothalamus

disorders such as in certain cases of Guillain–Barré syndrome, both of which also may present with low CSF hcrt-1 levels (Mignot et al. 2002). In conclusion, GFAP may be useful as an additional disease biomarker in patients with narcolepsy.

4.2 Changes in Hypocretin Neuronal Expression with Normal Aging in the Human Hypothalamus

It has been found that Hcrt concentration in the cerebral spinal fluid (CSF) of infants increased from birth until 2–4 postnatal months and then decreases throughout childhood and puberty (Feneberg et al. 2013). However, Aran et al. (2012) found no change in CSF Hcrt concentration between birth and 4 years of age. In older humans, the number of Hcrt neurons was unchanged between 50 and 90 of age (Kanbayashi et al. 2002; Fronczek et al. 2005). There was 24 % decrease in the number of Hcrt neurons in adults compared with infants and children (Fronczek et al. 2012). This may contribute to changes in sleep regulation during development and with aging. Animal studies have shown that decreased Hcrt expression is correlated with changes in sleep regulation with aging (Hunt et al. 2015; Brownell and Conti 2010).

4.3 Molecules Co-expressed in Hypocretin Neurons

Hcrt deficiency is directly involved in several neurological disorders. Dynorphin, glutamate and secretogranin II are found co-localized within Hcrt cells (Sawai et al. 2010; Chou et al. 2001; Bayer et al. 2002). There are several receptors and transporters that are expressed in Hcrt cells such as 5HT1-A, adenosine A1-R, GABA A alpha 3, GABA A epsilon, GABA B, group III metabotropic glutamate receptors, leptin R, Y4-R, vGlut1 and vGlut2 (Torrealba et al. 2003; Collin et al. 2002; Thakkar et al. 2002; Bäckberg et al. 2003, 2004; Moragues et al. 2003; Acuna-Goycolea et al. 2004; Håkansson et al. 1999; Rosin et al. 2003; Campbell et al. 2003). Acetylcholinesterase E, STAT-3, Narp and neuroglobin (Chou et al. 2004; Reti et al. 2002; Håkansson et al. 1999; Hundahl et al. 2008) are also found in Hcrt cells. Hcrt cell loss also causes the deficiency of these molecules. Their role in the symptoms of narcolepsy is not known.

5 Conclusions

The identification of hypocretin/orexin deficiency as the cause of human narcolepsy and the potential role of hypocretin peptides in other neurological disorders has sparked interest in the pathophysiology of the Hcrt system. As narcolepsy is found in about 1 of 2000 humans, it is the third most prevalent type of neurodegenerative disease, behind Alzheimer's (the most common) and Parkinson's (which affects about 1 in 1000 humans) (Dorsey et al. 2005), but more prevalent than Huntington's or amyotropic lateral sclerosis (each about 1 in 5000).

Acknowledgements Tissue specimens were obtained from the Human Brain and Spinal Fluid Resource Center, West Los Angeles Healthcare Center, Los Angeles, CA. This study was supported by the Medical Research Service of the Department of Veterans Affairs and NIH grants NS14610 and MH064109 to J.M.S.

References

Acuna-Goycolea C, Li Y, Van Den Pol AN (2004) Group III metabotropic glutamate receptors maintain tonic inhibition of excitatory synaptic input to hypocretin/orexin neurons. J Neurosci 24:3013–3022

Aran A, Shors I, Lin L, Mignot E, Schimmel MS (2012) CSF levels of hypocretin-1 (orexin-A) peak during early infancy in humans. Sleep 35:187–191

Bäckberg M, Collin M, Ovesjö ML, Meister B (2003) Chemical coding of GABA(B) receptor-immunoreactive neurones in hypothalamic regions regulating body weight. J Neuroendocrinol 15:1–14

Bäckberg M, Ultenius C, Fritschy JM, Meister B (2004) Cellular localization of GABA receptor alpha subunit immunoreactivity in the rat hypothalamus: relationship with neurones containing orexigenic or anorexigenic peptides. J Neuroendocrinol 16:589–604

Bayer L, Mairet-Coello G, Risold PY, Griffond B (2002) Orexin/hypocretin neurons:chemical phenotype and possible interactions with melanin-concentrating hormone neurons. Regul Pept 104:33–39

Blouin AM, Thannickal TC, Worley PF, Baraban JM, Reti IM, Siegel JM (2005) Narp immunostaining of human hypocretin (orexin) neurons: loss in narcolepsy. Neurology 65:1189–1192

Brownell SE, Conti B (2010) Age- and gender-specific changes of hypocretin immunopositive neurons in C57Bl/6 mice. Neurosci Lett 472:29–32

Campbell RE, Smith MS, Allen SE, Grayson BE, Ffrench-Mullen JM, Grove KL (2003) Orexin neurons express a functional pancreatic polypeptide Y4 receptor. J Neurosci 23:1487–1497

Chemelli RM, Willie JT, Sinton CM, Elmquist JK, Scammell T, Lee C, Richardson JA, Williams SC, Xiong Y, Kisanuki Y, Fitch TE, Nakazato M, Hammer RE, Saper CB, Yanagisawa M (1999) Narcolepsy in orexin knockout mice: molecular genetics of sleep regulation. Cell 98:437–451

Chou TC, Lee CE, Lu J, Elmquist JK, Hara J, Willie JT, Beuckmann CT, Chemelli RM, Sakurai T, Yanagisawa M, Saper CB, Scammell TE (2001) Orexin (hypocretin) neurons contain dynorphin. J Neurosci 21: RC168

Chou TC, Rotman SR, Saper CB (2004) Lateral hypothalamic acetylcholinesterase-immunoreactive neurons co-express either orexin or melanin concentrating hormone. Neurosci Lett 370:123–126

Collin M, Bäckberg M, Onnestam K, Meister B (2002) 5-HT1A receptor immunoreactivity in hypothalamic neurons involved in body weight control. NeuroReport 13:945–951

Crocker A, España RA, Papadopoulou M, Saper CB, Faraco J, Sakurai T, Honda M, Mignot E, Scammell TE (2005) Concomitant loss of dynorphin, NARP, and orexin in narcolepsy. Neurology 65:1184–1188

de Lecea L, Kilduff TS, Peyron C, Gao X, Foye PE, Danielson PE, Fukuhara C, Battenberg EL, Gautvik VT, Bartlett FS 2nd, Frankel WN, van den Pol AN, Bloom FE, Gautvik KM,

Sutcliffe JG (1998) The hypocretins: hypothalamus-specific peptides with neuroexcitatory activity. Proc Natl Acad Sci 95:322–327
Dorsey ER, Constantinescu R, Thompson JP, Biglan KM, Holloway RG, Kieburtz K, Marshall FJ, Ravina BM, Schifitto G, Siderowf A, Tanner CM (2005) Projected number of people with Parkinson disease in the most populous nations, 2005 through 2030. Neurology 68:384–386
Eng LF, Ghirnikar RS, Lee YL (2000) Glial fibrillary acidic protein: GFAP-thirty-one years (1969–2000). Neurochem Res 25:1439–1451
Feneberg E, Steinacker P, Lehnert S, Böhm B, Mayer G, Otto M (2013) Elevated glial fibrillary acidic protein levels in the cerebrospinal fluid of patients with narcolepsy. Sleep Med 14:692–694
Fronczek R, Lammers GJ, Balesar R, Unmehopa UA, Swaab DF (2005) The number of hypothalamic hypocretin (orexin) neurons is not affected in Prader-Willi syndrome. J Clin Endocrinol Metab 2005(90):5466–5470
Fronczek R, Overeem S, Lee SY, Hegeman IM, van Pelt J, van Duinen SG, Lammers GJ, Swaab DF (2007) Hypocretin (orexin) loss in Parkinson's disease. Brain 130:1577–1585
Fronczek R, Van Geest S, Frölich M, Overeem S, Roelandse FWC, Lammers GJ, Swaab DF (2012) Hypocretin (orexin) loss in Alzheimer's disease. Neurobiol Aging 33:1642–1650
Gerashchenko D, Murillo-Rodriguez E, Lin L, Xu M, Hallett L, Nishino S, Mignot E, Shiromani PJ (2003) Relationship between CSF hypocretin levels and hypocretin neuronal loss. Exp Neurol 84:1010–1016
Håkansson M, de Lecea L, Sutcliffe JG, Yanagisawa M, Meister B (1999) Leptin receptor- and STAT3-immunoreactivities in hypocretin/orexin neurones of the lateral hypothalamus. J Neuroendocrinol 11:653–663
Hundahl CA, Kelsen J, Dewilde S, Hay-Schmidt A (2008) Neuroglobin in the rat brain (II): co-localisation with neurotransmitters. Neuroendocrinology 88:183–198
Hunt NJ, Rodriguez ML, Waters KA, Machaalani R (2015) Changes in orexin (hypocretin) neuronal expression with normal aging in the human hypothalamus. Neurobiol Aging 36:292–300
Kanbayashi T, Yano T, Ishiguro H, Kawanishi K, Chiba S, Aizawa R, Sawaishi Y, Hirota K, Nishino S, Shimizu T (2002) Hypocretin-1 (orexin-A) levels in human lumbar CSF in different age groups: infants to elderly persons. Sleep 25:337–339
Lin L, Faraco J, Li R, Kadotani H, Rogers W, Lin X, Qiu X, de Jong PJ, Nishino S, Mignot E (1999) The sleep disorder canine narcolepsy is caused by a mutation in the hypocretin (orexin) receptor 2 gene. Cell 98:365–376
Mignot E, Lammers GJ, Ripley B, Okun M, Nevsimalova S, Overeem S, Vankova J, Black J, Harsh J, Bassetti C, Schrader H, Nishino S (2002) The role of cerebrospinal fluid hypocretin measurement in the diagnosis of narcolepsy and other hypersomnias. Arch Neurol 59:1553–1562
Moragues N, Ciofi P, Lafon P, Tramu G, Garret M (2003) GABAA receptor epsilonsubunit expression in identified peptidergic neurons of the rat hypothalamus. Brain Res 967:285–289
Nishino S, Ripley B, Overeem S, Lammers GJ, Mignot E (2000) Hypocretin (orexin) deficiency in human narcolepsy. Lancet 355:39–40
Oka Y, Inoue Y, Kanbayashi T, Kuroda K, Miyamoto M, Miyamoto T, Ikeda A, Shimizu T, Hishikawa Y, Shibasaki H (2006) Narcolepsy without cataplexy: 2 subtypes based on CSF hypocretin-1/orexin-A findings. Sleep 29:1439–1443
Peyron C, Tighe DK, van den Pol AN, de Lecea L, Heller HC, Sutcliffe JG, Kilduff TS (1998) Neurons containing hypocretin (orexin) project to multiple neuronal systems. J Neurosci 18:9996–10015
Peyron C, Faraco J, Rogers W, Ripley B, Overeem S, Charnay Y, Nevsimalova S, Aldrich M, Reynolds D, Albin R, Li R, Hungs M, Pedrazzoli M, Padigaru M, Kucherlapati M, Fan J, Maki R, Lammers GJ, Bouras C, Kucherlapati R, Nishino S, Mignot E (2000) A mutation in a case of early onset narcolepsy and a generalized absence of hypocretin peptides in human narcoleptic brains. Nat Med 9:991–997

Reti IM, Reddy R, Worley PF, Baraban JM (2002) Selective expression of Narp, a secreted neuronal pentraxin, in orexin neurons. J Neurochem 82:1561–1565

Rosin DL, Weston MC, Sevigny CP, Stornetta RL, Guyenet PG (2003) Hypothalamic orexin (hypocretin) neurons express vesicular glutamate transporters VGLUT1 or VGLUT2. J Comp Neurol 465:593–603

Sakurai T, Amemiya A, Ishii M, Matsuzaki I, Chemelli RM, Tanaka H, Williams SC, Richardson JA, Kozlowski GP, Wilson S, Arch JR, Buckingham RE, Haynes AC, Carr SA, Annan RS, McNulty DE, Liu WS, Terrett JA, Elshourbagy NA, Bergsma DJ, Yanagisawa M (1988) Orexins and orexin receptors: a family of hypothalamic neuropeptides and G protein-coupled receptors that regulate feeding behavior. Cell 92:573–585

Sawai N, Ueta Y, Nakazato M, Ozawa H (2010) Developmental and aging change of orexin-A and -B immunoreactive neurons in the male rat hypothalamus. NeurosciLett 468:51–55

Thakkar MM, Winston S, McCarley RW (2002) Orexin neurons of the hypothalamus express adenosine A1 receptors. Brain Res 944:190–194

Thannickal TC, Moore RY, Nienhuis R, Ramanathan L, Gulyani S, Aldrich M, Cornford M, Siegel JM (2000) Reduced number of hypocretin neurons in human narcolepsy. Neuron 27:469–474

Thannickal TC, Siegel JM, Nienhuis R, Moore RY (2003) Pattern of hypocretin (orexin) soma and axon loss, and gliosis, in human narcolepsy. Brain Pathol 13:340–351

Thannickal TC, Lai YY, Siegel JM (2007) Hypocretin (orexin) cell loss in Parkinson's disease. Brain 130:1586–1595

Thannickal TC, Nienhuis R, Siegel JM (2009) Localized loss of hypocretin (orexin) cells in narcolepsy without cataplexy. Sleep 32:993–998

Torrealba F, Yanagisawa M, Saper CB (2003) Colocalization of orexin a and glutamate immunoreactivity in axon terminals in the tuberomammillary nucleus in rats. Neuroscience 119:1033–1044

Trivedi P, Yu H, MacNeil DJ, Van der Ploeg LH, Guan XM (1998) Distribution of orexin receptor mRNA in the rat brain. FEBS Lett 438:71–75

Orexin and Circadian Influences in Sleep and Psychiatric Disorders: A Review of Experimental and Computational Modelling Studies

Alok Joshi, Mino D.C. Belle, KongFatt Wong-Lin and Hugh D. Piggins

Abstract Psychiatric disorders such as unipolar depression have complex pathologies, which include disruptions in circadian and sleep-wake cycles. At the neurochemical level, psychiatric diseases can also be accompanied by changes in neuromodulator systems such as orexin/hypocretin and the monoamines. Indeed, for decades the monoamine hypothesis of depression has been instrumental in driving discoveries and developments of antidepressant drugs. Recent preclinical and clinical advancement strongly suggests that neuropeptides such as orexin can play an important part in the pathophysiology of depression. Due to the complexity and extensive connectedness of neurobiological systems, understanding the biological causes and mechanisms of psychiatric disorders present major research challenges. In this chapter, we review experimental and computational studies investigating the complex relationship between orexinergic, monoaminergic, circadian oscillators, and sleep-wake neural circuitry. Our main aim is to understand how these physiological systems interact and how alteration in any of these factors can contribute to the behaviours commonly observed in depressive patients. Further, we examine how modelling across different levels of neurobiological organization enables insight into these interactions. We propose that a multiscale systems approach is necessary to understand the complex neurobiological systems whose dysfunctions are the underlying causes of psychiatric disorders. Such an approach could illuminate future treatments.

A. Joshi (✉) · M.D.C. Belle · H.D. Piggins
Faculty of Life Sciences, University of Manchester, Manchester, M13 9PT, UK
e-mail: alok.joshi@manchester.ac.uk

K. Wong-Lin
Intelligent Systems Research Centre, University of Ulster, Magee Campus, Derry, BT48 7JL, Northern Ireland, UK

© Springer International Publishing Switzerland 2015
T. Sakurai et al. (eds.), *Orexin and Sleep*,
DOI 10.1007/978-3-319-23078-8_16

Keywords Psychiatric disorders · Sleep · Neuromodulators · Orexin (hypocretin) · Monoamines · Circadian rhythms · Computational models

1 Introduction

Psychiatric disorders such as depression, schizophrenia, and bipolar disorder are multifaceted illnesses that present our society with significant challenges and burdens. In 2010, the Global Burden of Disease Study has rated mental and substance use disorders as the principal causes of global health burden (Whiteford et al. 2013). The trend is likely to continue and by 2030, it is estimated that among brain diseases, unipolar depressive disorder will be the largest cause of disability (Mathers and Loncar 2006). Such forecasts demonstrate the perniciousness of these disorders and also raise questions concerning current treatment strategies.

Many factors including neuromodulators and circadian rhythms as well as the patterns and quality of sleep contribute to the pathophysiology of psychiatric disorders (Germain and Kupfer 2008; Pandi-Perumal et al. 2009; Hasler 2010; Murray and Harvey 2010; Wulff et al. 2012). Individually, these factors have been discussed and reviewed extensively (Delgado 2000; Tsuno et al. 2005; Germain and Kupfer 2008; Murray and Harvey 2010; Wulff et al. 2012), while their potential interactions have received relatively little attention (Fig. 1). A comprehensive consideration of these factors and their interactions for psychiatric disorders is beyond the scope of this review. Therefore, the focus of this chapter will be on understanding how monoamines, orexin (also called hypocretin), sleep, and circadian oscillator(s) mutually interact, and how the disruption in one or more of these factors might contribute to the certain behavioural phenotypes that are altered in unipolar depressive disorder. In some cases, the relationships with other psychiatric disorders (schizophrenia or bipolar disorder) are also briefly discussed. While considering these interactions, emphasis is placed on the circadian system, especially the way it regulates the activity of specific neuromodulators and behavioural states, and how its disruption influences these interactions to potentially contribute to specific symptoms such as mood and sleep problems commonly observed in psychiatric disorders.

The organization of the chapter is as follows. The role of monoamines and orexin in psychiatric disorders are first discussed, and then the role of orexin along with other neuromodulators in the regulation of the sleep-wake cycle will be discussed. This is followed by a discussion of the regulatory functions of the main circadian pacemaker in the suprachiasmatic nuclei (SCN) which coordinates and drives daily activity in other semi-autonomous oscillators in the brain (such as lateral habenula, LHb) and controls circadian timing in some neuromodulatory systems (orexin, monoamines) and sleep. Then a computational modelling perspective will be presented. Finally, there will be a discussion on key unanswered questions and identification of future research directions.

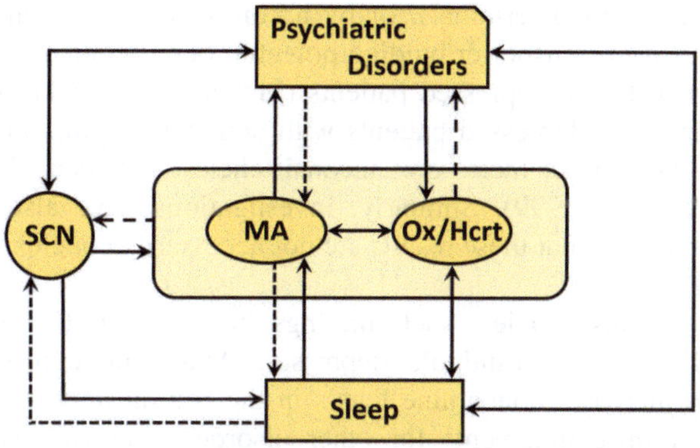

Fig. 1 Complex interactions of the master circadian oscillator in the suprachiasmatic nuclei (*SCN*), neuromodulators (e.g. orexin/hypocretin Ox/Hcrt; monoamines *MA*), sleep and psychiatric disorders. *Circle:* master circadian oscillator; *Oval:* neuromodulators; *Snip single corner rectangle:* psychiatric disorders; *Rectangle:* sleep; *Solid arrows:* known interactions; *Dashed arrows:* tentative interactions

2 Neuromodulators and Psychiatric Disorders

Results from many studies have indicated that altered circadian rhythms, disrupted sleep and abnormal basal levels of neuromodulators are associated with the clinical status of patients experiencing depression (Wehr et al. 1983; Ruhe et al. 2007; Wirz-Justice 2008; Germain and Kupfer 2008). Many of these characteristics are captured by the monoamine imbalance hypothesis which posits that perturbation in the basal levels of monoamines such as serotonin (5-HT), norepinephrine/noradrenaline (NE/NA), and dopamine (DA) can trigger depression (Nemeroff 1998). These monoamines are mainly synthesized at different brain regions. For example, 5-HT is synthesized in the raphe nuclei, NE in the locus coeruleus (LC), and DA in the substantia nigra and ventral tegmental area (VTA) (Grzanna and Molliver 1980; Park et al. 1999; Anzalone et al. 2012). These monoamine producing neurons project to many parts of the brain and are involved in the regulation of a wide range of behaviours (Leibowitz and Shor-Posner 1986; Cools et al. 2008; Aston-Jones et al. 2000; Bromberg-Martin et al. 2010). To understand how alteration in monoaminergic levels influences depression, many studies have focused on the expression and functionality of the monoaminergic receptors and transporters at the source and projection sites (Klimek et al. 1997; Rajkowska 2000; Arango et al. 2001; Boldrini et al. 2005). For example, depressed subjects have decreased 5-HT transporter, altered 5-$HT_{1A/2A}$ binding sites, and lower level of plasma tryptophan and metabolites (Quintana 1992; Owens and Nemeroff 1994; Malison et al. 1998; Drevets et al. 1999; Mintun et al. 2004). In the case of NE, altered adrenoreceptor (α_2) densities and lower NE levels have been observed in brain sites innervated by NE neurons, including the cortex (Ordway et al. 2003; Valdizán et al. 2010; Lanni et al. 2009; Moret and Briley 2011). Such changes may arise from reduced NE innervation, reduction in NE transporter activity in NE-containing terminals, or by a combination of both (Klimek et al. 1997). For DA, lower transporter binding potential (a correlate of receptor density) in striatum is reported in depressed patients (Meyer et al. 2001). Similar findings were also observed in depressed patients with anhedonia symptoms (Sarchiapone et al. 2006). However, there are inconsistencies between different studies (Laasonen-Balk et al. 1999). Similarly, investigations have also measured DA metabolites in the CSF, but these results are not necessarily conclusive (Dunlop and Nemeroff 2007).

Despite the inconsistencies, such findings have led to the development of pharmaceutical treatments of unipolar depression. Many antidepressant compounds are designed to increase monoamine levels in the central nervous system (CNS). The strategy for drug treatments for other disorders can vary since the symptomatology can be diverse and dynamic (Bowden 2005; Walderhaug et al. 2011). Current research focuses on other novel treatment options include the exploration of peptidergic ligands. Clinical and preclinical studies are suggestive of the roles of neuropeptide mechanisms (e.g. orexin, vasopressin, galanin, corticotropin-releasing hormone, neuropeptide Y, relaxin-3 and substance P) in the pathophysiology of psychiatric disorders, including depression (Den Boer 2006; Madaan and Wilson 2009; Nollet and Leman 2013). Indeed, emerging drug targets for the treatment of psychiatric disorders are compounds that signal via neuropeptide receptors (Fang et al. 2014; Smith et al. 2014).

Among the neuropeptides potentially involved in the symptomatology of psychiatric disorders are the orexins (Ox; also called the hypocretins or hcrt). These neuropeptides occur in two forms, Ox-A and Ox-B, and are produced by neurons in the lateral, perifornical and dorsomedial areas of the hypothalamus (Sakurai et al. 1998; de Lecea et al. 1998). These Ox-containing neurons send projections to many brain regions (Peyron et al. 1998), including key regions that regulate arousal and motivational states such as the dorsal raphe nucleus (DRN), LC, arcuate nucleus, VTA, tuberomammillary nucleus (TMN), basal forebrain (BF) and laterodorsal and pedunculopontine tegmental nucleus (LDT/PPT) (Tsujino and Sakurai 2009). Through these and other neural connections, Ox play key roles in the regulation of important behaviours and physiological functions such as sleep-wake cycle, energy homeostasis, addiction, endocrine function, reward seeking, and emotional behaviour (de Lecea et al. 2006; Sakurai 2006, 2007, 2010, 2014; Tsujino and Sakurai 2009; Aston-Jones et al. 2009; López et al. 2010). Interestingly, many of these are disrupted in depression (Drevets 2001), suggesting that insights into Ox signaling may lead to a better understanding of the biological basis of depression. Indeed, in the last few years, there have been many studies in humans and animals investigating the role of Ox in depression as well as in other psychiatric disorders.

For example, it has been known that Ox levels can exhibit circadian variation and depressed patients manifest lower amplitude circadian variation in Ox levels in the cerebrospinal fluid (CSF) than do control individuals. Treatment with sertraline (a selective serotonin reuptake inhibitor (SSRI) anti-depressant) decreased mean Ox levels, suggesting that elevated Ox signaling contributes to the depressive symptoms (Salomon et al. 2003). Such a putative interaction appears complex since other studies have reported that following attempted suicide the CSF Ox levels are decreased (Brundin et al. 2007a, b). In a later follow-up study (6–12 months after the attempt), CSF Ox levels among suicide survivors were relatively high compared to those measured immediately following suicide attempt (Brundin et al. 2009). In another investigation, Ox levels were reportedly unaltered in manic, control and depressed patients (Schmidt et al. 2010). Thus the effects of Ox in depression in humans remain inconclusive.

The relationship between Ox and depression has been also explored in animal models. For example, in the Flinders Sensitive Line (FSL), a rat genetic model of depression in which the animals are more susceptible to cholinergic agonists, the number of Ox neurons are higher than in control rats (Overstreet 1986; Overstreet et al. 2005; Mikrouli et al. 2011). Similarly, a 20 % increase in the number of detectable Ox neurons in lateral hypothalamus area was recently reported in a rodent model where high dosage of corticosterone treatment results in depressive like symptoms (Jalewa et al. 2014b). These findings suggest that during depression, either the number of Ox neurons increases or Ox synthesis increases, such that Ox-containing neurons are more readily available in the hypothalamus. In support for this interpretation, central infusion of Ox stimulates the hypothalamic–pituitary–adrenal (HPA) axis (Kuru et al. 2000), and hyperactivity of the HPA axis is often associated with depression (Vreeburg et al. 2009). Interestingly, attenuating Ox signaling in rodents through treatment with the Ox receptor antagonist almorexant improves HPA functioning and decreases behavioural measures of depression (Nollet et al. 2012). In contrast, exogenous treatments with Ox-A can have anti-depressant properties (Ito et al. 2008), suggesting that reduced Ox levels are associated with depression.

Ox is also potentially involved in other psychiatric disorders. For example, Ox signaling influences attentional and cognitive activities and has been linked with schizophrenia (Deutch et al. 2005; Lambe et al. 2007; Fukunaka et al. 2007; Poirier et al. 2010). Interestingly, higher plasma Ox levels are associated with improved symptoms in schizophrenia (Chien et al. 2015). There is also robust evidence indicating that a reduction in Ox neurons can lead to narcolepsy (Lin et al. 1999; Chemelli et al. 1999; Peyron et al. 2000). Since psychiatric patients frequently develop sleep disorders, this raises the possibility that changes in Ox signaling contributes to such arousal and wake-rest disruptions. This is explored in the next section.

3 Neuromodulators and Sleep

In mice, direct injection of Ox-A into the brain promotes wakefulness and suppresses non-rapid eye movement (NREM) and rapid eye movement (REM) sleep (Mieda et al. 2011), while similar treatment with a dual Ox_1/Ox_2 receptor antagonist promotes sleep and reduces locomotion (Mang et al. 2012). Thus, the activity of Ox neurons and Ox release varies across the sleep-wake cycle. Indeed, during the wake state, Ox neurons in rodents are very active (Lee et al. 2005b; Marston et al. 2008; Estabrooke et al. 2001), coinciding with the time when Ox release is maximal (Deboer et al. 2004). As noted earlier, Ox neurons innervate arousal-promoting nuclei in the pons, midbrain, posterior hypothalamus and forebrain, and since the main action of Ox is excitatory, it is interpreted that Ox activates neurons in these nuclei to drive wakefulness. Such neurons are major sources of acetylcholine, histamine, serotonin, and norepinephrine, all of which are known to contribute to arousal states. Indeed, the activity of these neurochemically identified neurons is higher during wakefulness than in sleep. Thus, during wakefulness, the outputs of arousal-promoting neurons act to inhibit the activity of NREM and REM promoting neurons in ventrolateral preoptic nucleus (VLPO) and LDT/PPT, respectively, and relay excitatory signals to the thalamus and cortical neurons (Saper et al. 2001; España and Scammell 2011).

During NREM sleep, neurons of the preoptic areas are activated, and these neurons act to inhibit the wake and REM-promoting neurons elsewhere in the brain. Thus, during this time, the firing in monoaminergic neurons decreases significantly and ceases completely in the cholinergic and Ox neurons (Saper et al. 2001; España and Scammell 2011). In addition to the preoptic neurons, interneurons (mainly GABAergic) in some of the arousal areas (LHA and BF) are also active during NREM phase and contribute in conveying inhibitory signals to the cortex to promote slow wave sleep (Manns et al. 2000; Hassani et al. 2009, 2010).

REM sleep occurs during the transition from NREM to waking states (Saper et al. 2001). Hallmark features of REM sleep include cortical activation and loss of muscle tone. Interestingly, during REM sleep, Ox neurons are not completely silenced but instead contribute to the phasic components of this brain state (Torterolo and Chase 2014). By contrast, during REM, cholinergic and VLPO neurons show high firing rate, while monoaminergic neurons cease firing (España and Scammell 2011). Thus an imbalance between cholinergic and monoaminergic discharge activity may contribute to the disruption between REM and NREM sleep which is very common among depression sufferers. Moreover, an imbalance between Ox and monoaminergic activity is believed to be a key factor in narcolepsy and depression (Brown et al. 2001; Feng et al. 2008).

These interpretations can be better understood by considering the two-process model of the dynamic regulation of sleep-wake cycle (Borbély 1982; Daan et al. 1984). It is widely believed that this regulation is achieved via two separate biological processes where process **S** is responsible for sleep homeostasis, and process **C** controls the circadian timing (via the circadian clock) (Fig. 2). Accordingly, the increase in sleep pressure that occurs during wakefulness subsequently declines during sleep. Studies have linked this pressure with extracellular adenosine levels (Sims et al. 2013; Huang et al. 2014). Adenosine forms from the degradation of adenine nucleotides, and during wakefulness its levels are believed to be increased in the BF and cortical regions (Huang et al. 2014). Thus, depending upon the adenosine concentration, process **S** sends timing signals to process **C** and regulates the sleep-wake cycle. So, the circadian clock not just controls the timing of process **S** but also interacts with various sleep stages, which may well decide the recovery

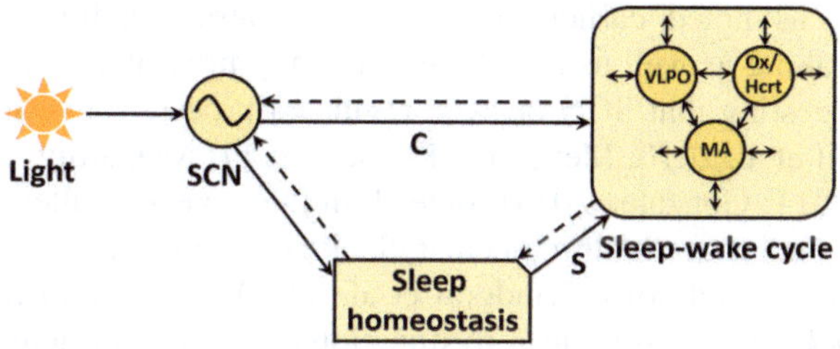

Fig. 2 A two-process model of sleep. The circadian clock in the SCN receives light inputs and sends the timing signals (**C**) to regions of the brain involved in sleep-wake (S-W) cycle and to a conceptual sleep homeostasis process which further regulates the balance between sleep and wake state. The S-W cycle is controlled by the complex interactions between sleep (ventrolateral preoptic nucleus VLPO) and wake promoting neurons (MA, Ox). *Circle*: circadian oscillator, sleep and wake promoting areas; *Snip single corner rectangle*: sleep homeostasis; *Solid and dashed arrows* as in Fig. 1 (Borbély 1982; Daan et al. 1984; Dijk and Lockley 2002)

pattern in instances of sleep deprivation. Interestingly, a study shows that activity of neurons in the brain's master circadian clock in the suprachiasmatic nuclei (SCN) is higher during the REM and wake states, and relatively lower during the NREM state (Deboer et al. 2003). There is further evidence that suggests that mutation, polymorphism or deletion of some of the clock genes disrupt sleep homeostasis and response to the sleep deprivation (Naylor et al. 2000; Toh et al. 2001; Laposky et al. 2005; Dijk and Lockley 2002). However, how changes in sleep-wake patterns affect the molecular mechanism of the clock still remains unknown (Deboer et al. 2003).

In the next section, we shall discuss how the circadian pacemaker in the SCN drives daily timing in different brain areas and regulates neuromodulator levels and sleep. We will also discuss how disruption of circadian clock timing is linked to specific symptoms commonly observed in psychiatric disorders.

4 Circadian Rhythms

Circadian rhythms are intrinsic near 24 h oscillations and pervade all aspects of physiology and behaviour. Neurons of the SCN contain the intracellular circadian clock of which the *Period* (*Per1-2*), *Cryptochrome* (*Cry1-2*), *Circadian Locomotor Output Cycles Kaput (CLOCK)* and *Brain and Muscle ARNT-like protein 1 (Bmal1)* genes and their protein products play important roles (Mohawk and Takahashi 2011). The synchronized activity of these autonomous cellular clocks enables the SCN as a whole to function as the master clock. In turn, the SCN receives information about environmental lighting directly from the retina, with daily variation in light entraining the SCN to the external world (Fig. 2). The SCN output then communicates these integrated timekeeping signals to the rest of the brain and body to control daily rhythms in sleeping and waking, metabolism, cognition, and mood (Piggins and Guilding 2011).

Intriguingly, as noted earlier, in many psychiatric conditions, alterations in sleeping and waking as well as mood are frequently present as part of the patients' symptoms, suggesting that SCN or SCN-regulated processes are disrupted in such disorders (Wulff et al. 2010; Menet and Rosbash 2011; McCarthy and Welsh 2012; Schnell et al. 2014; Gonzalez 2014). Indeed, in mice, genetic alterations that affect components of the intracellular molecular clock are implicated as murine models of human psychiatric conditions (Landgraf et al. 2014). For example, the Δ19 mutation of the *clock* gene, results in a dysfunctional CLOCK protein and the mouse manifests >24 h rhythms or lacks circadian rhythms as well as hyperactivity and other hallmarks of mania (Vitaterna et al. 1994; McClung et al. 2005; Roybal et al. 2007; Sidor et al. 2015). Such a mutation directly affects the SCN clock, but since circadian clock genes are also expressed in other brain areas as well as peripheral tissues, local physiology can be affected in such regions and organs. In the case of the Δ19 clock mutant mouse, mania-like behaviours are attributable to dysfunction of dopamine neurons in the VTA. This appears to be due to changes in gene regulation normally under the control of CLOCK (Mukherjee et al. 2010). Another mutation that influences the stability of CRY proteins also lengthens the period of circadian rhythms, which is accompanied by changes in anxiety (Keers et al. 2012).

In humans, some mutations in clock genes are associated with the sleep-wake cycle. Of note, treatment of bipolar disorder can include lithium which exerts some of its actions via GSK-3beta, a key regulator of the intracellular molecular clock (Iitaka et al. 2005). Indeed, in animal studies, lithium lengthens the period of the molecular clock rhythms and locomotor behavior (Welsh and Moore-Ede 1990; Li et al. 2012).

In rodents the activity of Ox neurons is under SCN circadian control (Zhang et al. 2004), with higher activity occurring during the behaviourally-active circadian night (Estabrooke et al. 2001; Marston et al. 2008). At this phase, Ox release in the brain is also at maximal (Deboer et al. 2004). Experimental destruction of the SCN or exposure to constant light, which attenuates SCN output, abolishes circadian variation in CSF levels of Ox, indicating that the day-night activity profile of this key arousal-promoting neurochemical is under SCN output control. Interestingly, the phase of the SCN circadian clockwork is also sensitive to feedback actions of arousal-promoting stimuli, particularly during the behaviourally-inactive circadian day (Hughes and Piggins 2012). During the day, arousal-promoting stimuli activate Ox neurons (Estabrooke et al. 2001; Marston et al. 2008; Webb et al. 2008) and suppress the SCN's electrical activity (van Oosterhout et al. 2012). Indeed, in SCN brain slice preparations Ox not only suppresses the electrical activity of SCN neurons, but also potentiates the phase-shifting capacity of another neurochemical correlate of arousal, neuropeptide Y (Belle et al. 2014).

Hence, alterations in the SCN circadian clock as well as circadian regulation of the brain and behavioural states can result in symptoms seen in psychiatric conditions. Since Ox-containing neurons project widely across the brain (Peyron et al. 1998; Nambu et al. 1999), including structures of the neural circadian system (McGranaghan and Piggins 2001; Backberg et al. 2002), Ox released during states of arousal can influence circadian timing in SCN and extra-SCN brain sites (Fig. 3). This may be particularly important in psychiatric conditions when the pattern of

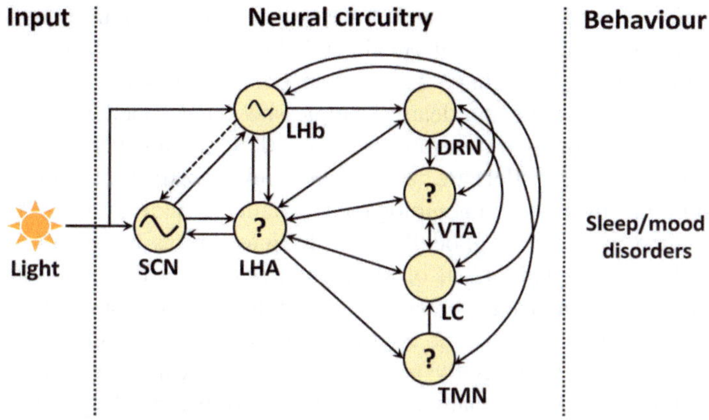

Fig. 3 Circadian clock, neuromodulators and their possible role in the etiology of sleep and mood disorders. Suprachiasmatic nuclei (SCN) with lateral habenula (LHb) receive the light inputs and regulate the activity of other brain areas that are important in S-W and mood regulation. SCN, LHb: master and semi-autonomous clocks. Lateral hypothalamus area (LHA), dorsal raphe nucleus (DRN), ventral tegmental area (VTA), locus coeruleus (LC), and tuberomamillary nucleus (TMN): major source of orexin/hypocretin, serotonin, dopamine, norepinephrine and histamine, respectively. '?': Either clock gene(s) are present in these areas (existence of intrinsic activity not yet confirmed) or their daily activity is indirectly regulated by the circadian clocks present in other brain regions (e.g. SCN and/or LHb). *Circle*: circadian oscillators (SCN, LHb), LHA and monoaminergic areas; *Note:* non-photic inputs are not shown (Jones and Moore 1977; Herkenham and Nauta 1979; Segal 1979; Deutch et al. 1986; Kalen et al. 1988; Ericson et al.1989; Herkenham and Nauta 1993; Peyron et al. 1998; Nambu et al. 1999; Abrahamson et al. 2001; Yoshida et al. 2001; Kim et al. 2004; Mileykovskiy et al. 2005; Lee et al. 2005a; McClung et al. 2005; Hattar et al. 2006; Guilding and Piggins 2007; Vertes and Linley 2008; Omelchenko et al. 2009; Tsujino and Sakurai, 2009; Schwartz et al. 2011; Goncalves et al. 2012; Watabe-Uchida et al. 2012; Stamatakis et al. 2013; Belle et al. 2014; Dorocic et al. 2014; Proulx et al. 2014; Root et al. 2014; Sakhi et al. 2014; Yu et al. 2014; Ogawa et al. 2014)

sleeping and waking is severely disturbed. Incorporating such knowledge when considering the treatments and the timing of treatments, therefore, has considerable therapeutic potential.

Thus, mutual interactions between circadian clock, neuromodulators and sleep are likely to be important in understanding the biological mechanisms of some aspects of psychiatric disorders. However, the behaviours due to these interactions are not intuitive. For example, 5-HT neurons in the DRN receive inputs from other monoaminergic neurons including the NE-containing neurons of the LC, histamine-containing neurons from TMN, as do DA neurons of the VTA (Kalen et al. 1988; Lee et al. 2005a; Dorocic et al. 2014; Ogawa et al. 2014). Additionally, the activities of these neuromodulators are also regulated by their auto-receptors. Thus, it is very difficult to assess the causal relationship (between activities or firing rate) among the different brain areas to establish precisely how these changes can cause or affect the severity of a symptom.

A promising approach towards assessing and understanding such complex interacting systems is to develop, simulate and analyze computational models. Some initial cognitive computational models have focused on improving the diagnosis of psychiatric disorders (Siegle 1999; Siegle and Hasselmo 2002). More recent cognitive computational models have focused on decision-making, reward and reinforcement learning aspects, e.g. learned helplessness and anhedonia, which are the core symptoms in depression (Dayan and Huys 2008; Gradin et al. 2011; Huys et al. 2013). However, these models are based on high-level abstractions and hence do not readily illuminate the underlying biological mechanisms underpinning the disorders. Moreover, there is a lack of modelling on the interactions between Ox and other neuromodulators, and their effects on sleep and depression. This is not surprising as substantial modelling and analysis work was more focused on associating Ox with the sleep-wake cycle, and in particular, its role in narcolepsy [e.g. (Behn et al. 2010)]. In the next section, we will briefly review some biologically based computational modelling studies on Ox, monoamines, circadian rhythms, and sleep-wake dynamics.

5 Neural Computational Models

The mathematical modelling and analysis of sleep-wake dynamics and circadian rhythms have a relatively long history, primarily due to the availability of behavioural and physiological (e.g. temperature and neuroendocrine) data (Czeisler 1978; Czeisler et al. 1980; Daan et al. 1984; Kawato et al. 1982; Kronauer et al. 1982, 1983; Strogatz and Carpenter 1986; Strogatz 1987; Winfree 1983). All these models address the sleep-wake dynamics and various autonomic circadian rhythms. There is a considerable similarity among these models (Kronauer et al. 1982; Kronauer et al. 1983), which typically consist of homogeneous neuronal populations, or coarse-grained population-averaged (firing-rate) type models (Wilson and Cowan 1972). The simplicity of the models allows tractability in the mathematical analyses and provides conceptual insights.

As various types of biological data become available, models of circadian rhythms become more physiologically detailed. At the molecular, genetic and protein levels, biochemical reactions and feedback loops are modelled to understand their regulations on the intrinsic circadian rhythms and their perturbations (Goldbeter 1995; Leloup and Goldbeter 2003; Forger and Peskin 2003). Physiologically, more realistic neuronal models of sleep-wake regulation have also begun, building on previously more abstract mathematical models (Behn et al. 2007; Phillips and Robinson 2007; Booth and Behn 2014). These models are typically considered at a single level of biological details.

At the neuronal level, a biophysical conductance-based model of the SCN neurons has successfully incorporated the available neuronal and synaptic properties in the SCN and can mimic the neuronal firing patterns, which are consistent with experiments (Sim and Forger 2007). Similar biophysical models for Ox neurons has also been developed with the minimal set of currents (Postnova et al. 2009; Carter et al. 2012) and model parameters were based on other neuronal types (Williams and Behn 2011). Despite the assumptions, such models could account for some aspects of sleep-wake transitions (Postnova et al. 2009), and the dynamical effects of Ox and dynorphin, a colocalized neuropeptide, on the Ox neurons (Williams and Behn 2011). In Carter et al. (2012), a more biophysical two-compartmental model of Ox neuron was also developed.

Neurocomputational models at the microcircuit level may offer a way to investigate heterogeneity effects and emergent properties, which are not feasible at the single neuronal modelling level. For example, Postnova et al. (2009) incorporated local glutamatergic neurons to provide local feedback to the Ox neurons to understand homeostasis in sleep-wake transitions. Then, Patriarca et al. (2012) extended this model into a multi-neuron model and demonstrated that sufficient diversity in the synapses is needed to provide efficient functioning of the homeostasis regulation of the sleep-wake cycle. In Gonze et al. (2005) the model incorporated coupling among 10,000 circadian oscillators and found that spontaneous synchronization is achieved through the "mean-field" coupling and the population can be entrained with light pulses. In Wong-Lin et al. (2012), a microcircuit model of the DRN with non-5-HT GABAergic neurons was proposed to link from single neuronal spiking behaviours of 5-HT and local non-5-HT neurons in the DRN to phasic neuronal activity observed in behaving animals. The microcircuit DRN model suggested inputs via inhibitory neurons, and predicted low frequency oscillations in the network, which is clearly emergent network behaviour.

To understand the overall behaviour of a system as complex as that shown in Fig. 3, one would need to extend beyond microcircuit modelling levels, and towards larger circuit models that include the interactions of multiple brain regions (Sorooshyari et al. 2015). For example, Behn et al. (2007) modelled the interactions among the sleep active (VLPO), wake active (LC, DR, TMN) and REM active (LDT/PPT) neural populations. The model could capture some of the features of mouse the sleep-wake behaviour (such as short-term awakening), and also predicts the mechanism for state transitions. Subsequent modelling studies showed that Ox could play an important part in the sustenance and stabilization of prolonged episodes of wake and sleep (Behn et al. 2008; Fulcher et al. 2014). Similarly, the role of Ox is also analyzed for other large-scale modelling work that reasonably reproduce sleep-wake timings (under normal and sleep deprived conditions), circadian influence on total sleep time, and rapid transition between sleep and wake states with the loss of Ox (Rempe et al. 2010).

Recently, there has been considerable interest in more focused modelling of Ox system's interactions with that of the monoamines (Joshi et al. 2011; Carter et al. 2012; Jalewa et al. 2014a). A common aim of these studies is to understand the possible roles of such interactions in sleep and depression (Joshi et al. 2011; Carter et al. 2012; Jalewa et al. 2014a; Mosqueiro et al. 2014). In one of these studies, the excitability of LC neurons is shown to be important for Ox-mediated transition in sleep-to-wake transition (Carter et al. 2012). In a different study, a mathematical neural circuit model with direct interactions between the DRN and LHA is developed (Joshi et al. 2011). The model's novel input-output functions for the DRN and LHA areas are derived from the relationship between the neural firing rate in that area and neuromodulator concentration level, which can be obtained directly from experiments. In addition, the dynamics of the release and uptake of the neuromodulators can, in principle, also be measured from experiments, e.g. voltammetry.

Even more complex models have been studied, where the models include not just across-region interactions, but also interactions with other local interneurons. For example, in Mosqueiro et al. (2014), inhibitory GABAergic interneurons are used to investigate the regulation of NE-producing LC neurons by Ox, in which the GABAergic neurons provide an indirect inhibitory connection from Ox to LC. This modelling work shows that the relatively fast $GABA_A$-mediated synapses are not sufficient to regulate LC activity. Similarly, the model in Jalewa et al. (2014a) has included more explicit interneurons, and for the first time, autoreceptors were included for the LHA-DRN interaction. More importantly, indirect connections are also considered in this model. The model demonstrates that LHA-DRN interactions are more stable if the indirect connections from 5-HT to GABAergic neurons in the LHA are strongly excitatory. Furthermore, it is found that faster 5-HT (e.g. via the 3A) receptor mediated timescales can quickly reset the Ox neuronal activities to baseline firing rate right after phasic 5-HT activation. Further, Kumar et al. (2012) have developed a mathematical model of REM-NREM dynamics that includes local GABAergic neurons, revealing the sensitive control of Ox on REM dynamics.

Taken together, these modeling studies have furthered our understanding of the relationship between Ox and the monoamines, and their roles in sleep-wake dynamics, providing mechanistic links to mood, cognition and psychiatric disorder. Despite the progress, integrated computational models that can link from cellular to behavioural levels have yet to be developed. Having such models would be of tremendous help in guiding future experiments across various levels.

6 Discussion

Many people worldwide suffer from psychiatric disorders. Depression is one psychiatric disorder with complex etiology and heterogeneous symptoms (Krishnan and Nestler 2008). Sleep anomalies and altered circadian rhythms are prevalent in depression (Germain and Kupfer 2008; Pandi-Perumal et al. 2009). Monoamine oxidase inhibitors (MAOI) are used for the clinical treatment of this atypical disorder (Thase et al. 1995) and are now augmented by drugs that target neuropeptide receptors, thought to be important in the etiology of depression (Saar et al. 2013; Nollet and Leman 2013; Yeoh et al. 2014).

In this chapter, we have discussed how Ox can interact with monoamines, circadian oscillators (e.g. SCN), and sleep wake-cycle. We raised some key issues, identified some of the unknown connections and attempted to understand how anomaly in one of the system can influence their complex interactions, which may contribute to certain behaviours that are common in psychiatric disorders.

Ox is a key neuropeptide which plays important roles in the regulation of a wide range of behaviours. Interestingly, Ox levels were inconsistent in different preclinical and clinical studies. There can be many reasons for such inconsistencies.

For example, Ox levels are regulated by the circadian clock and if not sampled at the same time in each study, considerable interstudy differences could emerge. Indeed, one investigation showed that depressed patients have dampened CSF Ox rhythms (Salomon et al. 2003). Similarly, circadian variation in serum serotonin and metabolites is also reported in depression (Pietraszek et al. 1991). Not surprisingly, there are other factors that may also contribute to inconsistency in measured brain neuromodulator levels. Some of them include heterogeneity among the animals (e.g. mice, rats), experimental protocols, small sample size and methods for measuring the neuromodulator levels (e.g. voltammetry, high-performance liquid chromatography or radioimmunoassay).

Another important issue is to conside whether Ox and other neuromodulator levels are regulated reciprocally. The last few years have witnessed a surge of interest in understanding Ox interactions with monoamines as these interactions can play important roles in the regulation of cognitive functions altered in many psychiatric disorders. Ox interaction with DA neurons may contribute to the better understanding of sleep related problems reported in schizophrenia, as anti-dopaminergic drugs actively lower the Ox CSF levels and promotes the NREM sleep in schizophrenics (Dalal et al. 2003). Also co-release of Ox with dynorphin in the VTA is important in motivational and reward-related behaviours (Muschamp et al. 2014) which are known to be disrupted in depression and schizophrenia. Similarly, knowledge of the interaction among Ox and serotonergic neurons is necessary for the better understanding of depression, as higher Ox levels in the hypothalamus are reported in a clomipramine (SSRI) induced rat model of depression (Feng et al. 2008). In some respects, there is much in common in these Ox-monoamine interactions. All these neuromodulators are active during arousal, and are also involved in some of the cognitive functions. The other similarity between Ox and other neuromodulators is the way Ox connects with them; interestingly Ox shares an excitatory-inhibitory feedback loop circuitry with these neuromodulators (e.g. serotonin, norepinephrine and dopamine). For example, Ox depolarizes these monoamine- containing neurons and conversely these neuromodulators can hyperpolarize Ox neurons in the LHA. Recently, experimental and modeling studies have contributed towards the better understanding of such interactions (Carter et al. 2012; Schone et al. 2014; Jalewa et al. 2014a). However, mutual interaction of all these arousal areas is neglected in the literature. This is because mutual interactions among the monoaminergic neurons add another layer of complexity, and practical implementation of these connections remains a considerable challenge.

Other than these complex interactions, Ox neurons also receive daily timing information from the SCN, and play a key role in the regulation of REM sleep (Kantor et al. 2009). However, the exact pathway via which SCN regulates Ox neurons and REM sleep is not known. One obvious possibility is that SCN and sleep homeostasis processes interact with sleep-wake states in accordance with the two-process model (Fig. 2). The other possibility is that the SCN directly regulates the sleep-wake cycle, and Ox plays a dual role where it stabilizes the sleep-wake switch and also regulates the sleep homeostasis process (Postnova et al. 2009). The excitatory drive of Ox neurons during wakefulness is due in part to excitatory feedback connections between Ox and glutamatergic neurons within the LHA. It is possible that synaptic efficacy of Ox neurons decreases due to their prolonged activity during wakefulness which is recovered during sleep.

Apart from Ox, the SCN clock can also influence the process of sleep homeostasis as this is disrupted in clock mutant mice (Naylor et al. 2000). However the exact relationship between Ox and the sleep homeostasis process remains elusive. Further, recent evidence indicating that clock genes influence the activity of DA neurons adds another level of complexity (McClung et al. 2005). This brings us to a question, whether a disrupted clock can influence the level of Ox or monoamines, and if so then what is the neural pathway of this dysregulation. One possibility is that the master clock in SCN first desynchronizes other local clocks (LHb, paraventricular nucleus) or the areas with clock genes. These areas then indirectly communicates with the other brain areas, affecting the electrical activities of their neurons (Fig. 3). On the other side, a change in the level of Ox or monoamines can alter circadian rhythms in other physiological activity, such as the profile in stress hormones and the HPA axis (Mazzocchi et al. 2001; Spinazzi et al. 2006; Ziolkowska et al. 2005).

Thus the role of Ox is not just limited to the regulation of the sleep-wake states but it is involved in the regulation of a wide range of behaviours and cognitive functions (Mahler et al. 2014). Most of these behaviours are determined by the complex interaction between subcortical and cortical areas (Jankowski and Sesack 2004; Onge et al. 2012; Pujara and Koenigs 2014). However, linking the neuronal activity with the observed behaviour is one of the greatest challenges in neuroscience.

To accomplish this difficult task, one important approach is to collate the experimental data and build computational models across multiple levels, from intracellular processes in individual neurons to network properties and behavioural states. For example, a concerted effort to systematically collect neurobiological data, such as electrophysiological properties of neurons, release-and-reuptake dynamics of the neuromodulators, and receptor-mediated currents at the target sites, would be vital towards developing biologically faithful computational models of neuromodulation. With such data available, an integrated model endowed with multiscales can be developed. The multiscale modelling approach would extract the essence at each scale or level and integrate this with processes happening at other scales or levels (Yamada and Forger 2010; Vasalou and Henson 2010; Qu et al. 2011; Dada and Mendes 2011). Such effort in developing a multiscale framework for neuromodulator systems, from intracellular biochemical reactions to behaviour, is currently undertaken by the authors—(Eckhoff et al. 2009, 2011; Wong-Lin et al. 2012; Wang and Wong-Lin 2013; Laviale et al. 2013; Nakamura and Wong-Lin 2014; Flower and Wong-Lin 2014; Cullen and Wong-Lin 2014). With the availability of such multiscale models, one can rapidly test hypothesis, and can make model predictions that can be verified by future experiments. Thus, computational models with pharmacological, imaging and optogenetic approaches can improve our understanding of the underlying neural mechanism(s) responsible for specific symptoms that are common in psychiatric disorders.

In summary, this chapter reviews studies related to Ox, monoamines, and circadian processes, and highlights some of their roles in psychiatric disorders. Further, we suggest the need to develop an integrated multiscale computational modelling framework that is based on systematic collection of experimental data. This will improve our understanding of how disruption in heterogeneous and highly connected neurochemically distinct neuronal populations and the circadian system contribute to symptoms in psychiatric conditions.

Acknowledgements This work was supported by the Human Frontier Science Program (AJ, HDP), BBSRC (MDCB, HDP), and Northern Ireland Functional Brain Mapping Facility (1303/101154803) funded by InvestNI and the University of Ulster, Innovate UK (102161), and The Royal Society—NSFC International Exchanges (KFW-L). The authors would like to thank Daniel B. Forger, Beatriz Baño Otálora, Joseph Timothy, Jaishree Jalewa, Christian Hölscher, T. Martin McGinnity and Girijesh Prasad for previous discussions in helping to generate some of the presented ideas.

References

Abrahamson E, Leak R, Moore R (2001) The suprachiasmatic nucleus projects to posterior hypothalamic arousal systems. Neuroreport 12:435–440

Anzalone A, Lizardi-Ortiz JE, Ramos M, De Mei C, Hopf FW, Iaccarino C, Halbout B, Jacobsen J, Kinoshita C, Welter M (2012) Dual control of dopamine synthesis and release by presynaptic and postsynaptic dopamine D2 receptors. J Neurosci 32:9023–9034

Arango V, Underwood MD, Boldrini M, Tamir H, Kassir SA, Hsiung S-C, Chen JJ, Mann JJ (2001) Serotonin 1A receptors, serotonin transporter binding and serotonin transporter mRNA expression in the brainstem of depressed suicide victims. Neuropsychopharmacology 25:892–903

Aston-Jones G, Rajkowski J, Cohen J (2000) Locus coeruleus and regulation of behavioral flexibility and attention. Prog Brain Res 126:165–182

Aston-Jones G, Smith RJ, Moorman DE, Richardson KA (2009) Role of lateral hypothalamic orexin neurons in reward processing and addiction. Neuropharmacology 56:112–121

Backberg M, Hervieu G, Wilson S, Meister B (2002) Orexin receptor-1 (OX-R1) immunoreactivity in chemically identified neurons of the hypothalamus: focus on orexin targets involved in control of food and water intake. Eur J Neurosci 15:315–328

Behn CG, Brown EN, Scammell TE, Kopell NJ (2007) Mathematical model of network dynamics governing mouse sleep-wake behavior. J Neurophysiol 97:3828–3840

Behn CD, Kopell N, Brown EN, Mochizuki T, Scammell TE (2008) Delayed orexin signaling consolidates wakefulness and sleep: physiology and modeling. J Neurophysiol 99:3090–3103

Behn CGD, Klerman EB, Mochizuki T, Lin S-C, Scammell TE (2010) Abnormal sleep/wake dynamics in orexin knockout mice. Sleep 33:297

Belle MD, Hughes AT, Bechtold DA, Cunningham P, Pierucci M, Burdakov D, Piggins HD (2014) Acute suppressive and long-term phase modulation actions of orexin on the mammalian circadian clock. J Neurosci 34:3607–3621

Boldrini M, Underwood MD, Mann JJ, Arango V (2005) More tryptophan hydroxylase in the brainstem dorsal raphe nucleus in depressed suicides. Brain Res 1041:19–28

Booth V, Behn CGD (2014) Physiologically-based modeling of sleep–wake regulatory networks. Math Biosci 250:54–68

Borbély AA (1982) A two process model of sleep regulation. Hum Neurobiol 1:195–204

Bowden CL (2005) A different depression: clinical distinctions between bipolar and unipolar depression. J Affect Disord 84:117–125

Bromberg-Martin ES, Matsumoto M, Hikosaka O (2010) Dopamine in motivational control: rewarding, aversive, and alerting. Neuron 68:815–834

Brown RE, Sergeeva O, Eriksson KS, Haas HL (2001) Orexin A excites serotonergic neurons in the dorsal raphe nucleus of the rat. Neuropharmacology 40:457–459

Brundin L, Bjorkqvist M, Petersen A, Traskman-Bendz L (2007a) Reduced orexin levels in the cerebrospinal fluid of suicidal patients with major depressive disorder. Eur Neuropsychopharmacol 17:573–579

Brundin L, Petersen A, Bjorkqvist M, Traskman-Bendz L (2007b) Orexin and psychiatric symptoms in suicide attempters. J Affect Disord 100:259–263

Brundin L, Bjorkqvist M, Traskman-Bendz L, Petersen A (2009) Increased orexin levels in the cerebrospinal fluid the first year after a suicide attempt. J Affect Disord 113:179–182

Carter ME, Brill J, Bonnavion P, Huguenard JR, Huerta R, de Lecea L (2012) Mechanism for Hypocretin-mediated sleep-to-wake transitions. Proc Natl Acad Sci U S A 109:E2635–E2644

Chemelli RM, Willie JT, Sinton CM, Elmquist JK, Scammell T, Lee C, Richardson JA, Williams SC, Xiong Y, Kisanuki Y, Fitch TE, Nakazato M, Hammer RE, Saper CB, Yanagisawa M (1999) Narcolepsy in orexin knockout mice: molecular genetics of sleep regulation. Cell 98:437–451

Chien Y-L, Liu C-M, Shan J-C, Lee H-J, Hsieh MH, Hwu H-G, Chiou L-C (2015) Elevated plasma orexin A levels in a subgroup of patients with schizophrenia associated with fewer negative and disorganized symptoms. Psychoneuroendocrinology 53:1–9

Cools R, Roberts AC, Robbins TW (2008) Serotoninergic regulation of emotional and behavioural control processes. Trends Cogn Sci 12:31–40

Cullen M, Wong-Lin K (2014) Analysis of a computational model of dopamine synthesis and release through perturbation. In: 2014 IEEE international conference on bioinformatics and biomedicine (BIBM), pp 1–7

Czeisler C (1978) Internal organization of temperature, sleep-wake, and neuroendocrine rhythms monitored in an environment free of time cues. Stanford University, Stanford

Czeisler CA, Weitzman E, Moore-Ede MC, Zimmerman JC, Knauer RS (1980) Human sleep: its duration and organization depend on its circadian phase. Science 210:1264–1267

Daan S, Beersma D, Borbély AA (1984) Timing of human sleep: recovery process gated by a circadian pacemaker. Am J Physiol-Regul Integr Comp Physiol 246:R161–R183

Dada JO, Mendes P (2011) Multi-scale modelling and simulation in systems biology. Integr Biol (Camb) 3:86–96

Dalal MA, Schuld A, Pollmacher T (2003) Lower CSF orexin A (hypocretin-1) levels in patients with schizophrenia treated with haloperidol compared to unmedicated subjects. Mol Psychiatry 8:836–837

Dayan P, Huys QJ (2008) Serotonin, inhibition, and negative mood. PLoS Comput Biol 4:e4

de Lecea L, Kilduff T, Peyron C, Gao X-B, Foye P, Danielson P, Fukuhara C, Battenberg E, Gautvik V, Bartlett FN (1998) The hypocretins: hypothalamus-specific peptides with neuroexcitatory activity. Proc Natl Acad Sci 95:322–327

de Lecea L, Jones BE, Boutrel B, Borgland SL, Nishino S, Bubser M, Dileone R (2006) Addiction and arousal: alternative roles of hypothalamic peptides. J Neurosci 26:10372–10375

Deboer T, Vansteensel MJ, Detari L, Meijer JH (2003) Sleep states alter activity of suprachiasmatic nucleus neurons. Nat Neurosci 6:1086–1090

Deboer T, Overeem S, Visser NA, Duindam H, Frolich M, Lammers GJ, Meijer JH (2004) Convergence of circadian and sleep regulatory mechanisms on hypocretin-1. Neuroscience 129:727–732

Delgado PL (2000) Depression: the case for a monoamine deficiency. J Clin Psychiatry 61(Suppl 6):7–11

Den Boer JA (2006) Looking beyond the monoamine hypothesis

Deutch AY, Goldstein M, Roth RH (1986) Activation of the locus coeruleus induced by selective stimulation of the ventral tegmental area. Brain Res 363:307–314

Deutch AY, Fadel J, Bubser M (2005) Dopamine-hypocretin/orexin interactions. In: Hypocretins. Springer, Berlin

Dijk D-J, Lockley SW (2002) Invited review: integration of human sleep-wake regulation and circadian rhythmicity. J Appl Physiol 92:852–862
Dorocic IP, Fürth D, Xuan Y, Johansson Y, Pozzi L, Silberberg G, Carlén M, Meletis K (2014) A whole-brain atlas of inputs to serotonergic neurons of the dorsal and median raphe nuclei. Neuron 83:663–678
Drevets WC (2001) Neuroimaging and neuropathological studies of depression: implications for the cognitive-emotional features of mood disorders. Curr Opin Neurobiol 11:240–249
Drevets WC, Frank E, Price JC, Kupfer DJ, Holt D, Greer PJ, Huang Y, Gautier C, Mathis C (1999) PET imaging of serotonin 1A receptor binding in depression. Biol Psychiatry 46:1375–1387
Dunlop BW, Nemeroff CB (2007) The role of dopamine in the pathophysiology of depression. Arch Gen Psychiatry 64:327–337
Eckhoff P, Wong-Lin KF, Holmes P (2009) Optimality and robustness of a biophysical decision-making model under norepinephrine modulation. J Neurosci 29:4301–4311
Eckhoff P, Wong-Lin K, Holmes P (2011) Dimension reduction and dynamics of a spiking neural network model for decision making under neuromodulation. SIAM J Appl Dyn Syst 10:148–188
Ericson H, Blomqvist A, Köhler C (1989) Brainstem afferents to the tuberomammillary nucleus in the rat brain with special reference to monoaminergic innervation. J Comp Neurol 281:169–192
España RA, Scammell TE (2011) Sleep neurobiology from a clinical perspective. Sleep 34:845
Estabrooke IV, McCarthy MT, Ko E, Chou TC, Chemelli RM, Yanagisawa M, Saper CB, Scammell TE (2001) Fos expression in orexin neurons varies with behavioral state. J Neurosci 21:1656–1662
Fang P, Min W, Sun Y, Guo L, Shi M, Bo P, Zhang Z (2014) The potential antidepressant and antidiabetic effects of galanin system. Pharmacol Biochem Behav 120:82–87
Feng P, Vurbic D, Wu Z, Hu Y, Strohl KP (2008) Changes in brain orexin levels in a rat model of depression induced by neonatal administration of clomipramine. J Psychopharmacol 22:784–791
Flower G, Wong-Lin K (2014) Reduced computational models of serotonin synthesis, release, and reuptake. IEEE Trans Biomed Eng 61:1054–1061
Forger DB, Peskin CS (2003) A detailed predictive model of the mammalian circadian clock. Proc Natl Acad Sci 100:14806–14811
Fukunaka Y, Shinkai T, Hwang R, Hori H, Utsunomiya K, Sakata S, Naoe Y, Shimizu K, Matsumoto C, Ohmori O, Nakamura J (2007) The orexin 1 receptor (HCRTR1) gene as a susceptibility gene contributing to polydipsia-hyponatremia in schizophrenia. Neuromolecular Med 9:292–297
Fulcher BD, Phillips AJ, Postnova S, Robinson PA (2014) A physiologically based model of orexinergic stabilization of sleep and wake. PLoS ONE 9:e91982
Germain A, Kupfer DJ (2008) Circadian rhythm disturbances in depression. Human Psychopharmacol Clin Exp 23:571–585
Goldbeter A (1995) A model for circadian oscillations in the Drosophila period protein (PER). Proc Roy Soc Lond Ser B Biol Sci 261:319–324
Goncalves L, Sego C, Metzger M (2012) Differential projections from the lateral habenula to the rostromedial tegmental nucleus and ventral tegmental area in the rat. J Comp Neurol 520:1278–1300
Gonzalez R (2014) The relationship between bipolar disorder and biological rhythms. J Clin Psychiatry 75:e323–e331
Gonze D, Bernard S, Waltermann C, Kramer A, Herzel H (2005) Spontaneous synchronization of coupled circadian oscillators. Biophys J 89:120–129
Gradin VB, Kumar P, Waiter G, Ahearn T, Stickle C, Milders M, Reid I, Hall J, Steele JD (2011) Expected value and prediction error abnormalities in depression and schizophrenia. Brain 134:1751–1764
Grzanna R, Molliver M (1980) The locus coeruleus in the rat: an immunohistochemical delineation. Neuroscience 5:21–40
Guilding C, Piggins HD (2007) Challenging the omnipotence of the suprachiasmatic timekeeper: are circadian oscillators present throughout the mammalian brain? Eur J Neurosci 25:3195–3216

Hasler G (2010) Pathophysiology of depression: do we have any solid evidence of interest to clinicians? World Psychiatry 9:155–161

Hassani OK, Lee MG, Henny P, Jones BE (2009) Discharge profiles of identified GABAergic in comparison to cholinergic and putative glutamatergic basal forebrain neurons across the sleep–wake cycle. J Neurosci 29:11828–11840

Hassani OK, Henny P, Lee MG, Jones BE (2010) GABAergic neurons intermingled with orexin and MCH neurons in the lateral hypothalamus discharge maximally during sleep. Eur J Neurosci 32:448–457

Hattar S, Kumar M, Park A, Tong P, Tung J, Yau KW, Berson DM (2006) Central projections of melanopsin-expressing retinal ganglion cells in the mouse. Journal of Comparative Neurology 497:326–349

Herkenham M, Nauta WJ (1979) Efferent connections of the habenular nuclei in the rat. J Comp Neurol 187:19-47

Herkenham M, Nauta WJ (1993) Afferent connections of the habenular nuclei in the rat. A horseradish peroxidase study, with a note on the fiber-of-passage problem. Neuroanatomy Springer 277–299

Huang ZL, Zhang Z, Qu WM (2014) Roles of adenosine and its receptors in sleep-wake regulation. Int Rev Neurobiol 119:349–371

Hughes AT, Piggins HD (2012) Feedback actions of locomotor activity to the circadian clock. Prog Brain Res 199:305–336

Huys QJ, Pizzagalli DA, Bogdan R, Dayan P (2013) Mapping anhedonia onto reinforcement learning: a behavioural meta-analysis. Biol Mood Anxiety Disord 3:12

Iitaka C, Miyazaki K, Akaike T, Ishida N (2005) A role for glycogen synthase kinase-3beta in the mammalian circadian clock. J Biol Chem 280:29397–29402

Ito N, Yabe T, Gamo Y, Nagai T, Oikawa T, Yamada H, Hanawa T (2008) Icv administration of orexin-A induces an antidepressive-like effect through hippocampal cell proliferation. Neuroscience 157:720–732

Jalewa J, Joshi A, McGinnity TM, Prasad G, Wong-Lin K, Holscher C (2014a) Neural circuit interactions between the dorsal raphe nucleus and the lateral hypothalamus: an experimental and computational study. PLoS ONE 9:e88003

Jalewa J, Wong-Lin K, McGinnity TM, Prasad G, Hölscher C (2014b) Increased number of orexin/hypocretin neurons with high and prolonged external stress-induced depression. Behav Brain Res 272:196–204

Jankowski MP, Sesack SR (2004) Prefrontal cortical projections to the rat dorsal raphe nucleus: ultrastructural features and associations with serotonin and gamma-aminobutyric acid neurons. J Comp Neurol 468:518–529

Jones BE, Moore RY (1977) Ascending projections of the locus coeruleus in the rat II. Autoradiographic study. Brain Res 127:25–53

Joshi A, Wong-Lin K, Mcginnity TM, Prasad G (2011) A mathematical model to explore the interdependence between the serotonin and orexin/hypocretin systems. In: Proceedings of conference on IEEE engineering in medicine and biology society, pp 7270–7273

Kalen P, Skagerberg G, Lindvall O (1988) Projections from the ventral tegmental area and mesencephalic raphe to the dorsal raphe nucleus in the rat. Exp Brain Res 73:69–77

Kantor S, Mochizuki T, Janisiewicz AM, Clark E, Nishino S, Scammell TE (2009) Orexin neurons are necessary for the circadian control of REM sleep. Sleep 32:1127–1134

Kawato M, Fujita K, Suzuki R, Winfree AT (1982) A three-oscillator model of the human circadian system controlling the core temperature rhythm and the sleep-wake cycle. J Theor Biol 98:369–392

Keers R, Pedroso I, Breen G, Aitchison KJ, Nolan PM, Cichon S, Nothen MM, Rietschel M, Schalkwyk LC, Fernandes C (2012) Reduced anxiety and depression-like behaviours in the circadian period mutant mouse afterhours. PLoS ONE 7:e38263

Kim MA, Lee HS, Lee BY, Waterhouse BD (2004) Reciprocal connections between subdivisions of the dorsal raphe and the nuclear core of the locus coeruleus in the rat. Brain research 1026:56–67

Klimek V, Stockmeier C, Overholser J, Meltzer HY, Kalka S, Dilley G, Ordway GA (1997) Reduced levels of norepinephrine transporters in the locus coeruleus in major depression. J Neurosci 17:8451–8458

Krishnan V, Nestler EJ (2008) The molecular neurobiology of depression. Nature 455:894–902

Kronauer RE, Czeisler CA, Pilato SF, Moore-Ede MC, Weitzman ED (1982) Mathematical model of the human circadian system with two interacting oscillators. Am J Physiol-Regul Integr Comp Physiol 242:R3–R17

Kronauer R, Czeisler C, Pilato S, Moore-Ede M, Weitzman E (1983) Mathematical representation of the human circadian system: two interacting oscillators which affect sleep. Sleep Dis Basic Clin Res 173–194

Kumar R, Bose A, Mallick BN (2012) A mathematical model towards understanding the mechanism of neuronal regulation of wake-NREMS-REMS states. PLoS ONE 7:e42059

Kuru M, Ueta Y, Serino R, Nakazato M, Yamamoto Y, Shibuya I, Yamashita H (2000) Centrally administered orexin/hypocretin activates HPA axis in rats. NeuroReport 11:1977–1980

Laasonen-Balk T, Kuikka J, Viinamaki H, Husso-Saastamoinen M, Lehtonen J, Tiihonen J (1999) Striatal dopamine transporter density in major depression. Psychopharmacology 144:282–285

Lambe EK, Liu RJ, Aghajanian GK (2007) Schizophrenia, hypocretin (orexin), and the thalamocortical activating system. Schizophr Bull 33:1284–1290

Landgraf D, McCarthy MJ, Welsh DK (2014) The role of the circadian clock in animal models of mood disorders. Behav Neurosci 128:344–359

Lanni C, Govoni S, Lucchelli A, Boselli C (2009) Depression and antidepressants: molecular and cellular aspects. Cell Mol Life Sci 66:2985–3008

Laposky A, Easton A, Dugovic C, Walisser J, Bradfield C, Turek F (2005) Deletion of the mammalian circadian clock gene BMAL1/Mop3 alters baseline sleep architecture and the response to sleep deprivation. Sleep 28:395–409

Laviale A, Weiss T, Veh RW, Mcginnity T, Maguire L, Wong-Lin K (2013) A single spiking neuronal model to account for the diverse spontaneous firing patterns of lateral habenula neurons. Society for Neuroscience, Washington DC

Lee HS, Lee BY, Waterhouse BD (2005a) Retrograde study of projections from the tuberomammillary nucleus to the dorsal raphe and the locus coeruleus in the rat. Brain Res 1043:65–75

Lee MG, Hassani OK, Jones BE (2005b) Discharge of identified orexin/hypocretin neurons across the sleep-waking cycle. J Neurosci 25:6716–6720

Leibowitz SF, Shor-Posner G (1986) Brain serotonin and eating behavior. Appetite 7:1–14

Leloup J-C, Goldbeter A (2003) Toward a detailed computational model for the mammalian circadian clock. Proc Natl Acad Sci 100:7051–7056

Li J, Lu WQ, Beesley S, Loudon AS, Meng QJ (2012) Lithium impacts on the amplitude and period of the molecular circadian clockwork. PLoS ONE 7:e33292

Lin L, Faraco J, Li R, Kadotani H, Rogers W, Lin X, Qiu X, de Jong PJ, Nishino S, Mignot E (1999) The sleep disorder canine narcolepsy is caused by a mutation in the hypocretin (orexin) receptor 2 gene. Cell 98:365–376

López M, Tena-Sempere M, Diéguez C (2010) Cross-talk between orexins (hypocretins) and the neuroendocrine axes (hypothalamic–pituitary axes). Front Neuroendocrinol 31:113–127

Madaan V, Wilson DR (2009) Neuropeptides: relevance in treatment of depression and anxiety disorders. Drug News Perspect 22:319–324

Mahler SV, Moorman DE, Smith RJ, James MH, Aston-Jones G (2014) Motivational activation: a unifying hypothesis of orexin/hypocretin function. Nat Neurosci 17:1298–1303

Malison RT, Price LH, Berman R, van Dyck CH, Pelton GH, Carpenter L, Sanacora G, Owens MJ, Nemeroff CB, Rajeevan N (1998) Reduced brain serotonin transporter availability in major depression as measured by [123 I]-2β-carbomethoxy-3β-(4-iodophenyl) tropane and single photon emission computed tomography. Biol Psychiatry 44:1090–1098

Mang GM, Durst T, Burki H, Imobersteg S, Abramowski D, Schuepbach E, Hoyer D, Fendt M, Gee CE (2012) The dual orexin receptor antagonist almorexant induces sleep and decreases orexin-induced locomotion by blocking orexin 2 receptors. Sleep 35:1625–1635

Manns ID, Alonso A, Jones BE (2000) Discharge profiles of juxtacellularly labeled and immunohistochemically identified GABAergic basal forebrain neurons recorded in association with the electroencephalogram in anesthetized rats. J Neurosci 20:9252–9263

Marston OJ, Williams RH, Canal MM, Samuels RE, Upton N, Piggins HD (2008) Circadian and dark-pulse activation of orexin/hypocretin neurons. Mol Brain 1:19

Mathers CD, Loncar D (2006) Projections of global mortality and burden of disease from 2002 to 2030. PLoS medicine 3:e442

Mazzocchi G, Malendowicz LK, Gottardo L, Aragona F, Nussdorfer GG (2001) Orexin A stimulates cortisol secretion from human adrenocortical cells through activation of the adenylate cyclase-dependent signaling cascade. J Clin Endocrinol Metab 86:778–782

McCarthy MJ, Welsh DK (2012) Cellular circadian clocks in mood disorders. J Biol Rhythms 27:339–352

McClung CA, Sidiropoulou K, Vitaterna M, Takahashi JS, White FJ, Cooper DC, Nestler EJ (2005) Regulation of dopaminergic transmission and cocaine reward by the Clock gene. Proc Natl Acad Sci U S A 102:9377–9381

McGranaghan PA, Piggins HD (2001) Orexin A-like immunoreactivity in the hypothalamus and thalamus of the Syrian hamster (*Mesocricetus auratus*) and Siberian hamster (*Phodopus sungorus*), with special reference to circadian structures. Brain Res 904:234–244

Menet JS, Rosbash M (2011) When brain clocks lose track of time: cause or consequence of neuropsychiatric disorders. Curr Opin Neurobiol 21:849–857

Meyer JH, Kruger S, Wilson AA, Christensen BK, Goulding VS, Schaffer A, Minifie C, Houle S, Hussey D, Kennedy SH (2001) Lower dopamine transporter binding potential in striatum during depression. NeuroReport 12:4121–4125

Mieda M, Hasegawa E, Kisanuki YY, Sinton CM, Yanagisawa M, Sakurai T (2011) Differential roles of orexin receptor-1 and -2 in the regulation of non-REM and REM sleep. J Neurosci 31:6518–6526

Mikrouli E, Wörtwein G, Soylu R, Mathé AA, Petersén Å (2011) Increased numbers of orexin/hypocretin neurons in a genetic rat depression model. Neuropeptides 45:401–406

Mileykovskiy BY, Kiyashchenko LI, Siegel JM (2005) Behavioral correlates of activity in identified hypocretin/orexin neurons. Neuron 46:787–798

Mintun MA, Sheline YI, Moerlein SM, Vlassenko AG, Huang Y, Snyder AZ (2004) Decreased hippocampal 5-HT 2A receptor binding in major depressive disorder: in vivo measurement with [18 F] altanserin positron emission tomography. Biol Psychiatry 55:217–224

Mohawk JA, Takahashi JS (2011) Cell autonomy and synchrony of suprachiasmatic nucleus circadian oscillators. Trends Neurosci 34:349–358

Moret C, Briley M (2011) The importance of norepinephrine in depression. Neuropsychiatric Dis Treat 7:9

Mosqueiro T, De Lecea L, Huerta R (2014) Control of sleep-to-wake transitions via fast aminoacid and slow neuropeptide transmission. New J Phys 16

Mukherjee S, Coque L, Cao JL, Kumar J, Chakravarty S, Asaithamby A, Graham A, Gordon E, Enwright JF, 3RD, Dileone RJ, Birnbaum SG, Cooper DC, Mcclung CA (2010) Knockdown of Clock in the ventral tegmental area through RNA interference results in a mixed state of mania and depression-like behavior. Biol Psychiatry 68:503–511

Murray G, Harvey A (2010) Circadian rhythms and sleep in bipolar disorder. Bipolar Disord 12:459–472

Muschamp JW, Hollander JA, Thompson JL, Voren G, Hassinger LC, Onvani S, Kamenecka TM, Borgland SL, Kenny PJ, Carlezon WA Jr (2014) Hypocretin (orexin) facilitates reward by attenuating the antireward effects of its cotransmitter dynorphin in ventral tegmental area. Proc Natl Acad Sci U S A 111:E1648–E1655

Nakamura K, Wong-Lin K (2014) Functions and computational principles of serotonergic and related systems at multiple scales. Front Integr Neurosci 8:23

Nambu T, Sakurai T, Mizukami K, Hosoya Y, Yanagisawa M, Goto K (1999) Distribution of orexin neurons in the adult rat brain. Brain Res 827:243–260

Naylor E, Bergmann BM, Krauski K, Zee PC, Takahashi JS, Vitaterna MH, Turek FW (2000) The circadian clock mutation alters sleep homeostasis in the mouse. J Neurosci 20:8138–8143
Nemeroff CB (1998) The neurobiology of depression. Sci Am-Am Edit 278:42–49
Nollet M, Leman S (2013) Role of orexin in the pathophysiology of depression: potential for pharmacological intervention. CNS Drugs 27:411–422
Nollet M, Gaillard P, Tanti A, Girault V, Belzung C, Leman S (2012) Neurogenesis-independent antidepressant-like effects on behavior and stress axis response of a dual orexin receptor antagonist in a rodent model of depression. Neuropsychopharmacology 37:2210–2221
Ogawa SK, Cohen JY, Hwang D, Uchida N, Watabe-Uchida M (2014) Organization of monosynaptic inputs to the serotonin and dopamine neuromodulatory systems. Cell reports 8:1105–1118
Omelchenko N, Bell R, Sesack SR (2009) Lateral habenula projections to dopamine and GABA neurons in the rat ventral tegmental area. Eur J Neurosci 30:1239–1250
Onge JRS, Stopper CM, Zahm DS, Floresco SB (2012) Separate prefrontal-subcortical circuits mediate different components of risk-based decision making. J Neurosci 32:2886–2899
Ordway GA, Schenk J, Stockmeier CA, May W, Klimek V (2003) Elevated agonist binding to α 2-adrenoceptors in the locus coeruleus in major depression. Biol Psychiatry 53:315–323
Overstreet DH (1986) Selective breeding for increased cholinergic function: development of a new animal model of depression. Biol Psychiatry 21:49–58
Overstreet DH, Friedman E, Mathé AA, Yadid G (2005) The Flinders Sensitive Line rat: a selectively bred putative animal model of depression. Neurosci Biobehav Rev 29:739–759
Owens MJ, Nemeroff CB (1994) Role of serotonin in the pathophysiology of depression: focus on the serotonin transporter. Clin Chem 40:288–295
Pandi-Perumal SR, Moscovitch A, Srinivasan V, Spence DW, Cardinali DP, Brown GM (2009) Bidirectional communication between sleep and circadian rhythms and its implications for depression: lessons from agomelatine. Prog Neurobiol 88:264–271
Park S, Harrold JA, Widdowson PS, Williams G (1999) Increased binding at 5-HT 1A, 5-HT 1B, and 5-HT 2A receptors and 5-HT transporters in diet-induced obese rats. Brain research 847:90–97
Patriarca M, Postnova S, Braun HA, Hernandez-Garcia E, Toral R (2012) Diversity and noise effects in a model of homeostatic regulation of the sleep-wake cycle. PLoS Comput Biol 8: e1002650
Peyron C, Tighe DK, van den Pol AN, de Lecea L, Heller HC, Sutcliffe JG, Kilduff TS (1998) Neurons containing hypocretin (orexin) project to multiple neuronal systems. J Neurosci 18:9996–10015
Peyron C, Faraco J, Rogers W, Ripley B, Overeem S, Charnay Y, Nevsimalova S, Aldrich M, Reynolds D, Albin R, Li R, Hungs M, Pedrazzoli M, Padigaru M, Kucherlapati M, Fan J, Maki R, Lammers GJ, Bouras C, Kucherlapati R, Nishino S, Mignot E (2000) A mutation in a case of early onset narcolepsy and a generalized absence of hypocretin peptides in human narcoleptic brains. Nat Med 6:991–997
Phillips AJ, Robinson PA (2007) A quantitative model of sleep-wake dynamics based on the physiology of the brainstem ascending arousal system. J Biol Rhythms 22:167–179
Pietraszek MH, Takahashi S, Takada Y, Ohara K, Inatomi H, Kondo N, Ohara K, Takada A (1991) Diurnal patterns of serotonin, 5-hydroxyindoleacetic acid, tryptophan and fibrinolytic activity in blood of depressive patients and healthy volunteers. Thromb Res 64:243–252
Piggins HD, Guilding C (2011) The neural circadian system of mammals. Essays Biochem 49:1–17
Poirier S, Legris G, Tremblay P, Michea R, Viau-Guay L, Merette C, Bouchard RH, Maziade M, Roy MA (2010) Schizophrenia patients with polydipsia and water intoxication are characterized by greater severity of psychotic illness and a more frequent history of alcohol abuse. Schizophr Res 118:285–291
Postnova S, Voigt K, Braun HA (2009) A mathematical model of homeostatic regulation of sleep-wake cycles by hypocretin/orexin. J Biol Rhythms 24:523–535
Proulx CD, Hikosaka O, Malinow R (2014) Reward processing by the lateral habenula in normal and depressive behaviors. Nat Neurosci 17:1146–1152

Pujara M, Koenigs M (2014) Mechanisms of reward circuit dysfunction in psychiatric illness: prefrontal-striatal interactions. Neuroscientist 20:82–95

Qu Z, Garfinkel A, Weiss JN, Nivala M (2011) Multi-scale modeling in biology: how to bridge the gaps between scales? Prog Biophys Mol Biol 107:21–31

Quintana J (1992) Platelet serotonin and plasma tryptophan decreases in endogenous depression. Clinical, therapeutic, and biological correlations. J Affect Disord 24:55–62

Rajkowska G (2000) Histopathology of the prefrontal cortex in major depression: what does it tell us about dysfunctional monoaminergic circuits? Prog Brain Res 126:397–412

Rempe MJ, Best J, Terman D (2010) A mathematical model of the sleep/wake cycle. J Math Biol 60:615–644

Root DH, Mejias-Aponte CA, Qi J, Morales M (2014) Role of glutamatergic projections from ventral tegmental area to lateral habenula in aversive conditioning. J Neurosci 34:13906–13910

Roybal K, Theobold D, Graham A, Dinieri JA, Russo SJ, Krishnan V, Chakravarty S, Peevey J, Oehrlein N, Birnbaum S, Vitaterna MH, Orsulak P, Takahashi JS, Nestler EJ, Carlezon WA Jr, Mcclung CA (2007) Mania-like behavior induced by disruption of CLOCK. Proc Natl Acad Sci U S A 104:6406–11

Ruhe HG, Mason NS, Schene AH (2007) Mood is indirectly related to serotonin, norepinephrine and dopamine levels in humans: a meta-analysis of monoamine depletion studies. Mol Psychiatry 12:331–359

Saar I, Lahe J, Langel K, Runesson J, Webling K, Jarv J, Rytkonen J, Narvanen A, Bartfai T, Kurrikoff K, Langel U (2013) Novel systemically active galanin receptor 2 ligands in depression-like behavior. J Neurochem 127:114–123

Sakhi K, Belle MD, Gossan N, Delagrange P, Piggins HD, (2014) Daily variation in the electrophysiological activity of mouse medial habenula neurones. J physiol 592:587–603

Sakurai T (2006) Roles of orexins and orexin receptors in central regulation of feeding behavior and energy homeostasis. CNS Neurol Disorders-Drug Targets (Formerly Curr Drug Targets-CNS Neurol Disorders) 5:313–325

Sakurai T (2007) The neural circuit of orexin (hypocretin): maintaining sleep and wakefulness. Nat Rev Neurosci 8:171–181

Sakurai T (2014) The role of orexin in motivated behaviours. Nat Rev Neurosci 15:719–731

Sakurai T, Amemiya A, Ishii M, Matsuzaki I, Chemelli RM, Tanaka H, Williams SC, Richardson JA, Kozlowski GP, Wilson S (1998) Orexins and orexin receptors: a family of hypothalamic neuropeptides and G protein-coupled receptors that regulate feeding behavior. Cell 92:573–585

Sakurai T, Mieda M, Tsujino N (2010) The orexin system: roles in sleep/wake regulation. Ann N Y Acad Sci 1200:149–161

Salomon RM, Ripley B, Kennedy JS, Johnson B, Schmidt D, Zeitzer JM, Nishino S, Mignot E (2003) Diurnal variation of cerebrospinal fluid hypocretin-1 (Orexin-A) levels in control and depressed subjects. Biol Psychiatry 54:96–104

Saper CB, Chou TC, Scammell TE (2001) The sleep switch: hypothalamic control of sleep and wakefulness. Trends Neurosci 24:726–731

Sarchiapone M, Carli V, Camardese G, Cuomo C, di Giuda D, Calcagni M-L, Focacci C, de Risio S (2006) Dopamine transporter binding in depressed patients with anhedonia. Psychiatry Res Neuroimaging 147:243–248

Schmidt FM, Brugel M, Kratzsch J, Strauss M, Sander C, Baum P, Thiery J, Hegerl U, Schonknecht P (2010) Cerebrospinal fluid hypocretin-1 (orexin A) levels in mania compared to unipolar depression and healthy controls. Neurosci Lett 483:20–22

Schnell A, Albrecht U, Sandrelli F (2014) Rhythm and mood: relationships between the circadian clock and mood-related behavior. Behav Neurosci 128:326–343

Schone C, Apergis-Schoute J, Sakurai T, Adamantidis A, Burdakov D (2014) Coreleased orexin and glutamate evoke nonredundant spike outputs and computations in histamine neurons. Cell Rep 7:697–704

Schwartz MD, Urbanski HF, Nunez AA, Smale L (2011) Projections of the suprachiasmatic nucleus and ventral subparaventricular zone in the Nile grass rat (*Arvicanthis niloticus*). Brain Res 1367:146–161

Segal M (1979) Serotonergic innervation of the locus coeruleus from the dorsal raphe and its action on responses to noxious stimuli. J Physiol 286:401–415

Sidor MM, Spencer SM, Dzirasa K, Parekh PK, Tye KM, Warden MR, Arey RN, Enwright JF, 3RD, Jacobsen JP, Kumar S, Remillard EM, Caron MG, Deisseroth K, Mcclung CA (2015) Daytime spikes in dopaminergic activity drive rapid mood-cycling in mice. Mol Psychiatry 5

Siegle GJ (1999) A neural network model of attention biases in depression. Prog Brain Res 121:407–432

Siegle GJ, Hasselmo ME (2002) Using connectionist models to guide assessment of psychological disorder. Psychol Assess 14:263–278

Sim CK, Forger DB (2007) Modeling the electrophysiology of suprachiasmatic nucleus neurons. J Biol Rhythms 22:445–453

Sims RE, Wu HH, Dale N (2013) Sleep-wake sensitive mechanisms of adenosine release in the basal forebrain of rodents: an in vitro study. PLoS ONE 8:e53814

Smith CM, Walker AW, Hosken IT, Chua BE, Zhang C, Haidar M, Gundlach AL (2014) Relaxin-3/RXFP3 networks: an emerging target for the treatment of depression and other neuropsychiatric diseases? Frontiers Pharmacol 5

Sorooshyari S, Huerta R, De_Lecea L (2015) A framework for quantitative modeling of neural circuits involved in sleep-to-wake transition. Name Frontiers Neurol 6:32

Spinazzi R, Andreis PG, Rossi GP, Nussdorfer GG (2006) Orexins in the regulation of the hypothalamic-pituitary-adrenal axis. Pharmacol Rev 58:46–57

Stamatakis AM, Jennings JH, Ung RL, Blair GA, Weinberg RJ, Neve RL, Boyce F, Mattis J, Ramakrishnan C, Deisseroth K (2013) A unique population of ventral tegmental area neurons inhibits the lateral habenula to promote reward. Neuron 80:1039–1053

Strogatz SH (1987) Open peer. Commentary the mathematical structure of the human sleep-wake cycle. In: Strogatz SH (1986) Lecture notes in biomathematics, vol 69. Springer, Berlin. J Biol Rhythms 2:317–329

Strogatz S, Carpenter G (1986) A comparative analysis of models of the human sleep-wake cycle. Lect Math Life Sci 19:1–38

Thase ME, Trivedi MH, Rush AJ (1995) MAOIs in the contemporary treatment of depression. Neuropsychopharmacology 12:185–219

Toh KL, Jones CR, He Y, Eide EJ, Hinz WA, Virshup DM, Ptacek LJ, Fu YH (2001) An hPer2 phosphorylation site mutation in familial advanced sleep phase syndrome. Science 291:1040–1043

Torterolo P, Chase MH (2014) The hypocretins (orexins) mediate the "phasic" components of REM sleep: a new hypothesis. Sleep Sci 7:19–29

Tsujino N, Sakurai T (2009) Orexin/hypocretin: a neuropeptide at the interface of sleep, energy homeostasis, and reward system. Pharmacol Rev 61:162–176

Tsuno N, Besset A, Ritchie K (2005) Sleep and depression. J Clin Psychiatry 66:1254–1269

Valdizán EM, Díez-Alarcia R, González-Maeso J, Pilar-Cuéllar F, García-Sevilla JA, Meana JJ, Pazos A (2010) α 2-Adrenoceptor functionality in postmortem frontal cortex of depressed suicide victims. Biol Psychiatry 68:869–872

van Oosterhout F, Lucassen EA, Houben T, Vanderleest HT, Antle MC, Meijer JH (2012) Amplitude of the SCN clock enhanced by the behavioral activity rhythm. PLoS ONE 7:e39693

Vasalou C, Henson MA (2010) A multiscale model to investigate circadian rhythmicity of pacemaker neurons in the suprachiasmatic nucleus. PLoS Comput Biol 6:e1000706

Vertes RP, Linley SB (2008) Efferent and afferent connections of the dorsal and median raphe nuclei in the rat. Serotonin and sleep: molecular, functional and clinical aspects Springer 4:69

Vitaterna MH, King DP, Chang AM, Kornhauser JM, Lowrey PL, McDonald JD, Dove WF, Pinto LH, Turek FW, Takahashi JS (1994) Mutagenesis and mapping of a mouse gene, clock, essential for circadian behavior. Science 264:719–725

Vreeburg SA, Hoogendijk WJ, van Pelt J, Derijk RH, Verhagen JC, van Dyck R, Smit JH, Zitman FG, Penninx BW (2009) Major depressive disorder and hypothalamic-pituitary-adrenal axis activity: results from a large cohort study. Arch Gen Psychiatry 66:617–626

Walderhaug E, Varga M, San Pedro M, Hu J, Neumeister A (2011) The role of the aminergic systems in the pathophysiology of bipolar disorder. In: Behavioral neurobiology of bipolar disorder and its treatment. Springer, Berlin

Wang DH, Wong-Lin K (2013) Comodulation of dopamine and serotonin on prefrontal cortical rhythms: a theoretical study. Front Integr Neurosci 7:54

Watabe-Uchida M, Zhu L, Ogawa SK, Vamanrao A, Uchida N (2012) Whole-brain mapping of direct inputs to midbrain dopamine neurons. Neuron 74:858–873

Webb IC, Patton DF, Hamson DK, Mistlberger RE (2008) Neural correlates of arousal-induced circadian clock resetting: hypocretin/orexin and the intergeniculate leaflet. Eur J Neurosci 27:828–835

Wehr TA, Sack D, Rosenthal N, Duncan W, Gillin JC (1983) Circadian rhythm disturbances in manic-depressive illness. Fed Proc 42:2809–2814

Welsh DK, Moore-Ede MC (1990) Lithium lengthens circadian period in a diurnal primate, *Saimiri sciureus*. Biol Psychiatry 28:117–126

Whiteford HA, Degenhardt L, Rehm J, Baxter AJ, Ferrari AJ, Erskine HE, Charlson FJ, Norman RE, Flaxman AD, Johns N (2013) Global burden of disease attributable to mental and substance use disorders: findings from the Global Burden of Disease Study 2010. Lancet 382:1575–1586

Williams KS, Behn CG (2011) Dynamic interactions between orexin and dynorphin may delay onset of functional orexin effects: a modeling study. J Biol Rhythms 26:171–181

Wilson HR, Cowan JD (1972) Excitatory and inhibitory interactions in localized populations of model neurons. Biophys J 12:1–24

Winfree AT (1983) Impact of a circadian clock on the timing of human sleep. Am J Physiol-Regul Integr Comp Physiol 245:R497–R504

Wirz-Justice A (2008) Diurnal variation of depressive symptoms. Dialogues Clin Neurosci 10:337–343

Wong-Lin K, Joshi A, Prasad G, McGinnity TM (2012) Network properties of a computational model of the dorsal raphe nucleus. Neural Netw 32:15–25

Wulff K, Gatti S, Wettstein JG, Foster RG (2010) Sleep and circadian rhythm disruption in psychiatric and neurodegenerative disease. Nat Rev Neurosci 11:589–599

Wulff K, Dijk D-J, Middleton B, Foster RG, Joyce EM (2012) Sleep and circadian rhythm disruption in schizophrenia. Br J Psychiatry 200:308–316

Yamada Y, Forger D (2010) Multiscale complexity in the mammalian circadian clock. Curr Opin Genet Dev 20:626–633

Yeoh JW, Campbell EJ, James MH, Graham BA, Dayas CV (2014) Orexin antagonists for neuropsychiatric disease: progress and potential pitfalls. Front Neurosci 8:36

Yoshida Y, Fujiki N, Nakajima T, Ripley B, Matsumura H, Yoneda H, Mignot E, Nishino S (2001) Fluctuation of extracellular hypocretin-1 (orexin A) levels in the rat in relation to the light–dark cycle and sleep–wake activities. J Neurosci 14:1075–1081

Yu X, Zecharia A, Zhang Z, Yang Q, Yustos R, Jager P, Vyssotski AL, Maywood ES, Chesham JE, Ma Y, Brickley SG, Hastings MH, Franks NP, Wisden W (2014) Circadian factor BMAL1 in histaminergic neurons regulates sleep architecture. Curr Biol 24:2838–2844

Zhang S, Zeitzer JM, Yoshida Y, Wisor JP, Nishino S, Edgar DM, Mignot E (2004) Lesions of the suprachiasmatic nucleus eliminate the daily rhythm of hypocretin-1 release. Sleep 27:619–627

Ziolkowska A, Spinazzi R, Albertin G, Nowak M, Malendowicz LK, Tortorella C, Nussdorfer GG (2005) Orexins stimulate glucocorticoid secretion from cultured rat and human adrenocortical cells, exclusively acting via the OX1 receptor. J Steroid Biochem Mol Biol 96:423–429

A New Class of Hypnotic Compounds for the Treatment of Insomnia: The Dual Orexin Receptor Antagonists

Jason M. Uslaner, John J. Renger, Paul J. Coleman and Chris J. Winrow

Abstract Insomnia is a widespread, debilitating disorder responsible for enormous individual and societal costs. Currently available pharmacologic treatments for insomnia, including allosteric modulators of the gamma-aminobutyric acid type A receptor (GABAA), such as zolpidem (Ambien®) and eszopiclone (Lunesta®), have undesirable effects that limit their tolerability and utility. The dual orexin receptor antagonists (DORAs), possessing novel mechanisms of action, have demonstrated efficacy in improving sleep latency and quantity in several preclinical species, healthy human volunteers, and patients with insomnia. Importantly, accumulating data suggest that DORAs may be better tolerated than allosteric modulators of the GABAA receptor with respect to cognitive impairment and motor side effects. A greater understanding of the differences between these drug classes is warranted. This chapter attempts to explain some of their key differences in mechanisms of action as well as describe areas where greater experimentation is warranted.

1 Introduction

> I've always envied people who sleep easily. Their brains must be cleaner, the floorboards of the skull well swept, all the little monsters closed up in a steamer trunk at the foot of the bed.
> — David Benioff, *City of Thieves*

> With insomnia, you're never really awake; but you're never really asleep.
> — Chuck Palahniuk, *Fight Club*

It is difficult to overstate the importance of a good night's sleep. Epidemiologic and experimental studies show that sleep restriction has deleterious consequences, including but not limited to impairments of mood, memory consolidation,

J.M. Uslaner (✉) · J.J. Renger · P.J. Coleman · C.J. Winrow
Merck Co. Inc, WP 46-100. Sumneytown Pike, P.O. Box 4, West Point, PA 19486, USA
e-mail: jason_uslaner@merck.com

© Springer International Publishing Switzerland 2015

T. Sakurai et al. (eds.), *Orexin and Sleep*,
DOI 10.1007/978-3-319-23078-8_17

attention/learning, creativity, athletic performance, inflammatory processes, clearance of neurotoxins, body weight/metabolism, and responsivity to stress, as well as hypertension, cancer, potential for accidents, and shortened life span. Despite the vast amount of research supporting the importance of a good night's sleep, getting too little sleep, in the United States, for example, is quickly becoming the norm rather than the exception (Foundation 2006). The National Institutes of Health recommend that children and teens receive 9–10 h and adults 7–8 h of sleep each night. However, approximately one third of adults report getting less than 6 h of sleep nightly on average while approximately two thirds of high-school students sleep less than 8 h on a typical school night (Schoenborn et al. 2004).

Lifestyle changes can often effectively improve the quantity and quality of sleep (Montgomery and Dennis 2004). For some individuals, however, healthy sleep remains elusive despite their best efforts to implement these changes. Characterized by difficulty falling asleep, staying asleep, and/or poor sleep quality, insomnia is the most common sleep disorder, associated with incidence rates of 4–24 %, depending on the population sampled and diagnostic criteria used (Ohayon 2002; Mai and Buysse 2008; Morin et al. 2009; Ohayon and Guilleminault et al. 2010; Roth et al. 2011). The individual burden of insomnia is significant, as are its societal and economic costs, the latter estimated at more than $60 billion annually in the United States alone (Balter and Uhlenhuth 1991; Kuppermann et al. 1995; Roth and Ancoli-Israel 1999; Hajak et al. 2011; Kessler et al. 2011). Individuals suffer from impaired daytime functioning, deficits in work and cognitive performance, and mood disturbances. The societal consequences include increased healthcare costs and resources, reduced work productivity, worker absenteeism, and increased risk of accidents (Kessler et al. 2011). Effective treatments for sleep disorders are of extreme importance.

2 Current Standard of Care for Treating Insomnia

As a result of insomnia's high prevalence and deleterious effects on functioning, well-tolerated treatments that are effective at promoting and maintaining sleep are a significant medical need. Approximately 4 % of the U.S. adult population over 20 years of age report taking a prescription sleep aid over the past 30 days (Chong et al. 2013). At the time of this writing, by far the most commonly prescribed pharmacologic treatments for insomnia are compounds that act as activators of the gamma-aminobutyric acid type A receptor (GABAA), including the non-benzodiazepines eszopiclone (Lunesta®) and zolpidem (Ambien®) (Roth et al. 2007).

Discovered well before GABAA had been identified and its mechanism of action understood, benzodiazepines and non-benzodiazepines act as allosteric modulators by binding to a site distinct from the endogenous ligand, GABA. Upon binding to GABAA, these compounds behave as positive allosteric modulators, increasing the ability of GABA to effectively open the chloride ion channel within the receptor, resulting in an influx of chloride ions and neuronal hyperpolarization (Mohler et al. 2002). As a result of decreasing the membrane potential of the neuron, the net effect of these compounds is to inhibit the propensity of neurons containing GABAA receptors to propagate action potentials.

Composed of five subunits, the GABAA receptor family exhibits exceptional diversity with heterologous combinations giving rise to a multitude of distinct subtypes. Specifically, seven subunit families (α_{1-6}, β_{1-3}, γ_{1-3}, δ, ε, θ, ρ_{1-3}) comprised of 18 subunits are found in the central nervous system (CNS). The different combinations of subunits produce heterogeneity in the channel kinetics, rate of desensitization, localization at and outside the synapse, and affinity for GABA (as well as exogenous ligands, including the hypnotics). In addition to the high level of structural diversity among the GABAA receptors, there is also a high level of diversity in CNS expression of the receptors and, therefore, the physiologic and behavioral consequences of positively modulating the receptor within each brain region. For example, the most common GABAA receptor, which contains the α_1 subunit, is found at very high levels in the cerebral cortex, throughout the hippocampus, amygdala, globus pallidus, ventral pallidum, caudate nucleus and putamen, nucleus accumbens, most subnuclei of the thalamus, and olfactory bulb, and throughout the cerebellum (Mohler et al. 2002). Importantly, it appears that receptors containing the α_1 subunit are responsible for the sedative hypnotic effects of the GABAA-positive allosteric modulators, as all these standard-of-care therapeutics bind to GABAA α_1 subunit-containing receptors, and the sedative effects of these compounds are lost in mice with the α_1 subunit genetically knocked out (Rudolph et al. 1999).

Besides the allosteric modulators of GABAA, a number of additional drugs with distinct mechanisms of action are used by patients with insomnia. These include compounds targeting serotonin signaling as well as drugs used for other indications (e.g., antidepressants and antipsychotics), over-the-counter treatments (e.g., diphenhydramine), and the recently approved melatonin receptor agonists. Because less is known about the mechanisms of action of these agents relative to the positive allosteric modulators for the GABAA receptor, this chapter will focus on comparing and contrasting the better understood allosteric modulators of GABAA with the dual orexin receptor antagonists (DORAs).

3 Orexin Receptor Antagonism—a Novel Mechanism for Treating Insomnia

In contrast to the discovery of GABAA positive allosteric modulators, which came prior to the identification of the GABAA receptor or the characterization of positive allosterism, the identification of drugs that block the orexin receptor represented a concerted effort based on an understanding of orexin genetics, and associated biology and function. The orexin peptides A and B (also referred to as hypocretin peptides) are generated from the same prepropeptide and were discovered simultaneously in 1998 by two different teams (de Lecea et al. 1998; Sakurai et al. 1998). The orexins and their receptors were subsequently characterized for their involvement in rodent and dog models of narcolepsy, as well as in human narcolepsy (Chemelli et al. 1999; Lin et al. 1999; Nishino et al. 2000). It was only a couple of years following these discoveries that the first selective orexin receptor antagonist was developed. The ensuing 15 years have seen a number of pharmaceutical companies dedicating resources to the development of orexin receptor antagonists (see accompanying chapters).

The distribution of orexinergic neurons and the orexin receptors 1 (OX1R) and 2 (OX2R) is much more spatially discrete relative to that of the GABAergic cells and GABAA receptors. For example, the human brain has approximately 70,000 orexin-synthesizing neurons (Thannickal et al. 2000), almost all of which can be found in two subnuclei of the hypothalamus, the lateral hypothalamic area and the posterior hypothalamus (Sakurai 2007). Orexin receptors (OX1R and OX2R) have overlapping and distinct localization expressed in key brain regions involved in arousal and vigilance (Marcus et al. 2001; Sakurai 2007). Importantly, orexin signaling rises during the normal active period sustaining wakefulness and falls silent during the normal sleep period (Gotter et al. 2013). In contrast, GABA is the major inhibitory neurotransmitter, synthesized ubiquitously throughout the brain. It has been estimated that GABAA is expressed in 20–50 % of the approximately 86 billion neurons in the human brain (Herculano-Houzel and Lent 2005; Herculano-Houzel 2012; Nutt and Malizia 2001). As a result of the much more restricted expression of orexin-producing neurons and orexin receptors, it was hypothesized that orexin receptor antagonists might promote sleep with far fewer side effects relative to globally acting GABAA-positive allosteric modulators.

Brisbare-Roche et al. (2007) were the first to demonstrate the sleep-promoting effects of an orexin receptor antagonist in humans. In their study, almorexant, a dual orexin receptor (OX1R and OX2R) antagonist (DORA), promoted sleep in rodents, dogs, and healthy humans (Brisbare-Roch et al. 2007). Although later shown to be active in patients with insomnia, development of almorexant was discontinued in 2011. A more recent DORA, suvorexant (Merck), has exhibited sleep-promoting effects in rodents, dogs, and humans, and is currently under evaluation by the U.S. Food and Drug Administration and other regulatory agencies after demonstrating efficacy in multiple phase 2 and 3 studies (Winrow et al. 2011; Herring et al. 2012; Michelson et al. 2013; Sun et al. 2013). Several other single or dual orexin receptor antagonists with sleep-promoting properties have also been identified.

It is now clear that DORAs are effective in promoting sleep. But how might DORAs differ from GABAA-positive allosteric modulators in a meaningful way for patients? The remainder of this chapter focuses on characterizing the differences between orexin receptor antagonism and allosteric modulation of GABAA with respect to (a) the nature of sleep produced by the two mechanisms of action, (b) motor and cognitive side effects, (c) arousability, and (d) interaction with alcohol. It is important to note that most of the work directly comparing these two mechanisms of action has been performed in preclinical species; how these findings will translate to humans is still largely unknown. As more clinical experience accumulates with orexin receptor antagonists, it will be crucial to further compare and contrast these two drug classes so that their potential benefits and liabilities can be better understood.

4 GABAA-Positive Allosteric Modulators and DORAs: Electroencephalographic Findings

Self-reported time to sleep onset, total sleep time, and wake time after sleep onset are three primary measures of sleep efficacy used in clinical studies. Both GABAA-positive allosteric modulators and DORAs impact these measures, helping individuals to fall asleep more rapidly and stay asleep. Despite these similarities, however, there are important distinctions in how each mechanism of action affects the state of the brain during sleep postdose.

Sleep can be characterized by changes in electrical activity in the brain, as measured by electrodes placed on or just below the scalp using electroencephalography (EEG). Additional measurements include eye movements using electrooculography and muscle activity using electromyography. Researchers employing these tools descriptively identified changes accompanying sleep as early as the 1930s (Loomis et al. 1936); more qualitative findings were published in the 1960s (Rechtschaffen and Kales 1968). At its simplest, these tools can be used to differentiate wakefulness from two different forms of sleep, non-rapid eye movement (NREM) (sometimes subdivided into non slow-wave and slow-wave sleep, or more subdivisions) and rapid eye movement (REM) sleep. More sophisticated analysis can further characterize distinct EEG signatures accompanying different phases of sleep.

Using EEG rather than the subjective measures of sleep described above, it has become evident that the electrophysiologic correlates of sleep induced by GABAA-positive allosteric modulators and DORAs differ significantly. Most clinical studies indicate that the former increase NREM (particularly light and slow-wave) sleep but decrease REM sleep (Brunner et al. 1991; Lancel 1999; Kanno et al. 2000; Dijk et al. 2010; Bettica et al. 2012), whereas the latter reportedly produce no effect on REM or NREM measures (Sun et al. 2013), or modest increases in REM with no change in NREM (slow-wave) sleep (Bettica et al. 2012; Herring et al. 2012). Although the function of REM and NREM sleep is controversial, less controversial is what subjects report when being woken from REM and NREM sleep. It is well documented that individuals report the most difficulty waking from NREM slow-wave sleep; when woken, they report feeling groggy and perform poorly on cognitive tests, a phenomenon referred to as sleep inertia (Lubin et al. 1976; Bonnet 1983). In contrast, when woken from REM sleep, individuals are more likely to report having been dreaming, and relative to slow-wave sleep, perform better on cognitive tasks and are more responsive in general (Broughton 1968; Scott 1968; Stones 1977; Dinges 1984). These findings might imply that there would be differences in capabilities upon awakening from

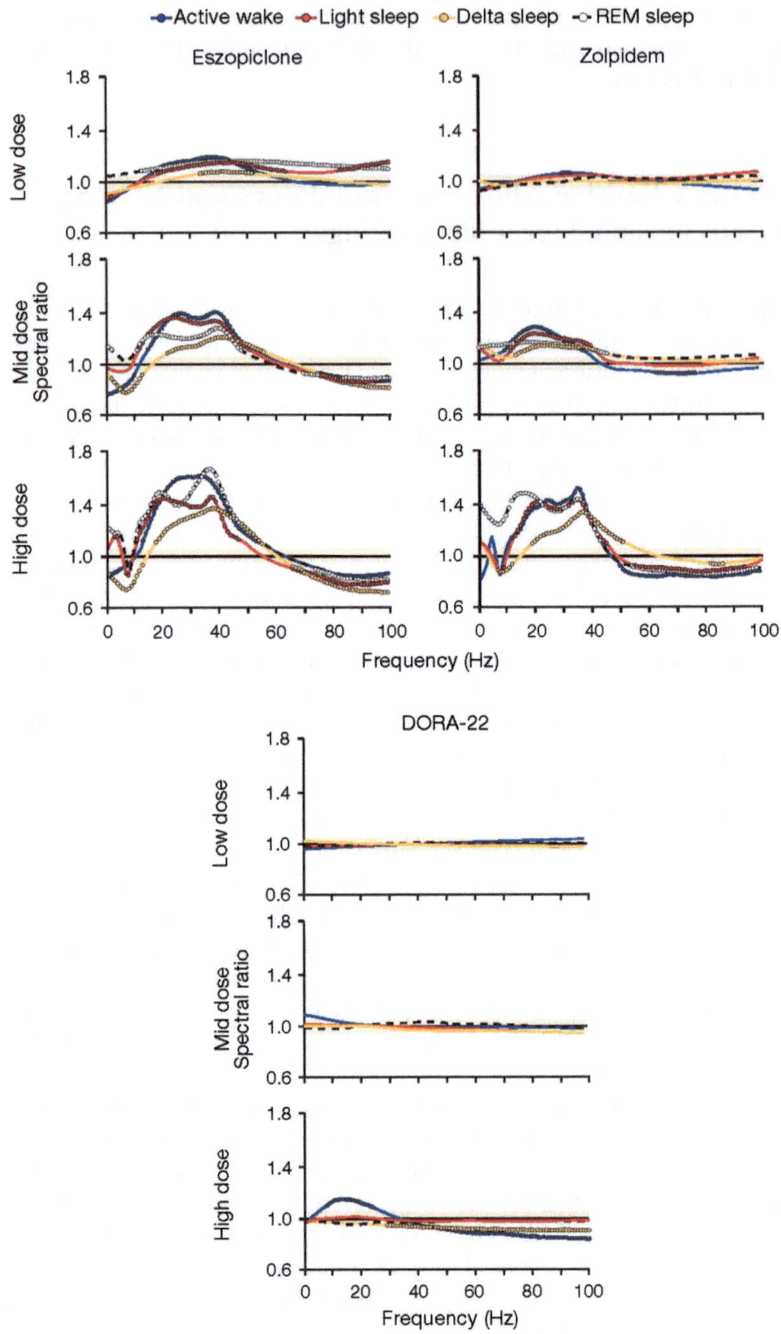
Active wake
Light sleep
Delta sleep
REM sleep
Eszopiclone
Zolpidem
Low dose
Mid dose
Spectral ratio
High dose
1.8
1.4
1.0
0.6
0
20
40
60
80
100
Frequency (Hz)
DORA-22

◀ **Fig. 1** Gamma aminobutyric acid (*$GABA_A$*) modulators (eszopiclone and zolpidem) dose-responsively alter electroencephalogram (*EEG*) spectral frequency of sleep/wake states in inactive-phase dosing, whereas DORA-22 produces sleep more akin to vehicle-treated animals. Shown are EEG spectral changes (as a ratio of treatment over vehicle) from 1 to 100 Hz within each sleep/wake state for each treatment at respective doses (low, mid, and high). Horizontal shaded area represents 1 (no change from vehicle) ±5 % as observed baseline dosing effect. Ninety-five percent confidence bounds were calculated, and *circles* indicate nonoverlapping areas above/below baseline dosing effect and 95 % confidence intervals for each sleep/wake state. All treatments and doses were collected as independent 3-day crossover study designs in male Sprague-Dawley rats (n = 16) during the inactive phase (ZT23). [Figure reproduced with permission from Fox et al. (2013)]

DORA or GABAA-modulator treatment, a topic that will be addressed under the discussion below on arousability.

The effects of DORAs versus GABAA-positive allosteric modulators on spectral frequency during each of the sleep stages have also been characterized. In a human study, it was shown that the DORA SB-649868 had negligible effect on spectral frequency during either REM or NREM sleep compared to placebo. In contrast, zolpidem (10 mg) significantly impacted spectral frequency during NREM sleep compared to placebo, increasing power density in very low frequencies (0.25–1 Hz) and decreasing it in frequencies between 2.25 and 11 Hz (Bettica et al. 2012). We extended these findings in a preclinical study in which the effects of several doses of eszopiclone, zolpidem, and the dual orexin receptor antagonist DORA-22 were compared in rats following administration during the active or inactive phase (Fox et al. 2013). Doses of each compound were chosen to produce equivalent effects on total sleep. Both eszopiclone and zolpidem dose-dependently disrupted the spectral frequencies of active wake, NREM sleep, and REM sleep relative to vehicle, independent of when either compound was dosed. In contrast, only the highest dose of DORA-22 produced modest changes in the spectral profile of REM when given in the active, but not inactive phase (Fig. 1). These findings in both humans and preclinical species suggest that the spectral frequency changes produced by GABAA-positive allosteric modulators are quite different from those observed during nonpharmacologically induced sleep, whereas the effects of DORAs are quite similar.

5 GABAA-Positive Allosteric Modulators and DORAs: Potential Differences in Cognitive and Behavioral Effects

As alluded to above, GABA is the major inhibitory neurotransmitter in the brain and the GABAA α_1 subunit–containing receptor, upon which the standard-of-care therapeutics interact, is found throughout the brain. Therefore, it is not surprising that these compounds produce a variety of effects in addition to impacting sleep. Most notably, compounds such as zolpidem and eszopiclone have been shown to produce cognitive disruption in human subjects, including deficits in attention and memory (Warot et al. 1987; Kuitunen et al. 1990; Balkin et al. 1992;

Berlin et al. 1993; Roehrs et al. 1994; Allain et al. 1995; Wesensten et al. 1995; Rush and Griffiths 1996; Verster et al. 2002; Leufkens et al. 2009), and motor disturbances, such as ataxia and loss of balance (Drover 2004; Allain et al. 2005; Zammit et al. 2008; de Haas et al. 2010). These findings are aligned with the high level of GABAA α_1 subunit-containing receptor found in the hippocampus and cortex, key regions involved in cognition, and the cerebellum and striatum, important for motor function. Many of these effects are also quite similar to those of ethanol, which also nonselectively impacts GABA receptor function.

Very few studies have directly compared the effects of GABAA-positive allosteric modulators versus DORAs on cognitive performance, but those that have suggest DORAs might be better tolerated. In a series of studies in rat and nonhuman primates, we evaluated the effects of zolpidem, eszopiclone, and DORA-22 on recognition memory in rats and working memory and attention in rhesus macaques (Uslaner et al. 2013). Compound testing was conducted shortly after dosing, while drug levels were at or near their maximal levels. The GABAA-positive allosteric modulators produced cognitive disruption in all three measures in both species at doses that were below or equal to those that promoted sleep. In contrast, DORA-22 increased sleep at doses 30-fold lower than the dose that impacted recognition memory, and no dose of DORA-22 tested impacted working memory or attention in rhesus macaques (Fig. 2). These findings are very similar to those of a more recent publication by Morairty et al. (2014), who demonstrated that, at doses producing equivalent levels of sleep, another DORA, almorexant, had no effect on spatial reference or spatial working memory in rats, whereas the effect of zolpidem was impairing. Finally, in rhesus macaques we have demonstrated that DORA-22 produced no next-day effects on working memory or attention, whereas diazepam produced next-day effects on both measures and eszopiclone produced effects on attention (Gotter et al. 2013). Importantly, in humans a dose of suvorexant five- or ten-fold greater than that necessary to promote sleep had effects on next-day subjective alertness (Sun et al. 2013). Future studies aimed at better characterizing the cognitive effects of these different mechanisms of action are warranted to determine the therapeutic window of each approach.

In addition to potential differences in the cognitive impact of these drugs in subjects who are awake, it appears that DORAs and the GABAA-positive allosteric modulators may also differentially impact the ability of salient stimuli to arouse the individual when asleep. Specifically, it is important for survival and safety to be able to wake to a meaningful stimulus, such as the calling of one's name, a fire alarm, or the sound of a predator, but to sleep through irrelevant stimuli, such as a bird chirping or background noise. We have demonstrated that dogs given vehicle or DORA-22 are much more likely to be awakened from sleep when exposed to a relevant, salient stimulus that has been previously paired with reward versus a nonrelevant, neutral stimulus (Tannenbaum et al. 2014). In stark contrast, rhesus macaques given eszopiclone or diazepam displayed no difference in their ability to awaken to a salient versus neutral stimulus, rarely waking to either (Tannenbaum et al. 2013) (Fig. 3). These results suggest that, at least in nonhuman primates, DORAs protect the ability to respond to salient stimuli when awake, whereas

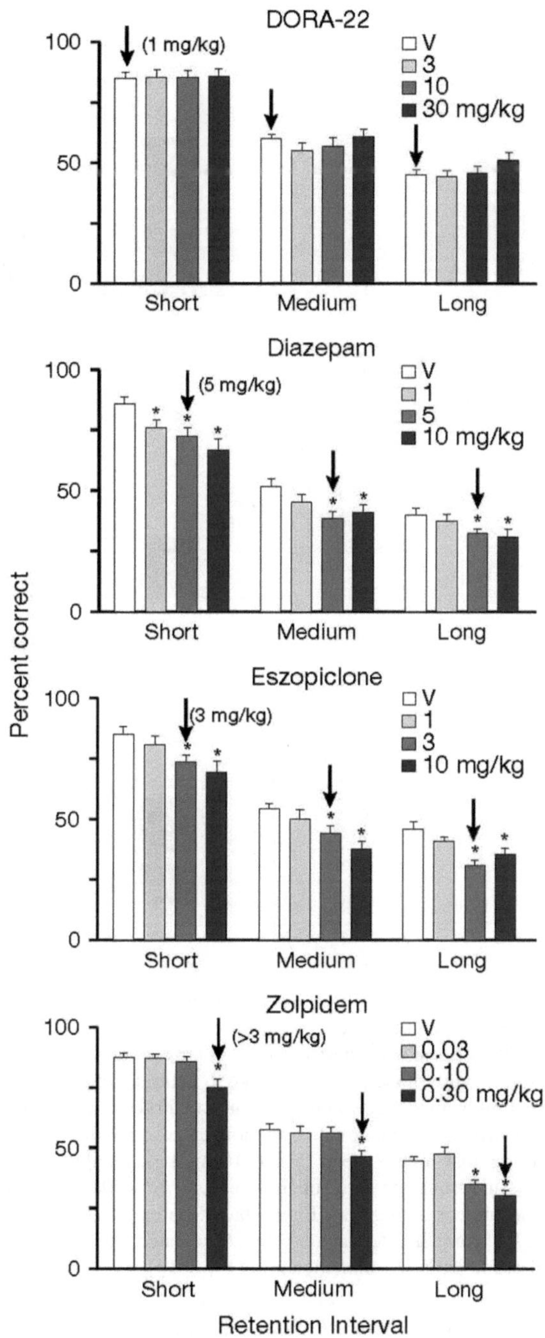

Fig. 2 Percentage correct on a working memory task, delayed match to sample, in rhesus monkeys. Animals were given stimuli and needed to remember them over a short, medium, or long duration. Eszopiclone (3 and 10 mg/kg), diazepam (1, 5, and 10 mg/kg), and zolpidem (0.3 mg/kg) impaired performance, whereas no dose of DORA-22 impacted performance (*asterisk* indicates significantly different from vehicle). The minimum effective doses to produce sleep as measured by electroencephalogram were as follows: DORA-22 1 mg/kg, eszopiclone 3 mg/kg, diazepam 5 mg/kg, and zolpidem >3 mg/kg (as indicted by *arrows*). [Figure reproduced with permission from Uslaner et al. (2013)]

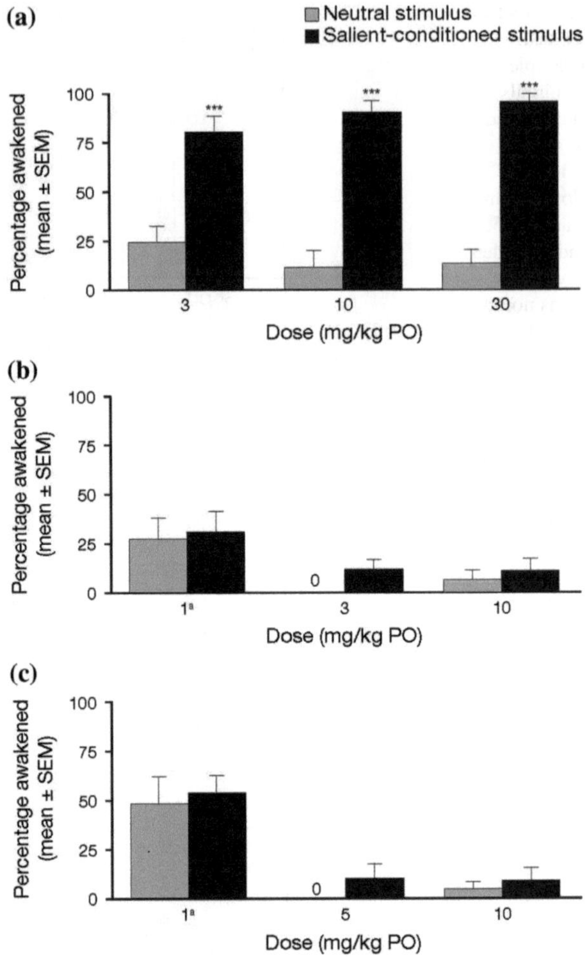

Fig. 3 Arousal from sleep to emotionally salient-conditioned or neutral acoustical stimuli at maximal nighttime drug exposure of **a** DORA-22, **b** eszopiclone, or **c** diazepam. **a** During DORA-22 sleep, rhesus monkeys (n = 12) woke to salient-conditioned acoustical stimuli significantly more than they woke to neutral stimuli. **b** During eszopiclone treatment, monkeys did not discriminate between salient and neutral stimuli; monkeys tended to sleep through both stimuli. **c** During diazepam treatment, monkeys did not discriminate between salient and neutral stimuli; at doses that induce sleep (5 and 10 mg/kg), monkeys slept through both stimuli. Data shown are mean ± standard error of the mean. $^{***}P < 0.001$. [a]Nonsedating daytime dose as measured by electroencephalogram. [Figure based on data from a study presented by Tannenbaum et al. at the 27th Annual Meeting of the Associated Professional Sleep Societies (APSS), June 2013, Baltimore, MD (Tannenbaum et al. 2013]

GABAA-positive allosteric modulators do not. Furthermore, these findings indicate that alternate arousal pathways responding to strong alerting signals remain intact while orexin signaling is silenced, as is observed when unmedicated subjects are awakened during normal sleep (and when endogenous orexin signaling is silent).

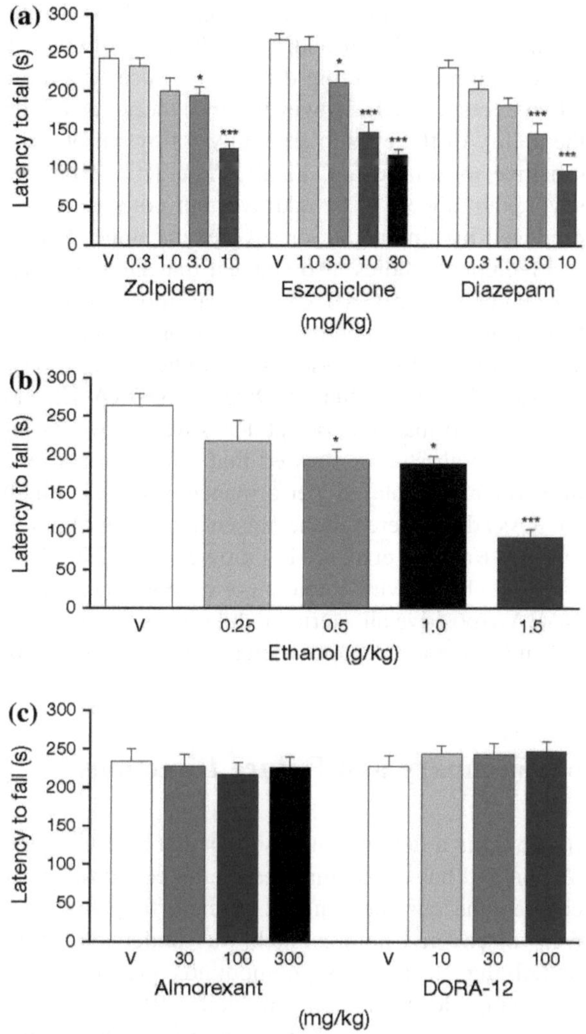

Fig. 4 Dose—response curve for latency to fall from rotarod for rats administered **a** GABAA receptor modulators, **b** ethanol, and **c** orexin receptor antagonists. Acute administration of zolpidem, eszopiclone, diazepam, and ethanol dose-dependently impaired rotarod performance. In contrast, the orexin receptor antagonists almorexant and DORA-12 did not impair rotarod performance after acute administration. Data shown are mean ± standard error of the mean. *$P < 0.05$, ***$P < 0.001$ versus vehicle-treated group. [Figure reproduced with permission from Ramirez et al. (2013)]

This could have important consequences, as experimental data in human subjects show that GABAA-positive allosteric modulators impair the ability to arouse to salient stimuli, such as a fire alarm (Johnson et al. 1987; Mendelson et al. 1988).

In addition to potentially differentiating with respect to cognitive performance, there is some evidence that the effects of these two mechanisms of action differ with respect to motor impairment. GABAA-positive allosteric modulators have been shown to produce motor disturbances, including loss of balance and an increased likelihood of falls, particularly in the elderly (Vermeeren 2004; Drover 2004; Allain et al. 2005; Zammit et al. 2008; de Haas et al. 2010). Far fewer clinical studies have characterized the motor effects of DORAs in humans. A very high dose of almorexant (1000 mg) was found to impact body sway; the effect was less pronounced than for zolpidem (10 mg), and it appeared to tolerate with repeated dosing (Hoever et al. 2010, 2012). In addition, whereas zolpidem produced reports of abnormal coordination and "feeling drunk," no dose of almorexant produced such an effect (Hoever et al. 2010). However, other measures of sensorimotor processing, such as the ability of the eyes to smoothly follow an object or saccade, appear to be more sensitive to almorexant. Finally, interactions between alcohol and almorexant or suvorexant appear to be additive, but not synergistic, to any of the clinical effects produced by almorexant or suvorexant alone (Hoch et al. 2013).

Preclinical studies also suggest that DORAs might have less impact on motor coordination than GABAA-positive allosteric modulators. Using the rotarod performance test, in which the ability of an animal to maintain its balance on a rotating accelerating rod is assessed, two studies (Steiner et al. 2011; Ramirez et al. 2013) reported that two different DORAs, DORA-12 and almorexant, did not disrupt rotarod performance in rats at doses well above those necessary to produce sleep. In contrast, both studies showed that zolpidem and eszopiclone produced significant impairment in rotarod performance (Fig. 4). Finally, both studies showed that DORAs administered in the presence of ethanol did not impair rotarod performance, whereas Ramirez et al. (2013) showed that ethanol produced synergistic impairment when combined with zolpidem or eszopiclone. The interaction between ethanol and GABAA-positive allosteric modulators is likely due to the fact that these compounds act on the same receptor, whereas DORAs act on a distinct set of neuronal circuits.

6 Summary and Future Directions

Insomnia is a debilitating disorder that affects a significant proportion of the population and has substantial health and economic consequences. As reviewed in this chapter, the current standard of care, the GABAa positive allosteric modulators, were discovered before much was understood regarding the neural mechanisms underlying sleep. These compounds produce a range of effects in addition to increasing sleep, including reducing REM and disrupting the spectral frequency observed during normal sleep, impairing cognition and disrupting motor coordination. In comparison, the DORAs have only recently been developed and as a result have less clinical characterization, but may offer some potential advantages. The discovery and development of orexin receptor antagonists was a concerted effort of rational drug design stemming from a core understanding of orexin biology, genetics, sleep physiology, and neuroanatomy. It was hypothesized that DORAs would promote sleep without disrupting spectral frequency and impairing cognition and motor coordination, and preclinical and clinical data appear consistent with this hypothesis. More clinical experience is necessary to better understand the effect of DORAs in humans and how they compare with the standard of care. Suvorexant is currently under evaluation by regulatory agencies for the treatment of insomnia, and if approved, would represent the first DORA available for patients suffering from this sleep disorder.

References

Allain H, Patat A et al (1995) Comparative study of the effects of zopiclone (7.5 mg), zolpidem, flunitrazepam and a placebo on nocturnal cognitive performance in healthy subjects, in relation to pharmacokinetics. Eur Psychiatry 10(Suppl 3):129s–135s

Allain H, Bentue-Ferrer D et al (2005) Postural instability and consequent falls and hip fractures associated with use of hypnotics in the elderly: a comparative review. Drugs Aging 22(9):749–765

Balkin TJ, O'Donnell VM et al (1992) Comparison of the daytime sleep and performance effects of zolpidem versus triazolam. Psychopharmacology 107(1):83–88

Balter MB, Uhlenhuth EH (1991) The beneficial and adverse effects of hypnotics. J Clin Psychiatry 52(Suppl):16–23

Berlin I, Warot D et al (1993) Comparison of the effects of zolpidem and triazolam on memory functions, psychomotor performances, and postural sway in healthy subjects. J Clin Psychopharmacol 13(2):100–106

Bettica P, Squassante L et al (2012) Differential effects of a dual orexin receptor antagonist (SB-649868) and zolpidem on sleep initiation and consolidation, SWS, REM sleep, and EEG power spectra in a model of situational insomnia. Neuropsychopharmacology 37(5):1224–1233

Bonnet MH (1983) Memory for events occurring during arousal from sleep. Psychophysiology 20 (1):81–87

Brisbare-Roch C, Dingemanse J et al (2007) Promotion of sleep by targeting the orexin system in rats, dogs and humans. Nat Med 13(2):150–155

Broughton RJ (1968) Sleep disorders: disorders of arousal? Enuresis, somnambulism, and nightmares occur in confusional states of arousal, not in "dreaming sleep". Science 159 (3819):1070–1078

Brunner DP, Dijk DJ et al (1991) Effect of zolpidem on sleep and sleep EEG spectra in healthy young men. Psychopharmacology 104(1):1–5

Chemelli RM, Willie JT et al (1999) Narcolepsy in orexin knockout mice: molecular genetics of sleep regulation. Cell 98(4):437–451

Chong Y, Fryer CD et al (2013) Prescription sleep aid use among adults: United States, 2005–2010. NCHS Data Brief 127:1–8

de Haas SL, Schoemaker RC et al (2010) Pharmacokinetics, pharmacodynamics and the pharmacokinetic/pharmacodynamic relationship of zolpidem in healthy subjects. J Psychopharmacol 24(11):1619–1629

de Lecea L, Kilduff TS et al (1998) The hypocretins: hypothalamus-specific peptides with neuroexcitatory activity. Proc Natl Acad Sci USA 95(1):322–327

Dijk DJ, James LM et al (2010) Sex differences and the effect of gaboxadol and zolpidem on EEG power spectra in NREM and REM sleep. J Psychopharmacol 24(11):1613–1618

Dinges DF (1984) The nature and timing of sleep. Trans Stud Coll Physicians Phila 6(3):177–206

Drover DR (2004) Comparative pharmacokinetics and pharmacodynamics of short-acting hypnosedatives: zaleplon, zolpidem and zopiclone. Clin Pharmacokinet 43(4):227–238

Foundation NS (2006) Sleep in America poll, Washington DC

Fox SV, Gotter AL et al (2013) Quantitative electroencephalography within sleep/wake states differentiates GABAA modulators eszopiclone and zolpidem from dual orexin receptor antagonists in rats. Neuropsychopharmacology 38(12):2401–2408

Gotter AL, Roecker AJ et al (2013a) Orexin receptors as therapeutic drug targets. Prog Brain Res 198:163–188

Gotter AL, Winrow CJ et al (2013b) The duration of sleep promoting efficacy by dual orexin receptor antagonists is dependent upon receptor occupancy threshold. BMC Neurosci 14:90

Hajak G, Petukhova M et al (2011) Days-out-of-role associated with insomnia and comorbid conditions in the America Insomnia Survey. Biol Psychiatry 70(11):1063–1073

Herculano-Houzel S (2012) Brains matter, bodies maybe not: the case for examining neuron numbers irrespective of body size. Ann N Y Acad Sci 1225:191–199

Herculano-Houzel S, Lent R (2005) Isotropic fractionator: a simple, rapid method for the quantification of total cell and neuron numbers in the brain. J Neurosci 25:2518–2521
Herring WJ, Snyder E et al (2012) Orexin receptor antagonism for treatment of insomnia: a randomized clinical trial of suvorexant. Neurology 79(23):2265–2274
Hoch M, Hay JL et al (2013) Dual orexin receptor antagonism by almorexant does not potentiate impairing effects of alcohol in humans. Eur Neuropsychopharmacol 23(2):107–117
Hoever P, de Haas S et al (2010) Orexin receptor antagonism, a new sleep-promoting paradigm: an ascending single-dose study with almorexant. Clin Pharmacol Ther 87(5):593–600
Hoever P, de Haas SL et al (2012) Orexin receptor antagonism: an ascending multiple-dose study with almorexant. J Psychopharmacol 26(8):1071–1080
Johnson LC, Spinweber CL et al (1987) Dose level effects of triazolam on sleep and response to a smoke detector alarm. Psychopharmacology 91(4):397–402
Kanno O, Sasaki T et al (2000) Comparison of the effects of zolpidem and triazolam on nocturnal sleep and sleep latency in the morning: a cross-over study in healthy young volunteers. Prog Neuropsychopharmacol Biol Psychiatry 24(6):897–910
Kessler RC, Berglund PA et al (2011) Insomnia and the performance of US workers: results from the America insomnia survey. Sleep 34(9):1161–1171
Kuitunen T, Mattila MJ et al (1990) Actions and interactions of hypnotics on human performance: single doses of zopiclone, triazolam and alcohol. Int Clin Psychopharmacol 5(Suppl 2):115–130
Kuppermann M, Lubeck DP et al (1995) Sleep problems and their correlates in a working population. J Gen Intern Med 10(1):25–32
Lancel M (1999) Role of GABAA receptors in the regulation of sleep: initial sleep responses to peripherally administered modulators and agonists. Sleep 22(1):33–42
Leufkens TR, Lund JS et al (2009) Highway driving performance and cognitive functioning the morning after bedtime and middle-of-the-night use of gaboxadol, zopiclone and zolpidem. J Sleep Res 18(4):387–396
Lin L, Faraco J et al (1999) The sleep disorder canine narcolepsy is caused by a mutation in the hypocretin (orexin) receptor 2 gene. Cell 98(3):365–376
Loomis AL, Harvey EN et al (1936) Brain potentials during hypnosis. Science 83(2149):239–241
Lubin A, Hord DJ et al (1976) Effects of exercise, bedrest and napping on performance decrement during 40 hours. Psychophysiology 13(4):334–339
Mai E, Buysse DJ (2008) Insomnia: prevalence, impact, pathogenesis, differential diagnosis, and evaluation. Sleep Med Clin 3(2):167–174
Marcus JN, Aschkenasi CJ et al (2001) Differential expression of orexin receptors 1 and 2 in the rat brain. J Comp Neurol 435(1):6–25
Mendelson WB, Martin JV et al (1988) Effects of flurazepam on sleep, arousal threshold, and the perception of being asleep. Psychopharmacology 95(2):258–262
Michelson D, Snyder E et al (2013) Safety and efficacy of suvorexant during 1-year treatment of insomnia with subsequent abrupt treatment discontinuation: a phase 3 randomised, double-blind, placebo-controlled trial. Lancet Neurol 13(5):461–471
Mohler H, Fritschy JM et al (2002) A new benzodiazepine pharmacology. J Pharmacol Exp Ther 300(1):2–8
Montgomery P, Dennis J (2004) A systematic review of non-pharmacological therapies for sleep problems in later life. Sleep Med Rev 8(1):47–62
Morairty SR, Wilk AJ et al (2014) The hypocretin/orexin antagonist almorexant promotes sleep without impairment of performance in rats. Front Neurosci 8:3
Morin CM, Belanger L et al (2009) The natural history of insomnia: a population-based 3-year longitudinal study. Arch Intern Med 169(5):447–453
Nishino S, Ripley B et al (2000) Hypocretin (orexin) deficiency in human narcolepsy. Lancet 355 (9197):39–40
Nutt DJ, Malizia AL (2001) New insights into the role of the GABA(A)-benzodiazepine receptor in psychiatric disorder. 179:390–396
Ohayon MM (2002) Epidemiology of insomnia: what we know and what we still need to learn. Sleep Med Rev 6(2):97–111

Ohayon MM, Guilleminault C et al (2010) Sleep epidemiology 30 years later: where are we? Sleep Med 11(10):961–962

Ramirez AD, Gotter AL et al (2013) Dual orexin receptor antagonists show distinct effects on locomotor performance, ethanol interaction and sleep architecture relative to gamma-aminobutyric acid-A receptor modulators. Front Neurosci 7:254

Rechtschaffen A, Kales A (1968) A manual of standardized terminology, techniques and scoring system for sleep stages of human subjects, Los Angeles: Brain Information Service/Brain Research Institute, University of California

Roehrs T, Merlotti L et al (1994) Sedative, memory, and performance effects of hypnotics. Psychopharmacology 116(2):130–134

Roth T, Ancoli-Israel S (1999) Daytime consequences and correlates of insomnia in the United States: results of the 1991 National Sleep Foundation Survey. II. Sleep 22(Suppl 2):S354–S358

Roth T, Franklin M et al (2007) The state of insomnia and emerging trends. Am J Manag Care 13 (5 Suppl):S117–S120

Roth T, Coulouvrat C et al (2011) Prevalence and perceived health associated with insomnia based on DSM-IV-TR; international statistical classification of diseases and related health problems, tenth revision; and research diagnostic criteria/international classification of sleep disorders, second edition criteria: results from the America Insomnia Survey. Biol Psychiatry 69(6):592–600

Rudolph U, Crestani F et al (1999) Benzodiazepine actions mediated by specific gamma-aminobutyric acid(A) receptor subtypes. Nature 401(6755):796–800

Rush CR, Griffiths RR (1996) Zolpidem, triazolam, and temazepam: behavioral and subject-rated effects in normal volunteers. J Clin Psychopharmacol 16(2):146–157

Sakurai T (2007) The neural circuit of orexin (hypocretin): maintaining sleep and wakefulness. Nat Rev Neurosci 8(3):171–181

Sakurai T, Amemiya A et al (1998) Orexins and orexin receptors: a family of hypothalamic neuropeptides and G protein-coupled receptors that regulate feeding behavior. Cell 92(4):573–585

Schoenborn CA, Adams PF et al (2004) Health behaviors of adults: United States, 1999-2001. Vital Health Stat 10(219):1–79

Scott EM (1968) Dreams of alcoholics. Percept Mot Skills 26(Suppl 3):1315–1318

Steiner MA, Lecourt H et al (2011) Differential effects of the dual orexin receptor antagonist almorexant and the GABA(A)-alpha1 receptor modulator zolpidem, alone or combined with ethanol, on motor performance in the rat. Neuropsychopharmacology 36(4):848–856

Stones MJ (1977) Memory performance after arousal from different sleep stages. Br J Psychol 68 (2):177–181

Sun H, Kennedy WP et al (2013) Effects of suvorexant, an orexin receptor antagonist, on sleep parameters as measured by polysomnography in healthy men. Sleep 36(2):259–267

Tannenbaum PL, Stevens J, Tye SJ, Fox SV, Binns J, Savitz AT, Kuduk SD, Uslaner JM, Winrow CJ, Renger JJ (2013) Dual orexin receptor antagonist sleep retains the ability to wake with emotionally salient acoustic stimuli yet sleep through neutral stimuli in dog and monkey models. Poster 0062 presented at the 27th Annual Meeting of the Associated Professional Sleep Societies, Baltimore, Maryland, USA, 1 June–5 June

Tannenbaum PL, Stevens J et al (2014) Orexin receptor antagonist-induced sleep does not impair the ability to wake in response to emotionally salient acoustic stimuli in dogs. Front Behav Neurosci 8:182

Thannickal TC, Moore RY et al (2000) Reduced number of hypocretin neurons in human narcolepsy. Neuron 27(3):469–474

Uslaner JM, Tye SJ et al 2013 Orexin receptor antagonists differ from standard sleep drugs by promoting sleep at doses that do not disrupt cognition. Sci Transl Med 5(179):179ra44

Vermeeren A (2004) Residual effects of hypnotics: epidemiology and clinical implications. CNS Drugs 18(5):297–328

Verster JC, Volkerts ER et al (2002) Residual effects of middle-of-the-night administration of zaleplon and zolpidem on driving ability, memory functions, and psychomotor performance. J Clin Psychopharmacol 22(6):576–583

Warot D, Bensimon G et al (1987) Comparative effects of zopiclone, triazolam and placebo on memory and psychomotor performance in healthy volunteers. Fundam Clin Pharmacol 1 (2):145–152

Wesensten NJ, Balkin TJ et al (1995) Effects of daytime administration of zolpidem versus triazolam on memory. Eur J Clin Pharmacol 48(2):115–122

Winrow CJ, Gotter AL et al (2011) Promotion of sleep by suvorexant-a novel dual orexin receptor antagonist. J Neurogenet 25(1–2):52–61

Zammit G, Wang-Weigand S et al (2008) Use of computerized dynamic posturography to assess balance in older adults after nighttime awakenings using zolpidem as a reference. BMC Geriatr 8:15

Pathway and Effect of Intranasal Orexin

Sara Lena Weinhold, Robert Göder and Paul Christian Baier

Abstract The neurotransmitter orexin has been shown to modulate wakefulness and sleep and other functions regarding cognition and the metabolic system. Because of the link between orexin-A deficiency and the disorder narcolepsy, effects of orexin on sleep and wakefulness are of particular interest for clinical purposes. Indeed, the transfer of results of animal studies to study protocols with humans is problematic. The utilized methods used in animals are not realizable in patients for ethical reasons, such as intracerebroventricular injection, possible side effects, and less effectiveness of intravenous and oral orexin. One way to avoid these problems is the intranasal administration of drugs. We present rare results of the relatively new procedure for administering orexin.

1 The Pathway of Intranasal Orexin

The manipulation of the central nervous orexinergic system by exogenous administration of orexin-A has been well studied in animal experiments. Whereas direct orexin-A modulates rapid eye movement sleep through activation of locus coeruleus neurons (Bourgin et al. 2000) and intracerebroventricular injections lead to relevant functional effects, this does not seem to be the case for intravenous administration which is limited by a tight blood brain barrier (Fujiki et al. 2003). Since direct administration to the central nervous system in humans is not a feasible option, the question arose whether the intranasal administration of this substance might be an option to bypass the blood brain barrier.

The intranasal administration of several substances has had a long tradition, starting with the use of intoxicants, such as cocaine and tobacco, and later continuing with the administration of drugs and proteins aiming at systemic rather

S.L. Weinhold (✉) · R. Göder · P.C. Baier
Department of Psychiatry and Psychotherapy, Centre for Sleep Medicine, University Kiel, Niemannsweg 147, 24105 Kiel, Germany
e-mail: s.weinhold@zip-kiel.de

© Springer International Publishing Switzerland 2015
T. Sakurai et al. (eds.), *Orexin and Sleep*,
DOI 10.1007/978-3-319-23078-8_18

than local effects (Born et al. 2002; Djupesland et al. 2014; Illum 2000; Thorne et al. 1995; Sakane et al. 1995). Evidence that molecules of different sizes can reach the central nervous system after intranasal administration is available for several species: Studies have shown functional effects of intranasal substance administration within one hour in rodents (Thorne et al. 2004; Ross et al. 2004, 2008; Sakane et al. 1995; Schipper et al. 1993; Schulz et al. 2004; Van den Berg et al. 2003, 2004; Wang et al. 2007; Westin et al. 2005, 2006), non-human primates (Balin et al. 1986; Deadwyler et al. 2007; Parker et al. 2005; Thorne et al. 2008), and humans (Van den Berg et al. 2003, 2004; Hallschmid et al. 2004; Benedict et al. 2011, 2007; Born et al. 2002; Hermens et al. 1992; Reger et al. 2008). As for orexin-A, there are animal studies with radio-labeled substance in mice (Hanson et al. 2004), rats (Dhuria et al. 2009a, b) and primates (Deadwyler et al. 2007) which have shown that this particular peptide can reach the central nervous system by the intranasal pathway.

Currently, there is strong evidence that intranasally administrated substances can indeed reach the brain, however the pathways are not fully understood. Some of the possible ways will be discussed below.

1.1 Vascular Pathway

There are two types of epithelia lining the inner walls of the nose. One is the ciliated respiratory epithelium lining approximately 90 % of the nasal cavity in humans, which warms and humidifies the inspired air and removes particulates, microorganisms, and allergens. The other is the olfactory epithelium whose blood vessel density is much lower and whose essential function is to host the olfactory receptor neurons and the so-called olfactory ensheathing cells, which permit olfactory sensation.

The most obvious way intranasally administered drugs can become systemically effective is by the parenteral uptake of substances through the highly vascularized respiratory mucosa which allows substances to enter the circulation. Thin walls of capillaries in the ciliated epithelium are semipermeable and have mostly 60–80 nm pores. In this way, smaller but also larger molecules can enter the systemic blood circulation. However, this path is limited by the fact that substances administered this way can only reach the brain after passing the blood brain barrier. Therefore, this pathway is particularly important for small, lipophilic molecules. Large and hydrophilic compounds, such as peptides and proteins like orexin-A, would probably not reach the brain sufficiently through this pathway (Dhuria et al. 2010). Furthermore, the distribution of substances via the systemic circulation is limited by renal and hepatic elimination, degradation by plasma proteases, and possible peripheral side effects.

However, there is increasing evidence for an alternative pathway for the delivery of intranasal substances through blood vessels but bypassing the systemic circulation: This could be by means of a perivasal space within the lymphatic system which is anatomically limited by the outer layer of the blood vessels and the basal membrane of the surrounding tissue (Pollock et al. 1997). This transcapillary transport seems not only to occur by directed diffusion but might also be driven by bulk flow due to arterial pulsation (Groothuis et al. 2007; Rennels et al. 1990). Evidence for this pathway and its relevance for the intranasal uptake of substances came from histological studies, in which insulin-like growth factor 1 (Thorne et al. 2004) and interferon-ß (Thorne et al. 2008) were found in high concentrations within the walls of cerebral vessel after intranasal application. Dhuria and colleagues (2009a) have made similar observations for orexin-A.

1.2 Pathway Along the Olfactory Bulb

The olfactory epithelium covers the posterior area of the nasal cavity. Apart from the olfactory receptor neurons, which are the chemoreceptive first neurons of the olfactory system, the olfactory epithelium consists of basal and supporting cells. The olfactory receptor neurons are bipolar cells whose apical dendrites terminate in the nasal cavity and whose axons extend in unmyelinated bundles (the so-called *fila olfactoria*) through the *lamina cribrosa* of the ethmoid bone. In the olfactory bulb they form the first synapses of the olfactory pathway. The pathway alongside the neurons of these axons is another possible mechanism of how substances might penetrate the central nervous system after intranasal administration. Evidence for this comes from observations that intranasally administered peptides, such as the glucagon-like-peptide-1 (Ross et al. 2004), the galanin-like peptide (Baker and Spencer 1986), interferon-ß (Dhuria et al. 2009b; Field et al. 2003), and orexin-A (Nonaka et al. 2008), already show high concentrations in the olfactory bulb within the first hour after intranasal administration.

Since intracellular transport of substances along the olfactory axons would take hours to days (Li et al. 2005a), this might not be the prevailing way uptake occurs. It is more likely that the substance transport to the brain occurs along the nervous cells extracellularly. Of importance here is a peculiarity of the transition between the peripheral and central nervous system in the olfactory system which is characterized by the presence of a special type of Schwann cells called olfactory neuron ensheathing cells. These cells play a special role in the axonal regeneration of the olfactory receptor neurons by enveloping them in a non-myelinating way (Li et al. 2005b; Luzzati et al. 2004a, b), and form a continuous perineural space from the olfactory epithelium to the olfactory bulb, filled with liquid. Along these perineural channels substances can rapidly reach the central nervous system, not only by directed diffusion but also via bulk flow. It has been shown that intranasally administered stem cells can reach the central nervous system along these perineural channels (Jiang et al. 2011; Danielyan et al. 2009).

Also other pathways, e.g. alongside the trigeminal nerve or the lymphatic system, are discussed. However, the abovementioned pathways, or most likely a combination thereof, seem to be a plausible explanation for the rapid central nervous system penetration of intranasally administered substances. In contrast to animals there is currently no proof in humans that intranasally administered orexin-A actually reaches the brain. However, there is functional evidence for the efficacy of intranasal administration, which will be discussed below.

2 The Effect of Intranasal Orexin-A

From animal studies we know that intranasal orexin-A reaches the brain within one hour (Hanson et al. 2004; Dhuria et al. 2009a; Deadwyler et al. 2007). The nasal delivery method shows higher effectiveness compared to intravenous orexin application despite a 10-fold lower dose, which has been shown by test performance, and PET scans in non-human monkeys (Deadwyler et al. 2007). There would be less non-central side effects after intranasal application as there are significant lower orexin concentrations in peripheral organs after intranasal administration compared to intravenous administration (Hanson et al. 2004). Furthermore, Deadwyler and colleagues (2007) found that different doses of intravenous orexin (up to 10.0 μg/kg) did not lead to an increase of this substance in the cerebrospinal fluid, whereas 1.0 μg/kg orexin administered directly to the nasal mucosa produced an significant increase 10 min after application.

2.1 The Effect of Intranasal Orexin on Cognitive Functions

The maximum activity of orexin-containing neurons during waking state is characterized by exploratory activity (Mileykovskiy et al. 2005; Estabrooke et al. 2001), supporting a role of orexin in attention and memory, e.g. remembering explored environment. Direct orexin injections to the brain have been shown to increase alertness in different species (i.e. Hagan et al. 1999; Piper et al. 2000; Bourgin et al. 2000). Indirect evidence for the role of orexin in cognitive functions has been demonstrated by memory impairment in rodents after selective inactivation of orexin receptors in the hippocampus (Akbari et al. 2006).

In addition, results of studies on intranasal orexin administration on cognitive function are available. Deadwyler and colleagues (2007) have studied the effects on short-term memory performance in non-human monkeys and found a reduction or reversal of deficits caused by sleep deprivation directly after (5–8 min) orexin-A application. In rested monkeys, orexin-A had only marginal effects in their study. Furthermore, they showed in a PET scan study that orexin-A reversed the changes caused by sleep deprivation on thalamic glucose metabolism in a brain region, which is highly susceptible to sleep deficits. Deadwyler and colleagues concluded from their results that orexin-A reverses deficits caused by sleep deprivation and does not influence cognitive performance as such.

3 The Effect of Intranasal Orexin-A on Attention in Humans

Apart from sleep dysregulation, neuropsychological impairments, attention deficit in particular, have been described in narcoleptic patients (Fulda and Schulz 2001). Although narcoleptics showed no differences compared to healthy controls in most short-term attention tasks, deficits in divided attention have been reported (also under stimulant medication) (Fulda and Schulz 2001; Rieger et al. 2003).

In our first experiments on intranasal orexin-A in humans, patients with narcolepsy and cataplexy were tested in the evening directly after intranasal orexin-A administration with the subtest "divided attention" of the test battery for attentional performance (TAP 2.1, Psytest, Herzogenrath, Germany). This test is very sensitive to impairment caused by sleep deprivation, and narcolepsy patients have been shown to be worse in this test (Rieger et al. 2003). However, we did not see any effect of intranasal orexin on divided attention directly after administration compared to placebo administered in this randomized and double-blind study (Baier et al. 2008).

In a second set of experiments in patients with narcolepsy and cataplexy, we administered intranasal orexin-A in the morning and performed attention testing with the same task at 10:40 a.m., 12:15 p.m., and 03:40 p.m. In this study, patients also participated in a "Maintenance of Wakefulness Test" (MWT) at 09:00 a.m., 11:00 a.m., 01:00 p.m., and 03:00 p.m. (see results of MWT below). In this second experiment intranasal orexin-A resulted in improved performance 220, 315, and 520 min after substance administration in the divided attention test. Patients had reduced mean reaction time and significantly fewer false reactions. The other parameters in this test showed no differences. The enhanced performance by orexin was independent of the time after administration in this study (Weinhold et al. 2014).

The results of our second study and the results in sleep-deprived primates demonstrate an effect of orexin-A on cognitive performance, an effect not seen in our first study with intranasal orexin-A in patients. This apparent contradiction might be explained as follows: The experiments from intranasal orexin-A in humans indicate that improvements in cognitive performance, in the divided attention test in particular, are indirectly mediated by stabilized sleep in daytime naps (see below) instead of being a direct consequence of the substance itself. Also, a wake-stabilizing effect, as seen in the non-human primates with an intact orexinergic system, could contribute to the effects on cognitive performance. If this were the case, the lack of effect on attention in our first study would be explained by the fact that the latency between drug administration and testing was with 5 min, too short to establish a measurable effect. In the animal experimental study of Deadwyler and colleagues (2007), testing was done only minutes after the drug administration but lasted longer.

3.1 The Effect of Intranasal Orexin on Sleep

After first research approaches of the role of orexin in weight and appetite regulation, especially motivated by the location of orexin-producing neurons in the hypothalamus, evidence for a central role of orexin in sleep–wake regulation has increased. The hypothesis of a wake-inducing effect by orexin results from intracerebral orexin injections, the observation of disturbed sleep wake cycles in animals with deficient orexin system and the finding of reduced or undetectable liquor orexin level in patients with narcolepsy and cataplexy (Dauvilliers et al. 2003; Heier et al. 2007; Knudsen et al. 2010; Mignot et al. 2002; Nishino et al. 2001). In subsequent studies, it seems that orexin-A as wake-inducing factor may be too simplistic: Patients with narcolepsy, for example, have the same total sleep duration over a 24-h period as do healthy subjects. Instead of being a disorder with an increased sleep demand, narcolepsy seems to be characterized by impairments in maintaining stable wakefulness during daytime on the one hand and in maintaining stable sleep in general and REM sleep in particular during nighttime on the other hand (Montplaisir et al. 1978; Lu et al. 2006; Weinhold et al. 2011).

To analyze the specific consequence of selective orexin cell loss in human narcolepsy and the reversibility by intranasal orexin substitution, we performed polysomnographic recordings and MWT after intranasal orexin-A in human narcolepsy subjects.

In the first study, we administered 435 nmol of human recombinant orexin-A intranasally before night sleep, preceded by an adaption night and followed by a second standard polysomnographic recording. We found intranasal orexin-A to have a clear REM-sleep-wake-stabilizing effect, as indicated by a reduction in direct wake to REM-sleep transitions. The number of total sleep stage changes did not differ. Other REM-sleep-modifying effects were a reduced REM-sleep duration and fewer sleep-onset REM periods. Apart from a marginal wakefulness-inducing effect in the first three hours after sleep onset, no other effects on wakefulness were seen. Furthermore, we observed an unexpected and marginal increase in slow wave sleep (Baier et al. 2011).

In a second study, patients received the same dose of intranasal orexin in the morning. The intranasal administration was preceded by an adaptation night and followed by MWT with sessions at 09:00 a.m., 11:00 a.m., 01:00 p.m. and 03:00 p.m. and a second full night of polysomnographic recording. As shown in our first study on nocturnal sleep, patients had a decreased REM-sleep duration and reduced wake-REM-sleep transitions also at daytime naps after intranasal orexin. The total number of sleep stage changes did not show any differences. Furthermore, marginal REM-sleep-modulating effects were an increase in REM-sleep latency and a reduction in sleep-onset REM periods. There were no effects on sleep latency and on wakefulness after sleep onset. In the night after the test day, patients in the orexin condition displayed an increase in N2 duration and a marginal reduction in the number of wake-REM-sleep transitions and in N3 latency. No significant results were seen for wake duration, REM-sleep duration, sleep onset or REM-sleep latency, total sleep time, and sleep efficiency (Weinhold et al. 2014).

These results provided first evidence of the functional effects of intranasal orexin-A on sleep in narcolepsy and cataplexy. The results are in line with the effects of intracerebral orexin-A administration and the REM-sleep increasing effect of a orexin receptor-2 antagonist in animals (Akanmu and Honda 2005; Bourgin et al. 2000; Kummangal et al. 2013) and with previous observations in animals of a REM-sleep-stabilizing effect of orexin (Piper et al. 2000; Watson et al. 2008). The observation of reduced wake-REM-sleep transitions within up to 24 h after a single dose of orexin suggests a long-lasting REM-sleep stabilizing effect. Comparable results on sleep were also found in an animal narcolepsy model within 24 h of intravenous orexin-A administration (John et al. 2000). Findings in animal studies clearly indicate that central nervous orexin-A administration promotes wakefulness (Hagan et al. 1999; Piper et al. 2000). A hypothetical explanation for the fact that this effect was missing in our studies could be due to instable REM-sleep in the night before testing, leading to stable sleep (instead of stable wakefulness) after intranasal orexin. Another explanation could be a dose-dependent effect. It may be that orexin leads to REM modulation in low doses and to wakefulness in high doses (Bourgin et al. 2000). Improved performance in the attention task could be indicative of a slight wake-promoting effect of intranasal orexin-A. This hypothesis is also supported by better cognitive function in sleep-deprived but not in rested monkeys (Deadwyler et al. 2007).

3.2 The Effect of Intranasal Orexin on Olfactory Function

Since olfaction is an important input for appetite regulation, food-seeking and motivational behavior, modulation of the olfactory system through orexin could link the involvement of orexin-A in weight and appetite regulation, energy consumption, and addictive behavior (Siegel 2004; Ganjavi and Shapiro 2007; Scammell and Saper 2007). Further evidence for the role of orexin in olfactory perception is seen in the mild olfactory dysfunction observed in narcolepsy. It is a very tempting hypothesis that the lack of orexin-A in narcolepsy patients explains their impaired olfactory performance (Stiasny-Kolster et al. 2007; Baier et al. 2008; Bayard et al. 2010; Buskova et al. 2010). Animal experiments support the assumption of olfactory signal processing in the olfactory bulb and in the cortex by centrally synthesized orexins (Caillol et al. 2003) and in the olfactory mucosa by locally synthesized ones (Gorojankina et al. 2007). Intracerebroventricular administration of orexin-A led to an improvement of olfactory detection performance (Julliard et al. 2007).

Apelbaum and colleagues (2005) studied the effect on mitral cell activity after intracerebroventricular orexin application and administration directly to the olfactory bulb. Both ways were found to induce a decrease in spontaneous activity in mitral cells. These results suggest enhanced olfactory performance to be a general orexin function and not only a local/specific effect and could underline a lower scale regulation of the olfactory system. This could allow the threshold for odor detection to be modified depending on the nutritional status. They also showed in this study that direct orexin application on the olfactory bulb led to higher effects compared to applications to the cerebral ventricle. The fact that similar reactions were observed independent of the locus of application, however, provides evidence that the effects of orexin are systemic and not local. The expected changes of responsiveness induced by food and non-food odors mediated by orexin could not be verified by this study (Apelbaum et al. 2005).

Because of experimental data it is not surprising that intranasal orexin-A affects the olfactory threshold in patients with narcolepsy and cataplexy directly after application. The significant decrease in the olfactory threshold scores compared to placebo reflects the results of animal studies (Julliard et al. 2007). It is reasonable to assume that the effects seen on olfactory threshold could be mediated either by a local effect of orexin-A in the nasal mucosa and/or by effects on the central processing of olfactory information (Baier et al. 2008). However, data of intracerebral orexin-A in animals that show similar results support the fact that the increased olfactory threshold after intranasal orexin-A was at least partially due to central orexin-A signaling.

4 Intranasal Orexin-A as Treatment Option in Narcolepsy

The therapy of narcolepsy is currently limited to symptomatic treatment. Substitution of orexin-A in narcolepsy patients is not possible at the moment. Animal experimental results of intracerebral application are convincing but because of the involved risks not feasible for the use in humans.

To better understand the effects of orexin we in the first place need studies using different doses and repeated applications. In our experiments, we used the 20-fold of weight-adapted intranasal orexin dose used in non-humans primates (Deadwyler et al. 2007) and twentieth of the intravenous dose used in dogs (Fujiki et al. 2003). Currently, there are no studies on different doses in humans. This would be very important in order to be able to assess the possibility of intranasal orexin-A administration in narcolepsy treatment. It would be of particular importance, if higher doses had a wake-promoting effect in patients, as indicated by results seen after intracerebral application in animals. Furthermore, the effects of intranasal orexin on other narcolepsy symptoms, such as cataplexies, are of similar relevance. This effect has also been seen in narcoleptic dogs after intravenous orexin administration. To investigate such effects in humans, we need to study multiple orexin administrations and longitudinal analyses, because cataplexies do not occur every day and are triggered by emotions (Mattarozzi et al. 2008).

Still, first insights into the effects of intranasal orexin seem to be promising but need further confirmation in order to be employed in narcolepsy treatment. In summary, intranasal orexin-A can effectively modulate central nervous functions in narcoleptic patients (Baier et al. 2008, 2011; Weinhold et al. 2014). Considering that daytime and sleep-associated symptoms are a major burden of narcoleptic patients, these data may bode well for future pharmacological approaches in the treatment of this disease.

References

Akanmu MA, Honda K (2005) Selective stimulation of orexin receptor type 2 promotes wakefulness in freely behaving rats. Brain Res 1048(1–2):138–145. doi:10.1016/j.brainres.2005.04.064

Akbari E, Naghdi N, Motamedi F (2006) Functional inactivation of orexin 1 receptors in CA1 region impairs acquisition, consolidation and retrieval in Morris water maze task. Behav Brain Res 173(1):47–52. doi:10.1016/j.bbr.2006.05.028

Apelbaum AF, Perrut A, Chaput M (2005) Orexin A effects on the olfactory bulb spontaneous activity and odor responsiveness in freely breathing rats. Regul Pept 129(1–3):49–61. doi:10.1016/j.regpep.2005.01.003

Baier PC, Weinhold SL, Huth V, Gottwald B, Ferstl R, Hinze-Selch D (2008) Olfactory dysfunction in patients with narcolepsy with cataplexy is restored by intranasal Orexin A (Hypocretin-1). Brain: J Neurol 131(Pt 10):2734–2741. doi:10.1093/brain/awn193

Baier PC, Hallschmid M, Seeck-Hirschner M, Weinhold SL, Burkert S, Diessner N, Goder R, Aldenhoff JB, Hinze-Selch D (2011) Effects of intranasal hypocretin-1 (orexin A) on sleep in narcolepsy with cataplexy. Sleep Med 12(10):941–946. doi:10.1016/j.sleep.2011.06.015

Baker H, Spencer RF (1986) Transneuronal transport of peroxidase-conjugated wheat germ agglutinin (WGA-HRP) from the olfactory epithelium to the brain of the adult rat. Exp Brain Res 63(3):461–473

Balin BJ, Broadwell RD, Salcman M, el-Kalliny M (1986) Avenues for entry of peripherally administered protein to the central nervous system in mouse, rat, and squirrel monkey. J Comp Neurol 251 (2):260–280. doi:10.1002/cne.902510209

Bayard S, Plazzi G, Poli F, Serra L, Ferri R, Dauvilliers Y (2010) Olfactory dysfunction in narcolepsy with cataplexy. Sleep Med 11(9):876–881. doi:10.1016/j.sleep.2010.07.004

Benedict C, Hallschmid M, Schmitz K, Schultes B, Ratter F, Fehm HL, Born J, Kern W (2007) Intranasal insulin improves memory in humans: superiority of insulin aspart. Neuropsychopharmacol: Official Publ Am Coll Neuropsychopharmacol 32(1):239–243. doi:10.1038/sj.npp.1301193

Benedict C, Brede S, Schioth HB, Lehnert H, Schultes B, Born J, Hallschmid M (2011) Intranasal insulin enhances postprandial thermogenesis and lowers postprandial serum insulin levels in healthy men. Diabetes 60(1):114–118. doi:10.2337/db10-0329

Born J, Lange T, Kern W, McGregor GP, Bickel U, Fehm HL (2002) Sniffing neuropeptides: a transnasal approach to the human brain. Nat Neurosci 5(6):514–516. doi:10.1038/nn849

Bourgin P, Huitron-Resendiz S, Spier AD, Fabre V, Morte B, Criado JR, Sutcliffe JG, Henriksen SJ, de Lecea L (2000) Hypocretin-1 modulates rapid eye movement sleep through activation of locus coeruleus neurons. J Neurosci (the official journal of the Society for Neuroscience) 20(20):7760–7765

Buskova J, Klaschka J, Sonka K, Nevsimalova S (2010) Olfactory dysfunction in narcolepsy with and without cataplexy. Sleep Med 11(6):558–561. doi:10.1016/j.sleep.2010.01.009

Caillol M, Aioun J, Baly C, Persuy MA, Salesse R (2003) Localization of orexins and their receptors in the rat olfactory system: possible modulation of olfactory perception by a neuropeptide synthetized centrally or locally. Brain Res 960(1–2):48–61

Danielyan L, Schafer R, von Ameln-Mayerhofer A, Buadze M, Geisler J, Klopfer T, Burkhardt U, Proksch B, Verleysdonk S, Ayturan M, Buniatian GH, Gleiter CH, Frey WH 2nd (2009) Intranasal delivery of cells to the brain. Eur J Cell Biol 88(6):315–324. doi:10.1016/j.ejcb.2009.02.001

Dauvilliers Y, Baumann CR, Carlander B, Bischof M, Blatter T, Lecendreux M, Maly F, Besset A, Touchon J, Billiard M, Tafti M, Bassetti CL (2003) CSF hypocretin-1 levels in narcolepsy, Kleine-Levin syndrome, and other hypersomnias and neurological conditions. J Neurol Neurosurg Psychiatry 74(12):1667–1673

Deadwyler SA, Porrino L, Siegel JM, Hampson RE (2007) Systemic and nasal delivery of orexin-A (Hypocretin-1) reduces the effects of sleep deprivation on cognitive performance in nonhuman primates. J Neurosci (the official journal of the Society for Neuroscience) 27(52):14239–14247. doi:10.1523/JNEUROSCI.3878-07.2007

Dhuria SV, Hanson LR, Frey WH 2nd (2009a) Intranasal drug targeting of hypocretin-1 (orexin-A) to the central nervous system. J Pharm Sci 98(7):2501–2515. doi:10.1002/jps.21604

Dhuria SV, Hanson LR, Frey WH 2nd (2009b) Novel vasoconstrictor formulation to enhance intranasal targeting of neuropeptide therapeutics to the central nervous system. J Pharmacol Exp Therapeutics 328(1):312–320. doi:10.1124/jpet.108.145565

Dhuria SV, Hanson LR, Frey WH 2nd (2010) Intranasal delivery to the central nervous system: mechanisms and experimental considerations. J Pharm Sci 99(4):1654–1673. doi:10.1002/jps.21924

Djupesland PG, Messina JC, Mahmoud RA (2014) The nasal approach to delivering treatment for brain diseases: an anatomic, physiologic, and delivery technology overview. Therapeutic delivery 5(6):709–733. doi:10.4155/tde.14.41

Estabrooke IV, McCarthy MT, Ko E, Chou TC, Chemelli RM, Yanagisawa M, Saper CB, Scammell TE (2001) Fos expression in orexin neurons varies with behavioral state. J Neurosci (the official journal of the Society for Neuroscience) 21(5):1656–1662

Field P, Li Y, Raisman G (2003) Ensheathment of the olfactory nerves in the adult rat. J Neurocytol 32(3):317–324. doi:10.1023/B:NEUR.0000010089.37032.48

Fujiki N, Yoshida Y, Ripley B, Mignot E, Nishino S (2003) Effects of IV and ICV hypocretin-1 (orexin A) in hypocretin receptor-2 gene mutated narcoleptic dogs and IV hypocretin-1 replacement therapy in a hypocretin-ligand-deficient narcoleptic dog. Sleep 26(8):953–959

Fulda S, Schulz H (2001) Cognitive dysfunction in sleep disorders. Sleep Med Rev 5(6):423–445. doi:10.1053/smrv.2001.0157

Ganjavi H, Shapiro CM (2007) Hypocretin/Orexin: a molecular link between sleep, energy regulation, and pleasure. J Neuropsychiatry Clin Neurosci 19(4):413–419. doi:10.1176/appi.neuropsych.19.4.413

Gorojankina T, Grebert D, Salesse R, Tanfin Z, Caillol M (2007) Study of orexins signal transduction pathways in rat olfactory mucosa and in olfactory sensory neurons-derived cell line Odora: multiple orexin signalling pathways. Regul Pept 141(1–3):73–85. doi:10.1016/j.regpep.2006.12.012

Groothuis DR, Vavra MW, Schlageter KE, Kang EW, Itskovich AC, Hertzler S, Allen CV, Lipton HL (2007) Efflux of drugs and solutes from brain: the interactive roles of diffusional transcapillary transport, bulk flow and capillary transporters. J Cereb Blood Flow and Metab: Official J Int Soc Cereb Blood Flow Metabolism 27(1):43–56. doi:10.1038/sj.jcbfm.9600315

Hagan JJ, Leslie RA, Patel S, Evans ML, Wattam TA, Holmes S, Benham CD, Taylor SG, Routledge C, Hemmati P, Munton RP, Ashmeade TE, Shah AS, Hatcher JP, Hatcher PD, Jones DN, Smith MI, Piper DC, Hunter AJ, Porter RA, Upton N (1999) Orexin A activates locus coeruleus cell firing and increases arousal in the rat. Proc Natl Acad Sci USA 96 (19):10911–10916

Hallschmid M, Benedict C, Schultes B, Fehm HL, Born J, Kern W (2004) Intranasal insulin reduces body fat in men but not in women. Diabetes 53(11):3024–3029

Hanson LR, Martinez PM, Taheri S, Kamsheh L, Mignot E, Frey WH 2nd (2004) Intranasal administration of hypocretin 1 (orexin A) bypasses the blood-brain barrier & targets the brain: a new strategy for the treatment of narcolepsy. Drug Delivery 4(4)

Heier MS, Evsiukova T, Vilming S, Gjerstad MD, Schrader H, Gautvik K (2007) CSF hypocretin-1 levels and clinical profiles in narcolepsy and idiopathic CNS hypersomnia in Norway. Sleep 30(8):969–973

Hermens WA, Belder CW, Merkus JM, Hooymans PM, Verhoef J, Merkus FW (1992) Intranasal administration of estradiol in combination with progesterone to oophorectomized women: a pilot study. Eur J Obstet Gynecol Reprod Biol 43(1):65–70

Illum L (2000) Transport of drugs from the nasal cavity to the central nervous system. Eur J pharmaceutical Sci: Official J Eur Fed Pharm Sci 11(1):1–18

Jiang Y, Zhu J, Xu G, Liu X (2011) Intranasal delivery of stem cells to the brain. Expert Opin Drug Deliv 8(5):623–632. doi:10.1517/17425247.2011.566267

John J, Wu MF, Siegel JM (2000) Systemic administration of hypocretin-1 reduces cataplexy and normalizes sleep and waking durations in narcoleptic dogs. Sleep Res Online: SRO 3(1):23–28

Julliard AK, Chaput MA, Apelbaum A, Aime P, Mahfouz M, Duchamp-Viret P (2007) Changes in rat olfactory detection performance induced by orexin and leptin mimicking fasting and satiation. Behav Brain Res 183(2):123–129. doi:10.1016/j.bbr.2007.05.033

Knudsen S, Jennum PJ, Alving J, Sheikh SP, Gammeltoft S (2010) Validation of the ICSD-2 criteria for CSF hypocretin-1 measurements in the diagnosis of narcolepsy in the Danish population. Sleep 33(2):169–176

Kummangal BA, Kumar D, Mallick HN (2013) Intracerebroventricular injection of orexin-2 receptor antagonist promotes REM sleep. Behav Brain Res 237:59–62. doi:10.1016/j.bbr.2012.09.015

Li Y, Field PM, Raisman G (2005a) Olfactory ensheathing cells and olfactory nerve fibroblasts maintain continuous open channels for regrowth of olfactory nerve fibres. Glia 52(3):245–251. doi:10.1002/glia.20241

Li Y, Li D, Raisman G (2005b) Interaction of olfactory ensheathing cells with astrocytes may be the key to repair of tract injuries in the spinal cord: the 'pathway hypothesis'. J Neurocytol 34 (3–5):343–351. doi:10.1007/s11068-005-8361-1

Lu J, Sherman D, Devor M, Saper CB (2006) A putative flip-flop switch for control of REM sleep. Nature 441(7093):589–594. doi:10.1038/nature04767

Luzzati V, Benoit E, Charpentier G, Vachette P (2004a) X-ray scattering study of pike olfactory nerve: elastic, thermodynamic and physiological properties of the axonal membrane. J Mol Biol 343(1):199–212. doi:10.1016/j.jmb.2004.08.029

Luzzati V, Vachette P, Benoit E, Charpentier G (2004b) X-ray scattering study of pike olfactory nerve: intensity of the axonal membrane, solution of the phase problem and electron density profile. J Mol Biol 343(1):187–197. doi:10.1016/j.jmb.2004.08.031

Mattarozzi K, Bellucci C, Campi C, Cipolli C, Ferri R, Franceschini C, Mazzetti M, Russo PM, Vandi S, Vignatelli L, Plazzi G (2008) Clinical, behavioural and polysomnographic correlates of cataplexy in patients with narcolepsy/cataplexy. Sleep Med 9(4):425–433

Mignot E, Lammers GJ, Ripley B, Okun M, Nevsimalova S, Overeem S, Vankova J, Black J, Harsh J, Bassetti C, Schrader H, Nishino S (2002) The role of cerebrospinal fluid hypocretin measurement in the diagnosis of narcolepsy and other hypersomnias. Arch Neurol 59 (10):1553–1562

Mileykovskiy BY, Kiyashchenko LI, Siegel JM (2005) Behavioral correlates of activity in identified hypocretin/orexin neurons. Neuron 46(5):787–798. doi:10.1016/j.neuron.2005.04.035

Montplaisir J, Billiard M, Takahashi S, Bell IR, Guilleminault C, Dement WC (1978) Twenty-four-hour recording in REM-narcoleptics with special reference to nocturnal sleep disruption. Biol Psychiatry 13(1):73–89

Nishino S, Ripley B, Overeem S, Nevsimalova S, Lammers GJ, Vankova J, Okun M, Rogers W, Brooks S, Mignot E (2001) Low cerebrospinal fluid hypocretin (Orexin) and altered energy homeostasis in human narcolepsy. Ann Neurol 50(3):381–388

Nonaka N, Farr SA, Kageyama H, Shioda S, Banks WA (2008) Delivery of galanin-like peptide to the brain: targeting with intranasal delivery and cyclodextrins. J Pharmacol Exp Therapeutics 325(2):513–519. doi:10.1124/jpet.107.132381

Parker KJ, Buckmaster CL, Schatzberg AF, Lyons DM (2005) Intranasal oxytocin administration attenuates the ACTH stress response in monkeys. Psychoneuroendocrinology 30(9):924–929. doi:10.1016/j.psyneuen.2005.04.002

Piper DC, Upton N, Smith MI, Hunter AJ (2000) The novel brain neuropeptide, orexin-A, modulates the sleep–wake cycle of rats. Eur J Neurosci 12(2):726–730

Pollock H, Hutchings M, Weller RO, Zhang ET (1997) Perivascular spaces in the basal ganglia of the human brain: their relationship to lacunes. J Anat 191(Pt 3):337–346

Reger MA, Watson GS, Green PS, Wilkinson CW, Baker LD, Cholerton B, Fishel MA, Plymate SR, Breitner JC, DeGroodt W, Mehta P, Craft S (2008) Intranasal insulin improves cognition and modulates beta-amyloid in early AD. Neurology 70(6):440–448. doi:10.1212/01.WNL.0000265401.62434.36

Rennels ML, Blaumanis OR, Grady PA (1990) Rapid solute transport throughout the brain via paravascular fluid pathways. Adv Neurol 52:431–439

Rieger M, Mayer G, Gauggel S (2003) Attention deficits in patients with narcolepsy. Sleep 26(1):36–43

Ross TM, Martinez PM, Renner JC, Thorne RG, Hanson LR, Frey WH 2nd (2004) Intranasal administration of interferon beta bypasses the blood-brain barrier to target the central nervous system and cervical lymph nodes: a non-invasive treatment strategy for multiple sclerosis. J Neuroimmunol 151(1–2):66–77. doi:10.1016/j.jneuroim.2004.02.011

Ross TM, Zuckermann RN, Reinhard C, Frey WH 2nd (2008) Intranasal administration delivers peptoids to the rat central nervous system. Neurosci Lett 439(1):30–33. doi:10.1016/j.neulet.2008.04.097

Sakane T, Akizuki M, Taki Y, Yamashita S, Sezaki H, Nadai T (1995) Direct drug transport from the rat nasal cavity to the cerebrospinal fluid: the relation to the molecular weight of drugs. J Pharm Pharmacol 47(5):379–381

Scammell TE, Saper CB (2007) Orexins: looking forward to sleep, back at addiction. Nat Med 13(2):126–128. doi:10.1038/nm0207-126

Schipper NG, Verhoef JC, De Lannoy LM, Romeijn SG, Brakkee JH, Wiegant VM, Gispen WH, Merkus FW (1993) Nasal administration of an ACTH(4-9) peptide analogue with dimethyl-beta-cyclodextrin as an absorption enhancer: pharmacokinetics and dynamics. Br J Pharmacol 110(4):1335–1340

Schulz C, Paulus K, Lehnert H (2004) Central nervous and metabolic effects of intranasally applied leptin. Endocrinology 145(6):2696–2701. doi:10.1210/en.2003-1431

Siegel JM (2004) Hypocretin (orexin): role in normal behavior and neuropathology. Annu Rev Psychol 55:125–148. doi:10.1146/annurev.psych.55.090902.141545

Stiasny-Kolster K, Clever SC, Moller JC, Oertel WH, Mayer G (2007) Olfactory dysfunction in patients with narcolepsy with and without REM sleep behaviour disorder. Brain:J Neurol 130(Pt 2):442–449. doi:10.1093/brain/awl343

Thorne RG, Emory CR, Ala TA, Frey WH 2nd (1995) Quantitative analysis of the olfactory pathway for drug delivery to the brain. Brain Res 692(1–2):278–282

Thorne RG, Pronk GJ, Padmanabhan V, Frey WH 2nd (2004) Delivery of insulin-like growth factor-I to the rat brain and spinal cord along olfactory and trigeminal pathways following intranasal administration. Neuroscience 127(2):481–496. doi:10.1016/j.neuroscience.2004.05.029

Thorne RG, Hanson LR, Ross TM, Tung D, Frey WH 2nd (2008) Delivery of interferon-beta to the monkey nervous system following intranasal administration. Neuroscience 152(3):785–797. doi:10.1016/j.neuroscience.2008.01.013

Van den Berg MP, Merkus P, Romeijn SG, Verhoef JC, Merkus FW (2003) Hydroxocobalamin uptake into the cerebrospinal fluid after nasal and intravenous delivery in rats and humans. J Drug Target 11(6):325–331. doi:10.1080/10611860310001640075

Van den Berg MP, Merkus P, Romeijn SG, Verhoef JC, Merkus FW (2004) Uptake of melatonin into the cerebrospinal fluid after nasal and intravenous delivery: studies in rats and comparison with a human study. Pharm Res 21(5):799–802

Wang Q, Chen G, Zeng S (2007) Pharmacokinetics of Gastrodin in rat plasma and CSF after i.n. and i.v. Int J Pharm 341(1–2):20–25. doi:10.1016/j.ijpharm.2007.03.041

Watson CJ, Soto-Calderon H, Lydic R, Baghdoyan HA (2008) Pontine reticular formation (PnO) administration of hypocretin-1 increases PnO GABA levels and wakefulness. Sleep 31(4):453–464

Weinhold SL, Seeck-Hirschner M, Nowak A, Göder R, Baier PB (2011) Wake–REM sleep transitions for measuring REM sleep disturbance: Comparison between narcolepsy, idiopathic hypersomnia and healthy controls. Sleep and Biological Rhythms 9:8

Weinhold SL, Seeck-Hirschner M, Nowak A, Hallschmid M, Goder R, Baier PC (2014) The effect of intranasal orexin-A (hypocretin-1) on sleep, wakefulness and attention in narcolepsy with cataplexy. Behav Brain Res 262:8–13. doi:10.1016/j.bbr.2013.12.045

Westin U, Piras E, Jansson B, Bergstrom U, Dahlin M, Brittebo E, Bjork E (2005) Transfer of morphine along the olfactory pathway to the central nervous system after nasal administration to rodents. Eur J Pharm Sci: Official J Eur Fed Pharm Sci 24(5):565–573. doi:10.1016/j.ejps.2005.01.009

Westin UE, Bostrom E, Grasjo J, Hammarlund-Udenaes M, Bjork E (2006) Direct nose-to-brain transfer of morphine after nasal administration to rats. Pharm Res 23(3):565–572. doi:10.1007/s11095-006-9534-z [illegible] (2016) [illegible] using [illegible] [illegible] [illegible] doi:10.[illegible]

[illegible]

Hypocretin (Orexin) Cell Transplantation as a New Therapeutic Approach in Narcolepsy

Oscar Arias-Carrión, Andrea Herrera-Solís, Alwin Poot-Aké, Ramsés Jiménez-Moreno and Eric Murillo-Rodríguez

Abstract Narcolepsy is a neurodegenerative disorder which main includes symptoms such as daytime sleepiness, sleep paralysis, hypnagogic hallucinations, disturbed nocturnal sleep as well as cataplexy. It is known that this disease is caused by deficiency of the neurotransmission system of the peptide named hypocretin (HCRT), also cited as orexin (OX). The ablation of HCRT/OX or HCRT/OX receptors in animal models has supported the understanding of the human narcolepsy. The current chapter describe the experimental model of narcolepsy in rats by pharmacological means. Targeting HCRT/OX neurons by the HCRT-2-saporin (HCRT2/SAP) toxin destroys the HCRT neurons. This experimental procedure promotes a significant diminution in number of HCRT/OX neurons as well as the endogenous levels of the peptide in the cerebrospinal fluid. We also discuss that HCRT2/SAP induces narcoleptic-like behaviour in rats as assessed by EEG/EMG means. Under this paradigm, here we present current evidence regarding the potential use of grafting HCRT neurons into lateral hypothalamus of lesioned rats to revert the sleep abnormalities observed in HCRT2/SAP animals.

Keywords Hypocretin/orexin neurons · Narcolepsy · Cell therapy · Sleep · Lateral hypothalamus

---

O. Arias-Carrión · A. Herrera-Solís
Unidad de Trastornos del Movimiento y Sueño (TMS),
Hospital General "Dr. Manuel Gea González", México, D.F., México

A. Poot-Aké · R. Jiménez-Moreno · E. Murillo-Rodríguez (✉)
Laboratorio de Neurociencias Moleculares e Integrativas,
Escuela de Medicina, División Ciencias de la Salud,
Universidad Anáhuac Mayab, Mérida, YUC, México
e-mail: eric.murillo@anahuac.mx

E. Murillo-Rodríguez
Grupo de Investigación en Envejecimiento, División Ciencias de la Salud,
Universidad Anáhuac Mayab, Mérida, YUC, México

E. Murillo-Rodríguez
Grupo de Investigación Desarrollos Tecnológicos para la Salud,
División de Ingeniería y Ciencias Exactas, Universidad Anáhuac Mayab, Mérida
YUC, México

© Springer International Publishing Switzerland 2015

T. Sakurai et al. (eds.), *Orexin and Sleep*,
DOI 10.1007/978-3-319-23078-8_19

1 Introduction

The Hypocretins (Hcrt; 1 and 2, also named as orexin A and B respectively) are two neuropeptides derived from the same precursor whose expression is restricted to a few thousand neurons of the lateral hypothalamus (Trivedi et al. 1998; Hervieu et al. 2001; Espana et al. 2005). Hcrt-1 consists of 33 amino acids whereas Hcrt-2 is a 28 amino acid molecule.

The Hcrt neurons are located between the rat fornix and the mammillothalamic tracts in the lateral hypothalamus (LH) from where hypocretins fiber project throughout the brain and spinal cord, including several areas implicated in regulations of the sleep/wake cycle (Trivedi et al. 1998; Hervieu et al. 2001; Peyron et al. 1998; Sakurai et al. 1998; Chemelli et al. 1999). The Hcrt fibers widely project throughout the brain and spinal cord and generally have excitatory effects on serotonergic (Hagan et al. 1999; Brown et al. 2002), noradrenergic (Bayer et al. 2002), histaminergic (Bayer et al. 2001), and cholinergic neurons in basal forebrain (Eggermann et al. 2001), laterodorsal tegmental nucleus (Takahashi et al. 2002) as well as thalamocortical neurons of thalamus (Bayer et al. 2002). Basically, Hcrt neurons regulate arousal and have been shown to be implicated in food reward and drug-seeking behavior (Sakurai 2007).

2 A Link Between Narcolepsy and Hypocretin Neurons

Narcolepsy is a debilitating neurological disease characterized by excessive daytime sleepiness, premature transitions to rapid eye movement sleep (named "sleep onset REM periods") and cataplexy (sudden bilateral skeletal muscle weakness without impairment of conscience) for review see (Siegel and Boehmer 2006) (Table 1).

The Hcrt was linked to human narcolepsy since it was discovered in a mutation in the Hcrt receptor in mice (Chemelli et al. 1999) and dogs (Lin et al. 1999). Once it was speculated that a deficiency in the Hcrt system might be the explanation for human narcolepsy, diverse groups explored this possible link. Indeed, Hcrt system was linked to human narcolepsy.

In a study of post-mortem brains of human narcoleptics, for instance, a massive reduction in the number of Hcrt-containing cells (85–98 %) was discovered compared with healthy controls (Peyron et al. 2000; Thannickal et al. 2000). On the other hand, narcoleptic patients have reduced cerebrospinal fluid (CSF) levels of Hcrt (Dalal et al. 2001; Kanbayashi et al. 2002; Mignot et al. 2002; Nishino et al. 2000; Ripley et al. 2001), a finding consistent with the loss of Hcrt-containing neurons as mentioned above. The CSF measurements of Hcrt provide a valuable diagnostic tool for narcolepsy, separating narcolepsy from other sleep and neurological disorders (Mignot et al. 2002).

Table 1 Summary of the symptoms of narcolepsy described in humans

Symptom	
Excessive daytime sleepiness (EDS)	A persistent sense of mental cloudiness, a lack of energy, a depressed mood, or extreme exhaustion. Some people experience memory lapses, and many have great difficulty maintaining their concentration while at school, work, or home
Cataplexy	Loss of muscle tone while you're awake. Muscle weakness affects part or all of your body. It is often triggered by sudden, strong emotions such as fear, anger, stress, excitement, or humor
Rapid entry into REM sleep	Patients may enter the REM or dream phase of sleep right after falling asleep, whereas most people take about 90 min to enter REM. Therefore, they experience the characteristics of REM sleep (vivid dreams and muscle paralysis) at the beginning of sleep, even if that sleep is during the day
Sleep paralysis	The temporary inability to move or speak while falling asleep or waking is similar to REM-induced inhibitions of voluntary muscle activity, lasting from a few seconds to minutes. After episodes end, people rapidly recover their full capacity to move and speak
Hallucinations	Unusually vivid hallucinations can occur when people are falling asleep, waking, or during sleep. Referred to as hypnagogic hallucinations when occurring during sleep onset and as hypnopompic hallucinations when occurring during waking
Disrupted nocturnal sleep	Most patients experience difficulties staying asleep or waking up repeatedly throughout the night
Obesity	After developing narcolepsy, many individuals suddenly gain weight
Other	Leg jerks, nightmares, and restlessness

3 Animal Models of Narcolepsy

There are experimental models that mimics the sleep disorder narcolepsy; for instance, Hcrt/orexin genes knockout mice (Chemelli et al. 1999), canines with a mutation in the Hcrt-2 receptor (Lin et al. 1999) or mice with a targeted destruction of the Hcrt neurons (Hara et al. 2001) that exhibit symptoms of narcolepsy (Fig. 1).

Recently, our group and other have generated an experimental model using a toxin which represents a very reliable procedure. This model consists of in a ribosome-inactivating protein saporin (SAP) (Stirpe et al. 1992) that is conjugated to the hypocretin/orexin receptor that binds ligand hypocretin-2/orexin-B (Hcrt-2) to lesion Hcrt receptor-bearing neurons. It is known that the LH contains a high concentration of Hcrt receptor mRNA (Trivedi et al. 1998) as well as immunoreactivity (Hervieu et al. 2001). This fact indicates of that presence of the Hcrt receptor on Hcrt neurons. When the Hcrt-2/SAP is injected into LH of rats, the toxin lesions Hcrt neurons, and produces behavioural symptoms that are characteristic of narcolepsy.

Comparative models of narcolepsy

Mouse	Dog	Rat	Human
*Cataplexy can be triggered by social interaction and/or positive emotions mutation in the Hcrt/OX receptor *Increase in REM sleep during the dark (active) period *Cataplexy in Hcrt/OX KO mice *Hcrt/OX KO mice shows intact circadian control of sleep, normal sleep homeostasis and an unchanged amount of wakefulness	*Emotional experiences, such as the presentation of food, water, or playing, are likely trigger cataplexy. *Narcoleptic dogs relatively inactive during daytime and shows an unclear rest/activity pattern, similar to that of human narcoleptics *Strong associations between narcolepsy and MHC class II HLA in humans *OXR2, was discovered present in narcolepsy *Mutation in the Hcrt-2 receptor	*OX KO revealed that rats show brief periods of peculiar behavior arrest during the dark period *Increase in REM sleep during the dark (active) period when Cataplexy in OX KO *Hcrt-2/SAP is injected into LH of rats, and the lesion destroys Hcrt neurons	*Excessive daytime *Sleepiness and cataplexy as primary symptoms *Positive for a specific class II HLA allele, HLA-DQB1*0602 *A massive reduction in the number of Hcrt-containing cells as well as Hcrt/OX ligand in cerebrospinal fluid

Fig. 1 Comparison of the available experimental models of nacolepsy: Mice, rat. dog and human. Please note the molecular, neurochemical, behavioral similarities among models. *Abbreviations HLA* Human leukocyte antigen; *Hcrt* hypocretin; *KO* knock out; *OX* orexin; *OX R2* orexin receptor 2; *REM* rapid eye movement sleep

The Hcrt-2/SAP binds to cells containing the Hcrt receptor but does not bind to cells that do not contain the Hcrt receptor. This indicates the specificity of the toxin. For instance, the diurnal rhythm of wakefulness (W) and slow wave sleep (SWS) in a rat lesion is attenuated. This is because of an increase in sleep during the lights-off period. Hcrt-2/SAP increases SWS over the 24 h period. Total time spent in SWS and rapid eye movement sleep (REMS) was found to correlate with a decline in number of Hcrt neurons.

Using this model, Gerashchenko et al. (2001) showed that Hcrt-2/SAP induced more SWS and REM sleep at night and multiple periods of abnormal behavioural arrest during purposeful behaviour. Indeed, this experimental model of narcolepsy provides a method of investigating the contribution of the Hcrt system to the regulation of the sleep-wake cycle and its relationship with narcolepsy. Gerashchenko et al. (2003) also reported that two concentrations (90 ng or 490 ng/0.5 μl) of the Hcrt-2/SAP injected directly to the lateral hypothalamus caused a significant Hcrt cell lost. Narcoleptic-like sleep behaviour was produced by both concentrations of this toxin (Gerashchenko et al. 2003).

As a conclusion, the use of Hcrt-2/SAP induced characteristic of narcolepsy, such as sleep fragmentation, sleep-onset REM sleep periods, increased NREM sleep and REM sleep time during the normally active lights-off period (Gerashchenko et al. 2001, 2003).

The relationship between Hcrt neuronal loss and changes in levels of the peptide was not known. Using the Hcrt-2/SAP to lesion neurons in the LH was found that there was a 50 % reduction in CSF Hcrt levels when 73 % of the number of Hcrt neurons was lost (Gerashchenko et al. 2003). The sleep deprivation method is used to test the homeostatic mechanism of sleep regulation. It was also found that in lesioned rats, the Hcrt levels were not increased by 6 h prolonged W, indicating that surviving neurons were not able to increase the output of Hcrt into CSF to compensate for the Hcrt neuronal loss.

In the same study, the authors showed that Hcrt levels in CSF measured at different times of the day-night cycle showed that control rats had significantly higher Hcrt levels at ZT0 (+72.9 %) compared to a different time point (ZT8). The rats with a 72 % Hcrt neuronal loss did not show a significant difference between ZT0 and ZT8. Surprisingly, it was found a significant correlation between Hcrt neurons and Hcrt levels at ZT0 (Gerashchenko et al. 2003).

The Hcrt neurons begin to appear on embryonic day 19 E19 and are fully developed by postnatal day 20 (Yamamoto et al. 2000; Van Den Pol et al. 2001). As reported by others (Fujiki et al. 2001; Yoshida et al. 2001), Hcrt display a diurnal rhythm in young animals.

Such a rhythm of Hcrt, however, has not been determined in aged animals. It was measured CSF Hcrt levels at 4-h intervals across a 24-h period to test the hypothesis that there was a decline in Hcrt levels in aged rats. In agreement with previous studies in young rats (Fujiki et al. 2001) peak levels of Hcrt were found at the end of the wake-active period (ZT0), and lowest levels occurred at the end of the sleep period (ZT12). This profile was present in young and old rats. The authors found, however, that the old rats had a significant reduction of Hcrt levels in CSF compared to the young rats. Moreover, the aged rats had significantly less CSF Hcrt levels compared to young rats as well (Desarnaud et al. 2004).

As mentioned above, CSF Hcrt levels are increased in response to prolonged W, we tested if Hcrt levels could be enhanced in aged animals after prolonged W. Rats were kept awake for 8 h to drive the activity of the Hcrt neurons, and CSF was collected in order to measure Hcrt. All groups of rats had significantly increased CSF Hcrt levels in response to 8 h of prolonged waking (Desarnaud et al. 2004), a finding that is consistent with other studies (Yoshida et al. 2001).

The overall CSF Hcrt levels after 8 h of prolonged W, however, were still lower in the old rats when compared to the young rats. The possible explanation for this phenomenon might be a decline in Hcrt-1 in old rats reflecting a diminution in prepro-Hcrt gene expression across the aging process. We therefore measured prepro-Hcrt mRNA levels in the posterior hypothalamus of young and old rats by Northern blot analysis. No difference in mRNA between age group was found (Desarnaud et al. 2004).

4 Narcolepsy: Current Treatment Options

What are the approaches that currently are used to treat narcolepsy? The goal of all therapeutic approaches in narcolepsy is to control the narcoleptic symptoms and to allow the patient to continue full participation in familial and professional daily activities. Pharmacological treatments for narcolepsy include the use of amphetamines and modafinil whereas cataplexy is treated with tricyclic antidepressants such as clomipramine (Mignot et al. 2002; Scammell 2003; Mignot and Nishino 2005).

Another possible experimental approach involves Hcrt cell replacement therapy. We have shown recently the use of cell Hcrt transplantation might represent a new approach to treat this disease (Arias-Carrión et al. 2004, 2006; Arias-Carrión and Murillo-Rodríguez 2014).

5 Cell Transplantation: A Future Therapy for Narcolepsy?

Neural transplantation is one of the most promising approaches for the treatment of Parkinson's disease (PD), a major neurodegenerative disorder with a prevalence as frequent as that of narcolepsy (Arias-Carrión et al. 2004). Neural transplantation involves implantation of living neuronal tissue into a host system. Several studies using animal models have demonstrated that grafted tissue survives, integrates within the host brain, and provides functional recovery following brain inventions (Drucker-Colín and Verdugo-Díaz 2004; Lindvall et al. 2004). This type of study for PD began in the latter half of the 1970s. Initially, DA neurons from animal fetuses were used as donors, and then paraneurons such as chromaffin cells were used (Drucker-Colín and Verdugo-Díaz 2004). Based on a large number of experimental animal studies, neural transplantation has been applied clinically (Drucker-Colín and Verdugo-Díaz 2004; Lindvall et al. 2004). Studies in patients with PD after intrastriatal transplantation of human fetal mesencephalic tissue, rich in postmitotic dopaminergic neurons, have provided proof of principle that neuronal replacement can work in the human brain (Lindvall et al. 2004). The grafted neurons survive and reinnervate the striatum for as long as 10 years despite an ongoing disease process, which destroys the patient's own dopaminergic neurons (Drucker-Colín and Verdugo-Díaz 2004). The grafts are able to normalize striatal dopamine release (Drucker-Colín and Verdugo-Díaz 2004) and to reverse the impairment of cortical activation underlying akinesia. Thus, grafted dopaminergic neurons can become functionally integrated into neuronal circuitries in the brain (Drucker-Colín and Verdugo-Díaz 2004; Lindvall et al. 2004). Several open-label trials have reported clinical benefit (for revision see Lindvall et al. 2004). Some patients have been able to withdraw from L-dopa treatment for several years and resume an independent life (Drucker-Colín and Verdugo-Díaz 2004; Lindvall et al. 2004). Beneficial effects have been demonstrated, and autopsy cases have shown that many transplanted cells were able to survive in the human brain for long periods. These findings contributed a great deal to the research in regeneration of the central nervous system.

Thus it would be interesting to know whether a graft of hypocretin neurons into a host brain could survive (Arias-Carrión et al. 2004, 2006; Arias-Carrión and Murillo-Rodríguez 2014). We demonstrated that Hcrt neurons suspension cells derived from posterior hypothalamus of 8–10-days old rat pups can survive when transplanted into the pons (a region of the brain that is innervated by hypocretin axons but where the hypocretin somata are not present) in adults rats (Arias-Carrión et al. 2004, 2006; Arias-Carrión and Murillo-Rodríguez 2014). In our preliminary study, we found that well-defined hypocretin-immunoreactive somata with processes and varicosities were present in the graft zone 36 days after implantation of the cell suspension, suggesting that hypocretin neurons obtained from rat pups can be grafted into an adult host brain (Arias-Carrión et al. 2004). These somata were similar in size and appearance to adult rat hypocretin-immunoreactive neurons.

Next, we investigated the time-course of survival of grafted Hcrt neurons into the pons of adult rats (Arias-Carrión et al. 2006). Control rats received a transplant that consisted of cells from the cerebellum where no Hcrt neurons are present. All adult host rats were sacrificed 1, 3, 6, 9, 12, 24, or 36 days after grafting. Immunohistochemistry was used to identify and count the presence of the Hcrt grafted neurons in the target area. The tally of Hcrt neurons present in the graft zone 1 day post-grafting was considered to be the baseline. From day 3 to 36 post-transplant there was a steady decline in the number of Hcrt neurons. We also noted that on day 36, the Hcrt neurons that survived in the pons had morphological features that were similar to mature Hcrt neurons in the adult lateral hypothalamus, suggesting that these neurons might be functionally active. Control rats that received grafts of cerebellar tissue did not show Hcrt neurons in the target area. These results demonstrate that there is a progressive decline in the number of transplanted neurons, but a significant percentage of Hcrt neurons do survive until day 36. Some evidence exists that hypocretin neuron-transplantation in rats somewhat diminished narcolepsy-like sleep behavior (Arias-Carrión and Murillo-Rodríguez 2014), but no studies in narcoleptic human subjects have been developed. These studies, however, highlights the potential use of transplants as a therapeutical tool in order to treat narcolepsy (Fig. 2).

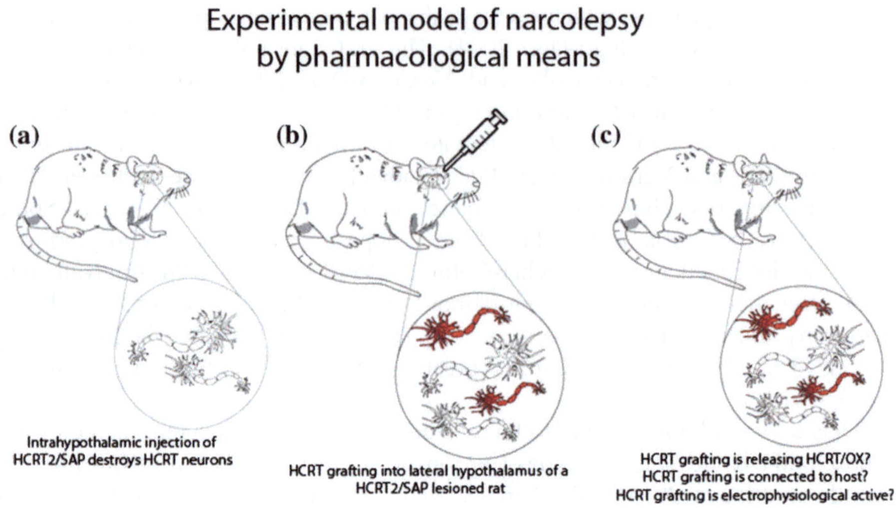

Fig. 2 The narolcpesy rat model by using HCRT2/SAP lesions establish a experimental model that resembles narcolepsy. Panel **a** shows that HCRT2/SAP toxin injected into lateral hypothalamus or rat destroys HCRT/OX neurons and promotes a diminution in HCRT/OX peptide levels in cerebrospinal fluid. This experimental model mimics narcolepsy as determined by EEG/EMG, neurochemical, and immunohistochemical measurements. On the other hand, grafting HCRT/OX neurons into lateral hypothalamus of HCRT2/SAP lesioned rats remain during 36 days post-transplant (Panel **b**; surviving neurons after HCRT2/SAP injection are indicated by drawing of neurons in *gray color* whereas grafted neurons are indicated by drawings of neurons in *color red*). Despite the behavioural improvement after HCRT/OX grafting in narcoleptic rats, it is unknown whether HCRT/OX transplant is active by releasing HCRT/OX ligand, or developing synapsis with host or if transplant is electrophysiologically active. These issues remain to be studied (color figure online)

6 Conclusions

Narcolepsy is a sleep disorder characterized by sleep attacks. Experimental models have been generated in order to understand the physiology of this disease. Recent evidence has concluded that narcolepsy in humans and animal models the result from the failure of cellular signalling mediated by Hcrt.

Recently, we have described that Hcrt cells are able to be transplanted with the aim to generate an alternative therapeutical tool to treat narcolepsy. We have showed here that grafted cells express the machinery for Hcrt release, and possess the morphological feature of an Hcrt neuron. In this context, we reported for the first time that transplantation of Hcrt neurons into the LH of Hcrt2/SAP-lesioned rats diminishes narcoleptic-like sleep behavior. The following points, however, have to be demonstrated in the future: (a) electrophysiological properties of fully mature Hcrt neuron; (b) grafted Hcrt neurons should re-establish a dense, functional Hcrt releasing terminal network; and (c) grafts have to become functionally integrated into host circuitries.

Acknowledgements Oscar Arias-Carrión is supported by CONACYT-BMBF 2013 (Grant 208132).

References

Arias-Carrión O, Murillo-Rodríguez E (2014) Effects of hypocretin/orexin cell transplantation on narcoleptic-like sleep behavior in rats. PLoS ONE 9(4):e95342

Arias-Carrión O et al (2004) Transplantation of hypocretin neurons into the pontine reticular formation: preliminary results. Sleep 27(8):1465–1470

Arias-Carrión O, Drucker-Colín R, Murillo-Rodríguez E (2006) Survival rates through time of hypocretin grafted neurons within their projection site. Neurosci Lett 404(1–2):93–97

Bayer L et al (2001) Orexins (hypocretins) directly excite tuberomammillary neurons. Eur J Neurosci 14(9):1571–1575

Bayer L et al (2002) Selective action of orexin (hypocretin) on nonspecific thalamocortical projection neurons. J Neurosci Official J Soc Neurosci 22(18):7835–7839

Brown RE, Sergeeva OA, Eriksson KS, Haas HL (2002) Convergent excitation of dorsal raphe serotonin neurons by multiple arousal systems (orexin/hypocretin, histamine and noradrenaline). J Neurosci Official J Soc Neurosci 22(20):8850–8859

Chemelli RM et al (1999) Narcolepsy in orexin knockout mice: molecular genetics of sleep regulation. Cell 98(4):437–451

Dalal MA et al (2001) Normal plasma levels of orexin A (hypocretin-1) in narcoleptic patients. Neurology 56(12):1749–1751

Desarnaud F et al (2004) The diurnal rhythm of hypocretin in young and old F344 rats. Sleep 27(5):851–856

Drucker-Colín R, Verdugo-Díaz L (2004) Cell transplantation for Parkinson's disease: present status. Cell Mol Neurobiol 24(3):301–316

Eggermann E et al (2001) Orexins/hypocretins excite basal forebrain cholinergic neurones. Neuroscience 108(2):177–181

Espana RA, Reis KM, Valentino RJ, Berridge CW (2005) Organization of hypocretin/orexin efferents to locus coeruleus and basal forebrain arousal-related structures. J Comp Neurol 481 (2):160–178

Fujiki N et al (2001) Changes in CSF hypocretin-1 (orexin A) levels in rats across 24 hours and in response to food deprivation. NeuroReport 12(5):993–997

Gerashchenko D et al (2001) Hypocretin-2-saporin lesions of the lateral hypothalamus produce narcoleptic-like sleep behavior in the rat. J Neurosci Official J Soc Neurosci 21(18):7273–7283

Gerashchenko D, Blanco-Centurión C, Greco MA, Shiromani PJ (2003a) Effects of lateral hypothalamic lesion with the neurotoxin hypocretin-2-saporin on sleep in Long-Evans rats. Neuroscience 116(1):223–235

Gerashchenko D et al (2003b) Relationship between CSF hypocretin levels and hypocretin neuronal loss. Exp Neurol 184(2):1010–1016

Hagan JJ et al (1999) Orexin A activates locus coeruleus cell firing and increases arousal in the rat. Proc Natl Acad Sci USA 96(19):10911–10916

Hara J et al (2001) Genetic ablation of orexin neurons in mice results in narcolepsy, hypophagia, and obesity. Neuron 30(2):345–354

Hervieu GJ, Cluderay JE, Harrison DC, Roberts JC, Leslie RA (2001) Gene expression and protein distribution of the orexin-1 receptor in the rat brain and spinal cord. Neuroscience 103(3):777–797

Kanbayashi T et al (2002) CSF hypocretin-1 (orexin-A) concentrations in narcolepsy with and without cataplexy and idiopathic hypersomnia. J Sleep Res 11(1):91–93

Lin L et al (1999) The sleep disorder canine narcolepsy is caused by a mutation in the hypocretin (orexin) receptor 2 gene. Cell 98(3):365–376

Lindvall O, Kokaia Z, Martinez-Serrano A (2004) Stem cell therapy for human neurodegenerative disorders-how to make it work. Nat Med 10(Suppl):S42–S50

Mignot E, Nishino S (2005) Emerging therapies in narcolepsy-cataplexy. Sleep 28(6):754–763

Mignot E et al (2002) The role of cerebrospinal fluid hypocretin measurement in the diagnosis of narcolepsy and other hypersomnias. Arch Neurol 59(10):1553–1562

Nishino S, Ripley B, Overeem S, Lammers GJ, Mignot E (2000) Hypocretin (orexin) deficiency in human narcolepsy. Lancet 355(9197):39–40

Peyron C et al (1998) Neurons containing hypocretin (orexin) project to multiple neuronal systems. J Neuroscience Official J Soc Neurosci 18(23):9996–10015

Peyron C et al (2000) A mutation in a case of early onset narcolepsy and a generalized absence of hypocretin peptides in human narcoleptic brains. Nat Med 6(9):991–997

Ripley B et al (2001) CSF hypocretin/orexin levels in narcolepsy and other neurological conditions. Neurology 57(12):2253–2258

Sakurai T (2007) The neural circuit of orexin (hypocretin): maintaining sleep and wakefulness. Nat Rev Neurosci 8(3):171–181

Sakurai T et al (1998) Orexins and orexin receptors: a family of hypothalamic neuropeptides and G protein-coupled receptors that regulate feeding behavior. Cell 92(4):573–585

Scammell TE (2003) The neurobiology, diagnosis, and treatment of narcolepsy. Ann Neurol 53(2):154–166

Siegel JM, Boehmer LN (2006) Narcolepsy and the hypocretin system–where motion meets emotion. Nature Clin Pract Neurol 2(10):548–556

Stirpe F, Barbieri L, Battelli MG, Soria M, Lappi DA (1992) Ribosome-inactivating proteins from plants: present status and future prospects. Biotechnology (N Y) 10(4):405–412

Takahashi K, Koyama Y, Kayama Y, Yamamoto M (2002) Effects of orexin on the laterodorsal tegmental neurones. Psychiatry Clin Neurosci 56(3):335–336

Thannickal TC et al (2000) Reduced number of hypocretin neurons in human narcolepsy. Neuron 27(3):469–474

Trivedi P, Yu H, MacNeil DJ, Van der Ploeg LH, Guan XM (1998) Distribution of orexin receptor mRNA in the rat brain. FEBS Lett 438(1–2):71–75

Van Den Pol AN, Patrylo PR, Ghosh PK, Gao XB (2001) Lateral hypothalamus: early developmental expression and response to hypocretin (orexin). J Comp Neurol 433(3):349–363

Yamamoto Y et al (2000) Postnatal development of orexin/hypocretin in rats. Brain Res Mol Brain Res 78(1–2):108–119

Yoshida Y et al (2001) Fluctuation of extracellular hypocretin-1 (orexin A) levels in the rat in relation to the light-dark cycle and sleep-wake activities. Eur J Neurosci 14(7):1075–1081

Orexin and Metabolism

Hiromasa Funato

Abstract Orexin has been confirmed to exhibit acute orexigenic effects, especially for palatable food. However, the loss of orexin renders mice susceptible to obesity, which suggests that orexin neurons function as a negative regulator of energy metabolism over an extended period of time. Thus, orexin exerts differential short-term and long-term effects on feeding behavior and energy metabolism. Orexin neurons are able to sense nutritional and energy status directly or indirectly through leptin, glucose, amino acids and other factors. Orexin neurons project to various brain regions from the forebrain to the spinal cord, which are involved in food intake, reward behavior and sympathetic nervous system activity; orexin's effects in these regions modulate food salience and energy expenditure. Orexin has also been reported to be involved in glucose and bone metabolism. Surprisingly, very little is currently known regarding the roles of the two orexin receptors and especially regarding the long-term effects on metabolism. The local action of orexin in the peripheral organs remains controversial.

1 Introduction

Orexin neurons are unique in their direct or indirect regulation of various aspects of metabolism, and, in turn, their activities are also regulated by metabolic signals. One of the most prominent features of orexin is its restricted expression in the lateral hypothalamic area (LHA); the LHA has been demonstrated as a feeding

H. Funato (✉)
Department of Anatomy, Toho University School of Medicine, 5-21-16 Omori-Nishi, Ota-ku, Tokyo 143-8540, Japan
e-mail: hiromasa.funato@med.toho-u.ac.jp; funato.hiromasa.km@u.tsukuba.ac.jp

H. Funato
International Institute for Integrative Sleep Medicine, University of Tsukuba, 1-1-1 Tennodai, Tsukuba, Ibaraki 305-8575, Japan

© Springer International Publishing Switzerland 2015
T. Sakurai et al. (eds.), *Orexin and Sleep*,
DOI 10.1007/978-3-319-23078-8_20

center because LHA lesioning suppressed feeding behavior (Anand and Brobeck 1951). Although increased attention has recently focused on the arcuate nucleus as a center of feeding behaviors, the LHA functions as an important downstream target of orexigenic agouti-related peptide (AgRP) neurons in the arcuate nucleus (Betley et al. 2013) and of the anorexigenic hormone leptin (Leinninger et al. 2009). Orexin neurons sense an internal state via nutrients, leptin, and glucose and subsequently modulate the neural activities of target sites via orexin receptor type 1 (OX1R) and type 2 (OX2R) through their broad projections from the forebrain to the spinal cord.

2 Feeding Behavior

2.1 Basal Feeding Behavior

Since orexin was first reported as an orexigenic neuropeptide (Sakurai et al. 1998), the acute orexigenic effect of orexin has since been repeatedly demonstrated. The central administration of orexin has been demonstrated to result in 2–3 g of food intake in the first 1–2 h post-injection during the light phase in rats (Sakurai et al. 1998; Haynes et al. 1999; Clegg et al. 2002; Tsuji et al. 2011), which is comparable with the food intake in the first hour after the intracerebroventricular injection of strong orexigenic peptides such as ghrelin (Nakazato et al. 2001) and neuropeptide Y (NPY) (Clark et al. 1984). In contrast, a selective OX1R antagonist suppressed food intake (Haynes et al. 2000; Rodgers et al. 2001). An opioid antagonist also reduced orexin-induced food intake (Clegg et al. 2002). Consistent with the acute orexigenic effects of orexin, fasting increased the number of c-Fos-positive orexin neurons in primates (Diano et al. 2003) and mice (Leinninger et al. 2011) and increased *orexin* mRNA (Sakurai et al. 1998; López et al. 2000). Orexin-deficient mice and orexin neuron-ablated mice exhibited a mild decrease in daily food intake (Hara et al. 2001, 2005; Sellayah et al. 2011).

However, when orexin was injected in the cerebral ventricle at the start of the dark period, food intake did not change (Haynes et al. 1999). Moreover, a chronic administration of orexin did not change daily food intake (Haynes et al. 1999; Yamanaka et al. 1999), which suggests that orexin's effects are short-term and apparent during the light period. The pharmacogenetic activation of orexin neurons during the light period increased food intake (Inutsuka et al. 2014), as well as the total wake time (Sasaki et al. 2011). Thus, the increased food intake induced by orexin during the light period may be accompanied by an increased wake time. However, it may be an oversimplification to consider increased food intake as the secondary consequence of the increased wakefulness induced by orexin because wake-promoting agents, such as modafinil and amphetamine, suppress food intake despite an increased wake time (Makris et al. 2004). Furthermore, the orexigenic effects of orexin are enhanced by fasting or exercise (Kotz et al. 2002; Thorpe et al.

2003, 2005), which suggests that the orexigenic effects of orexin are altered depending on the energy balance.

2.2 *Reward-Based Feeding*

Switching from normal chow to palatable diets, such as a high-fat diet, increases daily food intake because the palatability of food may overcome the homeostatic energy need. Increasing evidence suggests that orexin plays a crucial role in this reward-based feeding. Orexin injected in the third ventricle increased the intake of high-fat but not low-fat diets when the experimental rats were allowed access both diets (Clegg et al. 2002). Interestingly, a low dose of orexin (below the threshold that exerts an orexigenic effect) increased operant behavior for high-fat food (Choi et al. 2010). Orexin-deficient mice reduced their consumption of saccharine (Shiuchi et al. 2009) and sucrose solutions (Matsuo et al. 2011). Thus, orexin is required to enhance the hedonic aspect of palatable food intake. Orexin neurons are consistently activated in the anticipation of palatable food (Harris et al. 2005; Petrovich et al. 2012) and by the consumption of a high-fat diet (Valdivia et al. 2014). The systemic injection of an OX1R antagonist decreased responses for food reward or palatable food (Borgland et al. 2009; Sharf et al. 2010; Steiner et al. 2013; Valdivia et al. 2014; Alcaraz-Iborra et al. 2014), while an OX2R antagonist did not (Piccoli et al. 2012). Thus, orexin-OX1R signaling may enhance the drive for reward-based feeding.

3 Energy Metabolism and Body Weight Regulation

3.1 *Murine Study*

The injection of orexin-A in the cerebral ventricle or the arcuate nucleus increased energy expenditure (Wang et al. 2003; Semjonous et al. 2009). These effects were independent of food digestion and nutrient absorption because food deprivation did not affect orexin-induced energy expenditure (Lubkin and Stricker-Krongrad 1998). Thus, the orexin-OX1R signaling acutely enhances energy expenditure. Regarding the chronic effects of orexin, orexin-deficient mice exhibited age-dependent mild obesity (Tsuneki et al. 2008) and were susceptible to high-fat diet-induced obesity (Sellayah et al. 2011). Similarly, orexin neuron-ablated mice exhibited age-dependent (Hara et al. 2001) and high-fat diet-induced obesity (Hara et al. 2005). Conversely, orexin overexpression mice were resistant to high-fat diet-induced obesity due to increased energy expenditure, which suggests that orexin functions as a negative regulator of body weight. The suppressive effects of orexin on high-fat diet-induced obesity are mediated by OX2R; the effects of orexin overexpression were abolished in OX2R-deficient mice, and an OX2R agonist suppressed high-fat diet-induced weight gain (Funato et al. 2009). Thus, orexin can increase energy output in the short-term and long-term, and an OX2R agonist could function as an anti-obesity drug.

3.2 *Narcoleptic Patients*

Narcoleptic patients tend to have an increased body mass index (BMI) (Schuld et al. 2000; Dahmen et al. 2001; Nishino et al. 2001; Kotagal et al. 2004). Narcoleptic patients with low levels of orexin in the cerebrospinal fluid (CSF) had a higher BMI compared with narcoleptic patients with normal levels of orexin in the CSF despite no difference between the two groups in sleep-related symptoms (Nishino et al. 2001). Importantly, narcoleptic patients without obesity (BMI < 30) exhibited a lower metabolic rate compared with control subjects with similar BMIs (Dahmen et al. 2009), which suggests that non-obese narcoleptics are susceptible to obesity; this is in accordance with the orexin-deficient animals previously described. Once the patients become obese, there are no differences in resting metabolic rate (Fronczek et al. 2008; Dahmen et al. 2009), leptin or ghrelin (Arnulf et al. 2006; Donjacour et al. 2013).

4 Glucose Metabolism

Studies on the acute effects of orexin on glucose metabolism have produced differing results. The administration of orexin-A in the cerebral ventricles in rats increased (Yi et al. 2009), decreased (Tsuneki et al. 2002), or did not change blood glucose (Haynes et al. 1999). Interestingly, orexin-A administration enhanced the glucose uptake of skeletal muscles via the β2-adrenergic receptor (Shiuchi et al. 2009). Regarding the chronic effects of orexin via gene modification, orexin-deficient mice exhibited age-dependent mild glucose tolerance deterioration (Tsuneki et al. 2008); consistently, orexin overexpression mice exhibited better glucose metabolism (Funato et al. 2009). However, when mice were fed a normal chow diet, the changes in glucose metabolism parameters of orexin gene-modified mice were very subtle; thus, the significance of these findings remains unclear. In contrast, when C57BL/6 mice were given a high-fat diet, they exhibited an apparent deterioration of glucose metabolism, which is a good model for diet-induced insulin resistance. Orexin overexpression mice were resistant to diet-induced insulin resistance (Funato et al. 2009). In this case, better glucose metabolism would be a consequence of a non-obese phenotype because obesity is a major deteriorating factor in glucose metabolism. Importantly, OX1R-deficient mice fed a high-fat diet exhibited better glucose metabolism despite severe obesity, which suggests that the loss of OX1R-signaling causes a beneficial effect on glucose regulation that is independent of adiposity (Funato et al. 2009). The loss of orexin consistently reduced the deterioration of glucose metabolism after streptozocin treatment to destroy pancreatic beta cells (Adeghate et al. 2010).

5 Cardiovascular System

The central application of orexin-A increased blood pressure and heart rate (Shirasaka et al. 1999; Monda et al. 2003, 2007). This effect was abolished by an adrenergic receptor antagonist (Shirasaka et al. 1999). Accordingly, the blood pressure of orexin-deficient mice was 10–15 mmHg lower compared with their wild-type littermates (Kayaba et al. 2003). Similarly, air-jet stress-induced elevations in blood pressure and heart rate were attenuated in conscious orexin/ataxin 3 transgenic mice (Zhang et al. 2006). The orexin neurons were ablated in these mice, which suggests that orexin physiologically enhances sympathetic tone. The injection of orexin-A in the diagonal band of the Broca area increased the firing rate of sympathetic nerves to the brown adipose tissue, as well as the temperature of the brown adipose tissue (Monda et al. 2004).

6 Bone Metabolism

Orexin-deficient and OX2R-deficient mice have recently been reported to have a low bone mass (Wei et al. 2014). The injection of an OX2R agonist in the cerebral ventricle enhanced bone formation via leptin (Wei et al. 2014). Although the detailed mechanism requires further examination, this study indicates that osteoporosis is another potential therapeutic target of an OX2R agonist.

7 Interface of Multiple Behaviors and Metabolism

Fasting and caloric restriction enhance resilience to depressive states, which may enable animals to increase the possibility of successful forage and survival. However, orexin-deficient mice did not exhibit anti-depressive behaviors due to caloric restriction (Lutter et al. 2008a). Caloric restriction increased plasma ghrelin, which decreased depressive behavior in a manner similar to caloric restriction (Lutter et al. 2008b). Interestingly, orexin-deficient mice that were depressed via chronic social defeat stress exhibited impaired glucose metabolism, which was accompanied by blunted Akt phosphorylation, thereby indicating disturbed insulin signaling (Tsuneki et al. 2013). Thus, orexin functions to integrate mood-related behaviors with metabolism pathways.

8 The Nutritional and Hormonal Regulation of Orexin Neurons

8.1 Leptin

Leptin, a peptide hormone secreted from adipocytes, functions as a peripheral signal that conveys a positive energy balance to suppress food intake. The leptin receptor is moderately expressed in the LHA (Scott et al. 2009). Leptin increases *orexin* mRNA (Tritos et al. 2001; Louis et al. 2010; Leinninger et al. 2011) and the c-Fos expression of orexin neurons (Cui et al. 2012). However, leptin-expressing neurons in the LHA are distinct from orexin neurons (Leinninger et al. 2009). Considering the prominent anorexigenic and anti-obesity effects of leptin, the upregulation of *orexin* mRNA by leptin (Tritos et al. 2001; Louis et al. 2010; Leinninger et al. 2011; Cui et al. 2012) is not consistent with an acute orexigenic effect, but with orexin's role as a negative energy regulator (Funato et al. 2009).

8.2 Glucose

The LHA contains glucosensing neurons that increase their firing rates in response to reduced glucose levels (Oomura et al. 1969). Insulin-induced hypoglycemia increased the number of c-Fos-positive orexin neurons (Moriguchi et al. 1999; Cai et al. 2001), as well as *orexin* mRNA (Griffond et al. 1999). These findings demonstrate that changes in the circulating glucose concentration may directly or indirectly modulate the activity of orexin neurons. Orexin neurons in brain slices or in dissociated culture were inhibited or activated in response to increased or decreased glucose concentrations, respectively. In contrast, insulin did not affect orexin neuron activity (Yamanaka et al. 2003; Tsujino et al. 2005). Furthermore, an increased concentration of glucose within a physiological level was confirmed to hyperpolarize and cease orexin neuron firing due to increased potassium conductance (Burdakov et al. 2006). 2-deoxyglucose-6-phosphate, which is not metabolized by phosphoglucose isomerase, inhibits the firing of orexin neurons in a manner similar to glucose (González et al. 2008). The responses of orexin neurons during glucose application, such as the selectivity for potassium ions, Goldman-Hodgkin-Katz rectification, and minimal inactivation, indicate a member of tandem pore domain potassium channels (Burdakov et al. 2006). Among 15 members of tandem pore domain potassium channels, deficiencies of TASK1, TASK3, TREK1, TREK2, or TAAK did not abolish glucose-induced the hyperpolarization of orexin neurons (Guyon et al. 2009; González et al. 2009).

Interestingly, intracellular pyruvate, lactate, and ATP suppressed glucose-induced hyperpolarization in a dose-dependent manner (Venner et al. 2011). Because pyruvate, lactate, and ATP represent intracellular energy or sources of energy, orexin neurons have been suggested to be less sensitive to extracellular glucose in the presence of high intracellular energy levels.

8.3 Amino Acids

Similar to energy balance, protein balance is homeostatically regulated. For example, a low-protein diet increases food intake, whereas a high-protein diet decreases food intake (Morrison et al. 2012). The hypothalamus can directly monitor amino acids to alter feeding behavior (Cota et al. 2006). The administration of amino acids systemically or directly in the hypothalamus activated orexin neurons (Karnani et al. 2011). The application of amino acids in brain slices resulted in the depolarization and increased firing of orexin neurons and also suppressed glucose-induced hyperpolarization, whereas fatty acids did not change the membrane potential or firing rate of orexin neurons (Karnani et al. 2011).

8.4 Other Factors

In the brain, the pH of the interstitial space is approximately 7.25 and fluctuates between 6.9 and 7.4 during hypo or hyperventilation, respectively (van Hulst et al. 2002). Acidosis and hypercapnia have been demonstrated to suppress the excitabilities of most neurons, whereas both conditions activate several groups of neurons in the brain stem to stimulate breathing (Richerson 2004). Surprisingly, increased CO_2 and decreased pH within a physiological range depolarized orexin neurons, which resulted in an increased firing rate (Williams et al. 2007). The inhalation of 10 % CO_2 gas increased c-Fos expression in orexin neurons in the perifornical and dorsomedial hypothalamic regions (Sunanaga et al. 2009). These findings suggest that orexin neurons are able to enhance level of vigilance and cardiovascular activity in case of emergency. Orexin-deficient mice consistently exhibited a smaller increase in respiratory volume during exposure to a hypercapnic environment during wakefulness compared with wild-type mice (Nakamura et al. 2007). A dual antagonist for OX1R and OX2R inhibited CO_2-induced ventilation (Li and Nattie 2010). Orexin-A activates retrotrapezoid neurons, which play an important role in respiratory chemoreflexes (Lazarenko et al. 2011). The respiratory response to hypercapnia decreased when an OX1R antagonist was injected in the retrotrapezoid nucleus (Dias et al. 2009). Moreover, Burdakov et al. hypothesized that orexin neurons function as conditional glucose sensors to integrate information regarding extracellular glucose, intracellular energy, and extracellular pH, based on their finding that a lower extracellular pH suppressed the glucose-activated currents of orexin neurons (Burdakov et al. 2006).

9 Neural Networks that Mediate the Metabolic Effects of Orexin

9.1 Arcuate Nucleus

Orexin neurons send their fibers to the arcuate nucleus (Peyron et al. 1998), which is a pair of small nuclei located in the base of the medial hypothalamus and contains orexigenic NPY/AgRP neurons and anorexigenic POMC neurons. The injection of orexin-A in the arcuate nucleus increased food intake (Muroya et al. 2004) and energy expenditure (Wang et al. 2003). However, one half of the arcuate nucleus has a volume less than 0.2 mm^3. Therefore, it is technically impossible to inject 0.5 μl (0.5 mm^3) (Muroya et al. 2004) or 0.2 μl (0.2 mm^3) (Wang et al. 2003) into the arcuate nucleus without spilling into the adjacent structures.

Recent advances in the optogenetic activation and trans-synaptic labeling of genetically defined neural groups have assisted in overcoming the technical limitations of a stereotaxic injection into a specific hypothalamic nucleus. The optogenetic activation of Vglut2-positive neurons in the LHA did not activate AgRP neurons (Krashes et al. 2014), which suggests that orexin neurons do not directly activate AgRP because orexin neurons are glutamatergic and contain Vglut2 (Rosin et al. 2003). Consistent with the lack of evidence for the direct activation of AgRP neurons by LHA neurons, the monosynaptic retrograde labeling of AgRP neurons found that 3 % of the labeled cells resided in the LHA (Krashes et al. 2014). However, considering the dense fiber projections of orexin neurons to the arcuate nucleus, it is possible that orexin neurons regulate AgRP neurons via volume transmission, which cannot be detected via retrograde trans-synaptic labeling using a rabies virus glycoprotein or via optogenetic activation. The application of orexin to AgRP neurons in a dissociation culture (Muroya et al. 2004) or in brain slices (van den Top et al. 2004) increased the intracellular calcium concentration and induced depolarization, respectively. AgRP neurons have consistently been reported to express OX2R (van den Top et al. 2004).

Orexin has also been reported to affect POMC neurons, but the results are not consistent. Orexin decreased the intracellular calcium concentration of POMC neurons in a dissociation culture (Muroya et al. 2004) and hyperpolarized the POMC neurons in brain slices (Ma et al. 2007), whereas Acuna-Goycolea and van den Pol reported the activation of POMC neurons using orexin (Acuna-Goycolea and van den Pol 2009). The AgRP or POMC neuron-specific loss of orexin receptors is required to clarify the neural circuits that regulate feeding behavior and energy metabolism.

Regarding the afferent fibers to orexin neurons, transgenic mice that expressed tetanus toxin in the orexin neurons, which is transported trans-synaptically in a retrograde direction, exhibited many arcuate neurons that projected directly or indirectly to orexin neurons (Sakurai et al. 2005). NPY, AgRP, αMSH or galanin-like peptide-immunopositive fibers made close contact with orexin neurons (Elias et al. 1998; Issa and Wang 2008). However, NPY, AgRP, αMSH and galanin-like peptide did not alter the calcium concentration of the orexin neurons (Tsujino et al. 2005).

9.2 LHA

In addition to orexin neurons, there are several neuronal subgroups in the LHA that have specific neurochemical characteristics and convey different metabolic signals. Cui et al. and Leinninger et al. reported that a group of GABAergic neurons in the LHA expressed the long form of the leptin receptor (LepRb), melanocortin 4 receptor (MC4R), and neurotensin (Leinninger et al. 2011; Cui et al. 2012). These GABAergic neurons are distinct from orexin neurons and MCH neurons (Leinninger et al. 2009). Leptin hyperpolarized the LepRb/MC4R GABAergic neurons, which may project to orexin neurons (Cui et al. 2012); thus, leptin can disinhibit orexin neurons, which is consistent with leptin-induced c-Fos upregulation in orexin neurons (Tritos et al. 2001; Louis et al. 2010; Leinninger et al. 2011; Cui et al. 2012). Although neurotensin is expressed in many brain regions, the colocalization of neurotensin and leptin receptors was recognized only in the LHA. The deletion of the leptin receptor specifically in neurotensin-expressing neurons resulted in the loss of fasting-induced c-Fos expression in orexin neurons (Leinninger et al. 2011). Importantly, neurotensin neuron-specific leptin receptor-deficient mice increased body weight and adiposity (Leinninger et al. 2011), which suggests that the activation of neurotensin neurons by leptin is required to maintain a stable body weight. Because neurotensin may activate rather than inhibit orexin neurons (Tsujino et al. 2005), the inhibitory action of GABAergic transmission would exert a dominant effect on orexin neurons. The characterization of neurotensin-positive neurons is complicated by the fast that neurotensin is also expressed in the vast majority of orexin neurons. Neurotensin may function to maintain the activity of orexin neurons that express neurotensin receptor-2 (Furutani et al. 2013).

9.3 Other Hypothalamic Regions

Consistent with a broad hypothalamic projection of orexin neurons, the local administration of orexin in different brain regions induces multiple behavioral and metabolic changes that are dependent on the site of orexin injection. The injection of orexin-A in the paraventricular nucleus (PVN) and the dorsomedial hypothalamus (DMH) (Dube et al. 1999) increased food intake. The injection of orexin-A in the PVN also enhanced gastric acid secretion (Chaleek et al. 2012). Interestingly, orexin-A injected in the ventromedial hypothalamus (VMH) did not increase food intake but increased glucose uptake by skeletal muscles via the sympathetic nervous system (Shiuchi et al. 2009).

9.4 Brain Regions Related to the Mesolimbic Dopaminergic Pathway

Projections of orexin neurons to the ventral tegmental area (VTA) are important in reward-based feeding (Harris et al. 2005). The VTA has reciprocal fiber connections with the nucleus accumbens and medial prefrontal cortex to exert reward-dependent behavior (Lammel et al. 2014). In rats, the injection of a Mu-opioid receptor agonist in the ventral medial prefrontal cortex increased food intake and enhanced the preference for a carbohydrate-rich diet (Mena et al. 2011), which was accompanied by an increased c-Fos expression in orexin neurons (Mena et al. 2013). Similarly, a Mu-opioid receptor agonist injected in the nucleus accumbens activated orexin neurons and increased high-fat diet intake, which was not exhibited by orexin-deficient mice and was blocked via the injection of an OX1R antagonist in the VTA (Zheng et al. 2007). The injection of orexin-A in the nucleus accumbens and the ventral pallidum enhanced general feeding behavior (Thorpe and Kotz 2005) and strengthened sucrose preference (Ho and Berridge 2013), respectively. An OX1R knock-down in the paraventricular thalamic nucleus suppressed the high-fat diet consumption of mice (Choi et al. 2012).

9.5 Brain Stem

Orexin-A injected in the fourth ventricle increased licking of a sucrose solution during one meal (Baird et al. 2009) and food intake of a chow diet (Parise et al. 2011). The injection of orexin-A in the solitary tract nucleus, which is a structure adjacent to the fourth ventricle, increased high-fat diet intake (Kay et al. 2014).

Regarding sympathetic nerve activity, orexin-A applied in the fourth ventricle increased body temperature as well as heart rate (Zheng et al. 2005). Orexin neurons project to the raphe pallidus of the medulla, which contains sympathetic premotor neurons that regulate heart rate and heat production of the brown adipose tissue. Orexin-A injection in the raphe pallidus increased brown adipocyte thermogenesis (Tupone et al. 2011). In contrast, orexin-A injected in the solitary tract nucleus decreased heart rate and blood pressure (de Oliveira et al. 2003; Ciriello et al. 2013).

In addition to the sympathetic action of orexin, orexin neurons also project to regions of the parasympathetic system, such as the motor nucleus of the vagal nerve in the nucleus ambiguous, which is a major determinant of parasympathetic activity to the heart. Orexin inhibits vagal motor neurons in the nucleus ambiguus via the activation of local inhibitory neurons in slice preparations (Dergacheva et al. 2005).

9.6 Spinal Cord and Peripheral Nervous System

Several peripheral organs have been identified as the targets of the central orexin system. Orexin neurons directly innervate and are able to activate sympathetic preganglionic neurons of the spinal cord (Antunes et al. 2001; van den Top et al. 2003). Using a pseudorabies virus, which retrogradely transmits through multiple synapses, orexin neurons have been identified as a higher-order neuron circuit for the salivary gland, masseter muscle (Pérez et al. 2011), white adipose tissue, liver (van den Top et al. 2003; Stanley et al. 2010), skeletal muscle, adrenal gland (Kerman et al. 2007), and diaphragm (Badami et al. 2010). Consistent with the connection to the masseter muscle, the cerebroventricular injection of orexin enhanced the activity of the masseter muscle during eating (Tsuji et al. 2011). In this manner, the orexin system may help an individual animal survive given difficulty and may contribute to foraging behavior by altering the sympathetic tone, nutrient utilization, ventilation, and exercise capacity.

10 Peripheral Actions of Orexin

In addition to the nervous system, orexin receptors are expressed in the peripheral tissues, such as the pancreatic islet (Kirchgessner and Liu 1999; Nowak et al. 2005; Funato et al. 2009; Adeghate et al. 2010), adrenal gland (Jöhren et al. 2004), white adipose tissue (Digby et al. 2006), and bone (Wei et al. 2014). Although the peripheral actions of orexin have attracted attention, there is no consensus due to inconsistent results between different research groups and the discrepancy between in vivo and in vitro studies. For example, the peripheral administration of orexin increased plasma insulin and insulin secretion from the islet (Nowak et al. 2005), decreased it (Ouedraogo et al. 2003), or did not induce a change (Tsuneki et al. 2002; Ehrström et al. 2004).

Orexin has recently been reported to enhance the differentiation of brown fat adipocytes (Sellayah et al. 2011) and suppress osteoblast differentiation in vitro (Wei et al. 2014). Similarly, orexin enhanced glucose uptake and suppressed the lipase expression of 3T3-L1 adipocytes, which resulted in lipid accumulation (Skrzypski et al. 2011) and promoted the differentiation of 3T3-L1 preadipocytes into mature adipocytes (Skrzypski et al. 2012). However, the role of orexin in adipocytes in vitro is opposite that of the obesity-prone phenotype of orexin-deficient mice.

11 Conclusions and Future Directions

Orexin functions as an integrative modulator of multiple metabolism pathways and behaviors. Therefore, chemical agents that target OX1R and/or OX2R would have a rich therapeutic potential for eating disorders, obesity and diabetes mellitus. There has been a wealth of research using orexin-related reagents and gene-modified mice because of orexin's multifaceted role. However, the experimental results have not been consistent. A meta-analysis is necessary to make firm conclusions regarding orexin's metabolic role; a recent analysis on MCH is one example (Takase et al. 2014). Moreover, to determine whether orexin functions locally and directly in the peripheral tissue, it is necessary to examine animals in which orexin receptors are specifically deficient in a particular peripheral tissue.

References

Acuna-Goycolea C, van den Pol AN (2009) Neuroendocrine proopiomelanocortin neurons are excited by hypocretin/orexin. J Neurosci 29:1503–1513. doi:10.1523/JNEUROSCI.5147-08.2009

Adeghate E, Fernandez-Cabezudo M, Hameed R et al (2010) Orexin-1 receptor co-localizes with pancreatic hormones in islet cells and modulates the outcome of streptozotocin-induced diabetes mellitus. PLoS ONE 5:e8587. doi:10.1371/journal.pone.0008587

Alcaraz-Iborra M, Carvajal F, Lerma-Cabrera JM et al (2014) Binge-like consumption of caloric and non-caloric palatable substances in ad libitum-fed C57BL/6 J mice: Pharmacological and molecular evidence of orexin involvement. Behav Brain Res 272:93–99. doi:10.1016/j.bbr.2014.06.049

Anand BK, Brobeck JR (1951) Hypothalamic control of food intake in rats and cats. Yale J Biol Med 24:123–140

Antunes VR, Brailoiu GC, Kwok EH et al (2001) Orexins/hypocretins excite rat sympathetic preganglionic neurons in vivo and in vitro. Am J Physiol Regul Integr Comp Physiol 281: R1801–R1807

Arnulf I, Lin L, Zhang J et al (2006) CSF versus serum leptin in narcolepsy: is there an effect of hypocretin deficiency? Sleep 29:1017–1024

Badami VM, Rice CD, Lois JH et al (2010) Distribution of hypothalamic neurons with orexin (hypocretin) or melanin concentrating hormone (MCH) immunoreactivity and multisynaptic connections with diaphragm motoneurons. Brain Res 1323:119–126. doi:10.1016/j.brainres.2010.02.002

Baird J-P, Choe A, Loveland JL et al (2009) Orexin-A hyperphagia: hindbrain participation in consummatory feeding responses. Endocrinology 150:1202–1216. doi:10.1210/en.2008-0293

Betley JN, Cao ZFH, Ritola KD, Sternson SM (2013) Parallel, redundant circuit organization for homeostatic control of feeding behavior. Cell 155:1337–1350. doi:10.1016/j.cell.2013.11.002

Borgland SL, Chang S-J, Bowers MS et al (2009) Orexin A/hypocretin-1 selectively promotes motivation for positive reinforcers. J Neurosci 29:11215–11225. doi:10.1523/JNEUROSCI.6096-08.2009

Burdakov D, Jensen LT, Alexopoulos H et al (2006) Tandem-pore K^+ channels mediate inhibition of orexin neurons by glucose. Neuron 50:711–722. doi:10.1016/j.neuron.2006.04.032

Cai XJ, Evans ML, Lister CA et al (2001) Hypoglycemia activates orexin neurons and by the nucleus of the solitary tract. Diabetes 50:105–112

Chaleek N, Kermani M, Eliassi A, Haghparast A (2012) Effects of orexin and glucose microinjected into the hypothalamic paraventricular nucleus on gastric acid secretion in conscious rats. Neurogastroenterol Motil 24:e94–e102. doi:10.1111/j.1365-2982.2011.01789.x

Choi DL, Davis JF, Fitzgerald ME, Benoit SC (2010) The role of orexin-A in food motivation, reward-based feeding behavior and food-induced neuronal activation in rats. Neuroscience 167:11–20. doi:10.1016/j.neuroscience.2010.02.002

Choi DL, Davis JF, Magrisso IJ et al (2012) Orexin signaling in the paraventricular thalamic nucleus modulates mesolimbic dopamine and hedonic feeding in the rat. Neuroscience 210:243–248. doi:10.1016/j.neuroscience.2012.02.036

Ciriello J, Caverson MM, McMurray JC, Bruckschwaiger EB (2013) Co-localization of hypocretin-1 and leucine-enkephalin in hypothalamic neurons projecting to the nucleus of the solitary tract and their effect on arterial pressure. Neuroscience 250:599–613. doi:10.1016/j.neuroscience.2013.07.054

Clark JT, Kalra PS, Crowley WR, Kalra SP (1984) Neuropeptide Y and human pancreatic polypeptide stimulate feeding behavior in rats. Endocrinology 115:427–429

Clegg DJ, Air EL, Woods SC, Seeley RJ (2002) Eating elicited by orexin-A, but not melanin-concentrating hormone, is opioid mediated. Endocrinology 143:2995–3000. doi:10.1210/endo.143.8.8977

Cota D, Proulx K, Smith KAB et al (2006) Hypothalamic mTOR signaling regulates food intake. Science 312:927–930. doi:10.1126/science.1124147

Cui H, Sohn J-W, Gautron L et al (2012) Neuroanatomy of melanocortin-4 receptor pathway in the lateral hypothalamic area. J Comp Neurol 520:4168–4183. doi:10.1002/cne.23145

Dahmen N, Bierbrauer J, Kasten M (2001) Increased prevalence of obesity in narcoleptic patients and relatives. Eur Arch Psychiatry Clin Neurosci 251:85–89

Dahmen N, Tonn P, Messroghli L et al (2009) Basal metabolic rate in narcoleptic patients. Sleep 32:962–964

De Oliveira CVR, Rosas-Arellano MP, Solano-Flores LP, Ciriello J (2003) Cardiovascular effects of hypocretin-1 in nucleus of the solitary tract. Am J Physiol Heart Circ Physiol 284:H1369–H1377. doi:10.1152/ajpheart.00877.2002

Dergacheva O, Wang X, Huang Z et al (2005) Hypocretin-1 (Orexin-A) facilitates inhibitory and diminishes excitatory synaptic pathways to cardiac vagal neurons in the nucleus ambiguus. J Pharmacol Exp Ther 314:1322–1327. doi:10.1124/jpet.105.086421.sor

Diano S, Horvath B, Urbanski HF et al (2003) Fasting activates the nonhuman primate hypocretin (orexin) system and its postsynaptic targets. Endocrinology 144:3774–3778. doi:10.1210/en.2003-0274

Dias MB, Li A, Nattie EE (2009) Antagonism of orexin receptor-1 in the retrotrapezoid nucleus inhibits the ventilatory response to hypercapnia predominantly in wakefulness. J Physiol 587:2059–2067. doi:10.1113/jphysiol.2008.168260

Digby JE, Chen J, Tang JY et al (2006) Orexin receptor expression in human adipose tissue: effects of orexin-A and orexin-B. J Endocrinol 191:129–136. doi:10.1677/joe.1.06886

Donjacour CEHM, Pardi D, Aziz NA et al (2013) Plasma total ghrelin and leptin levels in human narcolepsy and matched healthy controls: basal concentrations and response to sodium oxybate. J Clin Sleep Med 9:797–803. doi:10.5665/sleep.3592

Dube MG, Kalra SP, Kalra PS (1999) Food intake elicited by central administration of orexins/hypocretins: identification of hypothalamic sites of action. Brain Res 842:473–477

Ehrström M, Näslund E, Levin F et al (2004) Pharmacokinetic profile of orexin A and effects on plasma insulin and glucagon in the rat. Regul Pept 119:209–212. doi:10.1016/j.regpep.2004.02.004

Elias CF, Saper CB, Maratos-flier E et al (1998) Chemically difined projections linking the mediobasal hypothalamus and the lateral hypothalamic area. J Comp Neurol 402:442–459

Fronczek R, Overeem S, Reijntjes R et al (2008) Increased heart rate variability but normal resting metabolic rate in hypocretin/orexin-deficient human narcolepsy. J Clin Sleep Med 4:248–254

Funato H, Tsai AL, Willie JT et al (2009) Enhanced orexin receptor-2 signaling prevents diet-induced obesity and improves leptin sensitivity. Cell Metab 9:64–76. doi:10.1016/j.cmet.2008.10.010

Furutani N, Hondo M, Kageyama H et al (2013) Neurotensin co-expressed in orexin-producing neurons in the lateral hypothalamus plays an important role in regulation of sleep/wakefulness states. PLoS ONE 8:e62391. doi:10.1371/journal.pone.0062391

González JA, Jensen LT, Fugger L, Burdakov D (2008) Metabolism-independent sugar sensing in central orexin neurons. Diabetes 57:2569–2576. doi:10.2337/db08-0548

González JA, Jensen LT, Doyle SE et al (2009) Deletion of TASK1 and TASK3 channels disrupts intrinsic excitability but does not abolish glucose or pH responses of orexin/hypocretin neurons. Eur J Neurosci 30:57–64. doi:10.1111/j.1460-9568.2009.06789.x

Griffond B, Risold P, Jacquemard C et al (1999) Insulin-induced hypoglycemia increases preprohypocretin (orexin) mRNA in the rat lateral hypothalamic area. Neurosci Lett 262:77–80

Guyon A, Tardy MP, Rovère C et al (2009) Glucose inhibition persists in hypothalamic neurons lacking tandem-pore K^+ channels. J Neurosci 29:2528–2533. doi:10.1523/JNEUROSCI.5764-08.2009

Hara J, Beuckmann CT, Nambu T et al (2001) Genetic ablation of orexin neurons in mice results in narcolepsy, hypophagia, and obesity. Neuron 30:345–354

Hara J, Yanagisawa M, Sakurai T (2005) Difference in obesity phenotype between orexin-knockout mice and orexin neuron-deficient mice with same genetic background and environmental conditions. Neurosci Lett 380:239–242. doi:10.1016/j.neulet.2005.01.046

Harris GC, Wimmer M, Aston-Jones G (2005) A role for lateral hypothalamic orexin neurons in reward seeking. Nature 437:556–559. doi:10.1038/nature04071

Haynes AC, Jackson B, Overend P et al (1999) Effects of single and chronic intracerebroventricular administration of the orexins on feeding in the rat. Peptides 20:1099–1105

Haynes AC, Jackson B, Chapman H et al (2000) A selective orexin-1 receptor antagonist reduces food consumption in male and female rats. Regul Pept 96:45–51

Ho C-Y, Berridge KC (2013) An orexin hotspot in ventral pallidum amplifies hedonic "liking" for sweetness. Neuropsychopharmacology 38:1655–1664. doi:10.1038/npp.2013.62

Inutsuka A, Inui A, Tabuchi S et al (2014) Concurrent and robust regulation of feeding behaviors and metabolism by orexin neurons. Neuropharmacology 85:451–460. doi:10.1016/j.neuropharm.2014.06.015

Issa EB, Wang X (2008) Sensory responses during sleep in primate primary and secondary auditory cortex. J Neurosci 28:14467–14480. doi:10.1523/JNEUROSCI.3086-08.2008

Jöhren O, Brüggemann N, Dominiak P (2004) Orexins (hypocretins) and adrenal function. Horm Metab Res 36:370–375. doi:10.1055/s-2004-814569

Karnani MM, Apergis-Schoute J, Adamantidis A et al (2011) Activation of central orexin/hypocretin neurons by dietary amino acids. Neuron 72:616–629. doi:10.1016/j.neuron.2011.08.027

Kay K, Parise EM, Lilly N, Williams DL (2014) Hindbrain orexin 1 receptors influence palatable food intake, operant responding for food, and food-conditioned place preference in rats. Psychopharmacology 231:419–427. doi:10.1007/s00213-013-3248-9

Kayaba Y, Nakamura A, Kasuya Y et al (2003) Attenuated defense response and low basal blood pressure in orexin knockout mice. Am J Physiol Regul Integr Comp Physiol 285:R581–R593. doi:10.1152/ajpregu.00671.2002

Kerman IA, Bernard R, Rosenthal D et al (2007) Distinct populations of presympathetic-premotor neurons express orexin or melanin-concentrating hormone in the rat lateral hypothalamus. J Comp Neurol 505:586–601

Kirchgessner AL, Liu M (1999) Orexin synthesis and response in the gut. Neuron 24:941–951

Kotagal S, Krahn LE, Slocumb N (2004) A putative link between childhood narcolepsy and obesity. Sleep Med 5:147–150. doi:10.1016/j.sleep.2003.10.006

Kotz CM, Teske JA, Levine JA, Wang C (2002) Feeding and activity induced by orexin A in the lateral hypothalamus in rats. Regul Pept 104:27–32

Krashes MJ, Shah BP, Madara JC et al (2014) An excitatory paraventricular nucleus to AgRP neuron circuit that drives hunger. Nature 507:238–242. doi:10.1038/nature12956

Lammel S, Lim BK, Malenka RC (2014) Reward and aversion in a heterogeneous midbrain dopamine system. Neuropharmacology 76:351–359. doi:10.1016/j.neuropharm.2013.03.019

Lazarenko RM, Stornetta RL, Bayliss DA, Guyenet PG (2011) Orexin A activates retrotrapezoid neurons in mice. Respir Physiol Neurobiol 175:283–287. doi:10.1016/j.resp.2010.12.003

Leinninger GM, Jo Y-H, Leshan RL et al (2009) Leptin acts via leptin receptor-expressing lateral hypothalamic neurons to modulate the mesolimbic dopamine system and suppress feeding. Cell Metab 10:89–98. doi:10.1016/j.cmet.2009.06.011

Leinninger GM, Opland DM, Jo Y-H et al (2011) Leptin action via neurotensin neurons controls orexin, the mesolimbic dopamine system and energy balance. Cell Metab 14:313–323. doi:10.1016/j.cmet.2011.06.016

Li A, Nattie E (2010) Antagonism of rat orexin receptors by almorexant attenuates central chemoreception in wakefulness in the active period of the diurnal cycle. J Physiol 588:2935–2944. doi:10.1113/jphysiol.2010.191288

López M, Seoane L, García MC et al (2000) Leptin regulation of prepro-orexin and orexin receptor mRNA levels in the hypothalamus. Biochem Biophys Res Commun 269:41–45. doi:10.1006/bbrc.2000.2245

Louis GW, Leinninger GM, Rhodes CJ, Myers MG (2010) Direct innervation and modulation of orexin neurons by lateral hypothalamic LepRb neurons. J Neurosci 30:11278–11287. doi:10.1523/JNEUROSCI.1340-10.2010

Lubkin M, Stricker-Krongrad A (1998) Independent feeding and metabolic actions of orexins in mice. Biochem Biophys Res Commun 253:241–245. doi:10.1006/bbrc.1998.9750

Lutter M, Krishnan V, Russo SJ et al (2008a) Orexin signaling mediates the antidepressant-like effect of calorie restriction. J Neurosci 28:3071–3075. doi:10.1523/JNEUROSCI.5584-07.2008

Lutter M, Sakata I, Osborne-Lawrence S et al (2008b) The orexigenic hormone ghrelin defends against depressive symptoms of chronic stress. Nat Neurosci 11:752–753. doi:10.1038/nn.2139

Ma X, Zubcevic L, Brüning JC et al (2007) Electrical inhibition of identified anorexigenic POMC neurons by orexin/hypocretin. J Neurosci 27:1529–1533. doi:10.1523/JNEUROSCI.3583-06.2007

Makris AP, Rush CR, Frederich RC, Kelly TH (2004) Wake-promoting agents with different mechanisms of action: comparison of effects of modafinil and amphetamine on food intake and cardiovascular activity. Appetite 42:185–195. doi:10.1016/j.appet.2003.11.003

Matsuo E, Mochizuki A, Nakayama K et al (2011) Decreased intake of sucrose solutions in orexin knockout mice. J Mol Neurosci 43:217–224. doi:10.1007/s12031-010-9475-1

Mena JD, Sadeghian K, Baldo BA (2011) Induction of hyperphagia and carbohydrate intake by μ-opioid receptor stimulation in circumscribed regions of frontal cortex. J Neurosci 31:3249–3260. doi:10.1523/JNEUROSCI.2050-10.2011

Mena JD, Selleck RA, Baldo BA (2013) Mu-opioid stimulation in rat prefrontal cortex engages hypothalamic orexin/hypocretin-containing neurons, and reveals dissociable roles of nucleus accumbens and hypothalamus in cortically driven feeding. J Neurosci 33:18540–18552. doi:10.1523/JNEUROSCI.3323-12.2013

Monda M, Viggiano A, De Luca V (2003) Parodoxical effect of orexin A: hypophagia induced by hyperthermia. Brain Res 961:220–228

Monda M, Viggiano A, Viggiano A et al (2004) Injection of orexin A into the diagonal band of Broca induces sympathetic and hyperthermic reactions. Brain Res 1018:265–271. doi:10.1016/j.brainres.2004.05.084

Monda M, Viggiano A, Viggiano A et al (2007) Sympathetic and hyperthermic reactions by orexin A: role of cerebral catecholaminergic neurons. Regul Pept 139:39–44. doi:10.1016/j.regpep.2006.10.002

Moriguchi T, Sakurai T, Nambu T et al (1999) Neurons containing orexin in the lateral hypothalamic area of the adult rat brain are activated by insulin-induced acute hypoglycemia. Neurosci Lett 264:101–104

Morrison CD, Reed SD, Henagan TM (2012) Homeostatic regulation of protein intake: in search of a mechanism. Am J Physiol Regul Integr Comp Physiol 302:R917–R928. doi:10.1152/ajpregu.00609.2011

Muroya S, Funahashi H, Yamanaka A et al (2004) Orexins (hypocretins) directly interact with neuropeptide Y, POMC and glucose-responsive neurons to regulate Ca^{2+} signaling in a reciprocal manner to leptin: orexigenic neuronal pathways in the mediobasal hypothalamus. Eur J Neurosci 19:1524–1534. doi:10.1111/j.1460-9568.2004.03255.x

Nakamura A, Zhang W, Yanagisawa M et al (2007) Vigilance state-dependent attenuation of hypercapnic chemoreflex and exaggerated sleep apnea in orexin knockout mice. J Appl Physiol 102:241–248. doi:10.1152/japplphysiol.00679.2006

Nakazato M, Murakami N, Date Y et al (2001) A role for ghrelin in the central regulation of feeding. Nature 409:194–198. doi:10.1038/35051587

Nishino S, Ripley B, Overeem S et al (2001) Low cerebrospinal fluid hypocretin (orexin) and altered energy homeostasis in human narcolepsy. Ann Neurol 50:381–388. doi:10.1002/ana.1130

Nowak KW, Strowski MZ, Switonska MM et al (2005) Evidence that orexins A and B stimulate insulin secretion from rat pancreatic islets via both receptor subtypes. Int J Mol Med 15:969–972. doi:10.3892/ijmm.15.6.969

Oomura Y, Ono T, Ooyama H, Wayner MJ (1969) Glucose and osmosensitive neurones in the rat hypothalamus. Nature 222:282–284

Ouedraogo R, Näslund E, Kirchgessner AL (2003) Glucose regulates the release of orexin-a from the endocrine pancreas. Diabetes 52:111–117. doi:10.2337/diabetes.52.1.111

Parise EM, Lilly N, Kay K et al (2011) Evidence for the role of hindbrain orexin-1 receptors in the control of meal size. Am J Physiol Regul Integr Comp Physiol 301:R1692–R1699. doi:10.1152/ajpregu.00044.2011

Pérez CA, Stanley SA, Wysocki RW et al (2011) Molecular annotation of integrative feeding neural circuits. Cell Metab 13:222–232. doi:10.1016/j.cmet.2010.12.013

Petrovich GD, Hobin MP, Reppucci CJ (2012) Selective Fos induction in hypothalamic orexin/hypocretin, but not melanin-concentrating hormone neurons, by a learned food-cue that stimulates feeding in sated rats. Neuroscience 224:70–80. doi:10.1016/j.neuroscience.2012.08.036

Peyron C, Tighe DK, van den Pol AN et al (1998) Neurons containing hypocretin (orexin) project to multiple neuronal systems. J Neurosci 18:9996–10015

Piccoli L, Micioni Di Bonaventura MV, Cifani C et al (2012) Role of orexin-1 receptor mechanisms on compulsive food consumption in a model of binge eating in female rats. Neuropsychopharmacology 37:1999–2011. doi:10.1038/npp.2012.48

Richerson GB (2004) Serotonergic neurons as carbon dioxide sensors that maintain pH homeostasis. Nat Rev Neurosci 5:449–461. doi:10.1038/nrn1409

Rodgers RJ, Halford JC, Nunes de Souza RL et al (2001) SB-334867, a selective orexin-1 receptor antagonist, enhances behavioural satiety and blocks the hyperphagic effect of orexin-A in rats. Eur J Neurosci 13:1444–1452. doi:10.1046/j.0953-816x.2001.01518.x

Rosin DL, Weston MC, Sevigny CP et al (2003) Hypothalamic orexin (hypocretin) neurons express vesicular glutamate transporters VGLUT1 or VGLUT2. J Comp Neurol 465:593–603. doi:10.1002/cne.10860

Sakurai T, Amemiya A, Ishii M et al (1998) Orexins and orexin receptors: a family of hypothalamic neuropeptides and G protein-coupled receptors that regulate feeding behavior. Cell 92:573–585

Sakurai T, Nagata R, Yamanaka A et al (2005) Input of orexin/hypocretin neurons revealed by a genetically encoded tracer in mice. Neuron 46:297–308. doi:10.1016/j.neuron.2005.03.010

Sasaki K, Suzuki M, Mieda M et al (2011) Pharmacogenetic modulation of orexin neurons alters sleep/wakefulness states in mice. PLoS ONE 6:e20360. doi:10.1371/journal.pone.0020360

Schuld A, Hebebrand J, Geller F, Pollmächer T (2000) Increased body-mass index in patients with narcolepsy. Lancet 355:1274–1275. doi:10.1016/S0140-6736(00)02101-2

Scott MM, Lachey JL, Sternson SM et al (2009) Leptin targets in the mouse brain. J Comp Neurol 514:518–532. doi:10.1002/cne.22025

Sellayah D, Bharaj P, Sikder D (2011) Orexin is required for brown adipose tissue development, differentiation, and function. Cell Metab 14:478–490. doi:10.1016/j.cmet.2011.08.010

Semjonous NM, Smith KL, Parkinson JRC et al (2009) Coordinated changes in energy intake and expenditure following hypothalamic administration of neuropeptides involved in energy balance. Int J Obes (Lond) 33:775–785. doi:10.1038/ijo.2009.96

Sharf R, Sarhan M, Brayton CE et al (2010) Orexin signaling via the orexin 1 receptor mediates operant responding for food reinforcement. Biol Psychiatry 67:753–760. doi:10.1016/j.biopsych.2009.12.035

Shirasaka T, Nakazato M, Matsukura S et al (1999) Sympathetic and cardiovascular actions of orexins in conscious rats. Am J Physiol Regul Integr Comp Physiol 277:R1780–R1785

Shiuchi T, Haque MS, Okamoto S et al (2009) Hypothalamic orexin stimulates feeding-associated glucose utilization in skeletal muscle via sympathetic nervous system. Cell Metab 10:466–480. doi:10.1016/j.cmet.2009.09.013

Skrzypski MT, Le T, Kaczmarek P et al (2011) Orexin A stimulates glucose uptake, lipid accumulation and adiponectin secretion from 3T3-L1 adipocytes and isolated primary rat adipocytes. Diabetologia 54:1841–1852. doi:10.1007/s00125-011-2152-2

Skrzypski M, Kaczmarek P, Le TT et al (2012) Effects of orexin A on proliferation, survival, apoptosis and differentiation of 3T3-L1 preadipocytes into mature adipocytes. FEBS Lett 586:4157–4164. doi:10.1016/j.febslet.2012.10.013

Stanley S, Pinto S, Segal J et al (2010) Identification of neuronal subpopulations that project from hypothalamus to both liver and adipose tissue polysynaptically. Proc Natl Acad Sci USA 107:7024–7029. doi:10.1073/pnas.1002790107

Steiner MA, Sciarretta C, Pasquali A, Jenck F (2013) The selective orexin receptor 1 antagonist ACT-335827 in a rat model of diet-induced obesity associated with metabolic syndrome. Front Pharmacol 4:165. doi:10.3389/fphar.2013.00165

Sunanaga J, Deng B-S, Zhang W et al (2009) CO_2 activates orexin-containing neurons in mice. Respir Physiol Neurobiol 166:184–186. doi:10.1016/j.resp.2009.03.006

Takase K, Kikuchi K, Tsuneoka Y et al (2014) Meta-analysis of melanin-concentrating hormone signaling-deficient mice on behavioral and metabolic phenotypes. PLoS ONE 9:e99961. doi:10.1371/journal.pone.0099961

Thorpe AJ, Kotz CM (2005) Orexin A in the nucleus accumbens stimulates feeding and locomotor activity. Brain Res 1050:156–162. doi:10.1016/j.brainres.2005.05.045

Thorpe AJ, Mullett MA, Wang C, Kotz CM (2003) Peptides that regulate food intake: regional, metabolic, and circadian specificity of lateral hypothalamic orexin A feeding stimulation. Am J Physiol Regul Integr Comp Physiol 284:R1409–R1417. doi:10.1152/ajpregu.00344.2002

Thorpe AJ, Teske JA, Kotz CM (2005) Physiological regulation of appetite: orexin A-induced feeding is augmented by caloric challenge. Am J Physiol Regul Integr Comp Physiol 289: R367–R372. doi:10.1152/ajpregu.00737.2004

Tritos NA, Mastaitis JW, Kokkotou E, Maratos-Flier E (2001) Characterization of melanin concentrating hormone and preproorexin expression in the murine hypothalamus. Brain Res 895:160–166

Tsuji T, Yamamoto T, Tanaka S et al (2011) Analyses of the facilitatory effect of orexin on eating and masticatory muscle activity in rats. J Neurophysiol 106:3129–3135. doi:10.1152/jn.01108.2010

Tsujino N, Yamanaka A, Ichiki K et al (2005) Cholecystokinin activates orexin/hypocretin neurons through the cholecystokinin A receptor. J Neurosci 25:7459–7469. doi:10.1523/JNEUROSCI.1193-05.2005

Tsuneki H, Sugihara Y, Honda R et al (2002) Reduction of blood glucose level by orexins in fasting normal and streptozotocin-diabetic mice. Eur J Pharmacol 448:245–252

Tsuneki H, Murata S, Anzawa Y et al (2008) Age-related insulin resistance in hypothalamus and peripheral tissues of orexin knockout mice. Diabetologia 51:657–667. doi:10.1007/s00125-008-0929-8

Tsuneki H, Tokai E, Sugawara C et al (2013) Hypothalamic orexin prevents hepatic insulin resistance induced by social defeat stress in mice. Neuropeptides 47:213–219. doi:10.1016/j.npep.2013.02.002

Tupone D, Madden CJ, Cano G, Morrison SF (2011) An orexinergic projection from perifornical hypothalamus to raphe pallidus increases rat brown adipose tissue thermogenesis. J Neurosci 31:15944–15955. doi:10.1523/JNEUROSCI.3909-11.2011

Valdivia S, Patrone A, Reynaldo M, Perello M (2014) Acute high fat diet consumption activates the mesolimbic circuit and requires orexin signaling in a mouse model. PLoS ONE 9:e87478. doi:10.1371/journal.pone.0087478

Van den Top M, Nolan MF, Lee K et al (2003) Orexins induce increased excitability and synchronisation of rat sympathetic preganglionic neurones. J Physiol 549:809–821. doi:10.1113/jphysiol.2002.033290

Van den Top M, Lee K, Whyment AD et al (2004) Orexigen-sensitive NPY/AgRP pacemaker neurons in the hypothalamic arcuate nucleus. Nat Neurosci 7:493–494. doi:10.1038/nn1226

Van Hulst RA, Hasan D, Lachmann B (2002) Intracranial pressure, brain PCO_2, PO_2, and pH during hypo- and hyperventilation at constant mean airway pressure in pigs. Intensive Care Med 28:68–73. doi:10.1007/s00134-001-1157-6

Venner A, Karnani MM, Gonzalez JA et al (2011) Orexin neurons as conditional glucosensors: paradoxical regulation of sugar sensing by intracellular fuels. J Physiol 589:5701–5708. doi:10.1113/jphysiol.2011.217000

Wang J, Osaka T, Inoue S (2003) Orexin-A-sensitive site for energy expenditure localized in the arcuate nucleus of the hypothalamus. Brain Res 971:128–134

Wei W, Motoike T, Krzeszinski JY et al (2014) Orexin regulates bone remodeling via a dominant positive central action and a subordinate negative peripheral action. Cell Metab 19:927–940. doi:10.1016/j.cmet.2014.03.016

Williams RH, Jensen LT, Verkhratsky A et al (2007) Control of hypothalamic orexin neurons by acid and CO_2. Proc Natl Acad Sci USA 104:10685–10690. doi:10.1073/pnas.0702676104

Yamanaka A, Sakurai T, Katsumoto T et al (1999) Chronic intracerebroventricular administration of orexin-A to rats increases food intake in daytime, but has no effect on body weight. Brain Res 849:248–252

Yamanaka A, Beuckmann CT, Willie JT et al (2003) Hypothalamic orexin neurons regulate arousal according to energy balance in mice. Neuron 38:701–713

Yi C, Serlie MJ, Ackermans MT et al (2009) A major role for perifornical orexin neurons in the control of glucose metabolism in rats. Diabetes 58:1998–2005. doi:10.2337/db09-0385

Zhang W, Sakurai T, Fukuda Y, Kuwaki T (2006) Orexin neuron-mediated skeletal muscle vasodilation and shift of baroreflex during defense response in mice. Am J Physiol Regul Integr Comp Physiol 290:R1654–R1663. doi:10.1152/ajpregu.00704.2005

Zheng H, Patterson LM, Berthoud H-R (2005) Orexin-A projections to the caudal medulla and orexin-induced c-Fos expression, food intake, and autonomic function. J Comp Neurol 485:127–142. doi:10.1002/cne.20515

Zheng H, Patterson LM, Berthoud H-R (2007) Orexin signaling in the ventral tegmental area is required for high-fat appetite induced by opioid stimulation of the nucleus accumbens. J Neurosci 27:11075–11082. doi:10.1523/JNEUROSCI.3542-07.2007

Orexin Regulates Glucose Homeodynamics with Daily Rhythm

Hiroshi Tsuneki, Tsutomu Wada and Toshiyasu Sasaoka

Abstract Daily rhythm of glucose metabolism is tightly controlled by circadian systems, and entrained to sleep/wake cycles. Chronic sleep disturbance causes insulin resistance and increased risk of type 2 diabetes. Thus, synchronizing the daily rhythm of glucose metabolism with biological clock, sleep/wake cycle, and feeding cycle is required for maintaining glucose homeostasis, although the coordination mechanism remains unclear. A hypothalamic neuropeptide, orexin, contributes to maintenance of glucose and energy homeostasis, in addition to stabilization of sleep/wake cycle. Here we summarized current knowledge about the role of hypothalamic orexin system in dynamic regulation of glucose homeostasis, namely, glucose homeodynamics at physiological and pathophysiological states. Importantly, the levels of orexin per se oscillate with a daily rhythm in the central nervous system under the control of biological clock. Based on this rhythm, orexin promotes to generate daily blood glucose oscillation through bidirectional regulation of hepatic glucose production, as well as sleep/wake cycle. The glucose oscillation not only helps adequate supply of energy at the time of awakening but also effectively prevents insulin resistance caused by aging, obesity, and depression. Therefore, we would like to propose a novel concept of chronotherapy to treat type 2 diabetes by targeting hypothalamic orexin system.

H. Tsuneki (✉) · T. Wada · T. Sasaoka (✉)
Department of Clinical Pharmacology, University of Toyama, 2630 Sugitani,
Toyama 930-0194, Japan
e-mail: htsuneki@pha.u-toyama.ac.jp

T. Sasaoka
e-mail: tsasaoka@pha.u-toyama.ac.jp

© Springer International Publishing Switzerland 2015

T. Sakurai et al. (eds.), *Orexin and Sleep*,
DOI 10.1007/978-3-319-23078-8_21

1 Introduction

Glucose is the main source of energy to survive. However, hyperglycemia that is seen in diabetic patients causes devastating abnormalities in many organs, including metabolic and cardiovascular systems. Therefore, blood glucose needs to be tightly controlled by both central nervous system and peripheral hormonal systems (Kosse et al. 2015). Basal blood glucose levels oscillate with a daily rhythm, according to sleep/wake cycles. The physiological blood glucose rhythm, independent of feeding rhythm, is mainly caused by daily changes in endogenous glucose production in the liver under the control of circadian clock system (Kalsbeek et al. 2010). Daily changes in blood glucose appear to be reasonable not only for adapting to daily changes in demand of energy, but also for preventing metabolic disorders. Chronic disruption of circadian rhythm by sleep disturbance or abnormal life-style, such as shift work, increases the risk of obesity and type 2 diabetes (Sahar and Sassone-Corsi 2012). However, the mechanisms underlying coordination of the circadian clock rhythm, sleep/wake cycle, and glucose metabolism had long been uncertain.

Orexin-A and -B are a pair of neuropeptides produced in the hypothalamus. The action of orexin is mediated by orexin receptor-1 (OX1R) and/or orexin receptor-2 (OX2R) (Sakurai 2014). The activity of orexin-producing neurons is under the control of circadian clock system located in the suprachiasmatic nuclei (SCN), and the levels of orexin A in cerebrospinal fluid (CSF) exhibit a daily rhythm, i.e., increase at active wake period and decrease at resting sleep period (Gotter et al. 2013). In parallel with its periodical expression, orexin maintains sleep/wake rhythm by stabilizing wakefulness at active period. Orexin also controls energy balance by regulating food intake, locomotor activity, and glucose metabolism (Tsuneki et al. 2012). Therefore, orexin may serve as a coordinator of functional link between circadian clock, sleep/wake, and glucose metabolism.

The purpose of this review is to summarize current knowledge about physiological role of orexin in maintenance of the glucose homeodynamics. Moreover, we described the functional significance of orexin at the pathophysiological states, such as aging, obesity, and depression, all of which cause the development of insulin resistance. Finally, we will discuss about the therapeutic potential of orexin agonists and antagonists for treating metabolic diseases associated with disturbance of circadian rhythm, including type 2 diabetes.

2 Physiological Role of Orexin in Regulation of Glucose Homeodynamics

2.1 Daily Activation of Hypothalamic Orexin System

Orexin-producing neurons, localized in lateral hypothalamus and adjacent perifornical area, have an intrinsic activity that promotes the spontaneous firing through transient receptor potential C channels (Burt et al. 2011). Orexin neurons are also regulated by multiple feedback mechanisms with local neurotransmitters (e.g., orexin, glutamate, GABA, melanin-concentrating hormone, thyrotropin-releasing hormone, and corticotropin-releasing factor) and several factors released from neighboring astrocytes (e.g., lactate and protons) (Burt et al. 2011). Moreover, the activity of orexin neurons is modulated by peripheral metabolic signals. For instance, orexin neurons are inhibited by glucose and leptin, whereas they are activated by ghrelin (Tsujino and Sakurai 2009; Tsuneki et al. 2010). Hypoglycemia associated with hypothalamic glucoprivation activates orexin neurons in perifornical area, leading to increase in adrenal sympathetic nerve activity for maintaining normal blood glucose levels (Korim et al. 2014; Otlivanchik et al. 2015). However, the glucose-sensing ability of orexin neurons is disappeared in some specific circumstances (Gao and Horvath 2014), namely, adaptation and context-dependent modulation (Kosse et al. 2015). In the process of adaptation, the glucosensing function is reinstated after sustained elevation of blood glucose levels, so that orexin neurons can continuously monitor the changes in ambient glucose levels. The context-dependent modulation occurs when the ambient levels of lactate and protons are increased, both of which are co-released from astrocytes in highly active neural circuits in brain (Kosse et al. 2015; Parsons and Hirasawa 2010). Thus, orexin neurons have enormous versatility to integrate both central and peripheral signals.

More importantly, the SCN circadian system controls the daily activation rhythm in hypothalamic orexin neurons via the dorsomedial hypothalamus, according to the 24-h light/dark cycles (Chou et al. 2003; Yoshida et al. 2006). Therefore, there is a remarkable daily rhythm in CSF levels of orexin, which is abolished by lesions of the SCN (Zhang et al. 2004). In addition, there is reciprocal relationship between the SCN and orexin system, where low and high concentrations of orexin A inhibit and activate the SCN neurons, respectively, for controlling the phase of neural activity (Klisch et al. 2009). On the other hand, when feeding schedule is restricted at specific time of the day, food anticipatory activity (FAA) is increased (Tahara and Shibata 2013). The daily rhythm of FAA is controlled by food-entrainable oscillators (FEO), independent of the SCN clock, i.e., light-entrainable oscillators (LEO), although the location of FEO remains to be identified. It has been reported that hypothalamic orexin neurons contributes to elicit FAA (Tahara and Shibata 2013). In this process, the orexin system is uncoupled from the SCN control, and instead preferentially entrained to food-cues under daily restricted feeding conditions (Jiménez et al. 2013; Petrovich et al. 2012; Sakurai 2014). Thus, hypothalamic orexin system can synchronize the timing of arousal to either the light/dark or feeding/fasting cycles, depending on the motivation levels towards food (Fig. 1).

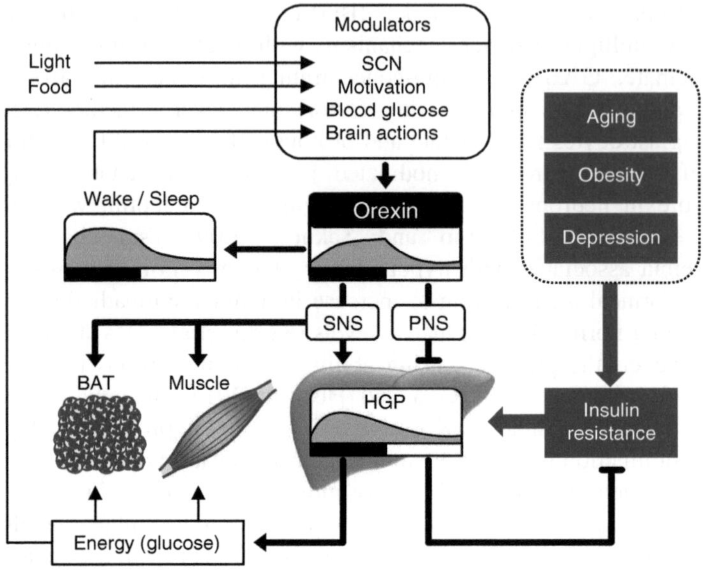

Fig. 1 Role of orexin in the regulation of glucose homeodynamics. Daily rhythm of the activation of hypothalamic orexin system is controlled by central biological clock located in suprachiasmatic nuclei (SCN), motivation towards food, blood glucose levels, and nutrient-consuming brain actions. The rhythmic activation of orexin contributes to not only stabilization of sleep/wake cycle but also generation of daily rhythm in hepatic glucose production (HGP) through the sympathetic (SNS) and parasympathetic nervous system (PNS). The resulting increase in blood glucose levels at the beginning of active period is useful for adapting to the growing demand for food. To support the food-searching behavior, orexin simultaneously enhances glucose consumption in skeletal muscle and thermogenesis in brown adipose tissue (BAT) through the SNS. Moreover, orexin-induced daily reduction of blood glucose is necessary for protecting the development of insulin resistance caused by aging, obesity and depression

2.2 Central Orexin Regulation of Glucose Homeodynamics

Glucose metabolism is regulated by both central and peripheral regulatory systems through their cooperative interactions (Schwartz et al. 2013). These systems are under the control of the SCN. Indeed, lesions of the SCN disrupted daily blood glucose oscillations (Kalsbeek et al. 2010). Hypothalamic orexin system plays crucial role in the central regulation of glucose metabolism through the autonomic nervous system (Tsuneki et al. 2010, 2012). Interestingly, it has been reported that orexin bidirectionally regulates the autonomic nervous system, depending on its concentration. For instance, low doses of orexin A suppressed renal- and white adipose tissue-sympathetic nerve activities, blood pressure, and lipolysis, whereas high doses of orexin A conversely elevated them (Shen et al. 2008; Tanida et al. 2006). Consistently, we found that orexin bidirectionally regulates hepatic glucose production via the autonomic nervous system, and generates daily rhythm of blood glucose (Tsuneki et al. 2015). In fact, daily glucose oscillation was disappeared in

Orexin$^{-/-}$ mice, whereas wild-type mice exhibited an obvious daily glucose rhythm peaked at the beginning of active wake period. Intracerebroventricular injection of orexin A at high dose promoted hepatic glucose production through the sympathetic nervous system, consistent with a previous report (Yi et al. 2009). In contrast, low dose of orexin A suppressed it through the parasympathetic nervous system (Tsuneki et al. 2015). These indicate that orexin-induced sequential activation of the sympathetic and parasympathetic nervous system promotes the generation of daily glucose oscillation via bidirectional changes in hepatic glucose production. Also, a previous study demonstrated that injection of orexin A into the ventromedial hypothalamus promoted insulin-induced glucose uptake and glycogen synthesis in skeletal muscle via the sympathetic nervous system in mice and rats (Shiuchi et al. 2009). Given that daily orexinergic activation starts at the beginning of active period, simultaneous induction of hepatic glucose production and muscle glucose metabolism by endogenous orexin may facilitate the energy flow from the liver to the skeletal muscle and support the food-searching behavior at the time of awakening (Fig. 1). Thus, hypothalamic orexin system appears to serve as a time-keeper to synchronize daily rhythm of glucose production and consumption with the sleep/wake rhythms for maintaining energy homeostasis throughout the day. Additional functional significances of the glucose-lowering effect of orexin at resting sleep period are mentioned below (see Sect. 3.1).

3 Role of Orexin in Prevention of Insulin Resistance

3.1 Aging

Advance of age is associated with insulin resistance and glucose intolerance (Barbieri et al. 2001). The causes of age-related insulin resistance include changes in body composition (increase in fat mass and decrease in lean mass), decline in physical activity, and increase in oxidative stress in the elderly. Moreover, abnormal endoplasmic reticulum (ER) stress response is promoted during aging (Brown and Naidoo 2012). ER stress response serves as an adaptive mechanism to cope with protein misfolding, but chronic ER stress causes deterioration of cellular functions, such as insulin resistance (Yalcin and Hotamisligil 2013). Excessive ER stress during aging is at least partly due to sleep disturbance, because quality of sleep declines along with aging, and because ER stress is augmented by disruption of normal sleep (Brown and Naidoo 2012). In addition, aging causes ER stress in orexin-producing neurons, leading to instability of wakefulness in mice (Naidoo et al. 2011). Orexin release and receptor activation in the brain also decrease during aging (Zink et al. 2014). Thus, a vicious cycle between perturbation of orexin system, sleep disturbance, and ER stress could promote the development of insulin resistance with aging. It should be noted, however, that orexin expression in the hypothalamus mainly declines during maturation (from adolescence to adulthood), compared to normal aging (from adulthood to old age) (Zink et al. 2014; Hunt et al.

2015). Moreover, daily rhythm of CSF orexin levels were largely maintained in aged rodents, compared to young controls (Desarnaud et al. 2004). Therefore, we consider that the endogenous orexin still contributes to various circadian regulations even in the elderly.

To know the significance of orexin on maintenance of glucose homeostasis during aging, we investigated age-related changes in glucose metabolism, using *Orexin*$^{-/-}$ mice at 3, 6, and 9 months old (Tsuneki et al. 2008, 2015). Impairment of insulin sensitivity occurred in the hypothalamus at 3 months old, in the liver at 6 months old, and in the skeletal muscle at 9 months old in male *Orexin*$^{-/-}$ mice, whereas there were no changes in the body weights. Consistently, hepatic glucose production was abnormally increased in *Orexin*$^{-/-}$ mice at 6 and 9 months old, whereas systemic insulin resistance and glucose intolerance occurred at 9 months old, when compared to age-matched wild-type mice. These indicate that orexin deficiency preferentially causes hepatic insulin resistance, leading to promotion of systemic insulin resistance. Thus, endogenous orexin action is required for preventing insulin resistance with aging, especially in the liver.

We further explored the mechanism underlying severe insulin resistance in the liver of *Orexin*$^{-/-}$ mice along with aging. In the liver, ER stress response serves to support metabolic reprogramming during transition from fasting to feeding state, in which protein synthesis is intensively reduced to adapt to metabolic changes (Deng et al. 2013). In general, these stress responses are rapidly terminated at postprandial state; however, when ER homeostasis is disrupted under the deleterious conditions, such as continuous nutrient overload, ER stress response is prolonged and causes insulin resistance and type 2 diabetes (Hotamisligil 2010; Lee and Ozcan 2014). In our study, *Orexin*$^{-/-}$ mice showed the disruption of daily blood glucose oscillation via hepatic glucose production at 3 months old, and thereafter at 6 months old, they exhibited abnormal ER stress responses in the liver upon refeeding after 24-h fasting, including an increase in phosphorylation levels of IRE1α and JNK, a major ER stress-related signaling. Thus, maintaining daily glucose rhythm by orexin appears to be beneficial for preventing ER dyshomeostasis. It has been reported that 'resting time' for ER is required for successful recovery from the state of ER stress (Eizirik et al. 2008). In fact, we observed that hepatic glucose production is daily suppressed at resting period under the control of endogenous orexin. This suppression may provide the resting time for ER, thereby preventing hepatic insulin resistance (Fig. 1).

It is well known that energy expenditure is reduced with aging, resulting in increase in body weight and fat volume. These abnormalities are considerably due to age-related impairment of thermogenesis in brown adipose tissue (Sellayah and Sikder 2014). Orexin promotes thermogenic function of brown adipose tissue through central pathways, including the rostral raphe pallidus-spinal intermediolateral nuclei-sympathetic nerve route (Morrison et al. 2014; Perez-Leighton et al. 2014). In addition, peripheral orexin action can directly promote the differentiation from committed brown preadipocyte to mature brown adipocyte (Sellayah et al. 2011). In fact, the cold-induced and diet-induced thermogenesis was impaired in *Orexin*$^{-/-}$ mice (Sellayah et al. 2011; Sellayah and Sikder 2012). Orexin injection daily for 14 days in aged mice (24-month-old male C57BL6 mice) increased the expression levels of uncoupling protein-1, a main regulator of thermogenesis in brown adipose tissue, and improved cold tolerance. Interestingly, glucose tolerance was also improved under this condition (Sellayah et al. 2011). Therefore, orexin is a crucial factor for preventing age-dependent deterioration of glucose and energy homeostasis.

3.2 *Obesity and Type 2 Diabetes*

Increased sympathetic activity predisposes obesity and metabolic syndrome (Lambert et al. 2010; Licht et al. 2013). Type 2 diabetic *db/db* mice showed a higher sympathetic tone with a reduced parasympathetic tone (Goncalves et al. 2009) and disrupted circadian rhythms of heart rate and locomotor activity (Su et al. 2008). In addition, type 2 diabetic subjects exhibited increased resting sympathetic nerve activity and blunted sympathetic response to oral glucose loading (Straznicky et al. 2012). Importantly, either hepatic sympathetic or parasympathetic denervation abolished daily rhythm of blood glucose, while no such effect was produced by complete denervation of both branches (Cailotto et al. 2008). Thus, the unbalanced autonomic nervous system in obese and diabetic states appears to deteriorate daily glucose homeodynamics. Moreover, since orexin-producing neurons belong to the glucose-inhibited neurons in the hypothalamus, the activity of orexin neurons and the expression levels of orexin were down-regulated by hyperglycemia in obese and diabetic animals, such as *ob/ob* and *db/db* mice (Yamamoto et al. 1999; Yamanaka et al. 2003). These raise a question whether or not orexin could induce the improvement of glucose metabolism impaired in obese and type 2 diabetic states. The answer is yes, because CAG/orexin transgenic mice on a high fat diet showed improvement of insulin sensitivity by an OX2R-dependent mechanism (Funato et al. 2009), whereas female *Orexin*$^{-/-}$ mice fed on a high fat diet exhibited remarkable obesity, glucose intolerance, and insulin resistance (Tsuneki et al. 2008). We further examined the effect of daily intracerebroventricular administration of orexin A on blood glucose in *db/db* mice (Tsuneki et al. 2015). Interestingly, nighttime administration of orexin A for 3 days, mimicking a physiological orexin secretion pattern, amplified daily glucose oscillation, leading to gradual reduction of blood glucose levels at daytime resting period. The glucose lowering effect in *db/db* mice was much greater than that in control mice. In contrast, neither daytime administration nor continuous administration of orexin A showed limited impacts on blood glucose levels in *db/db* mice. The mechanism underlying the glucose-lowering effect of orexin A administered at nighttime active period can be explained by suppression of hepatic glucose production at the subsequent daytime resting period in *db/db* mice. Therefore, we thought that the amplification of daily endogenous orexin action by the timely supplementation of orexin agonist effectively reconstitutes normal daily rhythm of hepatic glucose production, thereby ameliorating metabolic disorders related to obesity and type 2 diabetes (Fig. 1).

What about the timely inhibition of orexin system by orexin antagonist? Currently, an orexin receptor antagonist, suvorexant, is available for clinical use to treat insomnia; however the effect of this drug on glucose metabolism remains unknown. An early study reported that the other orexin receptor antagonist SB-334867-A exerted anti-obese and anti-diabetic effects when repetitively injected into genetically obese *ob/ob* mice at daytime for 2 weeks (Haynes et al. 2002). The mechanisms behind these beneficial effects have been explained by reduced adiposity and increased energy expenditure through the improvement of brown adipose tissue thermogenesis. We also assume, from the viewpoint of circadian rhythm, that administration of SB-334867-A at resting state may strengthen normal sleep/wake and feeding cycles, which can improve glucose metabolism in the liver and skeletal muscle.

Although high-fat feeding promotes the development of obesity and insulin resistance, time-restricted feeding strongly prevented the metabolic disorders at least through improvement of circadian clock gene expressions in the liver of high-fat-fed C57BL/6 mice (Hatori et al. 2012; Sherman et al. 2012) and *db/db* mice (Kudo et al. 2004). It should be reminded that feeding cycle is a robust entrainer of the daily orexinergic activation rhythm, as mentioned above. These raise a possibility that time-restricted feeding is a non-pharmacological regimen to alleviate obesity and type 2 diabetes via the activation of hypothalamic orexin system.

Taken together, impairment of daily orexin action under hyperglycemic conditions is an exacerbating factor for obesity and type 2 diabetes. Therefore, chronopharmacological approach to the treatment of such metabolic diseases, using orexin agonist and/or antagonist, appears to be promising as a new therapeutic strategy.

3.3 Depression

Several meta-analyses indicate that major depressive disorders are associated with insulin resistance and type 2 diabetes (Mezuk et al. 2008; Kan et al. 2013). This notion has been supported by animal studies. For instance, high fat diet-induced obese mice exhibited depressive behavior (Yamada et al. 2011). Acute psychological stress by inescapable foot shock rapidly caused impaired glucose tolerance and hepatic insulin resistance in mice (Li et al. 2013). Chronic social defeat stress, a mouse model of major depression, caused insulin resistance and lipid dysregulation in mice (Chuang et al. 2010). Thus, it appears that there is a synergistic relationship between major depression and metabolic disorders, including insulin resistance.

Most of the orexin-elicited behavioral changes are triggered by increased motivation under various physiological and psychological conditions, including hunger and reward-associated stimuli (Mahler et al. 2014; Sakurai 2014). The orexin levels in the human amygdala increased during positive emotion, social interaction and anger in addition to arousal (Blouin et al. 2013). Although many clinical observations demonstrate the implication of dysfunctional orexinergic activity in depression, the pathophysiological relationship between them is complicated and remains unclear (Nollet and Leman 2013). In fact, a behavioral test using *OX1R*$^{-/-}$ and *OX2R*$^{-/-}$ mice indicated that OX1R and OX2R signalings produced pro-depressant and anti-depressant effects, respectively (Scott et al. 2011), and SB-334867, a selective OX1R antagonist, exerted depressant or anti-depressant effects in mice, depending on the experimental conditions (Scott et al. 2011; Deats et al. 2014). Moreover, both hypoactivity and hyperactivity of orexin system have been reported to predispose depression (Nollet and Leman 2013). Therefore, there may be optimal levels of orexinergic activation for alleviating depression.

We asked whether or not central orexin system has some role in the regulation of glucose metabolism in depressive state. To address this question, we employed *Orexin*$^{-/-}$ mice subjected to chronic social defeat stress, and compared the efficacy of glucose metabolism with wild-type controls that received the same psychological stress (Tsuneki et al. 2013). The chronic social defeat stress caused depression-like behavior in both wild-type and *Orexin*$^{-/-}$ mice, suggesting that orexin deficiency alone does not affect the onset of depression. Instead, the defeat stress caused hyperinsulinemia without changing fasting blood glucose levels in *Orexin*$^{-/-}$ mice, and as a result, the HOMA-IR (homeostasis model assessment for insulin resistance) was increased. In pyruvate tolerance test to evaluate hepatic gluconeogenic activity reflecting hepatic insulin sensitivity, the defeat stress caused excessive glucose elevation in *Orexin*$^{-/-}$ mice compared to wild-type mice. Consistently, hepatic insulin signaling was severely impaired in *Orexin*$^{-/-}$ mice after the defeat stress exposure. These results indicate that endogenous orexin action is required for maintaining hepatic insulin sensitivity under prolonged stressful condition. Therefore, we propose that it is hypothalamic orexin system that prevents the development of type 2 diabetes under depressive condition.

4 Perspectives

In human, CSF levels of orexin A exhibits daily rhythm peaked at the end of active period (Salomon et al. 2003). The loss of orexin-producing neurons in the hypothalamus is the main cause of narcolepsy, a sleep disorder with excessive daytime sleepiness in human. The orexin-deficient narcolepsy is associated with dysregulation of the autonomic nervous system (Huda et al. 2013) and increased risk of obesity and type 2 diabetes (Schuld et al. 2000; Kok et al. 2003; Jennum et al. 2013). Orexin-deficient narcoleptic patients with cataplexy showed metabolic alterations, including insulin resistance, which is independent of body mass index in a case control study (Poli et al. 2009), although there are some conflicting reports (Beitinger et al. 2012; Donjacour et al. 2014). It is of note that sleep disturbance, such as poor sleep quality and insomnia, increases the risk of insulin resistance in healthy humans (Van Cauter 2011), and daily rhythms of glucose regulation are impaired in patients with type 2 diabetes and diabetic animals (Huang et al. 2011). Therefore, we anticipate that the timely activation and inactivation of endogenous orexin action may be a valuable, novel chronotherapeutic strategy to treat type 2 diabetes. Further study is required to clarify whether accentuation of daily sleep/wake rhythm by administration of orexin agonists at the time of arousal and/or orexin antagonists just before sleeping can improve metabolic defects in diabetic state.

Orexin also contributes to bone remodeling via central and peripheral regulation (Wei et al. 2014). Interestingly, these regulations are bidirectional in mice: central orexin action via OX2R increased bone mass by reducing serum levels of leptin, a suppressor of bone formation, whereas peripheral orexin action via OX1R in the bone caused bone loss by reducing local expression of ghrelin, an inducer of osteoblastogenesis. The central action is predominant, because global deletion and overexpression of orexin caused decrease and increase in bone formation, respectively, in mice. Therefore, the adverse effects of orexin agonists and antagonists on skeletal homeostasis should not be overlooked for future clinical use.

Calorie restriction strongly activates hypothalamic orexin system, leading to suppression of the depression-like behavior in mice exposed to chronic social defeat stress (Lutter et al. 2008), and as a result, the defeat stress no longer deranged hepatic glucose metabolism under this condition (Tsuneki et al. 2013). It has also been reported that bright light at the wake-up time activates hypothalamic orexin neurons, leading to amelioration of depression-like behaviors in the grass rat (Adidharma et al. 2012; Deats et al. 2014). Bright light therapy for patients with sleep disorders and/or depression, therefore, may be at least partly mediated by hypothalamic orexin system. In addition, since bright light stimulates LEO in the SCN, the rhythmic orexin regulation of glucose homeodynamics may be strengthened by bright light therapy. Thus, such non-pharmacological approaches for enhancing daily orexin action may be options for therapy to prevent metabolic diseases associated with depression.

5 Conclusion

As summarized in this review, hypothalamic orexin system has the ability to regulate both daily sleep/wake and blood glucose rhythms, according to circadian clock rhythm. The timely action of orexin enhances the glucose oscillation via autonomic nervous system. The resulting increase and decrease in blood glucose are beneficial for supplying energy at the active period and for preventing postprandial hyperglycemia at the resting period, respectively. In the pathophysiological states, such as aging, obesity and depression, the time-keeping actions of orexin can effectively prevent insulin resistance, especially in the liver. Therefore, we conclude that hypothalamic orexin system plays pivotal role in maintenance of the glucose homeostasis.

Acknowledgments This work was supported by JSPS KAKENHI Grant Number 24591317, 24390056, 15K09380 and 15K15599.

References

Adidharma W, Leach G, Yan L (2012) Orexinergic signaling mediates light-induced neuronal activation in the dorsal raphe nucleus. Neuroscience 220:201–207

Barbieri M, Rizzo MR, Manzella D, Paolisso G (2001) Age-related insulin resistance: is it an obligatory finding? The lesson from healthy centenarians. Diab Metab Res Rev 17:19–26

Beitinger PA, Fulda S, Dalal MA, Wehrle R, Keckeis M, Wetter TC, Han F, Pollmächer T, Schuld A (2012) Glucose tolerance in patients with narcolepsy. Sleep 35:231–236

Blouin AM, Fried I, Wilson CL, Staba RJ, Behnke EJ, Lam HA, Maidment NT, Karlsson KÆ, Lapierre JL, Siegel JM (2013) Human hypocretin and melanin-concentrating hormone levels are linked to emotion and social interaction. Nat Commun 4:1547

Brown MK, Naidoo N (2012) The endoplasmic reticulum stress response in aging and age-related diseases. Front Physiol 3:263

Burt J, Alberto CO, Parsons MP, Hirasawa M (2011) Local network regulation of orexin neurons in the lateral hypothalamus. Am J Physiol Regul Integr Comp Physiol 301:R572–R580

Cailotto C, van Heijningen C, van der Vliet J, van der Plasse G, Habold C, Kalsbeek A, Pévet P, Buijs RM (2008) Daily rhythms in metabolic liver enzymes and plasma glucose require a balance in the autonomic output to the liver. Endocrinology 149:1914–1925

Chou TC, Scammell TE, Gooley JJ, Gaus SE, Saper CB, Lu J (2003) Critical role of dorsomedial hypothalamic nucleus in a wide range of behavioral circadian rhythms. J Neurosci 23:10691–10702

Chuang JC, Cui H, Mason BL, Mahgoub M, Bookout AL, Yu HG, Perello M, Elmquist JK, Repa JJ, Zigman JM, Lutter M (2010) Chronic social defeat stress disrupts regulation of lipid synthesis. J Lipid Res 51:1344–1353

Deats SP, Adidharma W, Lonstein JS, Yan L (2014) Attenuated orexinergic signaling underlies depression-like responses induced by daytime light deficiency. Neuroscience 272:252–260

Deng Y, Wang ZV, Tao C et al (2013) The Xbp1s/GalE axis links ER stress to postprandial hepatic metabolism. J Clin Invest 123:455–468

Desarnaud F, Murillo-Rodriguez E, Lin L, Xu M, Gerashchenko D, Shiromani SN, Nishino S, Mignot E, Shiromani PJ (2004) The diurnal rhythm of hypocretin in young and old F344 rats. Sleep 27:851–856

Donjacour CE, Aziz NA, Overeem S, Kalsbeek A, Pijl H, Lammers GJ (2014) Glucose and fat metabolism in narcolepsy and the effect of sodium oxybate: a hyperinsulinemic-euglycemic clamp study. Sleep 37:795–801

Eizirik DL, Cardozo AK, Cnop M (2008) The role for endoplasmic reticulum stress in diabetes mellitus. Endocr Rev 29:42–61

Funato H, Tsai AL, Willie JT, Kisanuki Y, Williams SC, Sakurai T, Yanagisawa M (2009) Enhanced orexin receptor-2 signaling prevents diet-induced obesity and improves leptin sensitivity. Cell Metab 9:64–76

Gao XB, Horvath T (2014) Function and dysfunction of hypocretin/orexin: an energetics point of view. Annu Rev Neurosci 37:101–116

Goncalves AC, Tank J, Diedrich A, Hilzendeger A, Plehm R, Bader M, Luft FC, Jordan J, Gross V (2009) Diabetic hypertensive leptin receptor-deficient db/db mice develop cardioregulatory autonomic dysfunction. Hypertension 53:387–392

Gotter AL, Winrow CJ, Brunner J, Garson SL, Fox SV, Binns J, Harrell CM, Cui D, Yee KL, Stiteler M, Stevens J, Savitz A, Tannenbaum PL, Tye SJ, McDonald T, Yao L, Kuduk SD, Uslaner J, Coleman PJ, Renger JJ (2013) The duration of sleep promoting efficacy by dual orexin receptor antagonists is dependent upon receptor occupancy threshold. BMC Neurosci 14:90

Hatori M, Vollmers C, Zarrinpar A, DiTacchio L, Bushong EA, Gill S, Leblanc M, Chaix A, Joens M, Fitzpatrick JA, Ellisman MH, Panda S (2012) Time-restricted feeding without reducing caloric intake prevents metabolic diseases in mice fed a high-fat diet. Cell Metab 15:848–860

Haynes AC, Chapman H, Taylor C, Moore GB, Cawthorne MA, Tadayyon M, Clapham JC, Arch JR (2002) Anorectic, thermogenic and anti-obesity activity of a selective orexin-1 receptor antagonist in ob/ob mice. Regul Pept 104:153–159

Hotamisligil GS (2010) Endoplasmic reticulum stress and the inflammatory basis of metabolic disease. Cell 140:900–917

Huang W, Ramsey KM, Marcheva B, Bass J (2011) Circadian rhythms, sleep, and metabolism. J Clin Invest 121:2133–2141

Huda MS, Mani H, Durham BH, Dovey TM, Halford JC, Aditya BS, Pinkney JH, Wilding JP, Hart IK (2013) Plasma obestatin and autonomic function are altered in orexin-deficient narcolepsy, but ghrelin is unchanged. Endocrine 43:696–704

Hunt NJ, Rodriguez ML, Waters KA, Machaalani R (2015) Changes in orexin (hypocretin) neuronal expression with normal aging in the human hypothalamus. Neurobiol Aging 36:292–300

Jennum P, Ibsen R, Knudsen S, Kjellberg J (2013) Comorbidity and mortality of narcolepsy: a controlled retro- and prospective national study. Sleep 36:835–840

Jiménez A, Caba M, Escobar C (2013) Food-entrained patterns in orexin cells reveal subregion differential activation. Brain Res 1513:41–50

Kalsbeek A, Yi CX, La Fleur SE, Fliers E (2010) The hypothalamic clock and its control of glucose homeostasis. Trends Endocrinol Metab 21:402–410

Kan C, Silva N, Golden SH, Rajala U, Timonen M, Stahl D, Ismail K (2013) A systematic review and meta-analysis of the association between depression and insulin resistance. Diabetes Care 36:480–489

Klisch C, Inyushkin A, Mordel J, Karnas D, Pévet P, Meissl H (2009) Orexin A modulates neuronal activity of the rodent suprachiasmatic nucleus in vitro. Eur J Neurosci 30:65–75

Kok SW, Overeem S, Visscher TL, Lammers GJ, Seidell JC, Pijl H, Meinders AE (2003) Hypocretin deficiency in narcoleptic humans is associated with abdominal obesity. Obes Res 11:1147–1154

Korim WS, Bou Farah L, McMullan S, Verberne AJ (2014) Orexinergic activation of medullary premotor neurons modulates the adrenal sympathoexcitation to hypothalamic glucoprivation. Diabetes 63:1895–1906

Kosse C, Gonzalez A, Burdakov D (2015) Predictive models of glucose control: roles for glucose-sensing neurones. Acta Physiol (Oxf) 213:7–18

Kudo T, Akiyama M, Kuriyama K, Sudo M, Moriya T, Shibata S (2004) Night-time restricted feeding normalises clock genes and Pai-1 gene expression in the db/db mouse liver. Diabetologia 47:1425–1436

Lambert GW, Straznicky NE, Lambert EA, Dixon JB, Schlaich MP (2010) Sympathetic nervous activation in obesity and the metabolic syndrome: causes, consequences and therapeutic implications. Pharmacol Ther 126:159–172

Lee J, Ozcan U (2014) Unfolded protein response signaling and metabolic diseases. J Biol Chem 289:1203–1211

Li L, Li X, Zhou W, Messina JL (2013) Acute psychological stress results in the rapid development of insulin resistance. J Endocrinol 217:175–184

Licht CM, de Geus EJ, Penninx BW (2013) Dysregulation of the autonomic nervous system predicts the development of the metabolic syndrome. J Clin Endocrinol Metab 98:2484–2493

Lutter M, Krishnan V, Russo SJ, Jung S, McClung CA, Nestler EJ (2008) Orexin signaling mediates the antidepressant-like effect of calorie restriction. J Neurosci 28:3071–3075

Mahler SV, Moorman DE, Smith RJ, James MH, Aston-Jones G (2014) Motivational activation: a unifying hypothesis of orexin/hypocretin function. Nat Neurosci 17:1298–1303

Mezuk B, Eaton WW, Albrecht S, Golden SH (2008) Depression and type 2 diabetes over the lifespan: a meta-analysis. Diabetes Care 31:2383–2390

Morrison SF, Madden CJ, Tupone D (2014) Central neural regulation of brown adipose tissue thermogenesis and energy expenditure. Cell Metab 19:741–756

Naidoo N, Zhu J, Zhu Y, Fenik P, Lian J, Galante R, Veasey S (2011) Endoplasmic reticulum stress in wake-active neurons progresses with aging. Aging Cell 10:640–649

Nollet M, Leman S (2013) Role of orexin in the pathophysiology of depression: potential for pharmacological intervention. CNS Drugs 27:411–422

Otlivanchik O, Le Foll C, Levin BE (2015) Perifornical hypothalamic orexin and serotonin modulate the counterregulatory response to hypoglycemic and glucoprivic stimuli. Diabetes 64:226–235

Parsons MP, Hirasawa M (2010) ATP-sensitive potassium channel-mediated lactate effect on orexin neurons: implications for brain energetics during arousal. J Neurosci 30:8061–8070

Perez-Leighton CE, Billington CJ, Kotz CM (2014) Orexin modulation of adipose tissue. Biochim Biophys Acta 1842:440–445

Petrovich GD, Hobin MP, Reppucci CJ (2012) Selective Fos induction in hypothalamic orexin/hypocretin, but not melanin-concentrating hormone neurons, by a learned food-cue that stimulates feeding in sated rats. Neuroscience 224:70–80

Poli F, Plazzi G, Di Dalmazi G, Ribichini D, Vicennati V, Pizza F, Mignot E, Montagna P, Pasquali R, Pagotto U (2009) Body mass index-independent metabolic alterations in narcolepsy with cataplexy. Sleep 32:1491–1497

Sahar S, Sassone-Corsi P (2012) Regulation of metabolism: the circadian clock dictates the time. Trends Endocrinol Metab 23:1–8

Sakurai T (2014) The role of orexin in motivated behaviours. Nat Rev Neurosci 15:719–731

Salomon RM, Ripley B, Kennedy JS, Johnson B, Schmidt D, Zeitzer JM, Nishino S, Mignot E (2003) Diurnal variation of cerebrospinal fluid hypocretin-1 (Orexin-A) levels in control and depressed subjects. Biol Psychiatry 54:96–104

Schuld A, Hebebrand J, Geller F, Pollmächer T (2000) Increased body-mass index in patients with narcolepsy. Lancet 355:1274–1275

Schwartz MW, Seeley RJ, Tschöp MH, Woods SC, Morton GJ, Myers MG, D'Alessio D (2013) Cooperation between brain and islet in glucose homeostasis and diabetes. Nature 503:59–66

Scott MM, Marcus JN, Pettersen A, Birnbaum SG, Mochizuki T, Scammell TE, Nestler EJ, Elmquist JK, Lutter M (2011) Hcrtr1 and 2 signaling differentially regulates depression-like behaviors. Behav Brain Res 222:289–294

Sellayah D, Bharaj P, Sikder D (2011) Orexin is required for brown adipose tissue development, differentiation, and function. Cell Metab 14:478–490

Sellayah D, Sikder D (2012) Orexin receptor-1 mediates brown fat developmental differentiation. Adipocyte 1:58–63

Sellayah D, Sikder D (2014) Orexin restores aging-related brown adipose tissue dysfunction in male mice. Endocrinology 155:485–501

Shen J, Tanida M, Yao JF, Niijima A, Nagai K (2008) Biphasic effects of orexin-A on autonomic nerve activity and lipolysis. Neurosci Lett 444:166–171

Sherman H, Genzer Y, Cohen R, Chapnik N, Madar Z, Froy O (2012) Timed high-fat diet resets circadian metabolism and prevents obesity. FASEB J 26:3493–3502

Shiuchi T, Haque MS, Okamoto S, Inoue T, Kageyama H, Lee S, Toda C, Suzuki A, Bachman ES, Kim YB, Sakurai T, Yanagisawa M, Shioda S, Imoto K, Minokoshi Y (2009) Hypothalamic orexin stimulates feeding-associated glucose utilization in skeletal muscle via sympathetic nervous system. Cell Metab 10:466–480

Straznicky NE, Grima MT, Sari CI, Eikelis N, Lambert EA, Nestel PJ, Esler MD, Dixon JB, Chopra R, Tilbrook AJ, Schlaich MP, Lambert GW (2012) Neuroadrenergic dysfunction along the diabetes continuum: a comparative study in obese metabolic syndrome subjects. Diabetes 61:2506–2516

Su W, Guo Z, Randall DC, Cassis L, Brown DR, Gong MC (2008) Hypertension and disrupted blood pressure circadian rhythm in type 2 diabetic db/db mice. Am J Physiol Heart Circ Physiol 295:H1634–H1641

Tahara Y, Shibata S (2013) Chronobiology and nutrition. Neuroscience 253:78–88

Tanida M, Niijima A, Shen J, Yamada S, Sawai H, Fukuda Y, Nagai K (2006) Dose-different effects of orexin-A on the renal sympathetic nerve and blood pressure in urethane-anesthetized rats. Exp Biol Med (Maywood) 231:1616–1625

Tsujino N, Sakurai T (2009) Orexin/hypocretin: a neuropeptide at the interface of sleep, energy homeostasis, and reward system. Pharmacol Rev 61:162–176

Tsuneki H, Murata S, Anzawa Y, Soeda Y, Tokai E, Wada T, Kimura I, Yanagisawa M, Sakurai T, Sasaoka T (2008) Age-related insulin resistance in hypothalamus and peripheral tissues of orexin knockout mice. Diabetologia 51:657–667

Tsuneki H, Tokai E, Nakamura Y, Takahashi K, Fujita M, Asaoka T, Kon K, Anzawa Y, Wada T, Takasaki I, Kimura K, Inoue H, Yanagisawa M, Sakurai T, Sasaoka T (2015) Hypothalamic orexin prevents hepatic insulin resistance via daily bidirectional regulation of autonomic nervous system in mice. Diabetes 64:459–470

Tsuneki H, Tokai E, Sugawara C, Wada T, Sakurai T, Sasaoka T (2013) Hypothalamic orexin prevents hepatic insulin resistance induced by social defeat stress in mice. Neuropeptides 47:213–219

Tsuneki H, Wada T, Sasaoka T (2010) Role of orexin in the regulation of glucose homeostasis. Acta Physiol (Oxf) 198:335–348

Tsuneki H, Wada T, Sasaoka T (2012) Role of orexin in the central regulation of glucose and energy homeostasis. Endocr J 59:365–374

Van Cauter E (2011) Sleep disturbances and insulin resistance. Diabet Med 28:1455–1462

Wei W, Motoike T, Krzeszinski JY, Jin Z, Xie XJ, Dechow PC, Yanagisawa M, Wan Y (2014) Orexin regulates bone remodeling via a dominant positive central action and a subordinate negative peripheral action. Cell Metab 19:927–940

Yalcin A, Hotamisligil GS (2013) Impact of ER protein homeostasis on metabolism. Diabetes 62:691–693

Yamada N, Katsuura G, Ochi Y, Ebihara K, Kusakabe T, Hosoda K, Nakao K (2011) Impaired CNS leptin action is implicated in depression associated with obesity. Endocrinology 152:2634–2643

Yamamoto Y, Ueta Y, Date Y, Nakazato M, Hara Y, Serino R, Nomura M, Shibuya I, Matsukura S, Yamashita H (1999) Down regulation of the prepro-orexin gene expression in genetically obese mice. Brain Res Mol Brain Res 65:14–22

Yamanaka A, Beuckmann CT, Willie JT, Hara J, Tsujino N, Mieda M, Tominaga M, Yagami KI, Sugiyama F, Goto K, Yanagisawa M, Sakurai T (2003) Hypothalamic orexin neurons regulate arousal according to energy balance in mice. Neuron 38:701–713

Yi CX, Serlie MJ, Ackermans MT, Foppen E, Buijs RM, Sauerwein HP, Fliers E, Kalsbeek A (2009) A major role for perifornical orexin neurons in the control of glucose metabolism in rats. Diabetes 58:1998–2005

Yoshida K, McCormack S, España RA, Crocker A, Scammell TE (2006) Afferents to the orexin neurons of the rat brain. J Comp Neurol 494:845–861

Zhang S, Zeitzer JM, Yoshida Y, Wisor JP, Nishino S, Edgar DM, Mignot E (2004) Lesions of the suprachiasmatic nucleus eliminate the daily rhythm of hypocretin-1 release. Sleep 27:619–627

Zink AN, Perez-Leighton CE, Kotz CM (2014) The orexin neuropeptide system: physical activity and hypothalamic function throughout the aging process. Front Syst Neurosci 8:211

Orexinergic Tone in Cardiorespiratory Regulation

Leszek Kubin

Abstract Orexin-containing cells are located in the posterior lateral hypothalamus and send axonal projections to all brain and spinal cord locations important for the control of sleep-wake states, breathing, blood pressure, heart rate and motor activity, as well as mood, attention and motivation. While their strategic location and widespread efferent connections suggest that orexin neurons have major and multifaceted roles, elimination of orexins or their receptors results in a relatively discrete and focused deficiency—narcolepsy/cataplexy. Orexin-deficient rodents have been relatively extensively used to assess the role of endogenous orexins in cardiorespiratory regulation. Orexin receptor antagonists have been developed with a delay after the discovery of orexins and their use for focal probing of the endogenous role of orexins in different brain regions has been limited. Despite intense research, it is still unclear under which conditions orexins are physiologically important for the many different roles ascribed to them based on their anatomical connectivity and effects of focal administration of synthetic orexins. Here, we review recent studies which aimed to elucidate the endogenous role of orexins in the control of breathing and blood pressure under different physiologic and pathophysiologic conditions.

Keywords CO_2 chemoreception · Hypertension · Motoneurons · Obstructive sleep apnea · Stress · Upper airway

1 Introduction

The discovery of the new excitatory peptides, orexins (also named hypocretins), and the finding that the neurons that synthesize these peptides are exclusively located in the posterior lateral hypothalamus and have extremely broad axonal projections to diverse regions of the brain and spinal cord (de Lecea et al. 1998;

L. Kubin (✉)
Department of Animal Biology 209E/VET, School of Veterinary Medicine, 3800 Spruce Street, Philadelphia, PA 19104-6046, USA
e-mail: lkubin@vet.upenn.edu

© Springer International Publishing Switzerland 2015

T. Sakurai et al. (eds.), *Orexin and Sleep*,
DOI 10.1007/978-3-319-23078-8_22

Peyron et al. 1998; Marcus et al. 2001; Date et al. 1999) inspired many hypotheses about the possible functions of this newly discovered system. The location of orexin cells in the perifornical region of the posterior hypothalamus and the realization that one primary target of their efferent projections comprises cell groups with well established role in the control of sleep-wake states (noradrenergic, histaminergic, cholinergic) suggested a major role of orexins in the control of sleep and/or energy balance. The ensuing finding that the absence of orexins in mice (Chemelli et al. 1999), or one of their receptors in dogs (Lin et al. 1999), resulted in symptoms typical of narcolepsy-cataplexy, a relatively rare in humans sleep disorder, was possibly the most groundbreaking discovery in sleep research since a formal identification of rapid eye movement (REM) sleep as a distinct behavioral state (Aserinsky and Kleitman 1953) 45 years earlier.

While the discovery that a dysfunction of the orexin system causes narcolepsy-cataplexy had a major impact on sleep research and sleep medicine, the relatively subtle nature of the alteration caused by a total loss of the orexin system was surprising given how powerful the neuroexcitatory actions of orexins were and how widespread were the efferent connections of orexin neurons. Extensive studies with focal application of synthetic orexins have fully confirmed their powerful effects on different control systems, including sleep-wake, cardiovascular, respiratory and motor (e.g., Antunes et al. 2001; Bourgin et al. 2000; Horvath et al. 1999; Huang et al. 2010; Yamuy et al. 2004; Young et al. 2005; see Shirasaka et al. 2002 and Shahid et al. 2012 for reviews of cardiorespiratory effects of local orexin microinjections). Yet, of the many attempts to uncover significantly altered baseline ventilatory parameters in animals with genetically eliminated orexin (prepro-orexin KO rodents) or with genetic deletion of orexin-synthesizing neurons (orexin/ataxin-3 rodents), all were negative, and measurements of arterial blood pressure yielded conflicting results. Limited experiments with systemic administration of orexin-1, orexin-2, or dual orexin receptor antagonists also did not reveal major effects of elimination of orexin effects on baseline cardiorespiratory paramenters (Iigaya et al. 2012; Lee et al. 2013; reviewed by Silvani and Dampney 2013). A partial explanation for this apparent discrepancy may be found in the results from recording of activity of orexin neurons across sleep-wake states (Lee et al. 2005; Mileykovskiy et al. 2005; Takahash et al. 2008). These studies revealed that orexin neurons have high levels of activity only during active wakefulness, whereas during quiet wakefulness their activity is very low. It then ceases altogether during slow-wave sleep (SWS), and then becomes intermittent and bursty during REM sleep. Accordingly, a plausible explanation for difficulties to detect endogenous effects of orexins is that physiologically relevant levels of spontaneous activity of orexin cells are attained only under various stimulated conditions. One caveat here is that, to further examine the validity of this interpretation, one would need to measure the time course of the effective concentrations of orexins at their release sites in relation to the intensity and pattern of orexin cell activity (cf., Kiyashchenko et al. 2002). The recently developed ability to selectively and precisely stimulate orexin neurons using optogenetic approaches should help resolve this question.

It currently appears that, with the exception of the still not fully understood mode of their protective action against narcolepsy-cataplexy, orexins exert their effects under stimulated conditions, rather than as a matter of baseline control at rest. This interpretation is supported by reports of significant endogenous orexin role in conditions such as high attention, during high anxiety and anxiety-related insomnia (Betschart et al. 2013; Stettner et al. 2011; Wu et al. 2011a), certain types of stress and in defense behaviors (Furlong et al. 2009; Nisimaru et al. 2013; Samson et al. 2007; see Kuwaki 2011 for review), in neurogenic arterial hypertension (Johnson et al. 2012; Lee et al. 2013; Li et al. 2013; Xiao et al. 2013), during chemical stimulation of breathing above normal levels (Dias et al. 2009; reviewed by Nattie and Li 2012; Kuwaki 2008), or in response to focal pharmacological activation of orexin neurons (Cerri and Morrison 2005; DiMicco et al. 1986; Iigaya et al. 2012; Kayaba et al. 2003; Lu et al. 2007; Rusyniak et al. 2012; Stettner and Kubin 2013).

The focus of this brief review is on the endogenous cardiorespiratory effects of behavioral or pharmacological activation of orexin neurons. Figure 1 shows the main descending projections of hypothalamic orexin neurons to the most relevant locations and cell groups in the brainstem and spinal cord. The scheme includes both direct projections to major targets established by means of standard anterograde and retrograde tracing studies, as well as selected indirect pathways that have been particularly well documented. The discussion conducted in the subsequent sections will refer to different components of the scheme shown in Fig. 1.

2 Endogenous Respiratory Effects of Orexins

Baseline ventilation in orexin knock out (KO) mice does not differ from that in the control, wild-type mice but KO animals exhibit reduced ventilatory response when breathing is stimulated by increased level of inspired CO_2. In rats, pharmacological activation of orexin and other adjacent neurons in the posterior, lateral hypothalamus activates hypoglossal motoneurons that in obstructive sleep apnea (OSA) patients need to be activated to protect the airway against collapse during sleep. However, while genetic association studies show increased incidence of sleep apnea in patients with narcolepsy-cataplexy, it is the hypoxic ventilatory response, rather than the ventilatory sensitivity to CO_2, that is attenuated in narcolepsy-cataplexy patients. There are several possible explanations for these not fully concordant findings.

2.1 Orexin Contribution to Central CO_2 Sensitivity

Stimulation of breathing by elevated inspiratory CO_2 increases expression of the early gene c-Fos in orexin and other adjacent hypothalamic neurons in both anesthetized (Johnson et al. 2012) and freely behaving mice (Sunanaga et al. 2009).

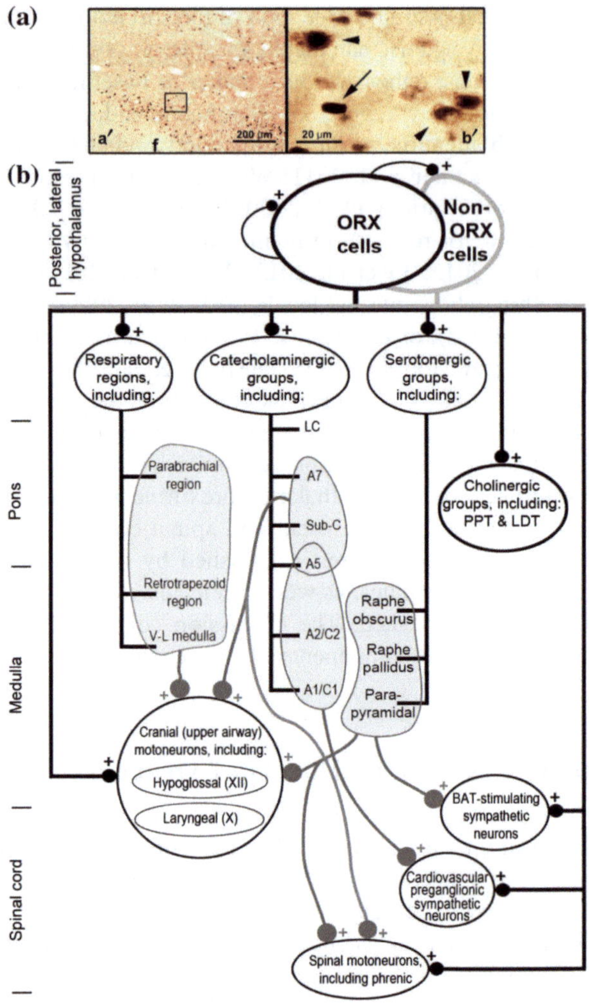

This indicates increased firing rates of these neurons. Since hypercapnic activation of breathing is associated with activation of multiple primary and accessory respiratory muscles and elicits anxiety caused by the feeling of dyspnea, the increased c-Fos expression does not necessarily indicate a primary respiratory role of orexin neurons. Rather, their activation may be secondary to those other aspects of experimental stimulation of breathing by hypercapnia. Nevertheless, it is evident from the scheme in Fig. 1b that orexin cell activation that occurs during hypercapnic states can facilitate generation of enhanced ventilatory effort.

Other evidence from rodents further supports a role of orexin in ventilatory activation specifically related to hypercapnia and central acidification. In arterially perfused brainstem-spinal cord juvenile rat preparation, respiratory rate response to

◀ **Fig. 1** Pathways transmitting activation from hypothalamic orexin neurons to cardiorespiratory outputs. **a** One method often used to activate orexin neurons involves local microinjections of $GABA_A$ receptor antagonists, such as bicuculline or gabazine, in anesthetized rodents. Compared to the anesthetized baseline state, such injections powerfully disinhibit orexin neurons, as indicated by massive expression and accumulation in their nuclei of c-Fos, a protein product of an early gene. The approach has been introduced prior to the discovery of orexin neurons as the means of producing cardiorespiratory, motor and electrocortical activation similar to that observed during a systemic defense response (e.g., Shekhar and DiMicco 1987). Posterior hypothalamic bicuculline or gabazine microinjections activate orexin neurons, as well as other local neurons that also may have widespread projections (Lu et al. 2007). In panel a′ (*left*), many cells located in the perifornical region of the posterior hypothalamus have c-Fos-stained nuclei following microinjection of 20 nl of 1 mM bicuculline in an urethane-anesthetized rat. Panel b′ shows an expanded frame marked in panel a′. *Arrowheads* point to selected c-Fos-stained nuclei (*black*) in orexin-positive neurons (*brown*) and *arrows* point to c-Fos-stained nuclei in non-orexin neurons (cell body not labeled). Unpublished data from the study of Lu et al. (2007) f—fornix. **b** Major descending projections from hypothalamic orexin neurons to the most relevant for cardiorespiratory regulation locations and cell groups in the brainstem and spinal cord. The scheme includes direct orexin cell projections to major targets, such as pontomedullary noradrenergic neurons (locus coeruleus—LC, A7, Sub-coeruleus region—Sub-C, A5, A2/C2 and A1/C1 groups) and medullary serotonergic neurons of the obscurus, pallidus and parapyramidal regions, as well as spinal and cranial motoneurons and preganglionic sympathetic neurons. Among the latter are those that control brown adipose tissue (BAT) and activate heat production and those responsible for vasoconstriction. Notably, spinal and cranial motoneurons and preganglionic sympathetic neurons can be activated directly by orexin-containing axon terminals and indirectly through aminergic cells that are themselves activated by orexin projections. Accordingly, orexinergic activation acts, in part, through enhancement of aminergic activation. Orexin neurons also send axons to numerous other pontomedullary regions with well-established roles in cardiorespiratory regulation. These include the pontine parabrachial region where the noradrenergic A7 group is located, the medullary retrotrapezoid region containing central chemosensory neurons, and other ventrolateral (V-L) medullary neurons that are part of the central respiratory rhythm and sympathetic tone generators. PPT & LDT—cholinergic neurons of the pedunculopontine tegmental and laterodorsal tegmental nuclei

hypercapnia is attenuated by bath application of orexin-1 receptor antagonist (Corcoran et al. 2010). This suggests that any endogenous orexin that might be present in the brainstem of this in vitro preparation acts at the brainstem level to facilitate transmission of hypercapnic stimulus to the central respiratory rhythm generating neurons. Such an action and interaction may occur in the ventrolateral medullary retrotrapezoid nucleus neurons which are known central receptors of brain acidification (Lazarenko et al. 2011). However, an alternative explanation for these findings would be that relevant orexin receptors are constitutively active and pH-sensitive in the conditions of these experiments.

A more detailed investigation of the sleep-wake state-dependent role of orexins in modulation of ventilatory sensitivity revealed that orexins facilitate the ventilatory response to CO_2 only during wakefulness (Nakamura et al. 2007). More specifically, the periods of wakefulness occurring during the dark phase when rodents are more active were associated with most pronounced endogenous orexinergic facilitation of hypercapnic ventilatory response and the ventrolateral medullary retrotrapezoid region was found to be an important site where such a state-dependent facilitation takes place (Dias et al. 2009). Orexin KO mice also had more frequent sleep-related central apneas (not associated with sighs) than wild-type mice, whereas hypoxic ventilatory response did not differ between the groups (Nakamura et al. 2007). Among these findings, the state-dependence of the effect of orexin on hypercapnic ventilatory response is consistent with the known state-dependence of orexin cell activity—activity present in wakefulness but absent, or nearly absent during sleep. However, given the absence of evidence for endogenous contribution of orexin to basal ventilation at rest, it would not be appropriate to conclude from the studies in which hypercapnia was used to stimulate breathing that a decrease in ventilation and the concomitant small increase in central CO_2 level that normally occur at the onset of SWS are due to silencing of orexin neurons. The increased expression of central apneas in orexin KO mice is more puzzling because orexin cells are typically silent in SWS in normal animals. Therefore, it is not clear how genetically induced absence of orexin and absence of orexin effects due to cell silencing would differentially contribute to the occurrence of central apneas during sleep. It is likely that a developmental reconfiguration of the respiratory system in orexin KO mice results in a more frequent expression of apneas in this mouse strain.

Importantly, when tested in vitro, orexin neurons have intrinsic pH chemosensitivity similar to noradrenergic neurons of the locus coeruleus and other neurons located in the pontomedullary reticular formation, and the level of highest pH sensitivity is around 7.2 (Williams et al. 2007). Combined with the known efferent projections of orexin neurons to aminergic neurons, as well as sympathetic and respiratory neurons, of the brainstem (Fig. 1b), this observation is supportive of a facilitatory role of orexins in ventilatory response to hypercapnia. Equally plausible, however, is the possibility that the primary role of hypercapnic activation of orexin neurons is to mediate the increased arousal and anxiety elicited by the hypercarbia-induced feeling of dyspnea (Williams et al. 2007). This function would be mediated by multiple suprapontine and ascending connections of orexin neurons that are outside of the scope of this discussion.

2.2 *Orexin Effects on the Brainstem Respiratory Network and Motoneurons Innervating the Upper Airway Muscles*

Orexin neurons send axonal projections to both cranial and spinal motoneurons. The role of these connections may be to facilitate motor activation in synchrony with the states of active wakefulness when orexin cells are more active than in other states. From the perspective of this review, it is particularly interesting whether endogenous orexinergic tone activates respiratory motoneurons, such as those innervating the diaphragm and intercostal muscles of the respiratory pump, and those that control the accessory respiratory muscles of the upper airway, such as hypoglossal (XII cranial nerve), facial (V) or laryngeal (X). Activation of upper airway motoneurons is especially important in OSA patients. In contrast to healthy persons, these patients require elevated levels of upper airway muscle activity in order to maintain the airway open. When this activity declines, as it typically does at sleep onset, the airway collapses (usually in the pharyngeal region), and the patient needs to awaken to resume breathing. Given the elevated activity level and prominent sleep-wake state-dependence of upper airway muscle activity in OSA patients, it is possible that endogenous orexinergic tone contributes to activation of upper airway motoneurons during wakefulness.

We and others have previously determined that norepinephrine- and serotonin-containing brainstem neurons provide a major component of the wakefulness related drive to XII motoneurons (Chan et al. 2006; Fenik et al. 2005). The relevance of this is that XII motoneurons innervate the muscles of the tongue which in OSA patients protect the airway from collapse by causing an anterior shift of the tongue which widens the lumen of the upper airway (Remmers et al. 1978; Sauerland and Harper 1976). As the scheme in Fig. 1b indicates, the wake-related activation of upper airway motoneurons may occur through their direct activation by axonal projections from hypothalamic orexin neurons and indirectly through orexinergic activation of brainstem aminergic neurons. To investigate these pathways, we used urethane-anesthetized rats in which orexin neurons are activated by local hypothalamic injections of $GABA_A$ receptor antagonists (bicuculline or gabazine). In the first study of this series, we determined that disinhibition of perifornical hypothalamic neurons, including the ones that express orexins (Fig. 1a), powerfully activated XII motoneurons and increased respiratory rate parallel to activation of cortical and hippocampal activity in a manner analogous to that in an aroused state (Lu et al. 2007). We then determined that a major component of the activation seen in XII motoneurons was preserved when noradrenergic and serotonergic excitatory effects on XII motoneurons were blocked at the level of the XII nucleus by local microinjections of a combination of α_1-adrenergic and serotonergic antagonists (Fenik et al. 2009). This suggested that a major activation of XII motoneurons in this model is mediated by direct projections from the perifornical hypothalamus to the XII motor nucleus, including the projections originating in orexin neurons (Fig. 1b). To test this, we compared the effects of hypothalamic gabazine injections on XII nerve activity before and after local microinjections of a dual orexin receptor antagonist, Almorexant (Brisbare-Roch et al. 2007), into the XII nucleus. Contrary to our expectation, Almorexant injected into the XII nucleus did not attenuate hypothalamic activation of XII motoneurons (Stettner and Kubin 2013). On the other hand, Almorexant injections placed into the same posterior hypothalamic site prior to gabazine injections delayed and attenuated the XII nerve and respiratory rate increases (as well as heart rate increases) (Stettner and Kubin 2013). This suggested that local, intrahypothalamic interactions mediated by orexin play an important, albeit only a partial, role in shaping of the response elicited by disinhibition of cells in the perifornical hypothalamus. A similar result was also obtained from the same animal model by others. Specifically, as in our study with local microinjections of Almorexant into the perifornical hypothalamus, systemic administrations of moderate doses of Almorexant (15 mg/kg) attenuated, but did not abolish, tachypneic and sympathoexcitatory effects of hypothalamic microinjections of bicuculline (Iigaya et al. 2012).

The following two points are worth deriving from the studies of the characteristic pattern of activation elicited by disinhibition of posterior hypothalamic neurons. First, there is to date no evidence that any component of the complex response elicited in this model can be entirely abolished by orexin antagonists regardless of whether they are administered locally or systemically. Rather, responses are either attenuated or unaffected. Second, local microinjections of the dual orexin receptor antagonist, Almorexant, into the perifornical region were sufficient to significantly attenuate both respiratory and cardiovascular effects of perifornical gabazine. Together, these findings suggest that the effects elicited by disinhibition of perifornical region neurons originate in orexin-containing and other neurons with overlapping projections and somewhat redundant functions. It also appears that orexin is an important excitatory peptide within the posterior lateral hypothalamus, and that its local role is to synchronize activation of orexin-containing and other local neurons who jointly ensure proper cardiorespiratory and behavioral activation, as needed by different environmental challenges. Accordingly, we would postulate that the perifornical hypothalamus may contain additional types of wake- and arousal-active neurons that interact with orexin neurons and generate complementary signals to those emanating from orexin neurons. A search for such neurons may prove to be a rewarding endeavor.

2.3 Association Between Orexin Dysfunction and OSA in Humans

OSA patients have reduced plasma levels of orexin (Busquets et al. 2004; Nishijima et al. 2003) which are then increased following treatment with nasal continuous positive airway pressure (nCPAP) (Sakurai et al. 2005). A plausible explanation of these findings is that the excessive sleepiness characteristically affecting OSA patients results in lower levels of orexin cell activity and, consequently, reduced plasma orexin levels. Once sleep is normalized by treatment with nCPAP, patients are less sleepy, more alert, and orexin cell activity levels rebounds. However, gene association studies also reveal a significant relationship between the genetic predisposition for narcolepsy-cataplexy and OSA. Specifically, a distinct polymorphism of prepro-orexin gene has been associated with OSA (Ahmed et al. 2012; Chen et al. 2012).

Interestingly, in contrast to findings in rats, narcolepsy-cataplexy patients with identified human leucocyte antigen (HLA)-DQB1*0602 allele that predisposes to narcolepsy-cataplexy do not exhibit reduced ventilatory sensitivity to hypercapnia. Instead, they have depressed ventilatory response to hypoxia (Han et al. 2010).

The discrepancy between the impairment of hypoxic ventilatory response in humans versus hypercapnic ventilatory response in mice in orexin deficiency is consistent with a complex and indirect, rather than a strictly deterministic, impact of orexins on ventilatory regulation.

3 Endogenous Cardiovascular Effects of Orexins

There are reports of both increased and decreased arterial blood pressure in orexin-deficient rodents when compared to their wild-type controls. The picture is complicated because, under physiologic conditions, blood pressure exhibits predictable and clinically significant dips on transition from wakefulness to SWS and then a relative increase, as well as increased variability, on transition from SWS to REM sleep. This pattern also appears to be affected by orexin deficiency. On the other hand, it is well established that arterial hypertension elicited by stress depends on increased activation of orexin neurons. Similarly, the spontaneous neurogenic hypertension depends on hyperactivity of orexin nerons. In both situations, there is also evidence of possible upregulation of orexin receptors and increased orexinergic innervation at brainstem sites relevant for the control of arterial blood pressure.

3.1 Baseline Blood Pressure Levels in Orexin Deficiency

In contrast to the absence of any effect of genetic elimination of orexin on baseline ventilation, it has been initially reported that orexin KO mice have significantly lower baseline blood pressure than their wild-type counterparts (Kayaba et al. 2003). For the systolic blood pressure, the magnitude of the difference was about 25 mmHg in anesthetized animals, and about 15 mmHg in chronically instrumented, behaving mice. A similar difference was also reported for the orexin/ataxin-3 mice (Zhang et al. 2006). A subsequent study using the orexin/ataxin-3 rats which included a more detailed analysis of blood pressure and heart rate separately during different phases of the circadian cycle and during transitions between different sleep-wake states also reported that hypocretin neuron-ablated rats had reduced blood pressure during wakefulness and all sleep stages, whereas heart rate measurements did not reveal any differences. Notably, blood pressure changes accompanying transitions between sleep-wake state were small, less than 4 mmHg, and not different between the orexin/ataxin-3 rats and their wild type controls (Schwimmer et al. 2010).

However, data from a series of reports from another group were not concordant with the finding of a reduced blood pressure in orexin deficient rodents. Instead, both orexin KO mice and orexin/ataxin-3 mice were reported to have higher blood pressure during SWS and REM sleep than the control mice of the same genetic background, whereas during wakefulness blood pressure did not differ among the three strains (Bastianini et al. 2011). As a result, blood pressure decrease (dip) that typically occurs on transition from wakefulness to non-REM sleep was attenuated in the orexin-deficient animals. In another study, neither the average blood pressure levels nor the blood pressure dips with entry into non-REM sleep differed between the orexin/ataxin-3 mice and control animals (Lo Martire et al. 2012). A third study from the same group devoted to the analysis of dynamic fluctuations of blood pressure and heart rate in different sleep-wake states in orexin deficient mice also concluded that the absence of orexin or orexin-containing neurons imparts only small and subtle changes on the central neural control of baseline blood pressure and heart rate across wake-sleep states (Silvani et al. 2012).

Data from human narcolepsy-cataplexy patients revealed that daytime blood pressure levels were similar in patients and control subjects, but the patients had a non-dipping night time blood pressure pattern (Grimaldi et al. 2012). Circadian rhythms of heart rate and blood pressure did not differ between patients and control subjects but systolic blood pressure during night time REM sleep was elevated and the average heart rate was significantly higher in the patient group across the entire 24 h period. The authors noted that the nocturnal blood pressure and heart rate patterns could be different in patients due to the characteristic for narcolepsy-cataplexy increase of sleep fragmentation and increased prevalence of arousals-causing periodic limb movements. Ultimately, a blunted night time blood pressure decrease in the patient group was significantly associated only with periodic limb movements. In another study, no differences were detected in mean systolic blood pressure or heart rate within different sleep-wake states between narcolepsy-cataplexy patients and control subjects, and only subtle differences were found in indices of blood pressure and heart rate variability, suggesting altered vago-sympathetic control and reduced bararoceptor reflex sensitivity in narcolepsy-cataplexy (Silvani et al. 2013). Similar to these studies, in a study focused exclusively of the heart rate, narcolepsy-cataplexy patients had a reduced transient heart rate acceleration associated with arousals (Sorensen et al. 2013). Notably, heart rate response was more attenuated in the patients with directly verified reduced plasma orexin levels and patients with cataplexy when compared to narcoleptic patients who had normal orexin levels or patients without cataplexy. Thus, orexin deficiency was associated with a significant attenuation of dynamic heart rate changes but, contrary to the generalization offered by the authors of the study, there was no evidence for a tonic contribution of the orexin system to modulation of cardiovascular function at rest.

The overall impression from the studies of cardiovascular baseline parameters in orexin deficiency is that the absence of orexin effects, has little if any effects.

3.2 Role of Orexins in Spontaneous Arterial Hypertension

Two studies have recently reported that elevated blood pressure in a commonly used rodent model of spontaneous neurogenic hypertension, the spontaneousy hypertensive rat (SHR), is dependent on orexinergic hyperactivity (Lee et al. 2013; Li et al. 2013). In the study of Li et al. (2013), the dual orexin receptor antagonist, Almorexant (250–300 mg/kg in oral gavage) profoundly reduced the mean arterial blood pressure by over 40 mmHg during the dark (active) period and by nearly 30 mmHg during the light (rest) period, whereas in the control rat strain, the normotensive Wistar-Kyoto rat, the effects of Almorexant were not significant. In the study of Lee et al. (2013), intracerebroventricular administration of orexin-2 receptor antagonist similarly reduced blood pressure and reduced heart rate in SHR, but not in Wistar-Kyoto rats, and local microinjections of the same antagonist into the ventrolateral medulla had stronger hypotensive effect in SHR than in the control rats. In contrast, orexin-1 receptor antagonist had no effect. Additionally, both studies revealed various biochemical markers consistent with upregulation of orexinergic transmission in the rostral ventrolateral medulla.

Thus, it appears that orexinergic hyperactivity, and/or increased sensitivity to orexin in medullary regions that control sympathetic vasoconstrictory activity, is an important mechanism of primary hypertension. It is also of note that the hypertension present in the SHR appears to depend on transmission in orexin-2 receptors. This is of note because various forms of stress-induced hypertension that are orexin-dependent are primarily mediated by orexin-1 receptors (discussed in the next section). This distinction is worth making because Wistar-Kyoto rats and their sub-strains are also used as a model of high vulnerability to stress (McAuley et al. 2009; Laitman et al. 2014). Accordingly, the contribution of different orexin receptors to hypertension in SHR and to stress-induced hypertension may help further refine our understanding of different states characterized by high anxiety.

3.3 *Role of Orexins in Cardiovascular Response to Stress*

Increased blood pressure and elevated heart rate represent major components of stress response. Given that most forms of acute stress are associated with high arousal, orexinergic tone is expected to increase under stressful conditions. This has been investigated in male Wistar rats for various forms of stress by quantifying c-Fos expression in orexin neurons and cerebrospinal fluid orexin levels (Furlong et al. 2009). It turned out that, when compared to quiet rest, orexinergic activation was strongest during active wakefulness and exploratory behaviors, moderately elevated following exposure to contextual fear, and negligible following acute restrain stress. Heart rate and blood pressure increases occurring during each of the conditions were measured with and without prior administration of the dual orexin receptor antagonist, Almorexant (100 and 300 mg/kg in oral gavage). Of the four tested conditions, only during exploratory behavior and exposure to fearful context, blood pressure elevation was partially reduced by Almorexant, and heart rate increase was sensitive to Almorexant only during the exploratory behavior. Thus, there is a selectivity of the behavioral conditions under which orexins contribute to cardiovascular response to stress. Nevertheless, in another study, plasma adrenocorticotropic hormone (ACTH) increase elicited by immobilization stress in male Sprague-Dawley rats was nearly abolished by intracerebroventricular administration of orexin-1 receptor antagonist (Samson et al. 2007).

In a rat model of hypertension induced by repeated exposure to electric shocks and noise, the protein levels of orexin and orexin-1 receptors were elevated in the rostral ventrolateral medulla and blood pressure elevation was partially reduced by either orexin-1 or orexin-2 receptor antagonists (Xiao et al. 2013). Also, in a rat model of cardiovascular activation and temperature increase acutely induced by amphetamine, pretreatment with orexin-1 receptor antagonist attenuated both hyperthermia and blood pressure increase (Rusyniak et al. 2012).

Contribution of orexin to cardiorespiratory and other markers of stress was also tested in conditions directly relevant for respiratory disorders. In rats exposed to high CO_2 levels to elicit a state of panic and high anxiety similar to that experienced by patients with chronic obstructive pulmonary disease (COPD) or asthma attacks, orexin-1 receptor antagonist attenuated the pressor response (Johnson et al. 2010, 2012). In another study on rats, a chronic partial tracheal airflow restriction was applied for seven weeks by surgical means to model airflow limitation in OSA. The intervention reduced sleep amounts, increased hypothalamic orexin mRNA and protein content, and caused growth retardation. Under these conditions, Almorexant (300 mg/kg) normalized sleep but induced breathing difficulties and dyspnea-like behavior, as deduced from video recordings showing exaggerated mouth opening during inspiration and significantly reduced respiratory rate (Tarasiuk et al. 2014). Thus, it appears that orexin hyperactivity contributes to the cardiorespiratory compensatory response to the stress of hypoventilation and dyspnea. As it is often the case, there is a concurrent maladaptive effect—the disruption of sleep. Further studies of the receptors and brain regions involved may help develop differential treatments for such complex conditions.

4 Conclusions

Based on the findings discussed in this review, the orexin system emerges as a powerful regulator of cardiorespiratory responses to challenges associated with states of high arousal, whereas contributions of orexins to baseline cardiorespiratory activity during quiet wakefulness and sleep is minimal. While broadly applicable, this generalization may be not fully accurate in conditions of partial chronic orexin deficiency or partial loss of orexin receptors. Under such conditions, a reconfiguration of other systems that normally interact with orexins (noradrenergic, serotonergic, cholinergic; see Fig. 1b) occurs and this likely alters the normal balance of orexin actions. Such an explanation was proposed to interpret the results of various pharmacological challenges in dogs lacking functional orexin-2 receptors (Wu et al. 2011b). The authors concluded that mutation of one orexin receptor caused long-term changes in the functioning of multiple brain systems. Such a reconfiguration also may explain conflicting findings about orexin contribution to baseline cardiorespiratory parameters in transgenic rodents with orexin deficiency.

The intense and growing interest in the development of effective orexin receptor antagonists is consistent with the notion that orexinergic hyperactivity, rather than hypoactivity, often leads to adverse health conditions (Betschart et al. 2013; Scammell and Winrow 2011). Studies of the role of the orexin system in cardiorespiratory regulation indicate that various forms of hypertension and respiratory disorders have components that depend on orexin hyperactivity. If so, they may be subject to treatments similar to such classic orexin-dependent disorders as insomnia and high anxiety.

Acknowledgments Our studies summarized in this review were supported by the following grants from the National Institutes of Health: HL-47600, HL-60287, HL-74385 and HL-116508.

References

Ahmed WA, Tsutsumi M, Nakata S et al (2012) A functional variation in the hypocretin neuropeptide precursor gene may be associated with obstructive sleep apnea syndrome in Japan. Laryngoscope 122:925–929

Antunes VR, Brailoiu GC, Kwok EH et al (2001) Orexins/hypocretins excite rat sympathetic preganglionic neurons in vivo and in vitro. Am J Physiol 281:R1801–R1807

Aserinsky E, Kleitman N (1953) Regularly occurring periods of eye motility, and concomitant phenomena, during sleep. Science 118:273–274

Bastianini S, Silvani A, Berteotti C et al (2011) Sleep related changes in blood pressure in hypocretin-deficient narcoleptic mice. Sleep 34:213–218

Betschart C, Hintermann S, Behnke D et al (2013) Identification of a novel series of orexin receptor antagonists with a distinct effect on sleep architecture for the treatment of insomnia. J Med Chem 56:7590–7607

Bourgin P, Huitrón-Reséndiz S, Spier AD et al (2000) Hypocretin-1 modulates rapid eye movement sleep through activation of locus coeruleus neurons. J Neurosci 20:7760–7765

Brisbare-Roch C, Dingemanse J, Koberstein R et al (2007) Promotion of sleep by targeting the orexin system in rats, dogs and humans. Nat Med 13:150–155

Busquets X, Barbé F, Barceló A et al (2004) Decreased plasma levels of orexin-A in sleep apnea. Respiration 71:575–579

Cerri M, Morrison SF (2005) Activation of lateral hypothalamic neurons stimulates brown adipose tissue thermogenesis. Neuroscience 135:627–638

Chan E, Steenland HW, Liu H, Horner RL (2006) Endogenous excitatory drive modulating respiratory muscle activity across sleep-wake states. Am J Respir Crit Care Med 174:1264–1273

Chemelli RM, Willie JT, Sinton CM et al (1999) Narcolepsy in orexin knockout mice: molecular genetics of sleep regulation. Cell 98:437–451

Chen W, Ye J, Han D et al (2012) Association of prepro-orexin polymorphism with obstructive sleep apnea/hypopnea syndrome. Am J Otolaryngol 33:31–36

Corcoran A, Richerson G, Harris M (2010) Modulation of respiratory activity by hypocretin-1 (orexin A) in situ and in vitro. Adv Exp Med Biol 669:109–113

Date Y, Ueta Y, Yamashita H et al (1999) Orexins, orexigenic hypothalamic peptides, interact with autonomic, neuroendocrine and neuroregulatory systems. Proc Natl Acad Sci USA 96:748–753 de Lecea L, Kilduff TS, Peyron C et al (1998) The hypocretins: hypothalamus-specific peptides with neuroexcitatory activity. Proc Natl Acad Sci USA 95:322–327

Dias MB, Li A, Nattie EE (2009) Antagonism of orexin receptor-1 in the retrotrapezoid nucleus inhibits the ventilatory response to hypercapnia predominantly in wakefulness. J Physiol 587:2059–2067

DiMicco JA, Abshire VM, Hankins KD et al (1986) Microinjection of GABA antagonists into posterior hypothalamus elevates heart rate in anesthetized rats. Neuropharmacology 25:1063–1066

Fenik VB, Davies RO, Kubin L (2005) REM sleep-like atonia of hypoglossal (XII) motoneurons is caused by loss of noradrenergic and serotonergic inputs. Am J Respir Crit Care Med 172:1322–1330

Fenik VB, Rukhadze I, Kubin L (2009) Antagonism of α_1-adrenergic and serotonergic receptors in the hypoglossal motor nucleus does not prevent motoneuronal activation elicited from the posterior hypothalamus. Neurosci Lett 462:80–84

Furlong TM, Vianna DM, Liu L, Carrive P (2009) Hypocretin/orexin contributes to the expression of some but not all forms of stress and arousal. Eur J Neurosci 30:1603–1614

Grimaldi D, Calandra-Buonaura G, Provini F et al (2012) Abnormal sleep-cardiovascular system interaction in narcolepsy with cataplexy: effects of hypocretin deficiency in humans. Sleep 35:519–528

Han F, Mignot E, Wei YC et al (2010) Ventilatory chemoresponsiveness, narcolepsy-cataplexy and human leukocyte antigen DQB1*0602 status. Eur Respir J 36:577–583

Horvath TL, Peyron C, Diano S et al (1999) Hypocretin (orexin) activation and synaptic innervation of the locus coeruleus noradrenergic system. J Comp Neurol 415:145–159

Huang SC, Dai YW, Lee YH et al (2010) Orexins depolarize rostral ventrolateral medulla neurons and increase arterial pressure and heart rate in rats mainly via orexin 2 receptors. J Pharmacol Exp Ther 334:522–529

Iigaya K, Horiuchi J, McDowall LM et al (2012) Blockade of orexin receptors with Almorexant reduces cardiorespiratory responses evoked from the hypothalamus but not baro- or chemoreceptor reflex responses. Am J Physiol 303:R1011–R1022

Johnson PL, Samuels BC, Fitz SD et al (2012) Activation of the orexin 1 receptor is a critical component of CO_2-mediated anxiety and hypertension but not bradycardia. Neuropsychopharmacology 37:1911–1922

Johnson PL, Truitt W, Fitz SD et al (2010) A key role for orexin in panic anxiety. Nat Med 16:111–115

Kayaba Y, Nakamura A, Kasuya Y et al (2003) Attenuated defense response and low basal blood pressure in orexin knockout mice. Am J Physiol 285:R581–R593

Kiyashchenko LI, Mileykovskiy BY, Maidment N et al (2002) Release of hypocretin (orexin) during waking and sleep states. J Neurosci 22:5282–5286

Kuwaki T (2008) Orexinergic modulation of breathing across vigilance states. Respir Physiol Neurobiol 164:204–212

Kuwaki T (2011) Orexin links emotional stress to autonomic functions. Auton Neurosci Bas Clin 161:20–27

Laitman BM, Gajewski ND, Mann G et al (2014) The alpha$_1$ adrenoceptor antagonist prazosin enhances sleep continuity in fear-conditioned Wistar-Kyoto rats. Progr Neuro-Psychopharmacol Biol Psych 49:7–15

Lazarenko RM, Stornetta RL, Bayliss DA, Guyenet PG (2011) Orexin A activates retrotrapezoid neurons in mice. Respir Physiol Neurobiol 175:283–287

Lee MG, Hassani OK, Jones BE (2005) Discharge of identified orexin/hypocretin neurons across the sleep-waking cycle. J Neurosci 25:6716–6720

Lee YH, Dai YW, Huang SC et al (2013) Blockade of central orexin 2 receptors reduces arterial pressure in spontaneously hypertensive rats. Exp Physiol 98:1145–1155

Li A, Hindmarch CC, Nattie EE, Paton JF (2013) Antagonism of orexin receptors significantly lowers blood pressure in spontaneously hypertensive rats. J Physiol 591:4237–4248

Lin L, Faraco J, Li R et al (1999) The sleep disorder canine narcolepsy is caused by a mutation in the hypocretin (orexin) receptor 2 gene. Cell 98:365–376
Lo Martire V, Silvani A, Bastianini S et al (2012) Effects of ambient temperature on sleep and cardiovascular regulation in mice: the role of hypocretin/orexin neurons. PLoS ONE 7:e47032. doi:10.1371/journal.pone.0047032
Lu JW, Fenik VB, Branconi JL et al (2007) Disinhibition of perifornical hypothalamic neurones activates noradrenergic neurones and blocks pontine carbachol-induced REM sleep-like episodes in rats. J Physiol 582:553–567
Marcus JN, Aschkenasi CJ, Lee CE et al (2001) Differential expression of orexin receptors 1 and 2 in the rat brain. J Comp Neurol 435:6–25
McAuley JD, Stewart AL, Webber ES et al (2009) Wistar-Kyoto rats as an animal model of anxiety vulnerability: support for a hypervigilance hypothesis. Behav Brain Res 204:162–168
Mileykovskiy BY, Kiyashchenko LI, Siegel JM (2005) Behavioral correlates of activity in identified hypocretin/orexin neurons. Neuron 46:787–798
Nakamura A, Zhang W, Yanagisawa M et al (2007) Vigilance state-dependent attenuation of hypercapnic chemoreflex and exaggerated sleep apnea in orexin knockout mice. J Appl Physiol 102:241–248
Nattie E, Li A (2012) Respiration and autonomic regulation and orexin. Progr Brain Res 198:25–46
Nishijima T, Sakurai S, Arihara Z, Takahashi K (2003) Plasma orexin-A-like immunoreactivity in patients with sleep apnea hypopnea syndrome. Peptides 24:407–411
Nisimaru N, Mittal C, Shirai Y et al (2013) Orexin-neuromodulated cerebellar circuit controls redistribution of arterial blood flows for defense behavior in rabbits. Proc Natl Acad Sci USA 110:14124–14131
Peyron C, Tighe DK, van den Pol AN et al (1998) Neurons containing hypocretin (orexin) project to multiple neuronal systems. J Neurosci 18:9996–10015
Remmers JE, DeGroot WJ, Sauerland EK, Anch AM (1978) Pathogenesis of upper airway occlusion during sleep. J Appl Physiol 44:931–938
Rusyniak DE, Zaretsky DV, Zaretskaia MV et al (2012) The orexin-1 receptor antagonist SB-334867 decreases sympathetic responses to a moderate dose of methamphetamine and stress. Physiol Behav 107:743–750
Sakurai S, Nishijima T, Takahashi S et al (2005) Low plasma orexin-A levels were improved by continuous positive airway pressure treatment in patients with severe obstructive sleep apnea-hypopnea syndrome. Chest 127:731–737
Samson WK, Bagley SL, Ferguson AV, White MM (2007) Hypocretin/orexin type 1 receptor in brain: role in cardiovascular control and the neuroendocrine response to immobilization stress. Am J Physiol 292:R382–R387
Sauerland EK, Harper RM (1976) The human tongue during sleep: electromyographic activity of the genioglossus muscle. Exp Neurol 51:160–170
Scammell TE, Winrow CJ (2011) Orexin receptors: pharmacology and therapeutic opportunities. Annu Rev Pharmacol Toxicol 51:243–266
Schwimmer H, Stauss HM, Abboud F et al (2010) Effects of sleep on the cardiovascular and thermoregulatory systems: a possible role for hypocretins. J Appl Physiol 109:1053–1063
Shahid IZ, Rahman AA, Pilowsky PM (2012) Orexin and central regulation of cardiorespiratory system. Vitam Horm 89:159–184
Shekhar A, DiMicco JA (1987) Defense reaction elicited by injection of GABA antagonists and synthesis inhibitors into the posterior hypothalamus in rats. Neuropharmacology 26:407–417
Shirasaka T, Kunitake T, Takasaki M, Kannan H (2002) Neuronal effects of orexins: relevant to sympathetic and cardiovascular functions. Regul Pept 104:91–95
Silvani A, Bastianini S, Berteotti C et al (2012) Control of cardiovascular variability during undisturbed wake-sleep behavior in hypocretin-deficient mice. Am J Physiol 302:R958–R964
Silvani A, Dampney RA (2013) Central control of cardiovascular function during sleep. Am J Physiol 305:H1683–H1692
Silvani A, Grimaldi D, Barletta G et al (2013) Cardiovascular variability as a function of sleep-wake behaviour in narcolepsy with cataplexy. J Sleep Res 22:178–184

Sorensen GL, Knudsen S, Petersen ER et al (2013) Attenuated heart rate response is associated with hypocretin deficiency in patients with narcolepsy. Sleep 36:91–98

Stettner GM, Kubin L (2013) Antagonism of orexin receptors in the posterior hypothalamus reduces hypoglossal and cardiorespiratory excitation from the perifornical hypothalamus. J Appl Physiol 114:119–130

Stettner GM, Kubin L, Volgin DV (2011) Antagonism of orexin 1 receptors eliminates motor hyperactivity and improves homing response acquisition in juvenile rats exposed to alcohol during early postnatal period. Behav Brain Res 221:324–328

Sunanaga J, Deng BS, Zhang W et al (2009) CO_2 activates orexin-containing neurons in mice. Respir Physiol Neurobiol 166:184–186

Takahash K, Lin JS, Sakai K (2008) Neuronal activity of orexin and non-orexin waking-active neurons during wake-sleep states in the mouse. Neuroscience 153:860–870

Tarasiuk A, Levi A, Berdugo-Boura N et al (2014) Role of orexin in respiratory and sleep homeostasis during upper airway obstruction in rats. Sleep 37:987–998

Williams RH, Jensen LT, Verkhratsky A et al (2007) Control of hypothalamic orexin neurons by acid and CO_2. Proc Natl Acad Sci USA 104:10685–10690

Wu MF, Nienhuis R, Maidment N et al (2011a) Cerebrospinal fluid hypocretin (orexin) levels are elevated by play but are not raised by exercise and its associated heart rate, blood pressure, respiration or body temperature changes. Arch Ital Biol 149:492–498

Wu MF, Nienhuis R, Maidment N et al (2011b) Role of the hypocretin (orexin) receptor 2 (Hcrt-r2) in the regulation of hypocretin level and cataplexy. J Neurosci 31:6305–6310

Xiao F, Jiang M, Du D et al (2013) Orexin A regulates cardiovascular responses in stress-induced hypertensive rats. Neuropharmacology 67:16–24

Yamuy J, Fung SJ, Xi M, Chase MH (2004) Hypocretinergic control of spinal cord motoneurons. J Neurosci 24:5336–5345

Young JK, Wu M, Manaye KF et al (2005) Orexin stimulates breathing via medullary and spinal pathways. J Appl Physiol 98:1387–1395

Zhang W, Sakurai T, Fukuda Y, Kuwaki T (2006) Orexin neuron-mediated skeletal muscle vasodilation and shift of baroreflex during defense response in mice. Am J Physiol 290: R1654–R1663

FSC
www.fsc.org
MIX
Papier aus verantwortungsvollen Quellen
Paper from responsible sources
FSC® C105338

If you have any concerns about our products,
you can contact us on
ProductSafety@springernature.com

In case Publisher is established outside the EU,
the EU authorized representative is:
Springer Nature Customer Service Center GmbH
Europaplatz 3, 69115 Heidelberg, Germany

Printed by Libri Plureos GmbH
in Hamburg, Germany